Lucia Lazarowski
Editor

Olfactory Research in Dogs

Springer

Editor
Lucia Lazarowski
Auburn University College of Veterinary Medicine
Auburn, AL, USA

ISBN 978-3-031-39369-3 ISBN 978-3-031-39370-9 (eBook)
https://doi.org/10.1007/978-3-031-39370-9

This Springer imprint is published by the registered company Springer Nature Switzerland AG
The registered company address is: Gewerbestrasse 11, 6330 Cham, Switzerland

Paper in this product is recyclable.

Contents

Fundamentals of Olfactory Function

Peripheral Olfactory Pathway Anatomy, Physiology, and Genetics

Melissa Singletary and Samantha Hagerty

Abstract

Various sensory systems provide environmental awareness and have evolved to meet species-specific needs. The earliest of these senses thought to be chemosensory in function include olfaction. Olfactory acuity varies greatly by species, with intra-species variations also notable. Broadly, a species' olfactory capability is categorized based on their overall olfactory development; species with a well-developed olfactory capability are macrosmatic (e.g. dogs, rats), compared to microsmatics (e.g. humans) with less developed sense of smell. A multitude of factors may influence this inter-species olfactory performance, which include anatomical and physiological variations across species. Despite these performance gradients, the foundational olfactory cytoarchitecture and cellular constituents appear to be fairly conserved across species. The field of olfaction has seen an increase in research since the discovery of olfactory receptors as G-protein coupled receptors (GPCRs) and part of the largest multi-gene family of receptors found in the mammalian gene repertoire by Linda Buck and Richard Axell in 1991 (Buck and Axel, Cell 65:175–187, 1991) winning them the Nobel Prize in 2004. However, the field remains understudied in relation to the other sensory systems and lacks significant comparative studies across species. The use of translational models, such as mice and rats, has served to provide the predominant foundational knowledge of olfactory sensory systems.

M. Singletary (✉)
Canine Performance Sciences, Auburn University College of Veterinary Medicine, Auburn, AL 36849, USA
e-mail: mas0028@auburn.edu

M. Singletary · S. Hagerty
Department of Anatomy, Physiology and Pharmacology, Auburn University College of Veterinary Medicine, Auburn, AL 36849, USA

L. Lazarowski (ed.), *Olfactory Research in Dogs*,
https://doi.org/10.1007/978-3-031-39370-9_1

This chapter will focus on the understanding of olfaction through an anatomical, physiological, and genetic foundation, highlighting the dog through the peripheral nervous system.

Keywords

Olfaction • Olfactory anatomy • Olfactory genetics • Peripheral olfactory system

1 Olfactory System Overview

Odors carry information regarding the past and the present including qualitative, quantitative, spatial, and temporal elements of a surrounding environment. Intraspecies and interspecies structural olfactory system variations reflect qualitative and quantitative differences, representing the complexity in anticipated and observed olfactory acuity. This overview will discuss these events in a comparative fashion across multiple species with emphasis on the dog.

Olfaction is one of two main chemosensory systems in mammals along with gustation. The olfactory system is considered the first sensory system to develop evolutionarily and has remained highly conserved across species, comprising the largest gene superfamily in mammals (Glusman et al. 2001; Reed 2004; Niimura 2012; Persuy et al. 2015). There are two main divisions of the olfactory system and, in species such as rodents, two lesser divisions. Of the two main divisions, the main olfactory epithelium (MOE) is responsible for the detection of volatile compounds and makes up the largest area of the olfactory system. Second, the vomeronasal organ (VNO) is primarily responsible for the detection of pheromones. The two less prominent components of the olfactory system in rodent species, in particular, include the Grueneberg's ganglion and septal organ (SO), or the Organ of Masera (Rodolfo-Maera 1943). Grueneberg's ganglion (GG) is spatially isolated from the MOE, located in the rostrodorsal aspect of the nasal cavity and is suggested to have thermosensory and chemosensory roles, with the capacity to sense coolness (Mamasuew et al. 2008) and alarm pheromones (Kikusui et al. 2001). While Grueneberg's ganglion has been identified in humans, its potential functional similarities to that in rodent counterspecies are not yet established (Gruneberg 1973). An isolated patch of neuroepithelium, known as the septal organ (SO), resides bilaterally on the caudoventral aspect of the nasal septum and is present in rodents and other mammalian species such as the koala (Kratzing 1984). The SO is thought to function as an airflow sensor (Ma 2010) and traditional chemosensory region similar to the MOE with a greatly reduced receptor repertoire (Grosmaitre et al. 2007). In dogs, there is a lack of evidence to support the presence of the SO or GG (Barrios et al. 2014). The VNO is adjacent to the nasal septum and is surrounded by cartilaginous walls. The anterior VNO duct is stratified squamous epithelium and the posterior is a simple columnar epithelium. In the dog all VNO receptors are VR-1 based (Quignon et al. 2006; Dzieciol et al. 2020).

Though there is controversy regarding the estimates of how many odors can be discriminated, it is recognized that there is an impressively large number of natural odors and odor combinations that can be discriminated by our sense of smell, and more so by macrosmatic species. The variety of odorants far outweighs the number of receptors, and the complex levels of cross-reactions result in a combinatorial coding, which leads to an indeterminate amount of possible odor perceptions (Malnic et al. 1999; Kurian et al. 2021). In rats, it has been shown that classification of mixtures according to the molar ratios of components occurs which allows for detection across a range of concentrations rather than specificity to a single previously learned concentration (Uchida and Mainen 2007).

Domestication in dogs has resulted in significant variation across breeds and divergence of characteristics from ancestral wolves. Domesticated dogs commonly demonstrate superior olfactory acuity to their human counterparts, and there are various foundational morphological and physiological underpinnings in support of this macrosmatic categorization. However, there are some reports that suggest the morphological characteristics of olfactory acuity of domestic dogs are diminished in relation to their ancestral wolves as measured by relative cribriform plate surface area size across 46 breeds of dogs (Bird et al. 2020, 2021). Significant neuroanatomical variations were noted across evaluation of 33 breeds by Hecht et al., noting that variations in specific neural networks were correlated with breed specialization within more recent phylogenetic terminal branches (Hecht et al. 2019).

Volatile compound properties, odor complexity, olfactory receptor odotype distribution, airflow rate, passage volume, mucosa surface area, and mucus solubility impact the odor availability, receptor interaction, and ultimately the odor precept. Various odorants, odorant combinations or varying concentrations result in activation of unique ensembles of odor receptors. Interactions including a range of receptor affinity agonism and antagonism contribute to the combinatorial coding at the odor receptor level (Malnic et al. 1999; Kurian et al. 2021). Binary and more complex mixtures in rodent model studies have shown complex and non-linear relationships. A recent study by Zak et al. (2020) explored the dynamic range of complex mixture in a freely breathing mouse model. Both olfactory sensory neuron (OSN) and glomerular imaging were conducted under an active sniffing evaluation across varying odor concentration and odor mixture presentations revealing a degree of mixture suppression corresponding to increased odor components (Zak et al. 2020).

Input is received and modified at multiple levels. The intricate balance of signal activation and inhibition is not well understood within the olfactory system regarding complex odor mixtures. From its source, an odor will undergo various conditions and interactions along its journey that will ultimately influence its availability and perception by an animal. Volatile availability is affected by the originating form and configuration influenced by factors such as surface area and surface moisture (Jorgensen 2009).

Further effects on odor availability come from the environment under which they are present, influenced by factors such as ambient temperature, humidity, air velocity, presence of airborne particulate matter, and barometric pressure (Kuehn et al. 2008; Ajmani et al. 2016; Qifan et al. 2017).

Each of these factors described thus far may promote or inhibit an odorant's movement toward engagement with the central nervous system and olfactory sensory pathway. The destination of a particular odorant may be entrapment, diversion, expulsion, degradation, metabolization, conversion, and varying affinity for direct or indirect receptor interaction within olfactory and non-olfactory sensory systems. When an odorant's journey results in interaction with an odor receptor of positive affinity and binding adequate for activation, it begins the translation of an environmental chemical signal into an electrical one through signal transduction within the olfactory sensory neuron.

2 Nasal Cavity

An animal's initial engagement with odor occurs at the entrance to the olfactory pathway in the nares. Upon entrance, the odorants traverse the intricate nasal passages where internal environmental factors influence their ultimate destination and interaction, such as passage diameter, air velocity, epithelial cytoarchitecture, mucous constituents, mucous viscosity, odor receptor distribution, and the air-flow path dynamics (Lawson et al. 2012; Challis et al. 2015; Rygg et al. 2017; Robert-Hazotte et al. 2019). The nares' morphological and physiological characteristics vary across species.

As the initial interface between odorant and anatomical structures, the morphology of the nares influences odorant delivery to chemosensory epithelia. While signal transduction processing at the receptor level is fairly conserved across species, the gross anatomical features vary greatly. Some aquatic species can have a blind-ended cavity with up to four up to four nostrils. Most mammals display bilateral nostrils supporting a dual functioning respiratory-olfactory design. An example of physical impact on odor delivery can be observed in some teleosts, wherein two of the four nostrils connect to form a one-way system of flow to the olfactory organ, facilitating a single-direction flow system for odorant delivery and transportation that allows for continuous undisrupted access to incoming odors (Niimura 2012). Additionally, the widely-separated and lateral positioning of the nares in some aquatic organisms has implicated roles in spatial and directional cues, particularly with low visibility.

It has been suggested that the dog as a species exhibits some of the largest morphological variation amongst land mammals due to wide breed variations (Ostrander 2012). In the dog, the nares represents a highly efficient design for external volatile sampling as examined through experimental fluid dynamic modeling and analysis of the functional aerodynamics associated with the sniff (Staymates et al. 2016; Settles et al. 2003; Craven et al. 2004, 2009). Nostril flexure along with nasal aperture anatomy controls air movement separately from inspiration to expiration (Settles et al. 2003). Sniffing activity results in air from within a 1–2 cm range being inhaled through the nostrils directed centrally. Upon exhalation the warmed and moistened air is expelled through the lateral slits or flaps that push the air ventrolaterally. This pattern of exhalation is suggested to enhance volatilization of nearby odorants, increasing the odorants within the inspiratory sampling field minimizing dilution, direct resampling or contamination of the central inhalational field, improving odor availability and sensitivity by the dog. Sniffing is a sinusoidal pattern that cycles through an inspiratory and expiratory phase at a frequency of up to 20 sniffs/minute at 4–7 Hz (Staymates et al. 2016). The sniffing pattern may change based on the task and can quickly change nostril selectivity based on biological relevance and odor-associations of the odor present (Siniscalchi et al. 2011, 2016; Cavelius et al. 2022). The sniffing cycle is influenced by the physiological properties of each nasal pathway separately. The highly vascularized tissue of the rostral nasal respiratory pathway cycles through a shifting lateralized congestion changing the airflow dynamics respectively through this selective tissue swelling (Friling et al. 2014). This tissue is sensitive to selective activation from the autonomic nervous system resulting in congestion.

Internally, once an odorant has passed the entrance at the nares, it will travel through the nasal vestibule, past the nasal meatuses onto the ethmoturbinates to reach the sensory cilia of the olfactory epithelium where the odorant will interact with receptors and initiate the cascade of events that lead to the recognition of an odor. The nasal cavity is divided into three main anatomical regions, which encompass the nasal vestibule, the respiratory, and the olfactory (Craven et al. 2007). Each piece and part of the pathway has a role to play and can be species-specific in form and function.

For example, the dog and rat have an elongated nasal passageway with a dedicated olfactory recess located posterior to the primary respiratory division. The longer nasal cavity of the dog may increase sensitivity and discrimination by the relatively longer time-of-flight comparison of odorant deposition compared to humans.

An evaluation of natural odor sampling behavior in a rat model indicated that sniff strength has a very limited role on shaping the primary odor precept (Cenier et al. 2013). However, sniffing in the canine model has shown that approximately 2.5 times more air is directed to the olfactory recess during active sniffing compared to quiet breathing and results in a corresponding increase in uptake of highly and moderately soluble odorants in the sensory field per unit of time (Rygg et al. 2017).

Upper respiratory and olfactory airways are structurally supported by the boney composition of the skull. The rostroventral surface of the passageway is formed by the bilateral incisive bones, with the lateral to dorsolateral walls formed by the maxillary bone, rostrodorsal surface by the nasal bone, and most caudal aspect formed by the ethmoid bone. The ethmoid bone is comprised of three regions, including the nasal septum extending directionally from caudal to rostral, dividing the midline of the nasal passages, the ethmoturbinates (also referred to as ethmoid labyrinth), and the perforated osseous cribriform plate. This plate consisting of numerous perforations through which axons traverse as the neuronal communication highway of odorant information directed toward the central nervous system from the olfactory epithelium. The caudoventral aspect of the nasal septum is formed by the vomer bone, an osseous structure partially making up the roof of the choanae, which form the openings of each nasal cavity to the associated nasopharyngeal meatus (Evans and de Lahunta 2013).

The most anterior portion of the nasal cavity through which odorants first enter is lined by a stratified squamous epithelium (Randall et al. 1987). Collectively called the vestibule, this passageway divides into meatuses that are also referred to as turbinals or concha. These curved structures, including the maxilloturbinal, nasoturbinal, and ethmoturbinals, extend inward from the lateral walls of the nasal cavity forming a fold-like appearance. The number of meatuses, turbinals, or concha vary between species, though in the canine middle vestibule, three main divisions are identified as the dorsal, vetral, and middle meatus (Moore 1981). They are functionally thought to be involved in directing airflow, reducing turbulence, and creating a sub-climate for inspired air to be warmed, humidified, and filtered (Menco and Morrison 2003).

In dogs, the turbinate complex includes multiple turbinates and concha. Six ectoturbinates are localized laterally with some projection into the frontal sinus. Four endoturbinates are localized medially forming the dorsal concha and middle concha with ectoturbinates I and II, respectively, then the ventral concha independently (Barrios et al. 2014). This complex scrollwork provides large surface area supporting an environmental chamber for air movement, heating, moisturizing, protection, and odor molecule transport.

On gross examination of the relative tissues in the nasal cavity, there is a distinct coloration change from sensory epithelia, which is yellow to brown, to respiratory epithelia, which is pink to red. From a histological perspective, the epithelial thickness and constituents vary significantly but have some overlap in the transition zones where the two epithelia meet.

In mammals, this passageway is dual functioning in the act of inhalation for both respiration and olfaction. It serves as a conduit for air carrying volatile odorants and oxygen molecules to sensory neurons and pulmonary organs, respectively. Infrastructure of the upper airway, including the nasal cavity, larynx, and trachea, are crucial for protection of more vulnerable lower airway regions. The epithelia lining the nasal septum is both sensory and non-sensory in nature, with the anterior-most aspect comprised of stratified squamous epithelium transitioning to non-sensory respiratory epithelium and ending in sensory olfactory epithelium in the posterior region. Cells of the respiratory epithelial tissue secrete high levels of defensins and antimicrobial peptides within the mucous layer lining. This mucous layer, which traps foreign particles from inhaled air, lies atop the apical surface of a stratified columnar epithelium covered in cilia with motile function allowing movement of the mucus to be directed away from the lower respiratory system. The mucous layer terminates in either the nasopharynx where it is discarded into the gastrointestinal tract or the nasal vestibule for removal by sneezing or manual discard. Mucosal microenvironments have been found to house a diverse range of immunologically protective microbes through actions such as competitive inhibition, antimicrobial peptide production, enzymatic degradation, and innate immune cell priming. Most evaluations of the respiratory tract have neglected to evaluate the olfactory mucosa as a separate system, though early studies suggest it is not equivalent (Francois et al. 2016).

Air containing odorant molecules is filtered, warmed, and humidified throughout the passage over respiratory epithelium and continues toward the folded labyrinth of ethmoturbinate structures in the most caudal aspect of the nasal cavity, which is lined with pseudostratified ciliated columnar olfactory epithelium. This epithelium is considered sensory due to the presence of ciliated olfactory sensory neurons, on which many transmembrane olfactory receptors (OR) reside. Here, odorants will interact with receptors that line the ciliary processes, resulting in a cascade of downstream signal transduction events that will be further detailed. An orthonasal route of odorant introduction is the classically described nasal passageway, through which both passive and active exposure occurs with the physical action of sniffing. Alternatively, in some cases, odorants may be introduced to olfactory epithelium retronasally when propagated from the oral cavity through the choanae in the caudodorsal nasopharynx to reach the nasal cavity and sensory structures for recognition (Small et al. 2005). However, it has been suggested that this retronasal olfaction is not significantly biologically relevant in dogs (Rygg et al. 2017). The main classical airflow in the dog is directed at higher velocities through the dorsal region of the left/right nasal cavity toward the olfactory recess curling back as it turns ventro-rostrallly, allowing a second pass over the olfactory epithelium. It is proposed that this air movement through the olfactory recess maximizes odorant exposure to the olfactory epithelium (Craven et al. 2007, 2009, 2010). This feature of the dedicated olfactory recess is lacking in humans and is thought to contribute to the higher olfactory acuity seen in dogs.

The separate nostrils, similar to other sensory structures such as ears and eyes, allow for a bilateral comparison between left and right-sided sensory input resulting in improved distinction of stimulus intensity and spatiotemporal parameters (Craven et al. 2010). Hemispheric olfactory lateralization has been suggested through behavioral studies (Webber et al. 1987; Sobel et al. 1999; Siniscalchi et al. 2008, 2011, 2016).

Olfactory lateralization starts with odorant delivery from two nostrils into two distinct and separate nasal chambers. Airflow patterns are mirrored on both left and right sides where the incoming air is split across a dorsally directed path representing 12–13% of inspired air by the dog, which travels caudally to enter the olfactory recess (Craven et al. 2010). The ventrally directed path moves the remaining inspired air caudally to the pharynx where it enters the pulmonary tract for gas exchange and respiration. The significant tissue structures that line this tract direct these airflow patterns creating turbulence, air particulate deposition patterns and microclimates within the nasal cavity (Rygg et al. 2017). In human studies, it is suggested that the absorption qualities, high or low, of an odor are varied correspondingly to airflow, low or high, resulting in optimal sensitivity respectively (Sobel et al. 1999).

3 Olfactory Neuroepithelia Functional Cytoarchitecture

The main olfactory epithelium (MOE), characterized histologically as pseudostratified ciliated columnar epithelium, is populated with both sensory and supportive cellular structures. Olfactory sensory neurons (OSN), also known as olfactory receptor neurons (ORN) and considered the parenchymal cell type of olfactory tissue, are true bipolar neurons. Extending from the soma toward the mucosal surface is a dendritic knob-like ending with approximately 20–30 non-motile cilia. Extending from the soma in the opposite direction the axon extends and terminates at the olfactory bulb (Morrison and Costanzo 1990). The dendritic processes allow for increased surface area of the receptive field for odorant sampling. More specific measurements have reported an estimated ciliary surface area of 23 cm^2 and knob density of 60,000 knobs/mm^2 (Menco and Jackson 1997; Doty 1998). OSN are surrounded by supporting (sustentacular) cells, whose primary function is to provide structural integrity to the surrounding epithelium. Supporting cells span from the mucosal surface to the basement membrane and are lined apically with microvilli. In mammals, the sustentacular cells have been classically shown to serve as an epithelial support structure and contribute to the microenvironment homeostasis through production of aqueous secretions and various cytokines. In more recent discoveries, these cells also show a schwann cell-like attribute through enwrapment of the olfactory sensory neuron dendrites (Liang 2020). Other principal cellular components within the olfactory mucosa include basal cells and Bowman's gland duct cells (Moran et al. 1982; Menco and Morrison 2003). Basal cells, which are found in the lower region of the epithelium near the basal lamina, have been found to act as progenitor cells for regenerating OSN following injury or natural maturation and turnover (Morrison and Costanzo 1989; Menco and Morrison 2003). The Bowman's glands are the primary secretory gland of the olfactory epithelium with some contribution from supporting cells, while goblet cells represent the principal secretory cells in non-sensory respiratory epithelium (Solbu and Holen 2012). The secretory composition from Bowman's glandular ducts is a diverse seromucous mixture of enzymes, acids, mucopolysaccharides, immune factors, antioxidants, antimicrobial proteins, and odorant binding proteins (OBP). The subepithelial region houses Bowman's gland acinar cells and glandular bulbs, Schwann cells, vasculature, connective tissue, and the OSN axonal fascicles (Ramon-Cueto and Avila 1998).

German shepherd olfactory mucosa spans 200 cm^2 while in the cocker spaniel only 67 cm^2, though significantly greater than the human at 5 cm^2 (Quignon et al. 2012; Galibert et al. 2016). There is considerable structural refinement present in the dog as compared to other species such as the sheep, with greater overall thickness of the olfactory epithelium and higher olfactory sensory neuron density within the cellular cytoarchitecture (60–80% of cells) estimating greater than 100 million OSN in the OE of the dog (Craven et al. 2010).

The ciliary dendritic knob, which protrudes into the mucous layer at the apical surface, is lined with many sensory receptors that interact with odorants, which are dissolved or actively transported by odorant binding proteins from incoming air flow. Following initial molecular events at the receptor level, generated signals travel through the axon towards the olfactory bulb for further processing. All OSN axons collectively form what is known anatomically as the first cranial nerve (CN I), or the olfactory nerve, which is classified as a special visceral afferent (SVA) fiber (Doty 1998).

A mouse model evaluating olfactory receptor neuron ciliary length showed a pattern of intrinsic arrangement suggested to enhance odor detection sensitivity and acuity (Challis et al. 2015). The length of cilia corresponds to the odorant response and odorant absorption in the nasal cavity (Challis et al. 2015). Cilia play a role in mammalian odor adaptation as one part of various complex mechanisms and feedback circuits within the olfactory pathway, which can be rapid or persistent in form (Zufall and Leinders-Zufall 2000).

Numerous factors contribute to olfactory acuity, which are observable interspecies and intraspecies. The number of olfactory receptors can change with age. A study by Kavoi et al. demonstrated dogs to have an overall increase in cilia olfactory receptor density from birth to adulthood while other species such as sheep showed a decrease in density with maturation, which may suggest that postnatal odor development is important to maximizing function (Kavoi and Jameela 2011).

Unlike most neurons, OSNs are continuously exposed to the external environment making them more vulnerable to physical, chemical, thermal, or pathogenic damage. Thus, the ability to repair and regenerate is a crucial characteristic of the olfactory system. Injury models in the olfactory system have demonstrated the stem cell function of basal cells (Graziadei and Graziadei 1979; Schwob et al. 1995), which can be further categorized into globose basal cells (GBC) or horizontal basal cells (HBC) based on their respective morphologies and activity level. GBC are continuously active stem cells responsible for routine turnover or acute maintenance of OSN (Schwob et al. 1995). Horizontal basal cells are generally quiescent and demonstrate mitotic activity only after severe and complete olfactory epithelial trauma where proliferation of all cell types is required (Leung et al. 2007; Mackay-Sim 2010; Suzuki et al. 2013). Maturing OSN gradually extend from the basement membrane through the pseudostratified neuroepithelial cell population, growing a new dendritic process that eventually reaches the mucosal surface and develops cilia. During this maturation process, an axon will navigate through the cribriform plate perforations reaching a glomerulus of the olfactory bulb and reestablishing a complete synaptic interface. The olfactory ensheathing cell (OEC) is essential in supporting the migration of the new axons to their target glomeruli exhibiting phenotypic and functional properties onto OSNs and basal cells in the OE to aide in axonal elongation (Ramon-Cueto and Avila 1998). Further investigation into the mechanism of these growth-promoting properties has shown a family of proteinases, matrix metalloproteinases (MMPs), to be critical for cell motility and ability to transverse extracellular matrix through support of the neurotrophic factors secreted by OEC (Ould-Yahoui et al. 2013).

The estimated lifespan of a given OSN differs by species and by environmental exposure amongst other factors. However, one estimate in mice maintained under laboratory conditions suggests a given OSN lifespan averages 30–90 days but some reported out to more than one year (Hinds et al. 1984; Kondo et al. 2010; Holl 2018). The neurogenerative capacity of individual olfactory sensory neurons allows for remarkably fast recovery periods of approximately 4–8 weeks in humans with some species-specific variations among mammals (Costanzo 1991).

4 Peri-receptor Environment

The unique mucosal surface of the olfactory epithelium with its cellular constituents functions in peri-receptor odorant and odor-receptor interaction modifications. Odorants brought to the olfactory epithelium may traverse the mucous layer interacting with microbiota, enzymes, proteins, cytokines, immune cells, and metals. Emerging work to establish the microbial composition of the olfactory system has demonstrated a core biome in mice of two dominant phyla, *Firmicutes* (30–70%) and *Bacteroidetes* (15–60%), followed by *Proteobacteria* (5–25%) and *Actinobacteria* (<10%) in descending abundance (Francois et al. 2016). In one evaluation of the human nasal mucosa, the phyla residing in olfactory-adjacent regions were markedly different in their respective ratios, the most prevalent being *Actinobacteria* (50%), followed by *Firmicutes* (24%), *Proteobacteria* (20%), and less than 3% from *Bacteroidetes* (Yan et al. 2013). Among a grouping of studies evaluating human fecal samples, the three dominant phyla representing approximately 75% of all microbiota include *Firmicutes*, *Bacteroidetes*, and *Actinobacteria* (Gibson and Roberfroid 2008). These results indicate a similar core biome, with a variation in the ratios of phyla isolated. There is significant individual variation, collection site variation, environmental factors, and collection methods that can account for this variation.

While the sensory apparatus itself, the olfactory receptor neuron, is unsurprisingly an important focus in the study of olfactory function, other factors within the peri-receptor environment also play a critical role in signal transduction. Along with the anatomical features discussed previously that can impact airflow dynamics, internal composition of the nasal mucous layer and chemical properties within it can influence odorant–receptor interactions. The nasal mucus layer covering both olfactory and respiratory epithelia ranges in thickness from 5 to 30 μm in the olfactory region and 5–12 μm in the respiratory region (Menco 1980). From a visual perspective, the sensory and non-sensory regions of nasal epithelia covered by a continuous layer of mucus have few distinguishable features. However, it is important to differentiate these regions as they represent two functionally distinct areas with varying contributory cells, microbial composition, and odorant concentration parameters. For example, the odorant partition coefficient between regions of nasal mucus has revealed quantifiable differences on odorant deposition patterns along sensory epithelia in mammals (Nagashima and Touhara 2010; Rygg et al. 2013).

There are conflicting results in the computational modeling studies that have examined the chromatographic theory. A combined electro-olfactogram and computational fluid dynamics model in the mouse by Coppola et al. in 2019 found no relationship between the EOG odor-responsive localized activation pattern and the corresponding odor sorption patterns (Coppola et al. 2019). However, in canine simulation models, highly soluble odors showed a concentrated deposition pattern along the rostral region of the olfactory recess through the dorsal meatus and nasal septum area compared to moderately soluble odor deposition patterns spread uniformly throughout the recess (Craven et al. 2010; Lawson et al. 2012).

Levels of enzymatic activity in the olfactory neuroepithelium have also been shown to influence odorant properties, as higher enzymatic activity resulted in increased active metabolites stemming from chemically-whole odorants (Nagashima and Touhara 2010). Enzyme-mediated transformation of odorants before receptor interaction as well as odorant removal by odor degrading enzymes (ODE) are unsurprisingly major factors that influence how often and in what ways odorant ligands meet olfactory receptors. Effects of odorant biotransformation into metabolites within the mucus layer were emphasized through reported variations in receptor responses between in vitro and in vivo mouse studies, wherein the only environmental difference was the presence or absence of an odorant-metabolite mixture prior to OR engagement (Nagashima and Touhara 2010). The key secretory source of mucus within the OE, the Bowman's gland, produces high concentrations of an essential metabolizing enzyme, cytochrome P450 (CYP), also found in the liver with comparable activity levels. Other relevant metabolizing enzymes secreted by the Bowman's glands that participate in degradation of xenobiotics include dehydrogenases, oxidases, reductases, carboxylesterases, epoxide hydrolases, uridine diphosphate glucuronyl transferase, glutathione S-transferase, rhodanese, angiotensin-converting enzyme, peptidases, kallikrein, and esterases (Kaliner 1991; Ohkubo et al. 1998; Mayer et al. 2009). These enzymatic factors play a complex role in odorant modification for both recognition as well as degradation and subsequent odor removal, as the mucus layer has a high turnover rate.

Furthermore, the presence of an extracellular signaling molecule and embryological inducer known as Sonic Hedgehog (Shh) has been established in nasal mucus where a dose-dependent relationship with olfactory function was reported (Henkin et al. 2017).

Along with odorant modification, some odorants are chaperoned through the mucus layer to receptor targets by odorant binding protein (OBP) (Pelosi 1996; Badonnel et al. 2009). These multi-functional proteins facilitate the delivery of volatile and hydrophobic odorants to their respective receptors. The quantity and variety of OBP types vary among mammalian species, with only one OBP established in humans while hundreds of variations have been identified in other mammals (Briand et al. 2002). OBPs belong to the lipocalins protein family, which consists of low molecular weight proteins with low ligand specificity, consistent with their relatively reversible binding properties (Pevsner and Synder 1990; Pelosi 1994).

Consistent with the tight linkage between immune function and mucosal surfaces found throughout the body where external and internal environments meet, the olfactory mucosa requires considerable multi-level defense mechanisms. Primary defenses include nasal secretions that contain immunoglobulins such as IgA, IgM, and IgG (Kaliner 1991). Pathogen-specific defenses include innate immune cells and B-lymphocytes, antimicrobial peptides, lysozyme protein, and lactoferrin (Mellert et al. 1992). There may also be defense-related roles of the microbiota populating olfactory epithelial tissue, though this remains to be directly established.

Other components of the peri-receptor environment include metals. Copper and zinc ions, as well as magnesium and calcium, specifically, have been reported in the nasal mucus. The estimated concentration of copper was 40 μM in mice and 16 $\mu g\ dL^{-1}$ in humans, while the reported levels of zinc, magnesium, and calcium in humans were 14 $\mu g\ dL^{-1}$, 1554 $\mu g\ dL^{-1}$, and 5303 $\mu g\ dL^{-1}$, respectively (Henkin et al. 2000). Relative concentrations of metallic content between whole nasal mucus and that covering specifically olfactory regions have not been compared. Although they are visually continuous, their distinct enzymatic and glandular contributions should be taken into consideration when conducting further evaluations. Similarly, the microbiota composition and influence within the olfactory system is not yet established as only limited studies report direct evaluation of the main olfactory epithelium and a few have discussed findings in the closely associated ethmoid recess in humans (Yan et al. 2013; Francois et al. 2016). Evaluations of the nasal cavity microbiome have focused on respiratory epithelium and pulmonary system considerations. While the role that microbes play in relation to social and behavioral cues through microbial odorant production within glands has been explored, continued work on more widespread composition and immunological roles across species is necessary.

A relatively recent field of study considering interactions between metals and the nervous system is called "metalloneurochemistry" (Lippard 2014). Primary metals of focus include zinc, iron, copper, and manganese as their composition makes them likely contributors of co-factors for various proteins and enzymes. Increased recognition of these metals in biologically-relevant environments and a need for further elucidation of their roles within the central nervous system have led to more directed effort toward characterization and investigative tools. Zinc ions, specifically, have been localized in areas of the CNS including the olfactory bulb, hippocampus, hypothalamus, and cerebral cortex in large concentrations. The concentration is mostly in the glomerular and granular layers and considered mobile within neurons (Frederickson and Danscher 1990; Sensi et al. 2009). Measurable quantities were also found in cerebrospinal fluid at an estimated concentration of 31.5 $mg\ L^{-1}$ (Agarwal and Henkin 1982). The observed abundance of zinc ions in such a highly protected bodily system is biologically considerable. Furthermore, reported reductions in some tissues during states of deprivation, particularly in peripheral sites such as bone, testes, and plasma, without a corresponding depletion from the CNS uniformly, suggest a significant role in normal homeostatic function (Jackson et al. 1982). Glutamate is an excitatory neurotransmitter utilized by OSN, mitral, and tufted secondary neurons of the olfactory bulb synaptic cleft. During glutaminergic neuronal signaling at this site, zinc is co-released by exocytosis, which may be contributing to the neurotransmitter recognition, though its target and functional role is not fully established. Presentations of hyposmia and anosmia have been associated with zinc deficiency, which further implicate its possible role in olfactory signal transduction.

Interest in understanding the role of various metals in receptor-level interactions, neurotransmission, and overall central nervous system physiology has steadily increased. As one of the most evolutionarily conserved subsets of the nervous system across species, the olfactory system is a relevant site to investigate such phenomena. A shuttlecock mechanism at the 4–5 transmembrane loop of olfactory receptors, which are classified as 7-transmembrane G-protein coupled receptors, (GPCR) has been proposed for the function of zinc or copper metals. The suggested mechanism is based on the principle of increased biochemical sensitivity to thiols and amines, which would assist activation of the olfactory receptor upon odorant binding and further indicate OR metalloprotein properties (Wang et al. 2003).

The interplay of metals and biology has a component established in neurobiology, with many implications in the sensory system of olfaction in the olfactory bulb and higher cognitive centers. Ionic zinc applied to the olfactory epithelium is inhibitory to olfaction and zinc salts are commonly used to induce anosmia in experimental settings (Smith 1938; Rowe and Smith 1972; Ishimaru et al. 2000). Zinc particles were identified in human and animal blood that was in the nanoscale (1–2 nm) and non-ionic which evoked significant enhancement in combination with odorant in the olfactory sensory neuron (Viswaprakash et al. 2006, 2009).

Subsequent ex vivo experiments in isolated, viable neuroepithelium from rat models demonstrated dose-dependent, specific, and reversible enhancement properties of zinc nanoparticles. In conjunction with effects observed at the epithelial level in rodents, further investigation into higher processing of olfactory information was conducted in an in vivo canine model. Functional magnetic resonance imaging (fMRI), a method that analyzes cognitive activity through relative levels of neuronal activation, was performed on awake and anesthetized canines exposed to odorants with and without zinc nanoparticles. Increased activity in brain regions involved in olfactory processing, specifically the olfactory bulb and hippocampus, was observed following zinc nanoparticle administration with odorants, which is consistent with previously shown signal intensity enhancement at the epithelial level. This complementary work suggests enhancement at the level of perception (Jia et al. 2016).

Microenvironments throughout biological systems are known to be colonized by diverse communities of microorganisms. The characterization of these communities is integral in understanding the interactions between the host and the colonies of microorganisms (Tress et al. 2017). Microbiomes are integral in the maintenance of homeostasis within individuals (Biswas et al. 2020). The microorganisms of the gut are intimately involved with digestion, provision of nutrients, social behavior, anxiety (Francois et al. 2016), immunological tolerance, and some autoimmune diseases. Disturbances in the commensal relationship with microbiome may lead to diseases (Wilson and Hamilos 2014) such as cancer, depression, and inflammatory and metabolic disorders (Koskinen et al. 2018). However, it is not known if community composition changes in the microbiome between health and disease are direct consequences of disease processes or if the changes themselves influence disease progression (Tress, et al. 2017). Additionally, the microbiome plays pivotal roles in epithelial development as demonstrated by germ-free rodents experiencing drastic deficits in intestinal epithelial development compared to typical rodents (Francois et al. 2016) and germfree zebrafish demonstrating deficits in olfactory epithelium (OE) development (Casadei et al. 2019).

Variation in host phylogeny, lifestyle, diet, physiology, and medical interventions can all have profound influences on the composition of the microbiome among individuals. Genetically identical rodents can have different microbiomes due to differences in factors of their environment including diet, litter, vendor, shipment, facilities, and exposures to microbes early in life (Chaves-Moreno et al. 2015). Additionally, the microbiota of the gut is known to impact the nasal region including the physiology of the olfactory epithelium (Francois et al. 2016; Isaiah et al. 2017; Thangaleela et al. 2022), feeding behavior and the hedonistic values of odors (Naudon et al. 2020). The epithelium of the upper airway aids in sensing the environment and regulating inspired air, linking the environment directly to the host immune system (Toppila-Salmi et al. 2015). Microbial communities act to aid in homeostasis, prevent pathogen attacks, and assist in immune regulation surviving within the host as mutualistic tenants or opportunistic pathogens (Thangaleela et al. 2022).

The bacterial community composition of the nasal cavity can mirror olfactory function. Microbiomes modulate olfactory epithelial physiology that can alter responses to odorant perception. Some bacteria produce strong-smelling compounds able to influence the perception of external environmental odors (Koskinen et al. 2018). However, describing the taxa can be challenging due to the fact that between 20 and 60% of the human microbiome is uncultivable (Wilson and Hamilos 2014) and low biomass samples subjected to marker gene amplification are prone to bias arising from over amplification over represention of some taxa (Knight et al. 2018).

Nasal cavity microbes aid in the maintenance of the physiology of olfactory epithelium and thereby maintaining olfactory capabilities (Biswas et al. 2020). The human nasal cavity is colonized by many taxa including Actinobacteria (50%), Firmacutes (28%), and Proteobacteria (14%). Normosmotics and hyposmotic humans have significantly different microbial communities within their nasal cavities. High proportions of taxa, such as the genus *Campylobacter*, appear to be significantly associated with a decrease olfactory ability (Koskinen et al. 2018) while reduction in *Corynebacterium* is linked to olfactory deficits. In humans grouped into anosmia, hyposmia, and normosmia, the nasal bacterial communities of those with anosmia have significantly reduced diversity than in humans with hyposmia and normosmia. Additionally, the relative abundances of *Streptococcus* and *Anaerococcus* were significantly more abundant in humans suffering from anosmia (Biswas et al. 2020).

When measured with 16S rDNA sequencing, the mouse OE microbial community consists of primarily of Bacteriodetes and Firmacutes with smaller proportions of Proteobacteria and Actinobacteria (Francois et al. 2016). In canines, Bacteriodetes, Firmacutes, Fusobacteria, and Tenericutes dominated the oral cavity while the nasal cavity principally contained Actinobacteria and Proteobacteria (Isaiah et al. 2017). The majority of studies examining the microbiomes of the nasal cavity have been performed with 16S marker gene amplification, which provides a low-resolution view of the microbial communities, and are prone to bias. Metagenomics, sequencing all of the microbial genomes from a sample, can be performed to achieve a higher taxonomic resolution to ascertain taxonomic and functional information of the microbe community (Knight et al. 2018). This information could prove to be vital in the treatment and prevention of diseases related to the respiratory and olfactory systems and provide a more comprehensive understanding of the function of the microbial community in relation to olfactory abilities.

5 Olfactory Receptors, Gene Families, Expression, and Evolution

The high density of olfactory sensory neurons represents a large receptive field for receiving stimuli. The olfactory system across species is a critical system well represented by the high number of OR genes present (Persuy et al. 2015). The OR gene origins can be traced back to the common ancestor of the phylum Chordata, demonstrating an important role throughout evolution in all vertebrate (Niimura 2012). This likely reflects the multiple and possible undiscovered roles of ORs and OR neurons. Physiologically, the olfactory system is involved in critical survival functions such as foraging, feeding, hunting, predator avoidance, spatial orientation, and stress. It also plays a role in the psychological processes of social behaviors, reproduction, and maternal bonding (Persuy et al. 2015).

The OR has been identified as a G-protein coupled receptor (GPCR) representing the largest family of GPCRs, the Rhodopsin family, and form 1–3% of the estimated mammalian gene repertoire (Glusman et al. 2001; Reed 2004). This approximately 300 amino acid protein receptor type is defined by its 7-transmembrane domains of approximately 20 amino acids each. The extracellular N-terminus exhibits higher polymorphisms and variability corresponding to a large diversity of ligand binding capacity and a more conserved C-terminus intracellular domain (Quignon et al. 2005). This receptor type operates to allow for signal amplification translating very minute detectable amounts of odorant stimuli. The functional importance of each transmembrane region varies relative to odor recognition and discrimination. There is a higher level of divergence noted in the transmembranes 3 (TM3), 4 (TM4), and 5 (TM5) (Liu et al. 2003).

The nomenclature system recently applied by the Vertebrate Gene Nomenclature Committee to the cow, horse, chimpanzee, and dog is the Mutual Maximum Similarity (MMS) algorithm. This system is a systematic classifier that assigns a human-centric nomenclature to olfactory receptor genes. In its evaluation, the chimpanzee showed ~87%, cow ~30%, and dog ~50% symbolic identity to humans, which suggested ecological, environmental, and imposed adaptive changes in the OR gene superfamily (Olender et al. 2020).

The environment is made up entirely of molecular information, allowing for a nearly unlimited number of odorant possibilities for the olfactory system to interpret. Given the unknown amount of odorant molecules that could be encountered and the relatively specific nature of odor perception amongst such complex mixtures, this chemosensory system that is equipped with a finite set of receptors is thought to employ a combinatorial coding mechanism (Malnic et al. 1999). In essence, a fixed capacity of olfactory machinery can interpret a vast array of odorant molecules through dynamic binding properties and topographically organized signal processing. The premise that only certain portions of an odorant molecule may be recognized by a given receptor type allows for this combinatorial mechanism by which one odorant may activate multiple receptors and one receptor may recognize multiple structurally similar odorants. Furthermore, a given odorant may elicit specific combinations of OR activation that leads to a distinct odor fingerprint. When the odor is presented in a mixture of others, the perceived scent may change. This concept is supported by the commonly reported observation that odor concentrations influence olfactory response and perception, some eliciting drastic changes, as increased presence of a given odorant allows for recognition by more receptors which may vary in excitability. Although the overall olfactory processing mechanism seems to remain fairly consistent across species, there are notable variations in the number and diversity of olfactory receptors between microsmatic and macrosmatic organisms. While humans, rodents, and dogs share functional aspects of odorant processing and perception, different genetic OR repertoire result in very different ranges of odor recognition and sensitivity. For example, mice have a receptor for detecting CO_2 that humans are lacking in their repertoire, rendering it odorless to humans (Hu et al. 2007). As with all phenotypes that exist in nature, selective pressures influence genotypic continuation in any given environment. The olfactory system works in direct response to its environment, so species inhabiting variable selective conditions may have evolved through a need for receptor diversity and others for quantity or regenerative capacity. Teleosts such as Zebrafish, for example, demonstrate fewer OR gene numbers than humans, though amongst those numbers is more OR diversity (Table 1) (Niimura 2012).

Phenotypic characterization through behavioral assessments has been the primary method for evaluation and selection in domestic dog breeding cohorts based on breeding goals. This may include trainability, olfactory acuity, size, or specific features as example. More recent efforts are underway in working dogs to improve selection through genotypic characterization and genome scanning of behavioral selection (Eyre et al. 2022).

Studies on olfactory receptor genes across multiple dog breeds have shown both high levels of uniformity, gene family level, and high levels of variation, lower gene levels (Issel-Tarver and Rine 1996; Olender et al. 2004; Quignon et al. 2005; Tacher et al. 2005; Robin et al. 2009; Chen et al. 2012; Derrien et al.

Table 1 Olfactory receptor genes species variance

Species	Approx OR genes	Identified % pseudogenes (%)	References
Human	860	50–60	Zozulya et al. (2001), Gilad et al. (2005)
Mouse	1000	23	Young and Trask (2002), Gilad et al. (2005)
Rat	1700	28	Young and Trask (2002), Gilad et al. (2005)
Zebrafish	133	12	Niimura (2012), Saraiva et al. (2015)
Dog	1100	20	Quignon et al. (2003), Niimura (2012)
Pig	1300	9	Nguyen et al. (2012)
Cat	1052	35	Galibert et al. (2016)
Gray short-tailed opossum (*Monodelphis domesitica*)	1492	20	Niimura and Nei (2007)
Platypus (*Monotremata*)	718	52	Niimura and Nei (2007)
Cow (*Cetartiodactyla*)	2119	46	Niimura and Nei (2007)
Primate (*Macaque*)	606	46	Niimura and Nei (2007)
Primate (*Chimpanzee*)	813	50–60	Gilad et al. (2005), Niimura (2012)
African elephant	4200	53	Niimura et al. (2014)
Chicken (*Gallus gallus*)	479	23	Steiger et al. (2009)
Zebra finch (*Taeniopygia guttata*)	553	40	Steiger et al. (2009)
Green anole (*Anolis carolinensis*)	156	27	Steiger et al. (2009)

2012, Ostrander 2012; Quignon et al. 2012; Rusyn et al. 2018). Genetic diversity of canine olfactory receptors comparing select breeds indicated differential effects by breed (Robin et al. 2009). Furthermore, breed clustering was found in good agreement with OR genotype clustering across multiple breeds, though not completely breed-specific (Galibert et al. 2016). Despite the significant variations across breeds, an analysis of the four olfactory receptor gene subfamilies in dogs across 26 breeds demonstrated stability in the number of genes in each subfamily (Issel-Tarver and Rine 1996). Specific alleles have been suggested to play a role in odor detection and odor recognition efficiency with specific single nucleotide polymorphisms (SNPs) (Lesniak et al. 2008). Furthermore, a study by Yang et al. suggests the selective breeding within Labrador retriever working dog populations has promoted SNP alleles of OR genes (Yang et al. 2022). Overall SNP distribution shows high heterogeneity across OR genes with more than 50% of OR genes disproportionally harboring the majority of SNPs. Also, across the dog breeds evaluated heterogeneity was observed with 35% of SNPs showing breed-specificity (Robin et al. 2009). Studies that focused on genetic variants of canine olfactory receptor genes have found a significant association with particular alleles such as *cOR52N9*, *cOR9S13*, and *OR10H1* linking single-nucleotide polymorphisms to odor detection capabilities (Lesniak et al. 2008; Yang et al. 2022). A series of studies using an olfactory epithelial brushing technique presented preliminary RNA profiling results in dogs that suggest the singular neuron olfactory receptor expression may be higher than in other species such as the rat and further demonstrated wide variations in expression which need further exploration to establish biological significance (Galibert et al. 2016; Azzouzi et al. 2022).

In a breed-select evaluation of 10 canine nasal biopsy samples, results suggest that up to 40% of the olfactory receptor gene repertoire may be silent and not expressed within the RNA profiles established (Azzouzi et al. 2022). However, other studies including mixed breeds suggest this rate to be much lower at only 14% of genes not detected (Saraiva et al. 2019). The overall ratios between both studies suggest a higher ratio of expression in dogs as compared to those in mice and humans and a large dynamic range was seen with some homologous subtypes highly abundant. Additionally, across all species analyzed (rat, mouse, dog, marmoset, macaque, and human), additional chemosensory receptors were expressed including trace amine-associated receptors (TAARs) and MS4A along with genetic markers for GUCY2D/GC-D+ or GUCY1B2+ OSNs. Of the TAAR gene family, TAAR5 was most abundant in humans and dogs, and of the MS4A gene family, MS4A7/Ms4a7 was most abundant in the marmoset, dog, and rat (Saraiva et al. 2019).

Comparisons of human, mouse, and canine olfactory receptor genes showed homologous gene clusters across various chromosomes between species showing close conservation of at least four olfactory receptor gene subfamilies (Carver et al. 1998). Between dogs and humans, there is a strongly conserved OR genomic distribution suggesting shared common mammalian ancestry, though subsequent expansion within the dog repertoire noted in emergence of canine-specific OR genes (Quignon et al. 2003).

Varying fractions of the OR genetic repertoire among all studied organisms are considered pseudogenes, or sequences that are non-functional due to the presence of frameshift mutations, nonsense mutations, or substantial deletions. Another possibility for incomplete fulfillment of OR genetic capacity occurs when there are partial sequences of intact genes, but they maintain the potential for completion without disruptive mutations. In studies evaluating the proportion of pseudogenes among functional genes, criteria for an OR gene to be considered functional are if the open reading frame starts at the initiation codon and ends at the stop codon. Additionally, truncated genes that maintain the capacity for complete sequencing without disruptive mutations are not included in pseudogene categories (Niimura

2012). Receptor diversity across species is evident as the percentages of known repertoire representing pseudogenes ranges from over 50% in some species such as humans to under 10% in other species such as pigs (Table 1) (Zozulya et al. 2001; Young and Trask 2002; Quignon et al. 2003; Gilad et al. 2005; Niimura and Nei 2007; Steiger et al. 2009; Nguyen et al. 2012; Niimura 2012; Niimura et al. 2014; Saraiva et al. 2015; Galibert et al. 2016). Canines have 1100 receptor genes, 75% considered intact or non-pseudo genes compared to humans with 802 odorant receptor genes and 48% considered intact or non-pseudo genes (Nei et al. 2008; Niimura et al. 2014). The relatively low pseudogene ratio and the breed-specific polymorphisms are suggestive of a low selective constraint relative to the high redundancy within the olfactory combinatorial code and pseudogenization process.

The OR sequences are further categorized into Class I and Class II genes. Class I genes are more conserved and represent receptor types that bind primarily hydrophilic odorants. Class I OR genes are further subcategorized based on whether they interact with airborne odorants (group α), water-soluble odorants (groups δ, ε, ζ, and η), or both (group β) (Niimura 2012). Class II OR genes represent those binding hydrophobic airborne odorants, and include only subgroup γ. Through olfactory receptor conservation amongst various species, mammals retain only Class I group α and Class II receptor genes, while Xenopus (amphibians) encode Class I and II genes. However, teleosts encode all subgroups of Class I without Class II genes (Table 2). Class I receptors may be referred to as "fish" genes due to the characteristic binding of hydrophilic odorants that are primarily water-soluble. However, a genetic repertoire to encode Class I receptors has been conserved among some non-aquatic and mammalian species that interact with only airborne odorants. In a phylogenic comparison of zebrafish to mouse OR Class I, it was demonstrated that the mice exhibit similar average pairwise identity to the zebrafish families with 27.3 ± 4.8%SD identity. A similar comparison of Class II OR between these species resulted in a 27.7 ± 5.5%SD identity match. They further showed that calculations comparing consensus sequences representing each family yielded similar results (Alioto and Ngai 2005). The conservation of these "fish" genes in non-aquatic species represents the evolutionarily tendency over time to develop pseudogenes in a particular class that has been more selectively limited. Some interpretations indicate that these seemingly unused yet conserved receptor types still have functional significance in humans (Niimura and Nei 2005a, b, c; Persuy et al. 2015).

Olfactory receptors (OR) of the MOE are only one of four olfactory receptor groups in mammals, though all share the common classification of being G-protein coupled receptors. Other receptor groups include vomeronasal receptors (V1R and V2R) and trace amine-associated receptors (TAARs). Distinct from the MOE, the vomeronasal organ (VNO) operates as an accessory olfactory system structure activating anatomically distinct neural pathways in response to pheromones. Not all mammals have an established VNO, as it is thought to have degenerated in some species such as humans. Unlike odorant transduction in the MOE, pheromones

Table 2 Olfactory receptor genes species variance in representative Class I and II groups

Species	α	Beta	γ	δ	ε	ζ	η
Human	58	–	329	–	–	–	–
Mouse	113	3	947	–	–	–	–
Rat	136	2	1121	–	–	–	–
Zebrafish		1	1	62	12	37	38
Dog	161	1	660	–	–	–	–
Primate (*Chimpanzee*)	64	–	335	–	–	–	–
Chicken (*Gallus gallus*)	10	–	290	–	–	–	–
Zebra finch (*Taeniopygia guttata*)	2	–	31	–	–	–	–
Green anole (*Anolis carolinensis*)	1	–	108	–	–	–	–
*Representative of intact OR genes	Steiger et al. (2009), Niimura (2012)						

are processed differently and engage with only two types of vomeronasal receptors (VR), V1R or V2R, which are expressed on microvilli of sensory neurons in the VNO that project their axons to the accessory OB (Korsching 2009). VR are far less diverse than OR found among the main olfactory epithelium, though the two subtypes share distinct evolutionary trajectories and are not thought to be universally conserved even in species with intact VNO. Dogs, cattle, and primates reportedly lack the V2R subtype while this gene family shows limited expression in rodents (Salazar et al. 1992; Young and Trask 2007).

A subset of VNO neurons (~1%) were found to express a novel group of functionally-related GPCRs in the vomeronasal organ, termed formyl peptide receptors (FPR) (Riviere et al. 2009). Expression of these receptors has also been established in cells of the immune system with demonstrated responses to formylated peptides and other proteins associated with inflammation (Le et al. 2007). FPR-expressing VNO neurons are thought to play a role in detection of infected tissues internally or among other species, as well as possible food contamination, based on the understanding that formyl peptides are released by bacteria (Riviere et al. 2009).

Among one subset of OSN known as GC-D neurons, which express guanylyl cyclase, the characteristic cAMP pathway common to most OSN in the MOE is absent while a cGMP-phosphodiesterase (PDE2A) pathway is displayed. This non-canonical pathway utilizes cGMP-sensitive cyclic nucleotide gated (CNG) channels versus the traditionally present cAMP-sensitive CNG channels. Information propagated to higher brain centers through GC-D neuron axons are found to synapse at glomeruli toward the caudal aspect of the olfactory bulb (Juilfs et al. 1997). Receptors demonstrating this transduction pathway were found to be sensitive to urinary peptides, uroguanylin, guanylin, and carbon dioxide in certain species (Leinders-Zufall et al. 2007; Sun et al. 2009).

Two basic principles of the olfactory system that underly signal transduction from OR to OB have been defined in most species. First, the "one-neuron-one-receptor" rule describes the concept that each individual OSN expresses a single functional OR type which interacts with a specific molecular range of odorants (Li et al. 2004). This suggests that a species can have any size of OR genetic repertoire, but only one of those specialized OR types is expressed on a given neuron. The second basic principle is that certain groupings of axons share target glomeruli, particularly OSNs expressing the same or similar ORs. Thus, combinations of receptors activated by related odorants propagate signals through axons that converge on a fixed set of glomeruli within the olfactory bulb (Korsching 2009). Consistent reports in agreement with the one neuron one receptor rule are most prevalent in zebrafish, though a few exceptions have been reported in goldfish (Speca et al. 1999; Sato et al. 2007).

The evaluation and categorization of receptor ligand pairs, deorphanization, have been performed to limited capacity across odorant receptors, with more limited studies in dogs due to epithelial access. However, some studies using in vitro cloning and expression have successfully deorphanized some of the receptor ligand pairs. This has resulted in the reinforcement that individual OR may respond to numerous odorants and that a single odorant may activate multiple OR types and additionally that odorants do not necessarily act in an additive fashion when applied in a mixture suggesting a non-additive complex receptor combinatorial code (Benbernou et al. 2007). Layering complexity into the odor mixture olfactory receptor interaction is the ligand dimerization, inhibition, and antagonism that can occur at a receptor (Oka et al. 2004; Katada et al. 2005).

Monogenic and Monoallelic expression of OR genes is achieved during the maturation and differentiation process of olfactory sensory neuron development (Monahan and Lomvardas 2015). Single genes are expressed in monoallelic fashion on a single olfactory sensory neuron, meaning the random expression of the maternal or paternal homologous allele for a given odorant receptor type occurs, which is a monoallelic expression of what can be a possible variant of the odor receptor gene type (Chess et al. 1994; Ishii et al. 2001; Mainland et al. 2014; Malnic et al. 2016). The receptor expressed guides not only selective odorant affinity and binding but also neuronal development guidance during axonal migration and establishment of neural circuitry with the respective glomeruli of the olfactory bulb (Feinstein et al. 2004; Feinstein and Mombaerts 2004).

Studies in rodent models have demonstrated a random distribution of a singly expressed OR across spatial zones within the olfactory epithelium resulting in an odor receptor expression pattern and zonular map (Mombaerts 1996; Mombaerts et al. 1996a, b; Zhu et al. 2022). Though there is more to learn regarding the full functionality of this zoning pattern and its presence across multiple species, it is suggested to aid in discrimination and may support the long-standing chromatographic theory (Mozell 1966; Ressler et al. 1993; Cenier et al. 2013; Scott et al. 2014; Secundo et al. 2014). However, recent modeling studies lack support for a corresponding odor deposition pattern aligning with higher affinity receptor subtypes within the epithelia (Coppola et al. 2019).

Graphical representations of signaling pathways, or "odor maps" are used to describe the combinations of specific OR-expressing neurons with their respective odorant receptive range. These species and odor-specific maps are key to understanding, and possibly engineering, the foundational molecular and neural mechanisms at play. At the core, odor maps use single-cell information about OR expression and connectivity to establish a larger scale view of combinatorial coding topographically. Early generated maps in rabbits, rodents, and Drosophila gave deeper insight into the consistency of molecular, anatomical, and functional organization and demonstrated that even species as distant as Drosophila share the basic olfactory system principles (Couto et al. 2005).

On average, OR proteins are 310 amino acids in length. Collective evaluation of these sequences reveals a degree of inter and intraspecies variability at specific locations. The respective variability or alignment is thought to indicate relatedness as well as functional significance in certain regions, particularly sites of odorant binding and G-protein activation. The GPCR family has shown very little divergence in key signaling characteristics. The OR as a member of this family, there are smiliarties at the genetic level associated with conservation of signaling molecules. The most notably conserved area is found where the third transmembrane region (TM3) meets the intracellular loop leading to transmembrane region 4 (TM4), in which the sequence of translated amino acids (aspartic acid–arginine–tyrosine) is shared amongst the rhodopsin-like GPCRs such as olfactory receptors (Rovati et al. 2007). This highly conserved region may be involved in the universal G-protein coupling mechanism observed in this family of receptors. In contrast, the diversity of odorant structures and receptor binding characteristics would likely implicate genetic variability at binding sites. Regions of hypervariability among olfactory receptor amino acid sequences are thought to be involved in ligand binding, and have been reported between TM3 and TM6 sites (Niimura 2012). Ligand-binding regions are more closely aligned between receptors of the same type, or orthologous receptors, that tend to interact with similar classes of odorants.

A unique feature of OR genes, and more broadly Rhodopsin-like GPCRs, is the lack of introns in their coding regions and additional 5′ untranslated exons found upstream of the coding region which can be alternatively spliced. These non-coding regions can generate multiple mRNA isoforms that are still translated into the same protein (Young et al. 2003; Niimura 2012).

Analyzing chromosomal distribution of OR genes is thought to reveal evolutionary relationships between organisms and respective conservation of specific sequences. While OR genes are found on almost all chromosomes, they are generally grouped in close proximity on individual chromosomes in what are termed 'clusters', or patterns in which fewer than 500 kb separates neighboring OR genes (Niimura and Nei 2003). Some genomic clusters contain only a few genes, while others contain over 100 OR genes; and the proportion of pseudogenes among these clusters varies and is thought to be associated with phylogenetic relationships. OR gene clusters often contain distantly related gene sequences while OR genes more closely related can be found on other chromosomes or clusters (Niimura and Nei

2003, 2005a, b, c). Functional OR genes that are phylogenetically related have been found in several chromosomal clusters, or tandem arrays, and a particular cluster often includes genes of multiple phylogenetic clades. Gene duplication alone, after which functional divergence or total inactivation may occur, does not sufficiently explain the apparent complexity of how phylogenetically related receptor genes are distributed across clusters. Other phenomena to explain OR evolution trends between tandem arrays of gene clusters have been explored and several explanations include repeated tandem duplication, chromosomal translocation, chromosomal inversion, unequal crossing over, and recombination. Gene duplication by unequal crossing over is thought to result in accumulation of mutations and increased diversity of the sequences of associated duplicates, which may influence rates of pseudogenization. In the case of a single fragmented cluster dispersing on different chromosomal regions through the process of chromosomal translocation, the observed distributions would be possible. A chromosome fission event at an OR gene cluster has been described in the great ape that may have generated human chromosomes 14 and 15 (Rudd et al. 2009) and suggests that several chromosomal rearrangements could have occurred at OR gene regions resulting in the shuffling of clusters consistent with observed genetic trends. Furthermore, recombination between different regions of chromosomes may result in complex dispersal of otherwise related clusters.

Copy number variations (CNVs), or variations seen in the genome structure itself which demonstrate many single nucleotide polymorphisms (SNPs), are common amongst OR genes (Waszak et al. 2010). These variations can lead to inactivation of genes or result in pseudogenes, which were previously reported to represent a range of less than 10% to over 50% of the total number of OR genes across species (Table 2). When a collection of individual functional OR genomes were analyzed, approximately 15% were affected by CNVs and 20% were affected by segregating pseudogenes (Waszak et al. 2010). Significant changes in genetic repertoire have been attributed to frequent gene duplications and individual rates of pseudogene formation as demonstrated in a broad scope analysis across species (Niimura 2012).

In species that are evolutionarily similar such as humans and chimpanzees, OR gene correlation seems to be significantly aligned (De la Cruz et al. 2009). Additionally, mouse and human OR gene clusters are generally well conserved although mice have a larger quantity of OR genes (Table 2). Although quantity of individual OR genes varies between these two species, the number of gene clusters that contain 5 or more OR genes (5+Clusters) is aligned. Therefore, while having comparable clusters, a single mouse OR gene cluster contains greater total numbers of OR genes than are observed in human cluster counterparts (Niimura and Nei 2005a, b, c). Variations of gene duplication or loss among individual lineages were studied across 13 placental mammals, in which intact OR genes were categorized into orthologous gene groups (OGGs) and sequentially compared. Among the 252 OGGs, the mean amino acid sequence identity was 81.3% and the median amino acid sequence identity was 82.1%. They found that more expanded gene lineages characterized by higher rates of gene duplication have evolved more dynamically in function between species. There were three OGGs with the highest amino acid sequence conservation coinciding with complete orthologous alignment across the evaluated species. This work concluded that the conservation of gene number and amino acid sequences of these OGGs may be indicative of functional importance common to placental mammals (Niimura et al. 2014).

6 Atypical Receptor Expression

Receptor presence is predominantly expressed within the olfactory epithelium, though more recent studies have demonstrated growing evidence of odor receptor expression in non-olfactory tissues. While the evolutionary purpose of ectopically expressed OR is relatively unknown, it is possible that they may serve important non-olfactory roles (De la Cruz et al. 2009). Supporting research has reported OR gene expression in atypical areas including the testis, where they had suggested involvement in sperm chemotaxis (Spehr et al. 2003). Additionally, OR expression in arterioles of the eye may be involved in detecting chemicals within the sensitive environment as an added alert mechanism for hazardous exposure (Pronin et al. 2014). Other reports have found OR expression in the kidney possibly implicating renin secretion and blood pressure regulation (Pluznick et al. 2009, 2013). Another atypical location with abundant OR expression was in enterochromaffin cells throughout portions of the gastrointestinal tract, where they are thought to function by inducing serotonin secretion. Also in this region but functionally distinct, OR expression in duodenal enterocytes was found to upregulate in response to high-fat diet in obese rats, which suggested a role in the dietary fat regulation and obesity susceptibility (Braun et al. 2007; Primeaux et al. 2013). OR localized in some areas of skeletal muscle had reported involvement in cell migration and muscle cell adhesion (Griffin et al. 2009). As a consensus, the most common functional attributes of OR in atypical regions include chemosensation, chemotaxis, and cell migration. Collectively, non-olfactory tissues in which OR have been found include the gastrointestinal tract, brain, kidney, muscle, placenta, eyes, and more, but the functional significance has not been established in each site. An important consideration is that OR proteins expressed in non-olfactory tissues appear to be regulated by different signaling factors than in traditionally-located nasal cavity (Persuy et al. 2015). OR gene expression regulation seems to differ in OSN of the main olfactory epithelium as compared to atypical sites observed. For example, pulmonary macrophages found to express OR genes were reported to respond to localized microbial infection. In this role, bacterially-released odorants appear to induce OR expression on proliferating macrophages that migration toward the site of infection (Li et al. 2013).

7 Comparative Olfactory Functional Variations

In fish species, the nose is comprised of a single olfactory organ known as the olfactory rosette, which lacks vomeronasal organ components. The olfactory epithelium of this organ consists of 3 olfactory neuron types, the most predominant of which are ciliated and microvillous cells, followed by crypt cells. Each innervates the same OB through fascicle bundles of olfactory nerves. The primary cells differ morphologically and spatially in the OE. While ciliated cells are found in the deeper basally located region of OE, microvillous cells are generally found in more superficial regions. Morphologically, ciliated OSN project long dendrite with extensive cilia processes while the microvillous OSN are equipped with shorter dendrites and topped with short microvilli. While a full molecular expression of receptors present in these cells has not been established (Korsching 2009), four main water-soluble odorant categories detected by teleosts have been suggested including amino acids, gonadal steroids, bile acids, and prostaglandins (Niimura 2012).

The olfactory transcriptome of zebrafish and mice was compared for analysis of evolutionary relationships, and results indicated strong conservation of cellular and molecular foundational mechanisms even though millions of years of evolutionary divergence separate the species. From an outside perspective, the single olfactory system of aquatic-dwelling teleosts would seem quite distinct from the dual vomeronasal and olfactory system of the air-breathing rodents. However, further focus on the molecular relationships between the two revealed a high degree of molecular conservation. For example, all but a single chemosensory receptor class-tested in mice were expressed in the zebrafish olfactory organ. While there were divergent classes of sensory neurons present, a closely correlated neural distribution and receptor gene repertoire were demonstrated (Saraiva et al. 2015). Zebrafish are phylogenetically found between mammals and insects/worms (Korsching 2009).

The insect olfactory system is evolutionarily independent from mammals, though many functional, pathway components, and signaling process correlations have been demonstrated. Odorant receptors in insects belong to the same family as gustatory receptors (Sato et al. 2008). These are 7-transmembrane proteins that are inversely embedded in the cell membrane and containing an N-terminus extension into the cytosol (Benton et al. 2006). Very distinct from GPCR classes, these receptors in OSN are ion channels that require heterodimerization to a secondary odorant receptor protein, termed Orco, to activate (Silbering and Benton 2010). These odorant receptors are present on hair-like protrusions from exoskeleton, or sensilla, of insect antennae. Their axons project to the antennal lobe, analogous to the OB in mammals, where a synaptic cleft with respective glomeruli is formed. Furthermore, the patterned code through which secondary neurons send information to what is known as the mushroom body is analogous to olfactory signal transduction to the piriform cortex in mammals (Vosshall 2001).

A series of events must occur to achieve the perception of odor. This sensation fundamentally requires an adequate stimulus with access to an accepting and available receptor that can interact within a supportive environment resulting in activation of signal transduction. Dogs are highly adapted as macrosmatic animals demonstrating higher olfactory acuity, which can be attributed to various factors including anatomical, physiological, and genetic considerations related to the peripheral olfactory pathway covered in this chapter.

Acknowledgements We would like to thank Dr. Lucia Lazarowski for the invitation to write this chapter. We also thank Dr. Xu Wang and Michelle Aono for their review and helpful comments on the chapter. Additionally, we thank Dr. Edward Morrison, Dr. Robert Judd, Dr. Vitaly Vodyanoy, and Dr. Iryna Sorokulova for their mentorship, input, and review of original supporting work.

References

Agarwal RP, Henkin RI (1982) Zinc and copper in human cerebrospinal fluid. Biol Trace Elem Res 4(2):117–124

Ajmani GS, Suh HH, Pinto JM (2016) Effects of ambient air pollution exposure on olfaction: a review. Environ Health Perspect 124(11):1683–1693

Alioto TS, Ngai J (2005) The odorant receptor repertoire of teleost fish. BMC Genomics 6:173

Azzouzi N, Guillory A-S, Chaudieu G, Galibert F (2022) Dog olfactory receptor gene expression profiling using samples derived from nasal epithelium brushing. Canine Med Genet 9(1):7

Badonnel K, Durieux D, Monnerie R, Grebert D, Salesse R, Caillol M, Baly C (2009) Leptin-sensitive OBP-expressing mucous cells in rat olfactory epithelium: a novel target for olfaction-nutrition crosstalk? Cell Tissue Res 338(1):53–66

Barrios AW, Sánchez-Quinteiro P, Salazar I (2014) Dog and mouse: toward a balanced view of the mammalian olfactory system. Front Neuroanat 8:106

Benbernou N, Tacher S, Robin S, Rakotomanga M, Senger F, Galibert F (2007) Functional analysis of a subset of canine olfactory receptor genes. J Hered 98(5):500–505

Benton R, Sachse S, Michnick S, Vosshall L (2006) Atypical membrane topology and heteromeric function of Drosophila odorant receptors in vivo. Chem Senses 31(5):A5–A5

Bianchi F, Flisi S, Careri M, Riboni N, Resimini S, Sala A, Conti V, Mattarozzi M, Taddei S, Spadini C, Basini G, Grolli S, Cabassi CS, Ramoni R (2019) Vertebrate odorant binding proteins as antimicrobial humoral components of innate immunity for pathogenic microorganisms. PLoS ONE 14(3):e0213545

Bird DJ, Hamid I, Fox-Rosales L, Van Valkenburgh B (2020) Olfaction at depth: cribriform plate size declines with dive depth and duration in aquatic arctoid carnivorans. Ecol Evol 10(14):6929–6953. https://doi.org/10.1002/ece3.6343

Bird DJ, Jacquemetton C, Buelow SA, Evans AW, Van Valkenburgh B (2021) Domesticating olfaction: dog breeds, including scent hounds, have reduced cribriform plate morphology relative to wolves. Anat Rec (Hoboken) 304(1):139–153. https://doi.org/10.1002/ar.24518

Biswas K, Wagner Mackenzie B, Ballauf C, Draf J, Douglas RG, Hummel T (2020) Loss of bacterial diversity in the sinuses is associated with lower smell discrimination scores. Sci Rep 10(1):16422

Braun T, Voland P, Kunz L, Prinz C, Gratzl M (2007) Enterochromaffin cells of the human gut: sensors for spices and odorants. Gastroenterology 132(5):1890–1901

Briand L, Eloit C, Nespoulous C, Bezirard V, Huet JC, Henry C, Blon F, Trotier D, Pernollet JC (2002) Evidence of an odorant-binding protein in the human olfactory mucus: location, structural characterization, and odorant-binding properties. Biochemistry 41(23):7241–7252

Bryche B, Baly C, Meunier N (2021) Modulation of olfactory signal detection in the olfactory epithelium: focus on the internal and external environment, and the emerging role of the immune system. Cell Tissue Res 384(3):589–605

Buck L, Axel R (1991) A novel multigene family may encode odorant receptors—a molecular-basis for odor recognition. Cell 65(1):175–187

Carver EA, Issel-Tarver L, Rine J, Olsen AS, Stubbs L (1998) Location of mouse and human genes corresponding to conserved canine olfactory receptor gene subfamilies. Mamm Genome 9(5):349–354

Casadei E, Tacchi L, Lickwar CR, Espenschied ST, Davison JM, Munoz P, Rawls JF, Salinas I (2019) Commensal bacteria regulate gene expression and differentiation in vertebrate olfactory systems through transcription factor REST. Chem Senses 44(8):615–630

Cavelius M, Brunel T, Didier A (2022) Lessons from behavioral lateralization in olfaction. Brain Struct Funct 227(2):685–696

Cenier T, McGann JP, Tsuno Y, Verhagen JV, Wachowiak M (2013) Testing the sorption hypothesis in olfaction: a limited role for sniff strength in shaping primary odor representations during behavior. J Neurosci 33(1):79–92

Challis RC, Tian H, Wang J, He J, Jiang J, Chen X, Yin W, Connelly T, Ma L, Yu CR, Pluznick JL, Storm DR, Huang L, Zhao K, Ma M (2015) An olfactory cilia pattern in the mammalian nose ensures high sensitivity to odors. Curr Biol 25(19):2503–2512

Chaves-Moreno D, Plumeier I, Kahl S, Krismer B, Peschel A, Oxley AP, Jauregui R, Pieper DH (2015) The microbial community structure of the cotton rat nose. Environ Microbiol Rep 7(6):929–935

Chen R, Irwin DM, Zhang YP (2012) Differences in selection drive olfactory receptor genes in different directions in dogs and wolf. Mol Biol Evol 29(11):3475–3484

Chess A, Simon I, Cedar H, Axel R (1994) Allelic inactivation regulates olfactory receptor gene expression. Cell 78(5):823–834

Coppola DM, Fitzwater E, Rygg AD, Craven BA (2019) Tests of the chromatographic theory of olfaction with highly soluble odors: a combined electro-olfactogram and computational fluid dynamics study in the mouse. Biology Open 8(10):bio047217

Costanzo RM (1991) Regeneration of olfactory receptor cells. Ciba Found Symp 160:233–242; discussion 243–238

Couto A, Alenius M, Dickson BJ (2005) Molecular, anatomical, and functional organization of the Drosophila olfactory system. Curr Biol 15(17):1535–1547

Craven BA, Paterson EG, Settles GS (2004) The fluid dynamics of canine olfaction: unique nasal airflow patterns as an explanation of macrosmia. J R Soc Interface 7(47):933–943

Craven BA, Neuberger T, Paterson EG, Webb AG, Josephson EM, Morrison EE, Settles GS (2007) Reconstruction and morphometric analysis of the nasal airway of the dog (Canis familiaris) and implications regarding olfactory airflow. Anat Rec 290(11):1325–1340

Craven BA, Paterson EG, Settles GS, Lawson MJ (2009) Development and verification of a high-fidelity computational fluid dynamics model of canine nasal airlow. J Biomech Eng 131:091002

Craven BA, Paterson EG, Settles GS (2010) The fluid dynamics of canine olfaction: unique nasal airflow patterns as an explanation of macrosmia. J R Soc Interface 7(47):933–943

De la Cruz O, Blekhman R, Zhang X, Nicolae D, Firestein S, Gilad Y (2009) A signature of evolutionary constraint on a subset of ectopically expressed olfactory receptor genes. Mol Biol Evol 26(3):491–494

Derrien T, Vaysse A, Andre C, Hitte C (2012) Annotation of the domestic dog genome sequence: finding the missing genes. Mamm Genome 23(1–2):124–131

Doty R (1998) Textbook of clinical neurology. Saunders, Philadelphia

Dzieciol M, Podgorski P, Stanczyk E, Szumny A, Woszczylo M, Pieczewska B, Nizanski W, Nicpon J, Wrzosek MA (2020) MRI features of the vomeronasal organ in dogs (Canis Familiaris). Front Vet Sci 7

Evans HE, de Lahunta A (2013) Miller's anatomy of the dog. E-book, Elsevier Health Sciences

Eyre AW, Zapata I, Hare E, Lee KMN, Bellis C, Essler JL, Otto CM, Serpell JA, Alvarez CE (2022) Genome scanning of behavioral selection in a canine olfactory detection breeding cohort. Sci Rep 12(1):14984

Feinstein P, Mombaerts P (2004) A contextual model for axonal sorting into glomeruli in the mouse olfactory system. Cell 117(6):817–831

Feinstein P, Bozza T, Rodriguez I, Vassalli A, Mombaerts P (2004) Axon guidance of mouse olfactory sensory neurons by odorant receptors and the beta2 adrenergic receptor. Cell 117(6):833–846

Francois A, Grebert D, Rhimi M, Mariadassou M, Naudon L, Rabot S, Meunier N (2016) Olfactory epithelium changes in germfree mice. Sci Rep 6:24687

Frederickson CJ, Danscher G (1990) Zinc-containing neurons in hippocampus and related CNS structures. Prog Brain Res 83:71–84

Friling L, Nyman HT, Johnson V (2014) Asymmetric nasal mucosal thickening in healthy dogs consistent with the nasal cycle as demonstrated by MRI and CT. Vet Radiol Ultrasound 55(2):159–165

Galibert F, Azzouzi N, Quignon P, Chaudieu G (2016) The genetics of canine olfaction. J Vet Behav 16:86–93

Gibson GR, Roberfroid MB (2008) Handbook of prebiotics. CRC Press, Boca Raton

Gilad Y, Man O, Glusman G (2005) A comparison of the human and chimpanzee olfactory receptor gene repertoires. Genome Res 15(2):224–230

Glusman G, Yanai I, Rubin I, Lancet D (2001) The complete human olfactory subgenome. Genome Res 11(5):685–702

Graziadei GA, Graziadei PP (1979) Neurogenesis and neuron regeneration in the olfactory system of mammals. II. Degeneration and reconstitution of the olfactory sensory neurons after axotomy. J Neurocytol 8(2):197–213

Griffin CA, Kafadar KA, Pavlath GK (2009) MOR23 promotes muscle regeneration and regulates cell adhesion and migration. Dev Cell 17(5):649–661

Grosmaitre X, Santarelli LC, Tan J, Luo M, Ma M (2007) Dual functions of mammalian olfactory sensory neurons as odor detectors and mechanical sensors. Nat Neurosci 10(3):348–354

Gruneberg H (1973) A ganglion probably belonging to the N terminalis system in the nasal mucosa of the mouse. Z Anat Entwicklungsgesch 140(1):39–52

Hecht EE, Smaers JB, Dunn WD, Kent M, Preuss TM, Gutman DA (2019) Significant neuroanatomical variation among domestic dog breeds. J Neurosci 39(39):7748–7758. https://doi.org/10.1523/JNEUROSCI.0303-19.2019

Henkin RIDA, Martin BM (2000) New frontiers in immunobiology. Kugler Publications; Library Research Associates [Distributor]; The Hague; Netherlands

Henkin RI, Hosein S, Stateman WA, Knoppel AB, Abdelmeguid M (2017) Improved smell function with increased nasal mucus sonic hedgehog in hyposmic patients after treatment with oral theophylline. Am J Otolaryngol 38(2):143–147

Hinds JW, Hinds PL, McNelly NA (1984) An autoradiographic study of the mouse olfactory epithelium: evidence for long-lived receptors. Anat Rec 210(2):375–383

Holl AM (2018) Survival of mature mouse olfactory sensory neurons labeled genetically perinatally. Mol Cell Neurosci 88:258–269

Hu J, Zhong C, Ding C, Chi Q, Walz A, Mombaerts P, Matsunami H, Luo M (2007) Detection of near-atmospheric concentrations of CO_2 by an olfactory subsystem in the mouse. Science 317(5840):953–957

Isaiah A, Hoffmann AR, Kelley R, Mundell P, Steiner JM, Suchodolski JS (2017) Characterization of the nasal and oral microbiota of detection dogs. PLoS ONE 12(9):e0184899

Ishii T, Serizawa S, Kohda A, Nakatani H, Shiroishi T, Okumura K, Iwakura Y, Nagawa F, Tsuboi A, Sakano H (2001) Monoallelic expression of the odourant receptor gene and axonal projection of olfactory sensory neurones. Genes Cells 6(1):71–78

Ishimaru T, Tsukatani T, Miwa T, Furukawa M (2000) Zinc modulates the electro-olfactogram of the frog. Auris Nasus Larynx 27(3):257–260

Issel-Tarver L, Rine J (1996) Organization and expression of canine olfactory receptor genes. Proc Natl Acad Sci USA 93(20):10897–10902

Jackson MJ, Jones DA, Edwards RHT (1982) Tissue zinc levels as an index of body zinc status. Clin Physiol 2(4):333–343

Jenkins EK, DeChant MT, Perry EB (2018) When the nose doesn't know: canine olfactory function associated with health, management, and potential links to microbiota. Front Vet Sci 5:56

Jia H, Pustovyy OM, Wang Y, Waggoner P, Beyers RJ, Schumacher J, Wildey C, Morrison E, Salibi N, Denney TS, Vodyanoy VJ, Deshpande G (2016) Enhancement of odor-induced activity in the canine brain by zinc nanoparticles: a functional MRI study in fully unrestrained conscious dogs. Chem Senses 41(1):53–67

Jorgensen E (2009) Emission of volatile compounds by seeds under different environmental conditions. Am Midl Nat 145:419–422

Juilfs DM, Fulle HJ, Zhao AZ, Houslay MD, Garbers DL, Beavo JA (1997) A subset of olfactory neurons that selectively express cGMP-stimulated phosphodiesterase (PDE2) and guanylyl cyclase-D define a unique olfactory signal transduction pathway. Proc Natl Acad Sci USA 94(7):3388–3395

Kaliner MA (1991) Human nasal respiratory secretions and host defense. Am Rev Respir Dis 144(3):S52–S56

Katada S, Hirokawa T, Oka Y, Suwa M, Touhara K (2005) Structural basis for a broad but selective ligand spectrum of a mouse olfactory receptor: mapping the odorant-binding site. J Neurosci 25(7):1806–1815

Kavoi BM, Jameela H (2011) Comparative morphometry of the olfactory bulb, tract and stria in the human, dog and goat. Int J Morphol 29(3):939–946

Kikusui T, Takigami S, Takeuchi Y, Mori Y (2001) Alarm pheromone enhances stress-induced hyperthermia in rats. Physiol Behav 72(1–2):45–50

Knight R, Vrbanac A, Taylor BC, Aksenov A, Callewaert C, Debelius J, Gonzalez A, Kosciolek T, McCall LI, McDonald D, Melnik AV, Morton JT, Navas J, Quinn RA, Sanders JG, Swafford AD, Thompson LR, Tripathi A, Xu ZZ, Zaneveld JR, Zhu Q, Caporaso JG, Dorrestein PC (2018) Best practices for analysing microbiomes. Nat Rev Microbiol 16(7):410–422

Kondo K, Suzukawa K, Sakamoto T, Watanabe K, Kanaya K, Ushio M, Yamaguchi T, Nibu K, Kaga K, Yamasoba T (2010) Age-related changes in cell dynamics of the postnatal mouse olfactory neuroepithelium: cell proliferation, neuronal differentiation, and cell death. J Comp Neurol 518(11):1962–1975

Korsching S (2009) The molecular evolution of teleost olfactory receptor gene families. Results Probl Cell Differ 47:37–55

Koskinen K, Reichert JL, Hoier S, Schachenreiter J, Duller S, Moissl-Eichinger C, Schopf V (2018) The nasal microbiome mirrors and potentially shapes olfactory function. Sci Rep 8(1):1296

Kratzing JE (1984) The anatomy and histology of the nasal cavity of the koala (Phascolarctos cinereus). J Anat 138(Pt 1):55–65

Kuehn M, Welsch H, Zahnert T, Hummel T (2008) Changes of pressure and humidity affect olfactory function. Eur Arch Otorhinolaryngol 265(3):299–302

Kurian SM, Naressi RG, Manoel D, Barwich A-S, Malnic B, Saraiva LR (2021) Odor coding in the mammalian olfactory epithelium. Cell Tissue Res 383(1):445–456

Lawson MJ, Craven BA, Paterson EG, Settles GS (2012) A computational study of odorant transport and deposition in the canine nasal cavity: implications for olfaction. Chem Senses 37(6):553–566

Le YY, Wang JMM, Liu XL, Kong Y, Hou XW, Ruan LF, Mou HW (2007) Biologically active peptides interacting with the G protein-coupled formylpeptide receptors. Protein Pept Lett 14(9):846–853

Leinders-Zufall T, Cockerham RE, Michalakis S, Biel M, Garbers DL, Reed RR, Zufall F, Munger SD (2007) Contribution of the receptor guanylyl cyclase GC-D to chemosensory function in the olfactory epithelium. Proc Natl Acad Sci USA 104(36):14507–14512

Lesniak A, Walczak M, Jezierski T, Sacharczuk M, Gawkowski M, Jaszczak K (2008) Canine olfactory receptor gene polymorphism and its relation to odor detection performance by sniffer dogs. J Hered 99(5):518–527

Leung CT, Coulombe PA, Reed RR (2007) Contribution of olfactory neural stem cells to tissue maintenance and regeneration. Nat Neurosci 10(6):720–726

Li J, Ishii T, Feinstein P, Mombaerts P (2004) Odorant receptor gene choice is reset by nuclear transfer from mouse olfactory sensory neurons. Nature 428(6981):393–399

Li JJ, Tay HL, Plank M, Essilfie AT, Hansbro PM, Foster PS, Yang M (2013) Activation of olfactory receptors on mouse pulmonary macrophages promotes monocyte chemotactic protein-1 production. PLoS ONE 8(11):e80148
Liang F (2020) Sustentacular cell enwrapment of olfactory receptor neuronal dendrites: an update. Genes 11(5):493
Lippard SJ (2014) Metalloneurochemistry: an overview and mobile zinc signaling in sensory perception. Abstracts Papers Am Chem Soc 248
Liu AH, Zhang X, Stolovitzky GA, Califano A, Firestein SJ (2003) Motif-based construction of a functional map for mammalian olfactory receptors. Genomics 81(5):443–456
Ma M (2010) Multiple olfactory subsystems convey various sensory signals. The neurobiology of olfaction. A. Menini, Boca Raton (FL)
Mackay-Sim A (2010) Stem cells and their niche in the adult olfactory mucosa. Arch Ital Biol 148(2):47–58
Mainland JD, Keller A, Li YR, Zhou T, Trimmer C, Snyder LL, Moberly AH, Adipietro KA, Liu WL, Zhuang H, Zhan S, Lee SS, Lin A, Matsunami H (2014) The missense of smell: functional variability in the human odorant receptor repertoire. Nat Neurosci 17(1):114–120
Malnic B, Hirono J, Sato T, Buck LB (1999) Combinatorial receptor codes for odors. Cell 96(5):713–723
Malnic B, Armelin-Correa L, Nagai M (2016) Monogenic and monoallelic expression of odorant receptors. Mol Pharmacol 90
Mamasuew K, Breer H, Fleischer J (2008) Grueneberg ganglion neurons respond to cool ambient temperatures. Eur J Neurosci 28(9):1775–1785
Mayer U, Kuller A, Daiber PC, Neudorf I, Warnken U, Schnolzer M, Frings S, Mohrlen F (2009) The proteome of rat olfactory sensory cilia. Proteomics 9(2):322–334
Mellert TK, Getchell ML, Sparks L, Getchell TV (1992) Characterization of the immune barrier in human olfactory mucosa. Otolaryngol Head Neck Surg 106(2):181–188
Menco M (1980) Qualitative and quantitative freeze-fracture studies on olfactory and respiratory epithelial surfaces of frog, ox, rat, and dog. IV. Ciliogenesis and ciliary necklaces (including high-voltage observations). Cell Tissue Res 212(1):1–16
Menco BPM, Jackson JE (1997) A banded topography in the developing rat's olfactory epithelial surface. J Comp Neurol 388(2):293–306
Menco BPM, Morrison EE (2003) Morphology of the mammalian olfactory epithelium: form, fine structure, function, and pathology. In: Doty RL (ed) Handbook of olfaction and gustation. Marcel Dekker, Inc., New York, pp 17–49
Mombaerts P (1996) Targeting olfaction. Curr Opin Neurobiol 6(4):481–486
Mombaerts P, Wang F, Dulac C, Chao SK, Nemes A, Mendelsohn M, Edmondson J, Axel R (1996a) Visualizing an olfactory sensory map. Cell 87(4):675–686
Mombaerts P, Wang F, Dulac C, Vassar R, Chao SK, Nemes A, Mendelsohn M, Edmondson J, Axel R (1996b) The molecular biology of olfactory perception. Cold Spring Harb Symp Quant Biol 61:135–145
Monahan K, Lomvardas S (2015) Monoallelic expression of olfactory receptors. Annu Rev Cell Dev Biol 31:721–740
Moore WJ (1981) The mammalian skull. Cambridge University Press, Cambridge; New York
Moran DT, Rowley JC 3rd, Jafek BW, Lovell MA (1982) The fine structure of the olfactory mucosa in man. J Neurocytol 11(5):721–746
Morrison EE, Costanzo RM (1989) Scanning electron microscopic study of degeneration and regeneration in the olfactory epithelium after axotomy. J Neurocytol 18(3):393–405
Morrison EE, Costanzo RM (1990) Morpholgy of the human olfactory epithelium. J Comp Neurol 297:1–13
Morrison EE, Costanzo RM (1992) Morphology of olfactory epithelium in humans and other vertebrates. Microsc Res Tech 23(1):49–61
Mozell MM (1966) The spatiotemporal analysis of odorants at the level of the olfactory receptor sheet. J Gen Physiol 50(1):25–41

Nagashima A, Touhara K (2010) Enzymatic conversion of odorants in nasal mucus affects olfactory glomerular activation patterns and odor perception. J Neurosci 30(48):16391–16398

Naudon L, Francois A, Mariadassou M, Monnoye M, Philippe C, Bruneau A, Dussauze M, Rue O, Rabot S, Meunier N (2020) First step of odorant detection in the olfactory epithelium and olfactory preferences differ according to the microbiota profile in mice. Behav Brain Res 384:112549

Nei M, Niimura Y, Nozawa M (2008) The evolution of animal chemosensory receptor gene repertoires: roles of chance and necessity. Nat Rev Genet 9(12):951–963

Nguyen DT, Lee K, Choi H, Choi MK, Le MT, Song N, Kim JH, Seo HG, Oh JW, Lee K, Kim TH, Park C (2012) The complete swine olfactory subgenome: expansion of the olfactory gene repertoire in the pig genome. BMC Genomics 13:584

Niimura Y (2012) Olfactory receptor multigene family in vertebrates: from the viewpoint of evolutionary genomics. Curr Genomics 13(2):103–114

Niimura Y, Nei M (2003) Evolution of olfactory receptor genes in the human genome. Proc Natl Acad Sci USA 100(21):12235–12240

Niimura Y, Nei M (2005a) Comparative evolutionary analysis of olfactory receptor gene clusters between humans and mice. Gene 346:13–21

Niimura Y, Nei M (2005b) Evolutionary changes of the number of olfactory receptor genes in the human and mouse lineages. Gene 346:23–28

Niimura Y, Nei M (2005c) Evolutionary dynamics of olfactory receptor genes in fishes and tetrapods. Proc Natl Acad Sci USA 102(17):6039–6044

Niimura Y, Nei M (2007) Extensive gains and losses of olfactory receptor genes in mammalian evolution. PLoS ONE 2(8):e708

Niimura Y, Matsui A, Touhara K (2014) Extreme expansion of the olfactory receptor gene repertoire in African elephants and evolutionary dynamics of orthologous gene groups in 13 placental mammals. Genome Res 24(9):1485–1496

Ohkubo K, Baraniuk JN, Hohman R, Merida M, Hersh LB, Kaliner MA (1998) Aminopeptidase activity in human nasal mucosa. J Allergy Clin Immunol 102(5):741–750

Oka Y, Omura M, Kataoka H, Touhara K (2004) Olfactory receptor antagonism between odorants. Embo J 23(1):120–126

Olender T, Fuchs T, Linhart C, Shamir R, Adams M, Kalush F, Khen M, Lancet D (2004) The canine olfactory subgenome. Genomics 83(3):361–372

Olender T, Jones TEM, Bruford E, Lancet D (2020) A unified nomenclature for vertebrate olfactory receptors. BMC Evol Biol 20(1):42

Ostrander EA (2012) Introduction. Mamm Genome 23(1):1–2

Ould-Yahoui A, Sbai O, Baranger K, Bernard A, Gueye Y, Charrat E, Clement B, Gigmes D, Dive V, Girard SD, Feron F, Khrestchatisky M, Rivera S (2013) Role of matrix metalloproteinases in migration and neurotrophic properties of nasal olfactory stem and ensheathing cells. Cell Transplant 22(6):993–1010

Pelosi P (1994) Odorant-binding proteins. Crit Rev Biochem Mol Biol 29(3):199–228

Pelosi P (1996) Perireceptor events in olfaction. J Neurobiol 30(1):3–19

Persuy MA, Sanz G, Tromelin A, Thomas Danguin T, Gibrat JF, Pajot Augy E (2015) Mammalian olfactory receptors: molecular mechanisms of odorant detection, 3D-modeling, and structure-activity relationships. Prog Mol Biol Transl Sci 130:1–36

Pevsner J, Synder SH (1990) Odorant binding protein: odorant transport function in the vertebrate nasal epithelium. Chem Senses 15(2):217–222

Pluznick JL, Zou DJ, Zhang X, Yan Q, Rodriguez-Gil DJ, Eisner C, Wells E, Greer CA, Wang T, Firestein S, Schnermann J, Caplan MJ (2009) Functional expression of the olfactory signaling system in the kidney. Proc Natl Acad Sci USA 106(6):2059–2064

Pluznick JL, Protzko RJ, Gevorgyan H, Peterlin Z, Sipos A, Han J, Brunet I, Wan LX, Rey F, Wang T, Firestein SJ, Yanagisawa M, Gordon JI, Eichmann A, Peti-Peterdi J, Caplan MJ (2013) Olfactory receptor responding to gut microbiota-derived signals plays a role in renin secretion and blood pressure regulation. Proc Natl Acad Sci USA 110(11):4410–4415

Primeaux SD, Braymer HD, Bray GA (2013) High fat diet differentially regulates the expression of olfactory receptors in the duodenum of obesity-prone and obesity-resistant rats. Dig Dis Sci 58(1):72–76

Pronin A, Levay K, Velmeshev D, Faghihi M, Shestopalov VI, Slepak VZ (2014) Expression of olfactory signaling genes in the eye. PLoS ONE 9(4):e96435

Qifan W, Jun S, Yang Z, Wanjun L (2017) Influence of environmental factors on volatile organic compound emissions from plywood tested by a rapid detection method. For Prod J 67(1–2):120–125

Quignon P, Kirkness E, Cadieu E, Touleimat N, Guyon R, Renier C, Hitte C, Andre C, Fraser C, Galibert F (2003) Comparison of the canine and human olfactory receptor gene repertoires. Genome Biol 4:R80

Quignon P, Giraud M, Rimbault M, Lavigne P, Tacher S, Morin E, Retout E, Valin AS, Lindblad-Toh K, Nicolas J, Galibert F (2005) The dog and rat olfactory receptor repertoires. Genome Biol 6(10):R83

Quignon P, Tacher S, Rimbault M, Galibert F (2006) The dog olfactory and vomeronasal receptor repertoires. Dog Genome 44:221–231

Quignon P, Rimbault M, Robin S, Galibert F (2012) Genetics of canine olfaction and receptor diversity. Mamm Genome 23(1–2):132–143

Ramon-Cueto A, Avila J (1998) Olfactory ensheathing glia: properties and function. Brain Res Bull 46(3):175–187

Randall HW, Bogdanffy MS, Morgan KT (1987) Enzyme histochemistry of the rat nasal mucosa embedded in cold glycol methacrylate. Am J Anat 179(1):10–17

Reed RR (2004) After the holy grail: establishing a molecular basis for Mammalian olfaction. Cell 116(2):329–336

Ressler KJ, Sullivan SL, Buck LB (1993) A zonal organization of odorant receptor gene expression in the olfactory epithelium. Cell 73(3):597–609

Riviere S, Challet L, Fluegge D, Spehr M, Rodriguez I (2009) Formyl peptide receptor-like proteins are a novel family of vomeronasal chemosensors. Nature 459(7246):574–577

Robert-Hazotte A, Schoumacker R, Semon E, Briand L, Guichard E, Le Quéré J-L, Faure P, Heydel J-M (2019) Ex vivo real-time monitoring of volatile metabolites resulting from nasal odorant metabolism. Sci Rep 9(1):2492

Robin S, Tacher S, Rimbault M, Vaysse A, Dréano S, André C, Hitte C, Galibert F (2009) Genetic diversity of canine olfactory receptors. BMC Genomics 10:21–21

Rodolfo-Maera T (1943) Su l'esistenza di un particolare organo olfacttivo nel setto nasale della cavia e di altri roditori. Arch Ital Anat Embryol

Rovati GE, Capra V, Neubig RR (2007) The highly conserved DRY motif of class A G protein-coupled receptors: beyond the ground state. Mol Pharmacol 71(4):959–964

Rowe FA, Smith WE (1972) Effects of peripherally induced anosmia on mating behavior of male mice. Psychon Sci 27(1):33–34

Rudd MK, Endicott RM, Friedman C, Walker M, Young JM, Osoegawa K, Program NCS, de Jong PJ, Green ED, Trask BJ (2009) Comparative sequence analysis of primate subtelomeres originating from a chromosome fission event. Genome Res 19(1):33–41

Rusyn I, Kleeberger SR, McAllister KA, French JE, Svenson KL (2018) Introduction to mammalian genome special issue: the combined role of genetics and environment relevant to human disease outcomes. Mamm Genome 29(1–2):1–4

Rygg AD, van Duin AC, Craven BA (2013) Molecular dynamics simulations of water/mucus partition coefficients for feeding stimulants in fish and the implications for olfaction. PLoS ONE 8(9):e72271

Rygg AD, Van Valkenburgh B, Craven BA (2017) The influence of sniffing on airflow and odorant deposition in the canine nasal cavity. Chem Senses 42(8):683–698

Salazar I, Barber PC, Cifuentes JM (1992) Anatomical and immunohistological demonstration of the primary neural connections of the vomeronasal organ in the dog. Anat Rec 233(2):309–313

Saraiva LR, Ahuja G, Ivandic I, Syed AS, Marioni JC, Korsching SI, Logan DW (2015) Molecular and neuronal homology between the olfactory systems of zebrafish and mouse. Sci Rep 5:11487

Saraiva LR, Riveros-McKay F, Mezzavilla M, Abou-Moussa EH, Arayata CJ, Makhlouf M, Trimmer C, Ibarra-Soria X, Khan M, Van Gerven L, Jorissen M, Gibbs M, O'Flynn C, McGrane S, Mombaerts P, Marioni JC, Mainland JD, Logan DW (2019) A transcriptomic atlas of mammalian olfactory mucosae reveals an evolutionary influence on food odor detection in humans. Sci Adv 5(7):eaax0396

Sato Y, Miyasaka N, Yoshihara Y (2007) Hierarchical regulation of odorant receptor gene choice and subsequent axonal projection of olfactory sensory neurons in zebrafish. J Neurosci 27(7):1606–1615

Sato K, Pellegrino M, Nakagawa T, Vosshall LB, Touhara K (2008) Insect olfactory receptors are heteromeric ligand-gated ion channels. Nature 452(7190):1002-U1009

Schwob JE, Youngentob SL, Mezza RC (1995) Reconstitution of the rat olfactory epithelium after methyl bromide-induced lesion. J Comp Neurol 359(1):15–37

Scott JW, Scott-Johnson PE (2002) The electroolfactogram: a review of its history and uses. Microsc Res Tech 58(3):152–160

Scott JW, Sherrill L, Jiang J, Zhao K (2014) Tuning to odor solubility and sorption pattern in olfactory epithelial responses. J Neurosci 34(6):2025–2036

Secundo L, Snitz K, Sobel N (2014) The perceptual logic of smell. Curr Opin Neurobiol 25:107–115

Sensi SL, Paoletti P, Bush AI, Sekler I (2009) Zinc in the physiology and pathology of the CNS. Nat Rev Neurosci 10(11):780-U738

Settles GS, Kester DA, Dodson-Dreibelbism LJ (2003) The external aerodynamics of canine olfaction. Sensors and sensing in biology and engineering. Springer, New York, pp 323–335

Silbering AF, Benton R (2010) Ionotropic and metabotropic mechanisms in chemoreception: 'chance or design'? EMBO Rep 11(3):173–179

Siniscalchi M, Quaranta A, Rogers LJ (2008) Hemispheric specialization in dogs for processing different acoustic stimuli. PLoS ONE 3(10):e3349

Siniscalchi M, Sasso R, Pepe AM, Dimatteo S, Vallortigara G, Quaranta A (2011) Sniffing with the right nostril: lateralization of response to odour stimuli by dogs. Anim Behav 82(2):399–404

Siniscalchi M, d'Ingeo S, Quaranta A (2016) The dog nose "KNOWS" fear: asymmetric nostril use during sniffing at canine and human emotional stimuli. Behav Brain Res 304:34–41

Small DM, Gerber JC, Mak YE, Hummel T (2005) Differential neural responses evoked by orthonasal versus retronasal odorant perception in humans. Neuron 47(4):593–605

Smith CG (1938) Changes in the olfactory mucosa and the olfactory nerves following intranasal treatment with one per cent zinc sulphate. Can Med Assoc J 39(2):138–140

Sobel N, Khan RM, Saltman A, Sullivan EV, Gabrieli JDE (1999) Olfaction—the world smells different to each nostril. Nature 402(6757):35–35

Solbu TT, Holen T (2012) Aquaporin pathways and mucin secretion of Bowman's glands might protect the olfactory mucosa. Chem Senses 37(1):35–46

Speca DJ, Lin DM, Sorensen PW, Isacoff EY, Ngai J, Dittman AH (1999) Functional identification of a goldfish odorant receptor. Neuron 23(3):487–498

Spehr M, Gisselmann G, Poplawski A, Riffell JA, Wetzel CH, Zimmer RK, Hatt H (2003) Identification of a testicular odorant receptor mediating human sperm chemotaxis. Science 299(5615):2054–2058

Staymates ME, MacCrehan WA, Staymates JL, Kunz RR, Mendum T, Ong T-H, Geurtsen G, Gillin GJ, Craven BA (2016) Biomimetic sniffing improves the detection performance of a 3D printed nose of a dog and a commercial trace vapor detector. Sci Rep 6:36876

Steiger SS, Kuryshev VY, Stensmyr MC, Kempenaers B, Mueller JC (2009) A comparison of reptilian and avian olfactory receptor gene repertoires: species-specific expansion of group gamma genes in birds. BMC Genomics 10:446

Sun LM, Wang HY, Hua J, Han JL, Matsunamic H, Luo MM (2009) Guanylyl cyclase-D in the olfactory CO_2 neurons is activated by bicarbonate. Proc Natl Acad Sci USA 106(6):2041–2046

Suzuki J, Yoshizaki K, Kobayashi T, Osumi N (2013) Neural crest-derived horizontal basal cells as tissue stem cells in the adult olfactory epithelium. Neurosci Res 75(2):112–120

Tacher S, Quignon P, Rimbault M, Dreano S, Andre C, Galibert F (2005) Olfactory receptor sequence polymorphism within and between breeds of dogs. J Hered 96(7):812–816

Thangaleela S, Sivamaruthi BS, Kesika P, Bharathi M, Chaiyasut C (2022) Nasal microbiota, olfactory health, neurological disorders and aging—a review. Microorganisms 10(7)

Toppila-Salmi S, van Drunen CM, Fokkens WJ, Golebski K, Mattila P, Joenvaara S, Renkonen J, Renkonen R (2015) Molecular mechanisms of nasal epithelium in rhinitis and rhinosinusitis. Curr Allergy Asthma Rep 15(2):495

Tress B, Dorn ES, Suchodolski JS, Nisar T, Ravindran P, Weber K, Hartmann K, Schulz BS (2017) Bacterial microbiome of the nose of healthy dogs and dogs with nasal disease. PLoS ONE 12(5):e0176736

Uchida N, Mainen ZF (2007) Odor concentration invariance by chemical ratio coding. Front Syst Neurosci 1:3

Viswaprakash N, Dennis JC, Samoylov AM, Josephson EM, Morrison EE, Vodyanoy V (2006) Enhancement of odorant-induced responses in olfactory receptor neurons by zinc nanoclusters. In: Society for neuroscience annual meeting. Society for Neuroscience, Atlanta, GA

Viswaprakash N, Dennis JC, Globa L, Pustovyy O, Josephson EM, Kanju P, Morrison EE, Vodyanoy V (2009) Enhancement of odorant-induced response in olfactory receptor neurons by zinc nanoparticles. Chem Senses 34:547–557

Vosshall LB (2001) The molecular logic of olfaction in Drosophila. Chem Senses 26(2):207–213

Wang J, Luthey-Schulten ZA, Suslick KS (2003) Is the olfactory receptor a metalloprotein? Proc Natl Acad Sci USA 100(6):3035–3039

Waszak SM, Hasin Y, Zichner T, Olender T, Keydar I, Khen M, Stutz AM, Schlattl A, Lancet D, Korbel JO (2010) Systematic inference of copy-number genotypes from personal genome sequencing data reveals extensive olfactory receptor gene content diversity. PLoS Comput Biol 6(11):e1000988

Webber RL, Jeffcoat MK, Harman JT, Ruttimann UE (1987) MR demonstration of the nasal cycle in the beagle dog. J Comput Assist Tomogr 11(5):869–871

Wilson MT, Hamilos DL (2014) The nasal and sinus microbiome in health and disease. Curr Allergy Asthma Rep 14(12):485

Yan M, Pamp SJ, Fukuyama J, Hwang PH, Cho DY, Holmes S, Relman DA (2013) Nasal microenvironments and interspecific interactions influence nasal microbiota complexity and S. aureus carriage. Cell Host Microbe 14(6):631–640

Yang M, Zhang H-X, Geng G-J, Wang F-J, Liu C-W, Liu J-L (2022) Artificial selection drives SNPs of olfactory receptor genes into different working traits in labrador retrievers. Genet Res 2022:8319396

Young JM, Trask BJ (2002) The sense of smell: genomics of vertebrate odorant receptors. Hum Mol Genet 11(10):1153–1160

Young JM, Trask BJ (2007) V2R gene families degenerated in primates, dog and cow, but expanded in opossum. Trends Genet 23(5):212–215

Young JM, Shykind BM, Lane RP, Tonnes-Priddy L, Ross JA, Walker M, Williams EM, Trask BJ (2003) Odorant receptor expressed sequence tags demonstrate olfactory expression of over 400 genes, extensive alternate splicing and unequal expression levels. Genome Biol 4(11):R71

Yu CR, Wu Y (2017) Regeneration and rewiring of rodent olfactory sensory neurons. Exp Neurol 287(Pt 3):395–408

Zak JD, Reddy G, Vergassola M, Murthy VN (2020) Antagonistic odor interactions in olfactory sensory neurons are widespread in freely breathing mice. Nat Commun 11(1):3350

Zhu KW, Burton SD, Nagai MH, Silverman JD, de March CA, Wachowiak M, Matsunami H (2022) Decoding the olfactory map through targeted transcriptomics links murine olfactory receptors to glomeruli. Nat Commun 13(1):5137

Zozulya S, Echeverri F, Nguyen T (2001) The human olfactory receptor repertoire. Genome Biol 2(6):RESEARCH0018

Zufall F, Leinders-Zufall T (2000) The cellular and molecular basis of odor adaptation. Chem Senses 25(4):473–481

Neuroanatomy and Neurophysiology of the Olfactory Signal Transduction Pathway

Melissa Singletary and Samantha Hagerty

Abstract

The sense of olfaction is surprisingly influential in many areas of the central nervous system that seem unrelated to how something smells. Perception of different odors can influence numerous biological functions: from reproduction to social interactions. Olfaction abnormalities have also been associated with various psychological conditions in humans, such as major depressive disorder and neurodegenerative diseases. This broad range of olfactory influence is giving researchers a new appreciation for the importance of sensory system. It is also unique in both neural organization and the environmental challenges that come from direct exposure of neurons to the external environment. This chapter will focus on predominantly the mammalian neuroanatomy and neurophysiology of the olfactory signal transduction pathway, with highlights on canine-specific attributes.

Keywords

Olfactory signal transduction • Olfactory neurophysiology • Olfactory neuroanatomy • Central olfactory pathway

M. Singletary (✉)
Canine Performance Science, Auburn University College of Veterinary Medicine, Al 36849 Auburn, US
e-mail: mas0028@auburn.edu

M. Singletary · S. Hagerty
Department of Anatomy, Physiology and Pharmacology, Auburn University College of Veterinary Medicine, Al 36849 Auburn, US

L. Lazarowski (ed.), *Olfactory Research in Dogs*,
https://doi.org/10.1007/978-3-031-39370-9_2

1 Olfactory Signal Transduction Pathway: Peripheral Nervous System

Olfactory receptors (OR) are chemosensory G-protein coupled receptors (GPCR) with three distinct, or heterotrimeric, G-proteins including alpha, beta, and gamma subunits. They represent the largest multi-gene receptor repertoire in mammalian genomes (Buck and Axel 1991a, b). While the repertoire of OR genes varies by species, the relative abundance in comparison to other receptor families is conserved across mammalian genomes. Olfactory sensory neurons (OSN) and Bowman's glands are the two main components that differentiate the areas of olfactory neuroepithelium from respiratory epithelia in the nasal cavity. At the apical surface, dendritic processes of an OSN are densely packed with approximately 1–50 non-motile cilia (Morrison and Costanzo 1990). The cilia express the classic 9 + 2 configuration, which is 9 pairs of microtubule doublets arranged in a concentric ring around a central pair of microtubules in the proximal segment but tapers to microtubule singlets more distally (Menco 1984; Williams et al. 2014). Though this configuration is typical of motile cilia, the OSN cilia are non-motile due to a lack of dynein arms. The heavy density of OR and signaling proteins are preferentially localized in the more distal end of the cilia where improved chance of stimulus interaction can occur (Menco 1984; Matsuzaki et al. 1999; Flannery et al. 2006; Jenkins et al. 2009), potentially increasing the sensitivity of odor detection.

A basic principle on which the olfactory system functions is the expression of only one receptor type on a given OSN also referred to as the one-receptor-one-neuron rule (Li et al. 2004). This monoallelic and monogenic expression provides stimulus specificity and discrimination. Odorants engage with these OR with varying degrees of affinity, setting off a cascade of enzymatic events (Buck and Axel 1991a, b; Firestein 2001).

The summation of potentials when multiple ORs are stimulated by the same odorant type result in an action potential when the threshold is reached (Firestein 2001). The combinatorial coding at the OR level is achieved by specific ORs being capable of odorant detection in a narrow spectrum or a broad spectrum while a specific odorant molecule can also activate a variety of ORs across a spectrum of intensity and attraction. The variations of OR sensitivity and specificity result in groups of odorant-specific activation that can be overlapping and allow for a more considerable combinatorial coding for more odorant detection possibilities (Persuy et al. 2015).

Odor intensity is directly influenced by odor concentration, which subsequently may correlate to the number of OSN stimulated, though conflating factors such as exposure duration or physiological state may limit interpretation as a solely linear relationship (Stevens 1960; Chastrette et al. 1998; Sirotin et al. 2015). There is a wide range of receptor tuning widths, described as the average number of activated glomeruli per single odorant, and non-linear responses to monomolecular, binary, similar molecular groups or complex odor mixtures. In mice, studies showed increased complexity from binary mixture to more complex odor mixtures involved increasing levels of antagonistic odor interactions (Zak et al. 2020). At the odorant–receptor interface, the series of molecular mechanisms involved in odorant recognition by the receptor is not yet fully defined. The subsequent signal cascade following receptor activation and cellular depolarization is better studied and established.

As with other electrically excitable cells, the intracellular environment of OSN is negative compared to the exterior space, though the unique characteristic of being exposed to the external environment requires these neurons to actively maintain a state of excitability under less than ideal conditions. The resting membrane potential of OSN that remain in holding states primed for activation is -65 mV with an activation threshold of approximately -45 mV (Firestein 2001). OR activation is initiated when the proper odorant ligand reaches the binding pocket of an appropriate GPCR, resulting in a conformational change in the heterotrimeric G-protein, G_{olf}, and guanine nucleotide exchange of GDP for GTP. This exchange prompts dissociation of the alpha subunit, $G_{\alpha olf,}$ from G_{β} and G_{γ} subunits. Specific sub-molecular events resulting in recognition and activation at the odorant-receptor interface are not fully elucidated, as several theories associated with vibrational, molecular, and biochemical properties are still explored yet not universally agreed upon (Turin 2002; Block et al. 2015; Hoehn et al. 2017). The considerable representation of OR among the mammalian gene repertoire emphasizes its evolutionary conservation and biological relevance (Buck and Axel 1991a, b). A characteristic of GPCR class receptors is the seven-transmembrane weaving pattern between intra and extracellular sides of the plasma membrane. Starting with an intracellular C-terminus, the protein transverses the membrane 7 times in loops leading to an extracellular N-terminus. Upon the conformational change resulting in the release of the $G_{\alpha olf}$ subunit, this subunit then interacts with adenylyl cyclase III (ACIII) enzyme allowing for the intracellular conversion of adenosine triphosphate (ATP) to cyclic adenosine monophosphate (cAMP) (Firestein 2001). Levels of cAMP increase to approximately 100 cAMP molecules per ACIII, allowing for cAMP molecules to bind to the transmembrane cyclic-nucleotide gated (CNG) ion channel which results in a conformational change and subsequent opening. CNG channel opening allows a selective influx of positively charged calcium (Ca^{2+}) and sodium (Na^{+}) into the cell. Concentration of CNG channels is estimated to exceed 2,000/μm^2, so the relative gradient change with incoming positive charge is considerable. The changing membrane potential is further facilitated by the opening of voltage-gated ion channels, specifically calcium-activated chloride channels, causing an efflux of negatively charged chloride (Cl^{-}) into the extracellular space (Firestein 2001). The opening of these voltage-gated ion channels is thought to occur via transmembrane protein TMEM16B, which are suggested to associate with the calcium-calmodulin precursor, calcium-free calmodulin apocalmodulin (apoCaM), under resting conditions (Yang et al. 2014). The combination of positive influx and negative outflux may allow for more rapid membrane depolarization as well as an environmental buffer to maintain proper resting states in less controlled conditions. As OSN are exposed to the external environment in the nasal cavity, the cilia on which the receptor sites are embedded may be more vulnerable to atypical extracellular ion concentrations through external environmental perturbations. Therefore, this dual ion buffer system allows for a unique fail-safe aiding in depolarization even when extracellular sodium levels may be low. As an electrically excitable cell, the high intracellular chloride concentration compared to the exterior environment is maintained as a mechanism for quick membrane potential depolarization (Firestein 2001). Chloride efflux has been reported to represent 80% of the primary depolarization in the OSN (Lowe and Gold 1993). The re-establishment of this Cl^- gradient is unsurprisingly critical in maintaining a buffer, and more recent work indicates ion transporter NKCC1as the main contributor (Haering et al. 2015). The NKCC1 ion transporter is a sodium, potassium, chloride symporter transporting in a 1:1:2 ratio, respectively (Haering et al. 2015). Redistribution of the cytosolic calcium to the extracellular space is also carried out by Na^+/ Ca^{2+} extrusion through the transporter NCKX4 (Stephan et al. 2012). The gradient re-establishment, or repolarization phase, initiates with temporary hyperpolarization in which the cell is unable to respond to a stimulus while membrane potential gradient is resetting.

There are several internal feedback mechanisms at play throughout this molecular cascade involving multiple ion channels. Calcium influx during depolarization acts directly on the associated ion channels through calcium-calmodulin binding, decreasing the ion channel sensitivity to cAMP and dampening the response, which elevates the threshold stimulus intensity required for subsequent excitation. Additionally, the regulator of G-protein signaling (RGS) reduces levels of cAMP production by direct inhibition of adenylyl cyclase III activity. Supporting evidence of this was demonstrated through experimental inhibition of RGS2, resulting in downregulation of signal transduction in the neuronal membrane (Sinnarajah et al. 2001). A similar multi-modal approach for response modulation to olfactory stimuli is seen in regulation of adaptation by adjusting OSN sensitivity (Kurahashi and Menini 1997; Reisert and Matthews 2000; Firestein 2001). Increased levels of cAMP by ACIII activate phosphokinase A (PKA), which phosphorylates the receptor, thereby inhibiting its continued function. Simultaneous action by odorant receptor kinase (ORK) functions through cAMP activation of the G-protein beta-gamma subunits. Furthermore, calcium-calmodulin activates phosphodiesterase (PDE) within the cytoplasm, which degrades cAMP to restore baseline levels (Firestein 2001). More recently, olfactory marker protein (OMP), generally considered a biomarker for neuronal maturity in OSN, has been implicated in modulation of the basal cAMP levels (Dibattista and Reisert 2016). OMP is not expressed in basal stem cells or developing OSN found in the lower region of OE but is present when functional activity is reached, which takes approximately one week in regenerating OSN (Kondo et al. 2010; Savya et al. 2019). The achievement of functional activity would be consistent with the ability to regulate and restore membrane potential, which is required for successful cell signaling.

Generated action potentials propagate through the basally extended axon where it terminates at a monosynaptic second neuron target in the olfactory bulb (OB). The OSN utilizes glutamate as its primary excitatory neurotransmitter for signal communication (Berkowicz et al. 1994). The OSN axons will functionally collate with similar OR expression and bundle into fascicles enwrapped in olfactory ensheathing cells (OEC) and traverse through the cribriform plate to enter the central nervous system.

2 Olfactory Signal Transduction Pathway: Central Nervous System

As first-order neurons, the OSN axons project from their soma within the olfactory epithelium to functionally respective glomeruli within the OB on second-order neurons. The various glomeruli are selective for individual OR genes allowing for collections of OSN axons expressing the same OR to bundle as axon fascicles and innervate selective glomeruli within the OB, passing along the signal from peripheral nervous system into the central nervous system (CNS) (Zhu et al. 2022). The OB will serve as the initial filter within the CNS aiding in discrimination, selectivity, and modification of odor sensitivities through modulation of odor noise (Jia et al. 2014).

There is significant variability in relative olfactory bulb size within the CNS across species. Some species show significantly larger relative bulb size, such as the shark and the dog, than other species such as humans. The OB volume for the dog was reported to be 0.18 ± 0.02 cm^3 while the humans' was 0.06 ± 0.01 cm^3 (Kavoi and Jameela 2011). The major cellular components of the olfactory bulb include mitral and tufted cell neurons, peri-glomerular cells, and granular cells. The site of synaptic connectivity between first-order OSN and second-order OB neurons occurs in the glomeruli of the glomerular layer. As many different OSN expressing individual OR alleles converge at shared glomeruli, there is significant input from the neuroepithelial level at these synaptic sites. In rabbit models, there are approximately 25,000 axons per glomeruli and about 1,800 glomeruli per olfactory bulb (Firestein 2001). The olfactory bulb can be divided into six distinct layers that are categorized based on cell types present and, particularly, where the cell bodies of such cells are located. OB glomeruli represent spatially encoded regions of incoming olfactory information as well as secondary processing for further projections. The basic OB divisions are olfactory nerve, glomerular, external plexiform, mitral cell plexiform, internal plexiform, and granule cell layers. First, the olfactory nerve layer is where OSN axonal projections enter the olfactory bulb with information obtained from the nasal epithelium. The second, termed glomerular layer, is the synaptic site where OSN release neurotransmitters for post-synaptic excitation by secondary neurons. These neurons are collectively called juxtaglomerular cells but are further subcategorized into periglomerular cells, external tufted cells, and superficial short-axon cells. The third layer is called the external plexiform layer, where primarily dendrodendritic synapses and tufted cells are found. The fourth, or mitral cell layer, as the terminology suggests is where mitral cell bodies are found although their dendritic processes extend to the glomeruli for OSN convergence. Next, the internal plexiform layer is characterized by little synaptic activity and few cells, through which axons from mitral or tufted cells pass. The innermost layer is known as the granular layer, where granular interneuron cells of the OB core function as stem cells (Sarnat and Yu 2016; Sarnat and Flores-Sarnat 2017; Sarnat et al. 2017).

The monoallelic and monogenic neuron expression also relates to associated specific receptor glomeruli. Located on the superficial layer of the olfactory bulb, these spherical glomeruli serve as the synaptic site between OSN axons and mitral and tufted cells. Mitral cells are the primary efferent projection cell that are also thought to play a role in post-synaptic signal modification, while tufted cells are also involved in signal reception and projection. Both are considered glutaminergic neurons as the excitatory neurotransmitter used is glutamate. Thousands of OR-specific olfactory sensory neurons terminate on a single glomerulus, with approximately 50 mitral and tufted cells involved in the post-synaptic response. Maturation of the olfactory bulb results in mitral neuron cell developmental changes, suggested by the increasing size of mitral cells with corresponding decrease in quantity from juvenile age to adulthood (Wei et al. 2008). Lateral interconnected mitral cells "refine" or "modify" the signal. This is thought to also be a mechanism to support discrimination of odors. Glomeruli are surrounded by periglomerular cells, which are interneurons that form dendrodendritic synapses between the olfactory signal-carrying neurons and associated cells, mainly for inhibitory purposes. Periglomerular cells consist primarily of inhibitory gabaminergic and dopaminergic neurons acting on NMDA receptors of the mitral and tufted projection cells within and between glomeruli (Ohm et al. 1990, 1991). The granular interneuron cells inhabiting the innermost layer also share inhibitory roles regulating mitral and tufted cells through dendrodendritic gabaminergic synapses (Hirata et al. 2006). An inverse reciprocal relationship between mitral and granular cells is evident, in which mitral to granular stimulation is excitatory while granular to mitral stimulation is inhibitory (Kosaka et al. 1985). An estimated 50–100 inhibitory granular cells are reported to interact with a given mitral cell (Hirata et al. 2006). Within the central nervous system, the olfactory bulb has the most robust presence of dendrodendritic synapses (Kaba and Keverne 1992; Hayashi 1999). Not only the OB size but also activity vary between species. Some studies in the dog have suggested that female dogs have more active olfactory bulbs and a suggested stronger long-term odor memory than those of males (Wei et al. 2017). This signal transduction pathway from first-order to second-order neuron has remained ipsilateral to the point of the OB. Post-OB the pathway complexity increases and is less well characterized with cross-over and extensive connections throughout the brain.

Olfactory signal information continues from olfactory bulb glomerular convergence through second-order neurons into the olfactory peduncle via the lateral olfactory tract and then into primary cortical olfactory areas. These areas include the basilar forebrain, limbic system, piriform lobe, lateral olfactory and parahippocampal gyri, anterior olfactory cortex, periamygdala, entorhinal cortex, and anterior cingulate cortex (Brunjes et al. 2011; Jia et al. 2014, 2016; Uemura 2015). From an evolutionary perspective, the limbic system that includes the OB, entorhinal cortex, hippocampus, and amygdala is generally considered a more primitive region of the brain associated with emotions and memories, which is relevant as it relates to olfactory processing (Kanter and Haberly 1990). The olfactory cortex consists of the anterior olfactory nucleus (AON), the tenia tecta, the olfactory tubercle (OT), the piriform cortex (PC), cortical amygdaloid nucleus, periamygdaloid cortex, and the entorhinal cortex (Price et al. 1991; Zelano and Sobel 2005). Signal processing occurs in the piriform cortex, where primary processing and assigning contextual information of odor sources is thought to occur (Haberly 2001). Conscious perception of the odor source may occur in the frontal lobe area of the neocortex (Ongur and Price 2000). More recent research using tandem diffusion tensor imaging (DTI) and the Klingler dissection method olfactory cortex network mapping in dogs revealed an extensive pathway to the occipital lobe in addition to the established cortical spinal tract, limbic system, piriform lobe, and entorhinal pathways (Andrews et al. 2022). The mapping and coding information from the OB is not retained within the piriform cortex, but rather has more plasticity. With repeated or varying exposures of odorants and odorant mixtures, the higher cognitive response has been shown to change and demonstrate plasticity that may result in unique odor perceptions under varying conditions. The trigeminal system engages with the olfactory pathway, providing ancillary information through odor activation of trigeminal sensory receptors producing somatic sensations related to temperature and nociception. This trigeminal activation can directly influence airflow and has even been shown to activate the piriform cortex (Hummel and Frasnelli 2019). Quantity and quality of odors are important to optimize olfactory perception, though there is much to learn about the complex neuroprocessing of odors.

Much of the projections throughout the brain are confined respectively through ipsilateral hemispheric communication, but there are contralateral communications that cross-over. Functional asymmetry in hemispheric processing of various information and associated behaviors has been noted across multiple species (Gunturkun et al. 2020; Vallortigara and Rogers 2020). There are suggestions that olfactory processing also is impacted by lateralization with detection, discrimination, and identification being differentially processed between the left and right hemispheres (Cavelius et al. 2022).

The AON is the most rostral region of the olfactory cortex (OC) and has a large number of commissural fibers that project contralateral and ipsilateral information to the piriform cortices (PC) (Brunjes et al. 2005; Zelano and Sobel 2005; Yan et al. 2008). The anterior commissure is one of two major regions of the brain where pathways for transfer of information between hemispheres occur. Olfactory allocortical structures transfer information across the midline at the anterior commissure. The average dog brain is 74600 mm^3 with the anterior commissure area representing 2.54 mm^2 (Ashwell 2016).

The olfactory sensory system does communicate with the thalamus but uniquely does not require communication with the thalamus as an intermediary prior to higher brain centers and has a direct connection to those areas (Ongur and Price 2000; Shepherd 2005; Kay and Sherman 2007). All other sensory systems require thalamocortical processing, but in regards to olfaction the thalamus is thought to hold a role in odor threshold (Challis et al. 2015). This direct connection to higher cortical areas is thought to contribute to the strength of odor-associated memories. Studies have suggested that the olfactory bulb cortex functions in a similar capacity to the sensory processing in the thalamus (Zelano and Sobel 2005; Sarnat and Flores-Sarnat 2017; Sarnat et al. 2017). Most OB output is carried toward the PC, where signals are projected to the dorsomedial nucleus of the thalamus, orbitofrontal cortex, and may have signal regulatory feedback roles as discussed in the OB. In contrast to the elaborate topographical spatial patterns found in the OB divisions, the PC is far less organized. Although there are three distinct layers, they appear highly associative without specific signal projections to particular target cells (Miyamichi et al. 2011; Wiegand et al. 2011). The entorhinal cortex receives input from multiple areas of the olfactory tract and connects to the hippocampus, so it is considered relevant in associating olfactory information with memories (Zelano and Sobel 2005). The amygdala, part of the limbic system, sends inputs into the hypothalamus while also feeding back onto the OB. This olfactory cortex region is thought to play a role in assigning emotion to odor profiles (Zelano and Sobel 2005; Good and Sullivan 2015). Among the olfactory cortex, the only region not known to directly feedback to the OB is the olfactory tubercle. The olfactory tubercle communicates with the dorsomedial thalamic nucleus as well as with the nucleus accumbens, ventral tegmentum, and pallidum which are thought to be the association of reward and motivation in olfaction (Heimer 2003; Ikemoto 2007). In dogs that perform odor-based detection tasks, the reward–behavior associations with the olfactory system include the caudate nucleus, the entorhinal cortex, and hippocampus related to development of odor memory, and emotional or motivational associations linked to the amygdala (Herrick 1933; Schoenbaum et al. 1999; Haberly 2001; Gottfried 2010; Wilson and Baietto 2011; Wilson and Sullivan 2011).

The olfactory cortex is so richly connected to other areas in the brain and many of these connections not only receive information but provide information to the olfactory sensory system as part of a feedback loop (Kay et al. 1996; Wilson 1998a, b; Zufall and Leinders-Zufall 2000). Feedback mechanisms from the olfactory and orbitofrontal cortex to the OB are thought to function as sources of adaptation and habituation. The source of adaptation can be either peripheral or central. Peripheral adaptation involves a decreased neural response in the pre-glomerular tract while central adaptation is characterized by a reduction in post-glomerular tract neural responses (Pellegrino et al. 2017). While the source of adaptation is mainly sensory fatigue, habituation is considered a reduced behavioral response resulting from repeated stimulation (Rankin et al. 2009). The collective phenomena have been further overlapped in some descriptions of adaptation as the neural basis of behavior response of habituation (Pellegrino et al. 2017). This adaptation can occur after as little as two repetitions of an intense odor stimulus where perception is diminished though the electrical activity at the OSN level is not necessarily diminished (Hummel et al. 1996; Hummel et al. 2006).

The multidimensional nature of this sense perceives odors that may have positive or negative hedonic value (the measure of pleasantness), may be repulsive or share in activation of the trigeminal system and may be the carrier of a biologically important message. As a primary sensory system for the dog, their behavior is significantly influenced by olfactory inputs processed in higher cortical areas (Siniscalchi 2016; Siniscalchi et al. 2016).

Various methods are used to evaluate neuroanatomical and functional neurophysiological processing including positron emission tomography scanning and functional magnetic resonance imaging (fMRI). The use of extensive training methods in dogs allows for voluntarily fMRI scans while awake and unrestrained (Karl et al. 2019; Strassberg et al. in press), allowing more research to explore the functional neurophysiological and cognitive processes. The limitations to these studies are the time and expense of maintaining a trained cohort of dogs for voluntary awake scanning to achieve biologically relevant scans that are unhindered by restraints or anesthesia, which can alter cognitive states and functional interpretation of the results (Thompkins 2016; Thompkins et al. 2016). One of the pioneer studies to scan awake unrestrained dogs showed the primary areas of the brain involved in odor processing, which included the OB and the olfactory cortex, were activated in both awake and anesthetized states at varying degrees, though higher cortical regions were seen predominantly in the awake state (Jia et al. 2014). As these higher cortical regions are involved in perception and identification through odor detection, discrimination, learning, and memory, this finding is consistent that an awake and alert state is necessary for maximal olfactory pathway activation (Siniscalchi 2016; Siniscalchi et al. 2016).

The superior behavioral olfactory acuity of macrosmatic species is attributable to multi-factorial traits. In the dog, there are a number of collective anatomical and physiological features that include skull nasal vestibule depth increasing odorant time of flight (Craven et al. 2010), larger total surface area of the olfactory neuroepithelium (Sjaastad et al. 2010), a dedicated olfactory recess for odorant processing (Craven et al. 2004), a greater number of functional odorant receptor genes with lower pseudogenization rate (Quignon et al. 2003; Olender et al. 2004), a higher density of olfactory sensory neurons present in the olfactory epithelium (Uemura 2015), a higher density of cilia present per olfactory sensory neuron (Uemura 2015), and an increase in OB volume relevant to the total brain weight (Reep et al. 2007; Kavoi and Jameela 2011; Uemura 2015). These unique features have positioned dogs as an excellent tool for uncovering the capacity and complexity of odor processing in mammals and a translational model for informing the underlying principles of odor learning and behavior.

Acknowledgements We would like to thank Dr. Lucia Lazarowski for the invitation to write this chapter. We also like to thank Michelle Aono for her for review and helpful comments on the chapter. Additionally, we thank Dr. Edward Morrison, Dr. Robert Judd, Dr. Vitaly Vodyanoy, and Dr. Iryna Sorokulova for their mentorship, input, and review of original supporting work.

References

Andrews EF, Pascalau R, Horowitz A, Lawrence GM, Johnson PJ (2022) Extensive connections of the canine olfactory pathway revealed by tractography and dissection. J Neurosci 42(33):6392–6407

Ashwell KWS (2016) Anterior commissure versus corpus callosum: a quantitative comparison across mammals. Zoology (jena) 119(2):126–136

Berkowicz DA, Trombley PQ, Shepherd GM (1994) Evidence for glutamate as the olfactory receptor cell neurotransmitter. J Neurophysiol 71(6):2557–2561

Block E, Jang S, Matsunami H, Sekharan S, Dethier B, Ertem MZ, Gundala S, Pan Y, Li S, Li Z, Lodge SN, Ozbil M, Jiang H, Penalba SF, Batista VS, Zhuang H (2015) Implausibility of the vibrational theory of olfaction. Proc Natl Acad Sci USA 112(21):E2766-2774

Brunjes PC, Illig KR, Meyer EA (2005) A field guide to the anterior olfactory nucleus (cortex). Brain Res Brain Res Rev 50(2):305–335

Brunjes PC, Kay RB, Arrivillaga JP (2011) The mouse olfactory peduncle. J Comp Neurol 519(14):2870–2886

Buck L, Axel R (1991a) A novel multigene family may encode odorant receptors—a molecular-basis for Odor recognition. Cell 65(1):175–187

Buck L, Axel R (1991b) A novel multigene family may encode odorant receptors: a molecular basis for odor recognition. Cell 65(1):175–187

Cavelius M, Brunel T, Didier A (2022) Lessons from behavioral lateralization in olfaction. Brain Struct Funct 227(2):685–696

Challis RC, Tian H, Wang J, He J, Jiang J, Chen X, Yin W, Connelly T, Ma L, Yu CR, Pluznick JL, Storm DR, Huang L, Zhao K, Ma M (2015) An Olfactory Cilia pattern in the mammalian nose ensures high sensitivity to Odors. Curr Biol 25(19):2503–2512

Chastrette M, Thomas-Danguin T, Rallet E (1998) Modelling the human olfactory stimulus-response function. Chem Senses 23(2):181–196

Craven BA, Paterson EG, Settles GS (2004) The fluid dynamics of canine olfaction: unique nasal airflow patterns as an explanation of macrosmia. J R Soc Interf 7(47):933–943

Craven BA, Paterson EG, Settles GS (2010) The fluid dynamics of canine olfaction: unique nasal airflow patterns as an explanation of macrosmia. J R Soc Interface 7(47):933–943

Dibattista M, Reisert J (2016) The odorant receptor-dependent role of olfactory marker protein in olfactory receptor neurons. J Neurosci 36(10):2995–3006

Firestein S (2001) How the olfactory system makes sense of scents. Nature 413(6852):211–218

Flannery RJ, French DA, Kleene SJ (2006) Clustering of cyclic-nucleotide-gated channels in olfactory cilia. Biophys J 91(1):179–188

Good KP, Sullivan RL (2015) Olfactory function in psychotic disorders: Insights from neuroimaging studies. World J Psychiatry 5(2):210–221

Gottfried JA (2010) Central mechanisms of odour object perception. Nat Rev Neurosci 11(9):628–641

Gunturkun O, Strockens F, Ocklenburg S (2020) Brain lateralization: a comparative perspective. Physiol Rev 100(3):1019–1063

Haberly LB (2001) Parallel-distributed processing in olfactory cortex: new insights from morphological and physiological analysis of neuronal circuitry. Chem Senses 26(5):551–576

Haering C, Kanageswaran N, Bouvain P, Scholz P, Altmuller J, Becker C, Gisselmann G, Waring-Bischof J, Hatt H (2015) Ion transporter NKCC1, modulator of neurogenesis in murine olfactory neurons. J Biol Chem 290(15):9767–9779

Hayashi Y (1999) Metabotropic glutamate receptor: its ligands and function in the olfactory system. Nihon Yakurigaku Zasshi 113(2):73–83

Heimer L (2003) A new anatomical framework for neuropsychiatric disorders and drug abuse. Am J Psychiatry 160(10):1726–1739

Herrick CJ (1933) The functions of the olfactory parts of the cerebral cortex. Proc Natl Acad Sci USA 19(1):7–14

Hirata T, Nakazawa M, Yoshihara S, Miyachi H, Kitamura K, Yoshihara Y, Hibi M (2006) Zinc-finger gene Fez in the olfactory sensory neurons regulates development of the olfactory bulb non-cell-autonomously. Development 133(8):1433–1443

Hoehn RD, Nichols DE, McCorvy JD, Neven H, Kais S (2017) Experimental evaluation of the generalized vibrational theory of G protein-coupled receptor activation. Proc Natl Acad Sci USA 114(22):5595–5600

Hummel T, Frasnelli J (2019) The intranasal trigeminal system. Handb Clin Neurol 164:119–134

Hummel T, Knecht M, Kobal G (1996) Peripherally obtained electrophysiological responses to olfactory stimulation in man: electro-olfactograms exhibit a smaller degree of desensitization compared with subjective intensity estimates. Brain Res 717(1–2):160–164

Hummel T, Mojet J, Kobal G (2006) Electro-olfactograms are present when odorous stimuli have not been perceived. Neurosci Lett 397(3):224–228

Ikemoto S (2007) Dopamine reward circuitry: two projection systems from the ventral midbrain to the nucleus accumbens-olfactory tubercle complex. Brain Res Rev 56(1):27–78

Jenkins PM, Zhang L, Thomas G, Martens JR (2009) PACS-1 mediates phosphorylation-dependent ciliary trafficking of the cyclic-nucleotide-gated channel in olfactory sensory neurons. J Neurosci 29(34):10541–10551

Jia H, Pustovyy OM, Waggoner P, Beyers RJ, Schumacher J, Wildey C, Barrett J, Morrison E, Salibi N, Denney TS, Vodyanoy VJ, Deshpande G (2014) Functional MRI of the olfactory system in conscious dogs. PLoS ONE 9(1):e86362

Jia H, Pustovyy OM, Wang Y, Waggoner P, Beyers RJ, Schumacher J, Wildey C, Morrison E, Salibi N, Denney TS, Vodyanoy VJ, Deshpande G (2016) Enhancement of Odor-induced activity in the canine brain by zinc nanoparticles: a functional MRI study in fully unrestrained conscious dogs. Chem Senses 41(1):53–67

Kaba H, Keverne EB (1992) Analysis of synaptic events in the mouse accessory olfactory bulb with current source-density techniques. Neuroscience 49(2):247–254

Kanter ED, Haberly LB (1990) NMDA-dependent induction of long-term potentiation in afferent and association fiber systems of piriform cortex in vitro. Brain Res 525(1):175–179

Karl S, Boch M, Viranyi Z, Lamm C, Huber L (2019) Training pet dogs for eye-tracking and awake fMRI. Behav Res Methods

Kavoi BM, Jameela H (2011) Comparative morphometry of the olfactory bulb, tract and stria in the human, dog and goat. Interan J Morphol 29(3):939–946

Kay LM, Lancaster LR, Freeman WJ (1996) Reafference and attractors in the olfactory system during odor recognition. Int J Neural Syst 7(4):489–495

Kay LM, Sherman SM (2007) An argument for an olfactory thalamus. Trends Neurosci 30(2):47–53

Kondo K, Suzukawa K, Sakamoto T, Watanabe K, Kanaya K, Ushio M, Yamaguchi T, Nibu K, Kaga K, Yamasoba T (2010) Age-related changes in cell dynamics of the postnatal mouse olfactory neuroepithelium: cell proliferation, neuronal differentiation, and cell death. J Comp Neurol 518(11):1962–1975

Kosaka T, Hataguchi Y, Hama K, Nagatsu I, Wu JY (1985) Coexistence of immunoreactivities for glutamate decarboxylase and tyrosine hydroxylase in some neurons in the periglomerular region of the rat main olfactory bulb: possible coexistence of gamma-aminobutyric acid (GABA) and dopamine. Brain Res 343(1):166–171

Kurahashi T, Menini A (1997) Mechanism of odorant adaptation in the olfactory receptor cell. Nature 385(6618):725–729

Li J, Ishii T, Feinstein P, Mombaerts P (2004) Odorant receptor gene choice is reset by nuclear transfer from mouse olfactory sensory neurons. Nature 428(6981):393–399

Lowe G, Gold GH (1993) Nonlinear amplification by calcium-dependent chloride channels in olfactory receptor cells. Nature 366(6452):283–286

Matsuzaki O, Bakin RE, Cai X, Menco BP, Ronnett GV (1999) Localization of the olfactory cyclic nucleotide-gated channel subunit 1 in normal, embryonic and regenerating olfactory epithelium. Neuroscience 94(1):131–140

Menco BP (1984) Ciliated and microvillous structures of rat olfactory and nasal respiratory epithelia. A study using ultra-rapid cryo-fixation followed by freeze-substitution or freeze-etching. Cell Tissue Res 235(2):225–241
Miyamichi K, Amat F, Moussavi F, Wang C, Wickersham I, Wall NR, Taniguchi H, Tasic B, Huang ZJ, He Z, Callaway EM, Horowitz MA, Luo L (2011) Cortical representations of olfactory input by trans-synaptic tracing. Nature 472(7342):191–196
Morrison EE, Costanzo RM (1990) Morpholgy of the human olfactory epithelium. J Comp Neurol 297:1–13
Ohm TG, Muller H, Braak E (1991) Calbindin-D-28k-like immunoreactive structures in the olfactory bulb and anterior olfactory nucleus of the human adult: distribution and cell typology–partial complementarity with parvalbumin. Neuroscience 42(3):823–840
Ohm TG, Muller H, Ulfig N, Braak E (1990) Glutamic-acid-decarboxylase-and parvalbumin-like-immunoreactive structures in the olfactory bulb of the human adult. J Comp Neurol 291(1):1–8
Olender T, Fuchs T, Linhart C, Shamir R, Adams M, Kalush F, Khen M, Lancet D (2004) The canine olfactory subgenome. Genomics 83(3):361–372
Ongur D, Price JL (2000) The organization of networks within the orbital and medial prefrontal cortex of rats, monkeys and humans. Cereb Cortex 10(3):206–219
Pellegrino R, Sinding C, de Wijk RA, Hummel T (2017) Habituation and adaptation to odors in humans. Physiol Behav 177:13–19
Persuy MA, Sanz G, Tromelin A, Thomas-Danguin T, Gibrat JF, Pajot-Augy E (2015) Mammalian olfactory receptors: molecular mechanisms of odorant detection, 3D-modeling, and structure-activity relationships. Prog Mol Biol Transl Sci 130:1–36
Price JL, Slotnick BM, Revial MF (1991) Olfactory projections to the hypothalamus. J Comp Neurol 306(3):447–461
Quignon P, Kirkness E, Cadieu E, Touleimat N, Guyon R, Renier C, Hitte C, Andre C, Fraser C, Galibert F (2003) Comparison of the canine and human olfactory receptor gene repertoires. Genome Biol 4:R80
Rankin CH, Abrams T, Barry RJ, Bhatnagar S, Clayton DF, Colombo J, Coppola G, Geyer MA, Glanzman DL, Marsland S, McSweeney FK, Wilson DA, Wu CF, Thompson RF (2009) Habituation revisited: an updated and revised description of the behavioral characteristics of habituation. Neurobiol Learn Mem 92(2):135–138
Reep RL, Finlay BL, DRB (2007) The limbic system in mammalian brain evolution. Brain Behav Evol 70:57-70
Reisert J, Matthews HR (2000) Adaptation-induced changes in sensitivity in frog olfactory receptor cells. Chem Senses 25(4):483–486
Sarnat HB, Flores-Sarnat L (2017) Olfactory development, part 2: neuroanatomic maturation and dysgeneses. J Child Neurol 32(6):579–593
Sarnat HB, Flores-Sarnat L, Wei XC (2017) Olfactory development, part 1: Function, from fetal perception to adult wine-tasting. J Child Neurol 32(6):566–578
Sarnat HB, Yu W (2016) Maturation and dysgenesis of the human olfactory bulb. Brain Pathol 26(3):301–318
Savya SP, Kunkhyen T, Cheetham CEJ (2019) Low survival rate of young adult-born olfactory sensory neurons in the undamaged mouse olfactory epithelium. J Bioenerg Biomembr 51(1):41–51
Schoenbaum G, Chiba AA, Gallagher M (1999) Neural encoding in orbitofrontal cortex and basolateral amygdala during olfactory discrimination learning. J Neurosci 19(5):1876–1884
Shepherd GM (2005) Perception without a thalamus: how does olfaction do it? Neuron 46:166–168
Siniscalchi M (2016) Olfaction and the canine brain. In: Jezierski T, Ensminger J, Papet LE (eds) Canine olfaction science and law. Press, Boca Raton, FL, USA, CRC, pp 31–37
Siniscalchi M, d'Ingeo S, Quaranta A (2016) The dog nose "KNOWS" fear: asymmetric nostril use during sniffing at canine and human emotional stimuli. Behav Brain Res 304:34–41
Sinnarajah S, Dessauer CW, Srikumar D, Chen J, Yuen J, Yllma S, Dennis JC, Morrison EE, Vodyanoy V, Kehrl JH (2001) RGS2 regulates signal transduction in olfactory neurons by attenuating activation of adenylyl cyclase III. Nature (london, UK) 409(6823):1051–1055

Sirotin YB, Shusterman R, Rinberg D (2015) Neural coding of perceived odor intensity. eNeuro 2(6)
Sjaastad OV, Sand O, Hove K (2010) Physiology of domestic animals, 2nd edn. Oslo, Scandinavian Veterindary Press
Stephan AB, Tobochnik S, Dibattista M, Wall CM, Reisert J, Zhao HQ (2012) The Na+/Ca2+ exchanger NCKX4 governs termination and adaptation of the mammalian olfactory response. Nat Neurosci 15(1):131-U164
Stevens SS (1960) The psychophysics of sensory function. Am Sci 48(2):226–253
Strassberg LR, Waggoner LP, Deshpande G, Katz JS (in press) Training dogs for awake, unrestrained functional magnetic resonance imaging. J Visual Exp
Thompkins A (2016) Investigating the dog-human social bond via behavioral and fMRI methodologies
Thompkins AM, Deshpande G, Waggoner P, Katz JS (2016) Functional Magnetic resonance imaging of the domestic dog: research, methodology, and conceptual issues. Comp Cogn Behav Rev 11:63–82
Turin L (2002) A method for the calculation of odor character from molecular structure. J Theor Biol 216(3):367–385
Uemura EE (2015) Fundamentals of canine neuroanatomy and neurophysiology, 1st edn. Danvers, MA, Blackwell
Vallortigara G, Rogers LJ (2020) A function for the bicameral mind. Cortex 124:274–285
Wei Q-G, Zhang H-H, Guo B (2008) Histological structure difference of dog's olfactory bulb between different age and sex. Zool Res 29:537–545
Wei Q-G, Zhang H-H, Ma S, Guo D (2017) Sex- and age-related differences in c-fos expression in dog olfactory bulbs. Acta Zoologica 98:370–376
Wiegand HF, Beed P, Bendels MH, Leibold C, Schmitz D, Johenning FW (2011) Complementary sensory and associative microcircuitry in primary olfactory cortex. J Neurosci 31(34):12149–12158
Williams CL, McIntyre JC, Norris SR, Jenkins PM, Zhang L, Pei Q, Verhey K, Martens JR (2014) Direct evidence for BBSome-associated intraflagellar transport reveals distinct properties of native mammalian cilia. Nat Commun 5:5813
Wilson AD, Baietto M (2011) Advances in electronic-nose technologies developed for biomedical applications. Sensors 11(1):1105–1176
Wilson DA (1998a) Habituation of odor responses in the rat anterior piriform cortex. J Neurophysiol 79(3):1425–1440
Wilson DA (1998b) Synaptic correlates of odor habituation in the rat anterior piriform cortex. J Neurophysiol 80(2):998–1001
Wilson DA, Sullivan RM (2011) Cortical processing of odor objects. Neuron 72(4):506–519
Yan Z, Tan J, Qin C, Lu Y, Ding C, Luo M (2008) Precise circuitry links bilaterally symmetric olfactory maps. Neuron 58(4):613–624
Yang T, Hendrickson WA, Colecraft HM (2014) Preassociated apocalmodulin mediates Ca2+-dependent sensitization of activation and inactivation of TMEM16A/16B Ca2+-gated Cl- channels. Proc Natl Acad Sci USA 111(51):18213–18218
Zak JD, Reddy G, Vergassola M, Murthy VN (2020) Antagonistic odor interactions in olfactory sensory neurons are widespread in freely breathing mice. Nat Commun 11(1):3350
Zelano C, Sobel N (2005) Humans as an animal model for systems-level organization of olfaction. Neuron 48(3):431–454
Zhu KW, Burton SD, Nagai MH, Silverman JD, de March CA, Wachowiak M, Matsunami H (2022) Decoding the olfactory map through targeted transcriptomics links murine olfactory receptors to glomeruli. Nat Commun 13(1):5137
Zufall F, Leinders-Zufall T (2000) The cellular and molecular basis of Odor adaptation. Chem Senses 25(4):473–481

Canine Olfactory Dysfunction

David C. Dorman

Abstract

This chapter considers the scientific evidence available examining the impact of disease on canine olfaction. The chapter is not intended to provide an exhaustive review of the topic, but rather key concepts are introduced to the reader. This chapter also considers the impact of aging and disease on human olfaction. Humans can serve as a useful model for scent detection dogs. Thus, it is anticipated that aging, nasal tumors, rhinitis, environmental exposures, and other disease states that affect olfaction in humans could also impair olfaction in dogs. There are also important differences in our ability to detect olfactory dysfunction in people and dogs. As evidenced by the recent COVID-19 pandemic, changes in human olfaction are often the result of self-reporting by an individual allowing for detection of subtle effects that precede other clinical signs. In contrast, assessment of olfactory function in dogs with either experimental or naturally occurring disease is uncommon especially in clinical settings.

Keywords

Canine • Olfaction • Hyposmia • Anosmia

1 Introduction

This chapter focuses on medical conditions that could affect olfaction in dogs. The chapter is not intended to provide an exhaustive review of the topic, but rather key concepts will be introduced to the reader. Whenever possible the reader has been directed to recent reviews for more information. Because our understanding of the

D. C. Dorman (✉)
Department of Molecular and Biomedical Sciences, College of Veterinary Medicine, North Carolina State University, Raleigh, NC, USA
e-mail: david_dorman@ncsu.edu

L. Lazarowski (ed.), *Olfactory Research in Dogs*,
https://doi.org/10.1007/978-3-031-39370-9_3

impacts of canine nasal disease on olfaction is limited, this chapter relies heavily on inferences drawn from the human literature. Additional information drawn from experimental studies using laboratory animals is also available to the interested reader. Unless otherwise noted, references to dogs or canine in this chapter refer to domesticated dogs (*Canis familiaris*).

1.1 Overview of Olfaction

Normal canine physiology and anatomy is discussed in Chap. 1 of this text. A recent review is also available (Jenkins et al. 2018). In brief, there are two major olfactory apparatuses in the nasal cavity: the main olfactory epithelium used to detect odorants (small organic molecules) and the vomeronasal organ, which predominantly responds to pheromones. This chapter focuses on the main olfactory system used to detect odorants. In dogs and other mammals, the sense of smell is dependent upon delivery of the odorant to sensory olfactory neurons found within the olfactory epithelium lining the dorsal or dorsoposterior nasal cavity (Dorman 2018). Other cells found in this pseudostratified epithelium include sustentacular cells that serve a glial-like support role as well as basal cells that serve as pluripotent stem cells. Olfactory receptor neurons have a lifetime of 2–4 weeks and are continually replaced by replication from underlying stem cells found in the basal cell layer (Yu and Wu 2017).

Unlike humans, dogs have an olfactory recess that is lined by olfactory epithelium and receives>10% of the inspired air during inspiration (Craven et al. 2010; Lawson et al. 2012). Sniffing increases air delivery to the canine olfactory recess and results in enhanced uptake of soluble odorants in this region (Rygg et al. 2017). Disease processes that obstruct nasal airflow, i.e., conductive disorders, could decrease air odorant delivery to the canine olfactory recess resulting in reduced olfactory abilities.

The olfactory epithelium is lined by a thin mucus sheet. Odorants in the air diffuse into this mucus sheet and subsequently bind to olfactory receptors located on cilia found on the olfactory receptor neurons. Mucosal odorant-binding proteins facilitate odorant–receptor binding (Heydel et al. 2013; Sun et al. 2018). Each olfactory neuron contains one type of odorant receptor that can detect a limited number of odorants (Quignon et al. 2012). Rapid metabolism of many odorants occurs within the olfactory mucosa. Cytochrome P450 isoforms, including CYP2A3/5/10/13 and CYP2G1 found in the olfactory epithelium are thought to contribute to the clearance of odorants (Heydel et al. 2013). Transport proteins including organic anion transporters, organic anion-transporting polypeptide, and divalent metal transporters are found in the mammalian nasal epithelium and may play a role in defense from xenobiotics (Burckhardt 2012; Jeong et al. 2022). Our understanding of how changes in the function of these proteins, receptors, and transporters may affect olfaction is incomplete. For example, humans have a single odor binding protein (OBPIIa) and polymorphisms in the OBPIIa gene have been associated with changes in olfactory performance. Different allelic frequencies have been seen in people with either a normal (normosmic) or decreased (hyposmic) sense of smell (Sollai et al. 2019). Whether similar effects could occur in dogs is unknown.

The sense of smell also depends upon normal signal transduction and message processing. Axonal projections from the olfactory neurons traverse the bony cribriform plate that separates the nasal and cranial cavities. Fractures of the human cribriform plate can sever olfactory nerve filaments resulting in a partial or total loss of smell (Gomez and Pickup 2022). It is likely that cribriform plate fractures (e.g., following head trauma) in dogs could adversely impact olfaction. Olfactory nerve projections enter the cranial cavity and synapse with mitral cells in the olfactory bulb. Olfactory impulses from the olfactory bulb can reach more distal sites within the brain. Neurodegeneration and other disease states affecting the central nervous system can result in an altered sense of smell in humans (Beecher et al. 2018; Walker et al. 2021) and would be anticipated to have similar effects in dogs. Plasticity in the olfactory system occurs as evidenced by the surprising finding of olfaction in some people without the presence of an anatomically recognizable olfactory bulb (Weiss et al. 2020).

1.2 Impact of Aging on Olfaction

Decreased olfactory function occurs frequently in people with incidence rates often exceeding 50% in people aged 65 years and older (Attems et al. 2015; Doty and Kamath 2014). Age-related changes within the nose, olfactory epithelium, olfactory bulb, olfactory cortex, and other brain structures occur and may contribute to this decline in olfactory function (Table 1).

It remains to be seen whether similar age-related changes also occur in dogs. Experimental studies in dogs examining age-related changes in either the olfactory system or olfaction remain limited. For example, Hirai and coworkers (1996) evaluated age-related changes in the olfactory system of dogs ranging in age from

Table 1 Selected changes in the aged human olfactory system

Change	References
Decreased size and number of patent foramina in the cribriform plate	Kalmey et al. (1998)
Olfactory epithelial metaplasia	Morrison and Costanzo (1990), Paik et al. (1992)
Decreased olfactory bulb size	Bhatnagar et al. (1987), Buschhüter et al. (2008)
Increased neurofibrillary tangles in the olfactory bulb	Kishikawa et al. (1990)
Reduced expression of phase I and phase II xenobiotic metabolizing enzymes	Getchell et al. (1993), Krishna et al. (1995)

10 to 19 years. They found that aging in dogs was associated with a decreased number of olfactory neurons and support cells and olfactory bulb changes including the presence of cerebrovascular amyloidosis, and ubiquitin deposits. Aging did not, however, increase the incidence of plaques in the canine olfactory bulb (Hirai et al. 1996). Increased plaques and A-beta protein deposits have been seen in the hippocampus and dentate gyrus of aged dogs (Czasch et al. 2006). In geriatric humans, a decreased number of cribriform plate foramen results in shearing of olfactory nerve projections. Loss of these projections may contribute to reduced olfactory ability and secondary degenerative changes in the olfactory epithelium seen in geriatric people (Kalmey et al. 1998). Some have theorized that increased surface and foramina areas of the cribriform plate are associated with enhanced olfactory abilities in a species (Bird et al. 2014). Wolves and coyotes have more complex cribriform plate morphologies when compared with domesticated dogs (Bird et al. 2021; Jacquemetton et al. 2021). Relative complexity of the cribriform plate does not differ significantly between scent breeds (beagle, bloodhound), breeds used in scent detection work (German shepherd, German short-haired pointer, golden retriever, Labrador retriever), and sight hounds (e.g., greyhound) (Bird et al. 2021). Behavioral data in dogs and wolves that would support the hypothesis that the more complex structure of the wolf cribriform plate versus the dog translates into enhanced olfactory abilities in this species is limited (Polgár et al. 2016). Likewise, anatomic studies evaluating age-related changes in cribriform plate morphology in dogs are lacking.

1.3 Lessons Learned from COVID

Olfactory dysfunction is one of the most prevalent symptoms seen in SARS-CoV-2 patients (Izquierdo-Dominguez et al. 2020; Najafloo et al. 2021). Multiple mechanisms have been proposed for how infection with SARS-CoV-2 leads to anosmia. One mechanism focuses on the role of angiotensin-converting enzyme 2 (ACE2) and transmembrane protease, serine 2 (TMPRSS2) as the functional receptors for SARS-CoV-2 (Bilinska and Butowt 2020) although other alternative mechanisms have been proposed (Hopkins et al. 2021). The prevailing hypothesis suggests that nasal sustentacular cells expressing ACE2 are initially infected with SARS-CoV-2. Sustentacular cell infection leads to secondary impairment of olfactory receptor neurons leading to olfactory dysfunction. Individuals infected with SARS-CoV-2 often develop olfactory dysfunction that precedes the onset of cough, fever, and other symptoms (Lechien et al. 2020). Klopfenstein et al. (2020) found that patients generally developed anosmia within four to five days of SARS-CoV-2 infection, with a duration of approximately nine days. Most patients in this retrospective study recovered within 28 days. Asymptomatic dogs cohabitating with SARS-CoV-2 infected people have had nasal swabs that yielded positive PCR with reverse transcription and serology results consistent with infection (Sit et al. 2020). It remains unknown whether SARS-CoV-2 infection in dogs has any effect on olfaction.

For the purposes of this chapter, it is important to consider how a diagnosis of olfactory dysfunction was made in patients infected with SARS-CoV-2. In most cases, the initial diagnosis was based on the patient's self-reporting rather than results from an olfactory function test (Hannum et al. 2020; Meng et al. 2020; Printza and Constantinidis 2020). Olfactory function tests including sniffing sticks (Bagnasco et al. 2021), the University of Pennsylvania Smell Identification Test (UPSIT), among others have been used in clinical studies with far fewer SARS-CoV-2 patients (Bagnasco et al. 2021; Boscolo-Rizzo et al. 2021; Hannum et al. 2020; Moein et al. 2020). This experience shows that self-reporting and olfactory function tests serve a critical role in the diagnosis of olfactory dysfunction in people. However, neither option is readily available for the assessment of dog olfactory abilities. Thus, we can anticipate that changes in olfaction in disease states will often go undiagnosed in our canine companions.

2 Disease States that Affect Olfaction

2.1 Overview of Nasal Disease in the Dog

Some systemic diseases can result in nasal effects. These include coagulopathies, thrombocytopenia, multiple myeloma, ehrlichiosis, thrombocytopenia, hypertension, and systemic infections (Cohn 2020). Primary nasal disease in dogs is common. Etiologies associated with primary nasal disease include nasal tumors, inflammatory rhinitis, fungal rhinitis (e.g., aspergillosis), periodontal disease, trauma, grass awns and other foreign bodies, amongst others (Cohn 2020; Meler et al. 2008; Plickert et al. 2014; Tasker et al. 1999). Chronic nasal disease in dogs is often caused by nasal tumors. Bloodhounds, Doberman pincher, Labrador retrievers, German shepherds and other long-nosed (dolichocephalic) breeds are more likely to develop nasal neoplasia than are mixed breed dogs, and large breed dogs have a higher incidence than smaller breeds (Mortier and Blackwood 2020). Dogs with foreign bodies or nasal mycosis typically occur in younger dogs.

2.2 Disease Syndromes Associated with Olfactory Dysfunction in Dogs

Few studies describing the impact of disease on functional or electrophysiologic changes in canine olfaction are available. Peterson and coworkers (1981) used bilateral surgical removal of the olfactory peduncle as an experimental model of anosmia in wolves. Two months later olfaction was tested in wolves that either underwent the olfactory pedunculotomy or a sham surgical procedure, as well as controls. Olfaction was tested using the animal's ability to find familiar food (deer meat) in an enclosure. Animals that underwent the olfactory pedunculotomy did not find or consume the deer meat within a 10 min test period while control or sham-operated animals readily found and consumed the deer meat.

Myers and coworkers (1988a, b) used a combination of electroencephalography (EEG) olfactometry, behavioral olfactometry, and electro-olfactography (EOG) to assess the effects of canine distemper virus (CDV) and canine parainfluenza virus (CPV) infection on olfaction in dogs. Behavioral and EEG olfactometry used eugenol and benzaldehyde as the test odorants. Dogs infected with CDV had altered EOG activity consistent with anosmia or hyposmia even months after the initial viral infection. Behavioral and EEG olfactometry indicated that all dogs infected with CPV were unresponsive to eugenol and benzaldehyde. Most dogs infected with CPV had rhinitis and olfactory epithelial atrophy. In contrast, dogs infected with CPV had fewer changes in their sense of smell. Dogs infected with CPV had normal EOG activity and lacked evidence of nasal pathology. Dogs infected with CPV had altered responses on both EEG olfactometry and the behavioral olfactometry tests using the two odorants. The concentration of odorant required for detection was increased in most dogs infected with CPV. Olfactory thresholds returned to normal after the disappearance of clinical signs in naturally infected dogs.

Houpt et al. (1978, 1982) placed a tracheostomy tube with an inflatable cuff in the nasal cavity of dogs to produce a reversible airflow obstruction and decreased olfactory function. They assessed olfaction using behavioral assays that required dogs to detect a pork sample that was buried under pine wood chips (Houpt et al. 1978) or a previously trained meat from a panel of four meats in a flavor-validation test (Houpt et al. 1982). Removal of the obstruction caused by the inflatable cuff restored olfactory function in these animals. These investigators also used nasal instillation of zinc sulfate as an alternative method of inducing anosmia in dogs (Houpt et al. 1978, 1982). This chemical for this purpose has an interesting history in the sensory literature. In the mid-1930s, intranasal administration of zinc sulfate was touted as a preventative for juvenile polio. Unfortunately, some children given zinc developed anosmia (Tisdall et al. 1938). Since then, zinc sulfate ablation of the olfactory mucosa has been extensively used as an experimental model of olfactory toxicity (Burd 1993). In the two studies performed by Houpt and coworkers, the use of zinc sulfate resulted in reduced responses on the behavioral assays of olfaction.

2.3 Disease Syndromes Associated with Olfactory Dysfunction in Humans

Surveys conducted prior to the advent of the COVID-19 pandemic found that nearly 5% of US adults who were 40 years or older reported some form of olfactory dysfunction (Hoffman et al. 2016). The same survey showed that by 80 years of age the prevalence of olfactory dysfunction approached 40% (Hoffman et al. 2016). Anosmia, hyposmia, and other forms of dysfunctional olfaction in people have multiple underlying causes (Boesveldt et al. 2017; Scangas and Bleier 2017). Disorders of olfaction in humans can be classified as either conductive (peripheral) or sensorineural (central) in origin (Scangas and Bleier 2017). Conductive

Table 2 Common causes of olfactory dysfunction of conductive and sensorineural origin in people

Etiology	Example
Conductive	Trauma
	Chronic rhinosinusitis (CRS)
	Nasal polyps
	Allergic rhinitis
	Nasal masses
Sensorineural	Upper respiratory infection
	Aging
	Environmental exposure
	Congenital
	Neurodegenerative diseases
	Medications

disorders develop when odorant delivery to the olfactory epithelium is restricted. Sensorineural disorders occur when there is decreased reception or processing of an olfactory stimulus. Sensorineural disorders can occur at any level of organization including altered function of olfactory receptors, olfactory neurons, or effects on the central nervous system. Table 2 provides the select examples of conductive and sensorineural disorders in people.

Many of these etiologies may be relevant for dogs—even though our knowledge of effects on canine olfaction is often inadequate. For example, canine parainfluenza virus is a highly contagious respiratory virus that has been associated with anosmia in dogs (Myers et al. 1988b). Other common respiratory viruses that can produce nasal effects in dogs include canine adenovirus-2, canine herpesvirus-1, canine influenza virus, canine reovirus, and canine coronaviruses (Sykes 2014). In dogs and people, nasal polyps can result in occlusion of the upper airway and extensive damage to nasal turbinates (Bottero et al. 2021). In people, nasal polyp formation is associated with chronic rhinosinusitis and altered olfaction (Kwah and Peters 2019; Marple et al. 2009; Stevens et al. 2016).

Canine cognitive dysfunction (CCD) is considered a canine analog of Alzheimer disease in people (Dewey et al. 2019). Brain lesions seen in geriatric dogs include atrophy, β-amyloid deposits, and neurofibrillary tangles that correlate with cognitive decline (Youssef et al. 2016). Increased β-amyloid deposits contribute to cerebrovascular amyloid angiopathy, a form of cerebrovascular disease. It remains unknown whether CCD is associated with olfactory dysfunction in dogs. In contrast, anosmia, hyposmia, and other olfactory disorders are commonly recognized in Alzheimer disease patients (Walker et al. 2021).

3 Environmental and Pharmaceutical Exposures

Environmental exposure of working dogs is a concern. Exposure to tobacco smoke and ambient air pollution impacts human olfactory function in people and may contribute to age-related declines in olfactory function (Ajmani et al. 2016; Ekström et al. 2022; Murphy et al. 2002). Dogs living in Mexico City, a city with high levels of particulate matter and other forms of ambient air pollution had chronic nasal inflammation, olfactory epithelial degeneration, and reactive astrocytosis in the olfactory bulb (Calderón-Garcidueñas et al. 2003).

Certain medications are also associated with abnormal olfaction in people (Gauvin et al. 2015; Thiermann and Buchbauer 2017). The possibility that a drug that affects olfaction in people could likewise impair performance of a scent detection dog should always be considered when using therapeutic agents in these animals. Few studies have examined the effect of drug treatments on the olfactory performance of scent detection dogs. For example, dogs given either dexamethasone or hydrocortisone plus desoxycorticosterone acetate (DOCA) had elevated olfactory detection threshold for benzaldehyde and eugenol in the absence of nasal pathology (Ezeh et al. 1992). Oral administration of metronidazole impaired the ability of trained dogs to detect ammonium nitrate and trinitrotoluene (Jenkins et al. 2016). In contrast, administration of doxycycline did not impair olfactory function of explosives detection dogs (Jenkins et al. 2016). Administration of naloxone following the administration of intravenous fentanyl to working dogs did not impair the ability of the dogs to detect universal detection calibrant (Essler et al. 2019). The impact of vaccination for *Bordetella bronchiseptica*, the pathogen associated with kennel cough in dogs, has been recently examined (Collins et al. 2022). These studies showed that the odor threshold for a universal detection calibrant was unaffected by the administration of either an oral or intranasal vaccine. However, the use of a combined regimen of an oral vaccine followed 28 days later by an intranasal vaccine resulted in a small but statistically significant increase in time to detect the calibrant (Collins et al. 2022).

4 Final Thoughts

This chapter has considered the scientific evidence available examining the impact of disease on canine olfactory dysfunction. The scant information available on this topic reflects several factors. Assessment of olfactory function in dogs with either experimental or naturally occurring disease is uncommonly performed. Interestingly, no mention of impaired olfaction occurs in a recent review of nasal disease (Cohn 2020). In addition, our ability to assess olfaction in dogs remains limited especially in clinical settings. Acute injury resulting in the loss of olfactory neurons can be repaired via replacement of lost neurons by pluripotent basal stem cells (Brenneman et al. 2002). Moreover, there is a significant amount of functional reserve and adaptation that can occur following injury. Studies performed in rats and other laboratory animal species with a well-developed sense of smell have shown that excessive damage to the olfactory epithelium is required before changes in olfactory function are observed (Hurtt et al. 1988; Owens et al. 1996).

The current state of the science may leave the reader with far more questions than answers but also points to the need for future studies. In the meantime, we should anticipate that disease states that affect the sense of smell in humans could also impair olfaction in dogs. This conclusion is bolstered by supportive data available from other mammalian species. For example, this chapter discussed age-related changes that occur in both the human and canine olfactory epithelium. Age-related declines in olfaction are well documented in humans but largely unknown in dogs due to a lack of studies. Studies performed with rats have shown age-related decreases in the number of olfactory receptor neurons found in the olfactory epithelium as well as reduced odorant-induced activity in this epithelium (Loo et al. 1996). Other studies have shown declines in the performance of aged rats on simple tests of olfaction or sniffing behaviors (Hlinák and Krejcí 1990; Luu et al. 2008). This broader knowledge base should be considered when answering the question of whether a disease state in dogs may affect their olfactory ability.

References

Ajmani GS, Suh HH, Pinto JM (2016) Effects of ambient air pollution exposure on olfaction: a review. Environ Health Perspect 124:1683–1693. https://doi.org/10.1289/EHP136

Attems J, Walker L, Jellinger KA (2015) Olfaction and aging: a mini-review. Gerontology 61(6):485–490

Bagnasco D, Passalacqua G, Braido F et al (2021) Quick olfactory Sniffin' Sticks Test (Q-Sticks) for the detection of smell disorders in COVID-19 patients. World Allergy Organ J 14(1):100497

Beecher K, St John JA, Chehrehasa F (2018) Factors that modulate olfactory dysfunction. Neural Regen Res 13(7):1151–1155

Bhatnagar KP, Kennedy RC, Baron G, Greenberg RA (1987) Number of mitral cells and the bulb volume in the aging human olfactory bulb: a quantitative morphological study. Anat Rec 218(1):73–87

Bird DJ, Amirkhanian A, Pang B, Van Valkenburgh B (2014) Quantifying the cribriform plate: influences of allometry, function, and phylogeny in Carnivora. Anat Rec (hoboken) 297(11):2080–2092

Bird DJ, Jacquemetton C, Buelow SA et al (2021) Domesticating olfaction: dog breeds, including scent hounds, have reduced cribriform plate morphology relative to wolves. Anat Rec (hoboken) 304(1):139–153

Bilinska K, Butowt R (2020) Anosmia in COVID-19: a bumpy road to establishing a cellular mechanism. ACS Chem Neurosci 11(15):2152–2155

Boesveldt S, Postma EM, Boak D et al (2017) Anosmia-A clinical review. Chem Senses 42(7):513–523

Boscolo-Rizzo P, Menegaldo A, Fabbris C et al (2021) Six-month psychophysical evaluation of olfactory dysfunction in patients with COVID-19. Chem Senses 46:bjab006

Bottero E, Mussi E, Raponi F, De Lorenzi D, Ruggiero P (2021) Diagnosis and outcome of nasal polyposis in 23 dogs treated medically or by endoscopic debridement. Can Vet J 62(7):736–742

Brenneman KA, Meleason DF, Sar M et al (2002) Olfactory mucosal necrosis in male CD rats following acute inhalation exposure to hydrogen sulfide: reversibility and the possible role of regional metabolism. Toxicol Pathol 30(2):200–208

Burckhardt G (2012) Drug transport by Organic Anion Transporters (OATs). Pharmacol Ther 136(1):106–130

Burd GD (1993) Morphological study of the effects of intranasal zinc sulfate irrigation on the mouse olfactory epithelium and olfactory bulb. Microsc Res Tech 24(3):195–213

Buschhüter D, Smitka M, Puschmann S et al (2008) Correlation between olfactory bulb volume and olfactory function. Neuroimage 42(2):498–502

Calderón-Garcidueñas L, Maronpot RR, Torres-Jardon R et al (2003) DNA damage in nasal and brain tissues of canines exposed to air pollutants is associated with evidence of chronic brain inflammation and neurodegeneration. Toxicol Pathol 31(5):524–538

Cohn LA (2020) Canine nasal disease: an update. Vet Clin North Am Small Anim Pract 50(2):359–374

Collins A, Bear RA, Mallikarjun A et al (2022) Effects of intranasal and oral *Bordetella bronchiseptica* vaccination on the behavioral and olfactory capabilities of detection dogs. Front Vet Sci 18(9):882424

Craven BA, Paterson EG, Settles GS (2010) The fluid dynamics of canine olfaction: unique nasal airflow patterns as an explanation of macrosmia. J R Soc Interface 7(47):933–943

Czasch S, Paul S, Baumgärtner W (2006) A comparison of immunohistochemical and silver staining methods for the detection of diffuse plaques in the aged canine brain. Neurobiol Aging 27(2):293–305

Dewey CW, Davies ES, Xie H, Wakshlag JJ (2019) Canine cognitive dysfunction: pathophysiology, diagnosis, and treatment. Vet Clin North Am Small Anim Pract 49(3):477–499

Dorman DC (2018) Olfactory system. In: McQueen CA (ed) Comprehensive toxicology, 3rd edn. Elsevier, New York, NY, pp 361–375

Doty RL, Kamath V (2014) The influences of age on olfaction: a review. Front Psychol 7(5):20

Ekström IA, Rizzuto D, Grande G et al (2022) Environmental air pollution and olfactory decline in aging. Environ Health Perspect 130(2):27005

Essler JL, Smith PG, Berger D et al (2019) A randomized cross-over trial comparing the effect of intramuscular versus intranasal naloxone reversal of intravenous fentanyl on odor detection in working dogs. Animals (Basel) 9(6):385

Ezeh PI, Myers LJ, Hanrahan LA, Kemppainen RJ, Cummins KA (1992) Effects of steroids on the olfactory function of the dog. Physiol Behav 51:1183–1187

Gauvin DV, Abernathy MM, Tapp RL, Yoder JD, Dalton JA, Baird TJ (2015) The failure to detect drug-induced sensory loss in standard preclinical studies. J Pharmacol Toxicol Methods 74:53–74

Getchell ML, Chen Y, Ding X et al (1993) Immunohistochemical localization of a cytochrome P-450 isozyme in human nasal mucosa: age-related trends. Ann Otol Rhinol Laryngol 102(5):368–374

Gomez J, Pickup S (2022) Cribiform plate fractures. In: StatPearls [Internet]. Treasure Island (FL): StatPearls Publishing

Hannum ME, Ramirez VA, Lipson SJ et al (2020) Objective sensory testing methods reveal a higher prevalence of olfactory loss in COVID-19-positive patients compared to subjective methods: A systematic review and meta-Analysis. Chem Senses 45(9):865–874

Heydel JM, Coelho A, Thiebaud N et al (2013) Odorant-binding proteins and xenobiotic metabolizing enzymes: implications in olfactory perireceptor events. Anat Rec (hoboken) 296(9):1333–1345

Hlinák Z, Krejcí I (1990) Spontaneous behaviour, habituation, and sniffing responsiveness of aged female rats. Act Nerv Super (praha) 32(2):87–94

Hirai T, Kojima S, Shimada A et al (1996) Age-related changes in the olfactory system of dogs. Neuropathol Appl Neurobiol 22(6):531–539

Hoffman HJ, Rawal S, Li CM, Duffy VB (2016) New chemosensory component in the U.S. National Health and Nutrition Examination Survey (NHANES): first-year results for measured olfactory dysfunction. Rev Endocr Metab Disord 17(2):221–240

Hopkins C, Lechien JR, Saussez S (2021) More that ACE2? NRP1 may play a central role in the underlying pathophysiological mechanism of olfactory dysfunction in COVID-19 and its association with enhanced survival. Med Hypotheses 146:110406

Houpt KA, Davis PP, Hintz HF (1982) Effect of peripheral anosmia in dogs trained as flavor validators. Am J Vet Res 43(5):841–843
Houpt KA, Shepherd P, Hintz HF (1978) Two methods for producing peripheral anosmia in dogs. Lab Anim Sci 28(2):173–177
Hurtt ME, Thomas DA, Working PK et al (1988) Degeneration and regeneration of the olfactory epithelium following inhalation exposure to methyl bromide: pathology, cell kinetics, and olfactory function. Toxicol Appl Pharmacol 94(2):311–328
Izquierdo-Dominguez A, Rojas-Lechuga MJ, Mullol J, Alobid I (2020) Olfactory dysfunction in the COVID-19 outbreak. J Investig Allergol Clin Immunol 30(5):317–326
Jacquemetton C, Drexler A, Kellerman G et al (2021) The impact of extreme skull morphology in domestic dogs on cribriform plate shape. Anat Rec (hoboken) 304(1):190–201
Jenkins EK, Lee-Fowler TM, Angle TC, Behrend EN, Moore GE (2016) Effects of oral administration of metronidazole and doxycycline on olfactory capabilities of explosives detection dogs. Am J Vet Res 77:906–912
Jenkins EK, DeChant MT, Perry EB (2018) When the nose doesn't know: Canine olfactory function associated with health, management, and potential links to microbiota. Front Vet Sci 5:56
Jeong SH, Jang JH, Lee YB (2022) Drug delivery to the brain via the nasal route of administration: exploration of key targets and major consideration factors. J Pharm Investig 24:1–34
Kalmey JK, Thewissen JG, Dluzen DE (1998) Age-related size reduction of foramina in the cribriform plate. Anat Rec 251(3):326–329
Kishikawa M, Iseki M, Nishimura M et al (1990) A histopathological study on senile changes in the human olfactory bulb. Acta Pathol Jpn 40(4):255–260
Klopfenstein T, Kadiane-Oussou NJ, Toko L et al (2020) Features of anosmia in COVID-19. Med Mal Infect 50(5):436–439
Krishna NS, Getchell TV, Dhooper N et al (1995) Age- and gender-related trends in the expression of glutathione S-transferases in human nasal mucosa. Ann Otol Rhinol Laryngol 104(10 Pt 1):812–822
Kwah JH, Peters AT (2019) Nasal polyps and rhinosinusitis. Allergy Asthma Proc 40(6):380–384
Lawson MJ, Craven BA, Paterson EG, Settles GS (2012) A computational study of odorant transport and deposition in the canine nasal cavity: implications for olfaction. Chem Senses 37(6):553–566
Lechien JR, Chiesa-Estomba CM, De Siati DR et al (2020) Olfactory and gustatory dysfunctions as a clinical presentation of mild-to-moderate forms of the coronavirus disease (COVID-19): a multicenter European study. Eur Arch Otorhinolaryngol 277(8):2251–2261
Loo AT, Youngentob SL, Kent PF, Schwob JE (1996) The aging olfactory epithelium: neurogenesis, response to damage, and odorant-induced activity. Int J Dev Neurosci 14(7–8):881–900
Luu TT, Pirogovsky E, Gilbert PE (2008) Age-related changes in contextual associative learning. Neurobiol Learn Mem 89(1):81–85
Marple BF, Stankiewicz JA, Baroody FM et al (2009) Diagnosis and management of chronic rhinosinusitis in adults. Postgrad Med 121(6):121–139
Meng X, Deng Y, Dai Z, Meng Z (2020) COVID-19 and anosmia: a review based on up to date knowledge. Am J Otolaryngol 41(5):102581
Meler E, Dunn M, Lecuyer M (2008) A retrospective study of canine persistent nasal disease: 80 cases (1998–2003). Can Vet J 49(1):71–76
Moein ST, Hashemian SM, Tabarsi P, Doty RL (2020) Prevalence and reversibility of smell dysfunction measured psychophysically in a cohort of COVID-19 patients. Int Forum Allergy Rhinol. 10(10):1127–1135
Morrison EE, Costanzo RM (1990) Morphology of the human olfactory epithelium. J Comp Neurol 297(1):1–13
Mortier JR, Blackwood L (2020) Treatment of nasal tumours in dogs: a review. J Small Anim Pract 61(7):404–415
Murphy C, Schubert CR, Cruickshanks KJ et al (2002) Prevalence of olfactory impairment in older adults. JAMA 288(18):2307–2312

Myers LJ, Hanrahan LA, Swango LJ, Nusbaum KE (1988a) Anosmia associated with canine distemper. Am J Vet Res 49(8):1295–1297
Myers LJ, Nusbaum KE, Swango LJ et al (1988b) Dysfunction of sense of smell caused by canine parainfluenza virus infection in dogs. Am J Vet Res 49(2):188–190 PMID: 2831762
Najafloo R, Majidi J, Asghari A et al (2021) Mechanism of anosmia caused by symptoms of COVID-19 and emerging treatments. ACS Chem Neurosci 12(20):3795–3805
Owens JG, James RA, Moss OR et al (1996) Design and evaluation of an olfactometer for the assessment of 3-methylindole-induced hyposmia. Fundam Appl Toxicol 33(1):60–70
Paik SI, Lehman MN, Seiden AM et al (1992) Human olfactory biopsy. The influence of age and receptor distribution. Arch Otolaryngol Head Neck Surg 118(7):731–738
Peterson EK, Letellier MA, Parsons JA et al (1981) Olfactory pedunculotomy induced anosmia in the wolf (*Canis lupus*). Physiol Behav 27(3):543–546
Plickert HD, Tichy A, Hirt RA (2014) Characteristics of canine nasal discharge related to intranasal diseases: a retrospective study of 105 cases. J Small Anim Pract 55(3):145–152
Polgár Z, Kinnunen M, Újváry D et al (2016) A test of canine olfactory capacity: comparing various dog breeds and wolves in a natural detection task. PLoS ONE 11(5):e0154087
Printza A, Constantinidis J (2020) The role of self-reported smell and taste disorders in suspected COVID-19. Eur Arch Otorhinolaryngol 277(9):2625–2630
Quignon P, Rimbault M, Robin S, Galibert F (2012) Genetics of canine olfaction and receptor diversity. Mamm Genome 23(1–2):132–143
Rygg AD, Van Valkenburgh B, Craven BA (2017) The Influence of sniffing on airflow and odorant deposition in the canine masal cavity. Chem Senses 42(8):683–698
Saltagi AK, Saltagi MZ, Nag AK et al (2021) Diagnosis of anosmia and hyposmia: a systematic review. Allergy Rhinol (providence). 5(12):21526567211026570
Scangas GA, Bleier BS (2017) Anosmia: differential diagnosis, evaluation, and management. Am J Rhinol Allergy 31(1):3–7
Sit THC, Brackman CJ, Ip SM et al (2020) Infection of dogs with SARS-CoV-2. Nature 586(7831):776–778
Sollai G, Melis M, Magri S et al (2019) Association between the rs2590498 polymorphism of Odorant Binding Protein (OBPIIa) gene and olfactory performance in healthy subjects. Behav Brain Res 17(372):112030
Stevens WW, Schleimer RP, Kern RC (2016) Chronic rhinosinusitis with nasal polyps. J Allergy Clin Immunol Pract 4(4):565–572
Sun JS, Xiao S, Carlson JR (2018) The diverse small proteins called odorant-binding proteins. Open Biol 8(12):180208
Sykes JE (2014) Canine viral respiratory infections. Canine Feline Infect Dis 170–181
Tasker S, Knottenbelt CM, Munro EA et al (1999) Aetiology and diagnosis of persistent nasal disease in the dog: a retrospective study of 42 cases. J Small Anim Pract 40(10):473–478
Thiermann F, Buchbauer G (2017) Influence of drugs on smelling capacity. A Rev Flavour Fragr J 32:147–158
Tisdall FF, Brown A, Defries RD (1938) Persistent anosmia following zinc sulfate spraying. J Pediatr 18(1):60–62
Walker IM, Fullard ME, Morley JF, Duda JE (2021) Olfaction as an early marker of Parkinson's disease and Alzheimer's disease. Handb Clin Neurol 182:317–329
Weiss T, Soroka T, Gorodisky L et al (2020) Human olfaction without apparent olfactory bulbs. Neuron 105(1):35-45.e5
Youssef SA, Capucchio MT, Rofina JE et al (2016) Pathology of the aging brain in domestic and laboratory animals, and animal models of human neurodegenerative diseases. Vet Pathol 53(2):327–348
Yu CR, Wu Y (2017) Regeneration and rewiring of rodent olfactory sensory neurons. Exp Neurol 287(Pt 3):395–408

Methodological Considerations in Dog Olfaction Research and Implementation

Behavioral Characteristics Associated with Detection Dog Success

Lucia Lazarowski and Bart Rogers

Abstract

A critical challenge to the canine detection industry is the identification of dogs with behavioral characteristics capable of fulfilling operational functions. This challenge is largely due to a lack of empirical evidence regarding the particular behavioral traits associated with long-term success in scent detection disciplines as well as reliable methods for identifying dogs possessing those traits. This chapter reviews behavioral characteristics universal to a range of detection tasks, highlighting aspects that may be unique to particular disciplines. We also discuss methods used to evaluate and select dogs for detection tasks, and make recommendations for future research needed to improve the selection process.

Keywords

Detection dogs • Behavior • Selection • Phenotyping

1 Introduction

Scent detection dogs play a critical role in homeland security (e.g., explosives and narcotics detection), forensics investigations (e.g., human remains detection, arson), disaster response (e.g., search and rescue), conservation efforts (e.g., detection of invasive or endangered species), among many other emerging areas. A critical challenge to the industry is the identification of dogs capable of fulfilling these operational functions. To succeed in an operational detection career, dogs must possess the physical and behavioral characteristics necessary for performing the task. While identifying dogs with the physical, structural, and medical

L. Lazarowski (✉) · B. Rogers
Canine Performance Sciences, Auburn University College of Veterinary Medicine, Auburn, AL 36849, USA
e-mail: lucia.lazarowski@auburn.edu

L. Lazarowski (ed.), *Olfactory Research in Dogs*,
https://doi.org/10.1007/978-3-031-39370-9_4

attributes (e.g., size, coat type, orthopedic soundness, etc.) needed to work in the field can be readily assessed through physical examinations and screenings of medical records, identifying dogs with the necessary behavioral attributes is a greater obstacle which remains a significant hurdle in the industry. This challenge is largely due to a lack of (1) empirical evidence regarding the particular behavioral traits associated with long-term success in scent detection disciplines, and (2) reliable methods for identifying dogs possessing those traits. Given that behavioral characteristics are greater determinants of successfully completing training and deploying in an operational role than non-behavioral characteristics (Graham and Gosling 2009; Sinn et al. 2010), there is a significant need for improving methods of behavioral selection. Furthermore, doing so would minimize the time, effort, and costs associated with developing, training, and procuring dogs, ultimately improving program efficiency, canine welfare, and industry outcomes.

Many of the behavioral characteristics that contribute to detection dog operational success are shared across the various sub-disciplines of detection (e.g., explosives, narcotics, search and rescue, human remains, etc.). Therefore, this chapter will review characteristics universal to a range of detection tasks, highlighting aspects with greater or less relevance to particular disciplines where applicable. For a more in-depth discussion on specific sub-disciplines of detection, we direct the reader to other published texts focusing on selecting dogs for explosives detection (Lazarowski et al. 2020); search and rescue (Schneider and Slotta-Bachmayr 2009); wildlife detection (Beebe et al. 2016; Jamieson et al. 2017); truffle detection (Čejka et al. 2022); and cadaver/human remains detection (Martin et al. 2020). We will also discuss methods used to evaluate and select dogs for detection tasks, and make recommendations for future research needed to improve the process.

2 Behavioral Characteristics

The behavioral characteristics desired of a detection dog vary somewhat across specific disciplines as a function of: (1) factors related to the operational environmental, (2) inherent aspects of the target odor, and (3) other demands specific to the particular task; however, some general characteristics are largely agreed to be universal for a successful and enduring detection career. For the purposes of this chapter, we will categorize these broad domains as (1) olfactory abilities, (2) motivational variables, and (3) emotional regulation. These domains will be discussed in further detail in the sections that follow. Because our focus is on those characteristics purported to have a genetic basis and therefore be relatively stable, our discussion will concentrate on phenotypic characterization; environmental effects of development and training are beyond the scope of this chapter, but are reviewed in detail elsewhere (Troisi et al. 2019; Lazarowski et al. 2021b).

2.1 Olfactory Behavior

Not surprisingly, olfactory-related abilities are critical to detection dog performance. Odor-detecting capabilities, such as sensitivity, specificity, and general olfactory acuity result from sensory and anatomical features of the olfactory system including head and nose shape, size of the nasal cavity, and number of olfactory receptor cells (Kokocinska-Kusiak et al. 2011). For example, natural (i.e., untrained) olfactory detection thresholds have been shown to vary between breeds originally bred for olfactory detection tasks (e.g., scent hounds) and those that were not (e.g., sight hounds) (Polgar et al. 2016). However, these nuances do not appear to have direct, appreciable impacts on the operational suitability of a detection dog, likely due to other aspects of detection dog performance resulting from a constellation of behavioral characteristics related to motivation, endurance, and trainability. In regard to the olfactory ability domain, olfactory behavior (i.e., tendencies to engage in odor-based search and specific search technique) will likely have a greater impact on detection performance than pure olfactory function.

One aspect of detection performance that varies in degree of importance as a function of specific detection discipline is the type of search technique. For example, some disciplines, such as tracking, benefit from a nose-to-the-ground search, detecting odors in footsteps, articles left behind, or fallen skin cells. For other disciplines, such as search and rescue, air-scenting with a nose-in-the-air search is more effective (Jones et al. 2004), where the dog is sampling odor molecules wafting in the air in the wake of a moving person, as is the case in person-borne improvised explosives device detection (PBIED). However, many odor detection disciplines utilize a combination of both techniques for the most efficient search. Regarding behavioral selection of detection dogs, although search technique can be trained to an extent, breeds originally selected for a particular type of search will show inherent preferences in their tendency to utilize a particular search technique, thus leading to more efficient training. For example, scent hounds are known for their superior tracking instincts, interrogating and pursuing primarily ground scent. By contrast, dogs bred for upland game hunting (e.g., pointers and setters) primarily engage in a nose-up air-scenting search to determine the direction of an odor source and locate a target without a scent trail to follow (Beebe et al. 2016). Capitalizing on breed-specific search techniques to accentuate certain capabilities for specific detection roles can be illustrated by the introduction of the German-Wirehaired pointer (GWP) breed to the Labrador retriever breeding population of Auburn University's Canine Performance Sciences program. The crossing of a GWP sire (from the Deutsch-Drahthaar lineage for which the standard for breeding was performance-based, which was not always upheld for GWPs bred in the United States due to the breed's rising popularity (DeRosa 2019)) to Labrador retriever dams was driven by the program's focus of producing dogs for PBIED, for which air-scenting is a desirable trait. This introduction resulted in generations of GWP X Labrador crosses exhibiting accentuated air-scenting (along with other traits related to the versatility of the GWP as a hunting breed), complemented by the sociability and other important aspects of trainability (discussed below) of the Labrador retriever (Lazarowski et al. 2018).

2.2 Motivational Drives

It is widely acknowledged that motivation to work is critical to the success of any working dog, enabling the endurance to continue performing their tasks in a range of conditions. For detection dogs, especially those trained to detect infrequently occurring targets, the motivation to continue searching with a low rate of encountering a target (and, subsequently, reinforcement), which is common in most disciplines, is critical to the ability to continue searching for long periods of time without a decline in performance. For example, an explosives detection dog must maintain the motivation to clear large venues or screen thousands of people with a low probability of encountering a target without a decline in search vigilance (Porritt et al. 2015); wildlife detection dogs must sustain endurance throughout long searches, often in harsh conditions, where there may be a low density of the target species. While there are tactics to maintain motivation, such as shorter duty cycles or planting finds (Porritt et al. 2015; Lazarowski et al. 2021c), selecting dogs with higher inherent motivation to work because they find it intrinsically reinforcing will lead to less frustration and better overall performance.

Despite the term 'drive' often not being operationally defined due to it referring to an internal construct that is difficult to characterize, leading to debate over the meaning and usefulness of the term (Cecil 2015), different types of motivational drives related to detection dog performance are commonly discussed. Perhaps the most important type for a detection dog is the 'hunt drive' (Cablk et al. 2006), or the inherent propensity to engage in persistent olfactory-based searching (Lazarowski et al. 2020). For some dogs, such a propensity is likely intrinsically reinforcing, resulting in higher stamina and endurance to continue searching for long periods of time despite challenges and distractions and without receiving any external reinforcement. Thus, while proper training can result in effective searching, selecting dogs with inherently strong hunt drives (i.e., breeds originally bred for such tasks) will result in more effective and efficient training. As above, this characteristic is particularly important for disciplines routinely requiring long durations of searches, in austere or challenging environments, with seldom reinforcement, such as explosives detection, wildlife detection, and search and rescue, but may not be as critical of a trait for disciplines requiring shorter or less complex searches in more static environments.

As mentioned, use of the term "drive" is often ambiguous and ill-defined. In a validation of a behavioral test predicting puppies' future selection for a detection career, Lazarowski et al. (2021a) rated puppies' performance on a series of tests including a search for several hidden targets and defined the concept of hunt drive as the dogs' "willingness and ability to investigate areas using its nose." Low scores were given for dogs that did not actively investigate areas but rather wandered aimlessly, and high scores reflected dogs that were constantly using their nose to methodically and efficiently clear areas without needing any handler prompting. Higher scores for this variable were related to other metrics of detection performance, such as air-scenting and measures related to reward motivation, and were predictive of future selection for a detection career as early as 3 months (Lazarowski et al. 2021a). However, further research is needed to determine how to best operationalize "hunt drive", and whether it represents a singular construct that is not redundant with others.

External reinforcers are also commonly used with detection dogs, the effectiveness of which depends on the dog's motivation to obtain the particular reinforcer used. First, dogs learn to detect target odors by pairing the odor with a reward; the stronger the dog's desire for that reward, the faster and more effective this conditioning will be. Next, dogs must be taught to perform an alert response to indicate the presence and location of a target, such as sitting (common with explosives and wildlife detection dogs), barking (common with search and rescue dogs), or performing a nose-hold (common with biomedical detection dogs). Such responses are typically shaped through operant conditioning, delivering a reward for performing the correct behavior. Generally, the stronger the desire for the reward, the more effective training will be. Further, for dogs without a strong inherent motivation to hunt for the sake of it, external reinforcers can be used to motivate dogs to engage in their task. However, note that for some dogs with high levels of arousal, highly desired rewards may increase arousal to an unproductive level that interferes with performance. For example, for detection dogs that strongly preferred a ball, the use of a ball as the reward during a cognitive task interfered with problem-solving abilities compared to when a desirable but relatively lower-value reward (a piece of food) was used (Krichbaum and Lazarowski 2022). Therefore, the particular reward used should be carefully chosen and dictated by the dog's arousal level and the demands of the task.

While food is inherently reinforcing for all animals, food rewards are not typically used in most detection dog disciplines due to the impracticalities of its use in the field (DeMatteo et al. 2019). Moreover, the reinforcing value of food at any given time is dependent on the dog's current state of hunger. Regardless, using food rewards can be effective in some contexts and for some disciplines, and is primarily used by the Bureau of Alcohol, Tobacco, Firearms, and Explosives in the training of their dogs (Oxley and Waggoner 2009). As mentioned, food may represent a lower-value reward for some dogs which may be useful in situations where maintaining low arousal is beneficial, such as shaping meticulous behaviors or tedious laboratory-based discrimination work. Interestingly, a genetic marker associated with food motivation and obesity was found in Labrador retrievers, and even more so in those selected for assistance roles, leading the authors to hypothesize a genetic underpinning of trainability (because assistance dogs are primarily trained with food reinforcers) important for these working dogs (Raffan et al. 2016).

For detection dogs, toys (tennis balls, Kong™, etc.) are more commonly used than food. The desire to retrieve and interact with a toy, usually in the context of interacting with a human, is often referred to as 'play drive' (Cablk et al. 2006), and has been reported to be important for explosives detection dogs (Lazarowski et al. 2018, 2020), wildlife detection dogs (Cablk et al. 2006), human remains detection dogs (Martin et al. 2020), truffle detection dogs (Čejka et al. 2022), and drug detection dogs (Ganitskaya et al. 2020). The motivation to play with objects is largely genetically based, stemming from the canine predatory sequence (Lazarowski et al. 2020). While domestic dogs' ancestor, the wolf, exhibits the full sequence of chasing prey until it is captured, dissected, and consumed, modern dog breeds exhibit partial predatory sequences with aspects enhanced or diminished as a result of selective breeding for different working roles (Udell 2014; Mehrkam et al. 2017). Though this behavior originates from predatory behavior towards prey, it is believed that selection has led to its translation to object-play with non-prey items and is unrelated to hunger (Burghardt 2003; Jamieson et al. 2017). Because of the strong genetic basis of the behavior, it is intrinsically rewarding and therefore can serve as a powerful reinforcer (that is, the toy represents an external reinforcer but stimulates intrinsically motivated behavior). Traditionally, especially in the security sector (i.e., explosives and narcotics detection), selection tests for detection dogs have emphasized this object-play trait, often measured as the dogs' desire to retrieve an object and maintain physical possession of it. However, as discussed in Lazarowski et al. (2020), too much emphasis on physical possession may be misleading when a dog prioritizes engaging in a search over engaging with the reward (i.e., completes a search and receives a reward, but eventually drops it to return to searching) which is misinterpreted as a negative quality. Arguably, a stronger hunt drive is more important than strong object-play drive in terms of the critical aspects of detection dog performance.

A related but somewhat different type of motivational drive discussed in relation to detection dog suitability is "prey drive", reflecting a dog's desire to chase (Cablk et al. 2006). While object-play motivation as described above is likely a remnant predatory behavior, it is important to distinguish from that which is elicited by and directed towards actual prey. For any detection dog that may encounter small animals (e.g., squirrels, birds, chipmunks) while working, prey drive could result in distraction and disruption of performance. Additionally, prey drive is of great concern in disciplines for which such behavior could lead to chasing or attacking the target, which could have devastating consequences if the dog harms the target species in the case of wildlife detection dogs (Cablk et al. 2006; DeMatteo et al. 2019), or a victim in the case of cadaver dogs (Martin et al. 2020). Prey drive could also put the dog in danger if the target species or other animals in the environment pose a threat to the dog. For example, in a python detection project conducted by our group in the Florida Everglades, it was critical that the dogs not chase or try to capture (1) the target pythons upon locating them, (2) other large predator species in the area such as bobcats, bears, and panthers, or (3) small animals into ambush zones of alligators and crocodiles. While it is unclear how and to what extent these play and prey mechanisms have diverged, in our experience with explosives detection dogs, prey chasing has not been an issue without any explicit training. We hypothesize that this is due to focusing the dogs' efforts towards the intended targets and rewards used in training from an early age, fulfilling and substituting any actual prey drive.

Little research has systematically examined the importance of these motivational variables in detection dog performance. In one study that assessed the US Transportation Security Administration (TSA) test for evaluating explosives detection dog suitability, physical possession of a toy, described as dogs' willingness to carry an object absent any external input (i.e., from a person), was not predictive of dogs' selection. However, a separate construct related to possession but that reflected the dog's desire to gain possession of a toy from a person (measured during a game of tug) was predictive (McGarrity et al. 2016). This suggests that it was the social nature of the interaction and not the interaction with the toy itself that was important. Similarly, the definition of play drive by Cablk et al. (2006) considers play specifically in the context of playing with a human (e.g., playing tug-of-war). Given that detection dogs primarily work as a dog-handler team, the relevance of social interaction in dogs' motivation to perform their tasks is not surprising.

Detection dogs work closely with people throughout all stages of their career; in training, dogs must be responsive to a trainer's commands and actions, and in operations, dogs must be able to effectively take guidance and direction from their handler. Higher levels of such "biddability" translates to easier training and better cooperation in the field (Morrill et al. 2022). Indeed, Čejka et al. (2022) suggest that truffle detection dogs that are highly cooperative and socially motivated require little to no external motivation to work. For detection dogs that work off-leash at a distance from the handler, such as search and rescue dogs, IED detection dogs, and some wildlife detection dogs, responsiveness to the handler is critical to success as well as the dog's safety (Lazarowski et al. 2020). In a study of cognitive and behavioral characteristics associated with success for IED detection dogs trained for directional control using hand signals, those that more readily responded to human gestures in a problem-solving task were more successful in their career (MacLean and Hare 2018). On the other hand, handler dependence is undesirable for detection dogs because they need to be able to make independent decisions, sometimes contradicting the handler. For example, due to dogs' superior olfactory sensitivity compared to humans, a dog may detect an odor that the handler is not aware of or believes has been cleared; the handler may command the dog to move in a different direction, putting the dog in conflict between the odor and the handler. In this scenario, the dog should defy the handler's command and be "obedient to odor." This concept of "intelligent disobedience" is well acknowledged in other working dog disciplines, such as guide dogs, where the dog must prioritize the handler's safety if, for example, the handler gives the dog a command to cross the road, unaware of oncoming traffic (Lazarowski et al. 2021b). This concept was demonstrated by Lazarowski et al. (2019), demonstrating that young candidate explosives detection dogs that prioritized olfactory cues when they conflicted with a human social cue were more likely to be selected for operational careers in the future.

Such responsiveness to humans has a genetic basis which can therefore be harnessed when selecting dogs for detection roles. Gácsi et al. (2009) differentiate between "cooperative workers" and "independent workers"; the former consists of breeds originally used for working cooperatively with a human partner, bred for their ability to work at a distance at times but still maintaining visual contact and cooperation (e.g., gundogs, herders); the latter consists of breeds used for headstrong independent tasks not involving a human partner (e.g., scent hounds, livestock guarding dogs, sled dogs). For this reason, despite the purported olfactory superiority of scent hounds such as bloodhounds, they are not commonly used in detection tasks involving a high level of cooperation with and responsiveness to a trainer/handler (Jamieson et al. 2017).

Another motivational aspect related to sociability is a dog's general desire to be around and in contact with people, termed "affability" by Wilsson and Sundgren (1997), also found to have a genetic basis. For example, Labrador retrievers scored higher on affability than German shepherds, again likely due to the genetic history of the breeds in relation to working closely with human partners (Wilsson and Sundgren 1997). While German shepherds are commonly used in detection roles, this is the result of their aptitude for police and patrol work (i.e., apprehension and protection) leading to their use in "dual-purpose" roles of both detection and protection out of efficiency, rather than any superior suitability specifically for detection work. The social nature of detection dogs is especially important for those working in areas of high pedestrian traffic, such as dogs screening passengers at airports or event attendees. While these roles require the dogs to be comfortable around people and not pose a risk to the public, the public perception of a dog's temperament (and therefore their comfort in getting close enough to the dog for screening) is an equal driver in selection. The importance of the public's acceptance of the dogs has purportedly led to TSA prioritizing selection for "floppy eared" breeds such as Labrador retrievers, German shorthaired pointers, and Vizslas over traditionally popular explosives detection "pointy eared" breeds such as German shepherds and Belgian Malinois (though the legitimacy of this claim has been contended). In terms of effects on behavior and training, interacting with people can be highly rewarding for some dogs and may be just as if not more reinforcing, and therefore motivating for work, as other external reinforcers, and likely explains the discrepancy in the predictive validity of the different types of possession (i.e., independent vs. during play) analyzed by McGarrity et al. (2016) in the TSA test. However, it is important that the dog's motivation to interact with people is not so high that it distracts the dog from working; engaging in the search should be more reinforcing than any competing available reinforcers such as people. Further, DeMatteo et al. (2019) caution that too strong of a bond between detection dog and handler can lead to instances of the dog playing the role of therapy dog when the handler is in need of support. Rather, they suggest that dogs that are more motivated by their reward than by their handler will be more reliable and flexible, able to work effectively with different handlers as needed.

All of the motivational variables described above likely make up the important construct of "trainability", or the speed and ease with which a dog learns a new task. Trainability as assessed by the validated Canine Behavioral Assessment and Research Questionnaire (Hsu and Serpell 2001) measures variables related to attention and responsiveness to commands and correction, interest in fetching toys, and the ability to ignore distractions in the environment. Therefore, the more motivated a dog is to respond to a person, engage in a given task, or obtain a reward, the more trainable they will be. For example, DeMatteo et al. (2019) argue that distractibility while working likely results from the dog's motivation for the reward not being strong enough to overcome competing sources of distraction. These characteristics have commonly been reported in the literature as important for detection dogs' performance; for example, Martin et al. (2020) list high focus and trainability as important traits for human remains detection dogs; Čejka et al. (2022) mention not losing interest due to engaging in a repetitive routine for truffle detection dogs; trainability as measured by the CBARQ is associated with success for search and rescue (Hare et al. 2018) and explosives detection dogs (Lazarowski et al. 2021a); and "desire to work" was shown to predict success for drug detection dogs (Maejima et al. 2007).

The characteristics motivating search behavior appear to starkly differ between dogs searching in operational environments, such as explosives detection or wildlife detection dogs, versus those working in laboratory settings such as biomedical detection dogs. For many detection dog breeding and training programs, such as Auburn University's Canine Performance Sciences program and the Penn Vet Working Dog Center, dogs utilized for laboratory-based detection tasks are often byproducts of procurement for operational roles; that is, dogs that lack the ability to work effectively in real-world settings, often due to reasons related to fear and anxiety towards the types of stimuli encountered in operational environments (discussed below), are still quite effective in their odor detection capabilities which are valuable for in-house detection research where the dog works in a comfortable, consistent environment. However, there may be additional performance characteristics that are unique to dogs capable of succeeding in the more tedious, nuanced task of biomedical detection, as evidenced by our experience with the significant attrition in dogs attempted for biomedical detection training. Gadbois and Reeve (2016) report that biomedical detection dogs must be carefully selected, with very few able to be successfully trained for the task. In our case, the incompatibility of dogs bred and selected for traditional detection tasks with biomedical detection is likely related to their high energy and arousal levels (Brady et al. 2018a), which does not translate well to detailed, repetitive discrimination tasks in a laboratory setting. Dogs that have been successful in our biological detection studies (i.e., discrimination of virus samples using a small carousel setup) were selected due to their calculated, methodical "micro search" detection techniques (Angle et al. 2016). Therefore, breeds bred for such type of searching behavior, such as flushing breeds that were designed to interrogate areas of high target probability up close (e.g., English cockers) (Spafford 2019), may be ideally suited for these tasks. In Chapter 12 of this text, Concha suggests that the major difference between dogs able to perform biomedical detection and other detection tasks is the need to compare multiple samples with similar odor profiles placed side by side (i.e., in a lineup or carousel), with several repetitions per session and only slight variability in the context. Along those lines, Gadbois et al. in Chap. 15 of this text suggest that the monotonous and artificial nature of laboratory-based discrete discrimination tasks may require even higher motivation than other types of search-based tasks, such as wildlife detection performed in natural environments, which may elicit more natural search behavior that is intrinsically reinforcing. Selecting for strong motivational drives is often piggybacked by high levels of arousal (Lazarowski et al. 2020), so we caution that selection for biomedical detection should strive to strike a balance of high motivation with low arousal. However, there is currently no published research systematically examining the search and motivational characteristics that may differ between dogs successful at laboratory-based and more traditional detection tasks; therefore, further research is needed to address these hypotheses.

2.3 Stress Resilience, Fear, and Anxiety

The tasks required of a detection dog and the environments in which they work are often challenging and unpredictable, which can lead to frustration and stress. Further, housing (e.g., kennels), transport, husbandry, routine veterinary exams, and other aspects of operations commonly encountered by detection dogs can be stressful. A lack of resilience to stressors can impact a detection dog's ability to perform effectively, lead to chronic stress, and impact its overall welfare and ultimately career longevity (Rooney et al. 2009, 2016). Resilience is likely multifaceted, but is generally agreed to reflect the ability to cope with and "bounce back" from negative experiences; in dogs, aspects of resilience include boldness, sociability, emotion regulation, inhibitory control, and adaptability to change (Tiira 2019). While early life experiences such as maternal care and environmental stressors contribute to the development of stress resilience, there is also a genetic component with some breeds such as the Labrador retriever, a popular detection dog breed, considered to be among the more resilient breeds (Tiira 2019).

The concept of stress resilience in regard to detection dog selection has been most commonly discussed for search and rescue dogs, due to the unpredictable and strenuous conditions in which they work. Schneider and Slotta-Bachmayr (2009) describe several psychological stressors leading to mental strain faced by search and rescue dogs such as transport by helicopter, sharing confined spaces with strangers or other dogs, and extensive searches in adverse environments (e.g., rubble piles and disaster sites). The authors report that during prolonged rubble searches, the change from beneficial stress (eustress) to negative (distress), measured by the stress hormone cortisol, occurred after 60–80 min of steady searching. Martin et al. (2020) also emphasizes the importance of the ability to deal with similar stressful situations for human remains detection dogs in disaster environments such as unstable surfaces from collapsed structures, smoke, and crawling in tight spaces. The degree of stress resilience needed will vary as a function of stressors routinely faced in the operational environment and therefore differs across detection disciplines; for example, urban search and rescue dogs responding to a large disaster will face many more stressors than a wilderness search and rescue dog working in an unpopulated, naturalistic area such as a forest or field.

A characteristic related to resilience that in many ways overlaps is that of fear and anxiety; however, here we refer to a dog's reaction to potentially fear-eliciting stimuli in the environment which is frequently reported as a primary reason for failure to complete training or qualify for operational roles across a range of working dog disciplines (Goddard and Beilharz 1984; Lazarowski et al. 2018, 2021a; Dollion et al. 2019). Fearfulness in detection dogs is often manifested as hesitation toward unfamiliar stimuli, including people, environments, sounds, and objects (Beebe et al. 2016). A dog's reaction to unfamiliar stimuli, which can be measured as approach or avoidance, can significantly impact its ability to work effectively. Many terms have been used to refer to this construct (though note some have overlapping definitions with resilience), including: nerve strength (Brownell and Marsolais 2000; Beebe et al. 2016), environmental soundness (Lazarowski et al. 2018)/sureness (Wilsson and Sinn 2012)/stability (McGarrity et al. 2016), emotional reactivity (Sherman et al. 2015), courage (Wilsson and Sundgren 1997), and sensitivity to aversives (Brady et al. 2018b). Again, the degree of acceptable reactivity will vary across disciplines, depending on the exposure typical of the operational environment. For example, fearfulness has been reported to be less important or not mentioned at all in discussions of the behavioral characteristics necessary for wildlife conservation dogs (Cablk et al. 2006; Beebe et al. 2016) and truffle detection dogs (Čejka et al. 2022), but is consistently reported as a critical aspect of performance for search and rescue (Brownell and Marsolais 2000; Hare et al. 2018) and explosives detection dogs (Rooney et al. 2004; McGarrity et al. 2016; Lazarowski et al. 2018), and has been shown to be predictive of future selection as an explosives detection dog as early as 3 months of age (Lazarowski et al. 2021a). Below we highlight a few specific fears pertinent to particular detection disciplines.

Fear of unfamiliar people can be detrimental to the effectiveness of any detection dog that will be working in urban environments or in the proximity of people, such as explosives detection dogs working in mass transit areas, event venues, or security checkpoints, particularly those performing passenger screening; narcotics or weapons detection dogs working in schools; contraband detection dogs screening luggage in airports; urban search and rescue dogs; and hospital infection detection dogs. Level of comfort around crowds/strangers will be less critical to the performance of dogs working only with a handler or small team of people, such as wildlife detection dogs working in fields/forests, forensics detection dogs (e.g., arson, cadaver) working a restricted investigation scene, bed bug detection dogs working an evacuated home, cargo screening dogs working in private areas of an airport, and biomedical detection dogs screening samples in a laboratory.

Confidence navigating a variety of surfaces, sometimes referred to as 'tactile nerve strength' (Brownell and Marsolais 2000), is one of the most commonly reported critical attributes of successful search and rescue and wildlife detection dogs alike, due to the need to navigate different types of environments. For example, urban search and rescue dogs must negotiate rubble piles and disaster sites that contain slippery, unstable, and rough surfaces, and wilderness rescue and wildlife detection dogs must traverse a variety of different terrain, such as thick vegetation and rocky ground (Brownell and Marsolais 2000; Schneider 2009; Hare et al. 2018; Martin et al. 2020); explosives, narcotics, and other types of detection dogs working in different buildings and venues must be comfortable searching on a variety of floorings such as carpet, concrete, grates, and slick floors.

Assessing this domain typically involves testing dogs' reactions to unfamiliar, unusual, or startling stimuli. Many detection dog programs assess dogs' confidence in operational scenarios by performing walk-throughs of real-world environments, assessing their ability to navigate effectively in the presence of people, machinery, noises, objects, and surfaces (Brownell and Marsolais 2000; McGarrity et al. 2016; Lazarowski et al. 2018). While more realistic, a disadvantage of testing in these settings is the inability to control the environment and standardize the test across dogs, which may result in unreliable outcomes. Another test format, often referred to as emotional reactivity testing, exposes dogs to a battery of startling and unusual stimuli and assesses reactions (response and recovery) to each (Sherman et al. 2015). While this testing format may seem artificial, standardization across dogs and timepoints can be ensured and research has shown it to be a valid and reliable method for evaluating dogs' fearfulness. For example, Sherman et al. (2015) found that explosives detection dogs' aggregate responses across a range of stimuli (e.g., unusual person, remote-controlled car, umbrella opening) in an emotional reactivity test correlated with an external measure of anxiety (open-field anxiety test), and that cortisol levels were higher after the test compared to baseline. Lazarowski et al. (2021a) validated a similar test designed for puppies in training for explosives detection, demonstrating consistency across different scorers, convergence with an external measure of anxiety (CBARQ), and predictive validity of future selection as early as 3 months of age. Therefore, emotional reactivity tests may be a more accurate, efficient method than more cumbersome and less standardized environmental testing.

3 Sourcing and Selecting Detection Dogs

An increasing challenge to the detection dog industry is the availability of dogs with the necessary behavioral attributes for reliable detection work, which has been further compounded during the COVID-19 pandemic. In the security sector, challenges in sourcing dogs arise from hurdles encountered during the government procurement process, as well as a lack of incentives for breeders to supply purpose-bred dogs for procurement (Leighton et al. 2018). Across all detection disciplines, identifying dogs with the necessary behavioral characteristics is difficult due to the high behavioral standards required of operational detection dogs as well as a general lack of reliable selection tools. While further research is needed to develop and validate reliable behavioral assessments for selecting detection dogs across a range of disciplines, recent progress has been made in demonstrating the effectiveness of behavioral tests (Sherman et al. 2015; Lazarowski et al. 2021a), survey-based methods (Hare et al. 2018, 2021), and cognitive measures (MacLean and Hare 2018) in identifying suitable detection dogs.

Traditionally, detection dogs have been sourced from breeding programs that specifically breed, raise, and train purpose-bred detection dogs. For example, Auburn University's Canine Performance Sciences program produces purpose-bred dogs primarily for explosives detection (Haney and Wilborn 2021), and the Penn Vet Working Dog Center produces dogs for a range of detection careers, primarily explosives detection and search and rescue (Hare et al. 2021). While selecting dogs from dedicated breeding programs is advantageous in ensuring medical soundness, breeding integrity, and assurance associated with the program or bloodline's reputation, obtaining dogs from such programs can be cost-prohibitive due to the need to offset the costs of dog housing, personnel, and veterinary care associated with breeding and puppy raising operations. Non-conventional methods of procurement such as sourcing dogs from shelters or community models (i.e., individuals volunteer to train and deploy with their privately owned dogs) are more common in search and rescue and wildlife detection disciplines (Byosiere et al. 2019). Programs such as Working Dogs for Conservation and the National Disaster Search Dog Foundation have had success selecting and training shelter dogs for detection careers; some of the behavioral characteristics that may have resulted in a dog being incompatible as a pet living in a home and relinquished to a shelter are often those that are desirable for a detection dog (Cablk et al. 2006; Byosiere et al. 2019). However, while more cost-effective, sourcing dogs from shelters requires a significant amount of time screening hundreds of dogs in order to find enough with the physical and behavioral characteristics necessary for detection careers; and volunteer-based models may not be as reliable as purpose-bred dogs and dedicated personnel.

4 Conclusions

Ultimately, the behavioral repertoire of a successful detection dog will depend on the operational environment the dog will be expected to work in (e.g., urban environments, wilderness, laboratories), the nature of the target to be detected (e.g., moving versus static targets), and the specific task (e.g., person-borne detection, static object screening, multiple sample discrimination). However, accurate and reliable identification of the behavioral characteristics relevant to detection dog operational success remains a persistent and significant challenge in the industry, and the majority of recommendations are based on anecdotal reports and experience. For significant progress to be made, there is a need for empirical research systematically examining behavioral characteristics, measured by objective methods, that are associated with detection dog performance outcomes (e.g., successful completion of training, certification, selection for an operational role, performance in the field, etc.). Determining which outcomes are valid measures of success, however, is a challenge in itself. For example, standards for training, selection, and certification vary across organizations and are often relatively subjective; further, such metrics represent more immediate outcomes and may not predict ultimate success in the field over longer periods of time (McGarrity et al. 2016; Lazarowski et al. 2021a). Efforts to improve behavioral selection should include standardization of the terminology and methods used to assess and select detection dogs; validation of such efforts through convergence with actual field success; and longitudinal assessments of predictive validity across dogs' careers in the field.

References

Angle TC, Passler T, Waggoner PL, Fischer TD, Rogers B, Galik PK, Maxwell HS (2016) Real-Time detection of a virus using detection dogs. Front VetY Sci 2:1–6. https://doi.org/10.3389/fvets.2015.00079

Beebe SC, Howell TJ, Bennett PC (2016) Using scent detection dogs in conservation settings: a review of scientific literature regarding their selection. Front VetY Sci 3:96. https://doi.org/10.3389/fvets.2016.00096

Brady K, Cracknell N, Zulch H, Mills DS (2018a) Factors associated with long-term success in working police dogs. Appl Anim Behav Sci 207:67–72. https://doi.org/10.1016/j.applanim.2018.07.003

Brady K, Cracknell N, Zulch H, Mills DS (2018b) A Systematic review of the reliability and validity of behavioural tests used to assess behavioural characteristics important in working dogs. Frontiers in Veterinary Science 5:103. https://doi.org/10.3389/fvets.2018.00103

Brownell DA, Marsolais M (2000) The Brownell-Marsolais scale: a proposal for the qualitative evaluation of SAR/disaster K9 candidates. Adv Rescue Technol 5:57–67

Burghardt WF (2003) Behavioral considerations in the management of working dogs. VetY Clin: Small Anim Pract 33:417–446

Byosiere S-E, Feng LC, Rutter NJ (2019) Factors That May Affect the Success of Scent Detection Dogs: Exploring Nonconventional Models of Preparation and Deployment. Comp. Cogn. & Behav. Rev. 14

Cablk ME, Heaton JS, Cablk1 ' ME, Heaton2 JS (2006) Accuracy and Reliability of Dogs in Surveying for Desert Tortoise (Gopherus agassizii). Source: Ecol Appl Ecol Appl 16:1926–1935. https://doi.org/10.1890/1051-0761(2006)016[1926:AARODI]2.0.CO;2

Cecil L (2015) Drive: Where it came from. Where it's gone to. What now? 25

Čejka T, Thomas PW, Oliach D, Stobbe U, Egli S, Tegel W, Centenaro G, Sproll L, Bagi I, Trnka M, Büntgen U (2022) Understanding the performance of truffle dogs. J VetY Behav 52–53:8–13. https://doi.org/10.1016/j.jveb.2022.04.002

DeMatteo KE, Davenport B, Wilson LE (2019) Back to the basics with conservation detection dogs: fundamentals for success. Wildlife Biology. https://doi.org/10.2981/wlb.00584

DeRosa AJ (2019) The difference between a deutsch-drahthaar and a german wirehaired pointer. In: Project Upland. https://projectupland.com/hunting-dogs/the-difference-between-a-deutsch-drahthaar-and-a-german-wirehaired-pointer-2/. Accessed 19 Dec 2022

Dollion N, Paulus A, Champagne N, St-Pierre N, St-Pierre É, Trudel M, Plusquellec P (2019) Fear/Reactivity in working dogs: An analysis of 37 years of behavioural data from the Mira Foundation's future service dogs. Appl Anim Behav Sci 221:104864. https://doi.org/10.1016/j.applanim.2019.104864

Gácsi M, McGreevy P, Kara E, Miklósi A (2009) Effects of selection for cooperation and attention in dogs. Behavioral and Brain Functions : BBF 5:31. https://doi.org/10.1186/1744-9081-5-31

Gadbois S, Reeve C (2016) The semiotic canine: Scent processing dogs as research assistants in biomedical and environmental research. Dog Behavior 2:26–32. https://doi.org/10.4454/DB.V2I3.43

Ganitskaya YuV, Feoktistova NYu, Vasukov DD, Surov AV (2020) Some behavioral features required for the selection of detection dogs. Biol Bull Russ Acad Sci 47:501–505. https://doi.org/10.1134/S1062359020050039

Goddard ME, Beilharz RG (1984) A factor analysis of fearfulness in potential guide dogs. Applied Animal Behaviour Sciencer Science 12:253–265. https://doi.org/10.1007/BF01071933

Graham LT, Gosling SD (2009) Temperament and personality in working dogs. In: Helton WS (ed) Canine ergonomics: The science of working dogs. CRC Press/Taylor & Francis, Boca Raton, pp 63–81

Haney PS, Wilborn RR (2021) Breeding program management. Veterinary Clinics of North America: Small Animal Practice 51:891–904. https://doi.org/10.1016/j.cvsm.2021.04.007

Hare E, Joffe E, Wilson C, Serpell J, Otto CM (2021) Behavior traits associated with career outcome in a prison Puppy-Raising program. Appl Anim Behav Sci 105218. https://doi.org/10.1016/j.applanim.2021.105218

Hare E, Kelsey KM, Serpell JA, Otto CM (2018) Behavior Differences Between Search-and-Rescue and Pet Dogs. Frontiers in Veterinary Science 5:1–5. https://doi.org/10.3389/fvets.2018.00118

Hsu Y, Serpell JA (2001) Development and validation of a novel method for evaluating behavior and temperament in guide dogs. Appl Anim Behav Sci 72:347–364. https://doi.org/10.1524/ncrs.2009.224.14.255

Jamieson LTJ, Baxter GS, Murray PJ (2017) Identifying suitable detection dogs. Applied Animal BehaviourScience 0–1. https://doi.org/10.1016/j.applanim.2017.06.010

Jones KE, Dashfield K, Downend AB, Otto CM (2004) Search-and-rescue dogs: An overview for veterinarians. J Am Vet Med Assoc 225:854–860

Kokocinska-Kusiak A, Woszczyło M, Zybala M, Maciocha J, Barłowska K, Dzieciol M (2011) Canine olfaction: physiology, behavior, and possibilities for practical applications. Int Libr Ethics, Law Technol 6:133–183. https://doi.org/10.1007/978-94-007-1330-7_6

Krichbaum S, Lazarowski L (2022) Reward type affects dogs' performance in the cylinder task. AB&C 9:287–297. https://doi.org/10.26451/abc.09.03.03.2022

Lazarowski L, Haney P, Brock J, Fischer T, Rogers B, Angle C, Katz J, Waggoner LP (2018) Investigation of the behavioral characteristics of dogs Purpose-Bred and prepared to perform vapor wake® detection of Person-Borne explosives. Front VetY Sci 5:50. https://doi.org/10.3389/fvets.2018.00050

Lazarowski L, Rogers B, Krichbaum S, Haney P, Smith JG, Waggoner P (2021a) Validation of a behavior test for predicting puppies ' suitability as detection dogs. Animals

Lazarowski L, Rogers B, Waggoner LP, Katz JS (2019) When the nose knows: ontogenetic changes in detection dogs' (Canis familiaris) responsiveness to social and olfactory cues. Anim Behav 153:61–68. https://doi.org/10.1016/j.anbehav.2019.05.002

Lazarowski L, Singletary M, Rogers B, Waggoner P (2021b) Development and training for working dogs. VetY Clin N Am Small Anim Pract 51:921–931. https://doi.org/10.1016/j.cvsm.2021.04.009

Lazarowski L, Waggoner LP, Krichbaum S, Singletary M, Haney P, Rogers B, Angle C (2020) Selecting dogs for explosives detection : behavioral characteristics. Front VetY Sci 7. https://doi.org/10.3389/fvets.2020.00597

Lazarowski L, Waggoner P, Hutchings B, Angle C, Porritt F (2021c) Maintaining long-term odor memory and detection performance in dogs. Appl Anim Behav Sci 105301. https://doi.org/10.1016/j.applanim.2021.105301

Leighton EA, Hare E, Waggoner LP, OttoCM, Thomas S (2018) A solution for the shortage of detection dogs: a detector dog center of excellence and a cooperative breeding program. Front VetY Sci 5:1–7. https://doi.org/10.3389/fvets.2018.00284

MacLean E, Hare B (2018) Enhanced selection of assistance and explosive detection dogs using cognitive measures. Frontiers in Veterinary Science 5:236. https://doi.org/10.3389/FVETS.2018.00236
Maejima M, Inoue-Murayama M, Tonosaki K, Matsuura N, Kato S, Saito Y, Weiss A, Murayama Y, Ito S (2007) Traits and genotypes may predict the successful training of drug detection dogs. Appl Anim Behav Sci 107:287–298. https://doi.org/10.1016/j.applanim.2006.10.005
Martin C, Diederich C, Verheggen C (2020) Cadaver Dogs and the Deathly Hallows—A Survey and literature review on selection and training procedure. Animals 10:1–19. https://doi.org/10.3390/ani10071219
McGarrity ME, Sinn DL, Thomas SG, Marti CN, Gosling SD (2016) Comparing the predictive validity of behavioral codings and behavioral ratings in a working-dog breeding program. Appl Anim Behav Sci 179:82–94. https://doi.org/10.1016/j.applanim.2016.03.013
Mehrkam LR, Hall NJ, Haitz C, Wynne CDL (2017). The Influence of Breed and Environmental Factors on Social and Solitary Play in Dogs (canis Lupus Familiaris). https://doi.org/10.3758/s13420-017-0283-0
Morrill K, Hekman J, Li X, McClure J, Logan B, Goodman L, Gao M, Dong Y, Alonso M, Carmichael E, Snyder-Mackler N, Alonso J, Noh HJ, Johnson J, Koltookian M, Lieu C, Megquier K, Swofford R, Turner-Maier J, White ME, Weng Z, Colubri A, Genereux DP, Lord KA, Karlsson EK (2022) Ancestry-inclusive dog genomics challenges popular breed stereotypes. Science 376:eabk0639. https://doi.org/10.1126/science.abk0639
Oxley J, Waggoner LP (2009) Detection of explosives by dogs. In: Asp Explos Detect. pp 27–40
Polgar Z, Kinnunen M, Ujvary D, Miklosi A, Gacsi M (2016) A test of canine olfactory capacity: comparing various dog breeds and wolves in a natural detection task. PLoS ONE. https://doi.org/10.1371/journal.pone.0154087
Porritt F, Shapiro M, Waggoner P, Mitchell E, Thomson T, Nicklin S, Kacelnik A (2015) Performance decline by search dogs in repetitive tasks, and mitigation strategies. Appl Anim Behav Sci 166:112–122. https://doi.org/10.1016/j.applanim.2015.02.013
Raffan E, Dennis RJ, O'Donovan CJ, Becker JM, Scott RA, Smith SP, Withers DJ, Wood CJ, Conci E, Clements DN, Summers KM, German AJ, Mellersh CS, Arendt ML, Iyemere VP, Withers E, Söder J, Wernersson S, Andersson G, Lindblad-Toh K, Yeo GSH, O'Rahilly S (2016) A Deletion in the canine POMC gene is associated with weight and appetite in Obesity-Prone labrador retriever dogs. Cell Metab 23:893–900. https://doi.org/10.1016/j.cmet.2016.04.012
Rooney N, Gaines S, Hiby E (2009) A practitioner's guide to working dog welfare. J VetY Behav: Clin Appl Res 4:127–134. https://doi.org/10.1016/j.jveb.2008.10.037
Rooney NJ, Bradshaw JWS, Almey H (2004) Attributes of specialist search dogs–a questionnaire survey of UK dog handlers and trainers. J Forensic Sci 49:300–306. https://doi.org/10.1520/JFS2003048
Rooney NJ, Clark CCA, Casey RA (2016) Minimizing fear and anxiety in working dogs: A review. J VetY Behav: Clin Appl Res 16:53–64. https://doi.org/10.1016/j.jveb.2016.11.001
Schneider, M., Slotta-Bachmayr, L. (2009) Physical and mental stress of SAR dogs during search work. In: Canine Ergon: Sci Work Dogs. CRC Press/Taylor & Francis
Sherman BL, Gruen ME, Case BC, Foster ML, Fish RE, Lazarowski L, DePuy V, Dorman DC (2015) A test for the evaluation of emotional reactivity in Labrador retrievers used for explosives detection. J VetY Behav: Clin Appl Res 10:94–102. https://doi.org/10.1016/j.jveb.2014.12.007
Sinn DL, Gosling SD, Hilliard S (2010) Personality and performance in military working dogs: Reliability and predictive validity of behavioral tests. Appl Anim Behav Sci 127:51–65. https://doi.org/10.1016/j.applanim.2010.08.007
Spafford M (2019) Should you consider a flushing breed for your next bird dog? In: Project Upland. https://projectupland.com/hunting-dogs/should-you-consider-a-flushing-breed-for-your-next-bird-dog-2/. Accessed 15 Dec 2022
Tiira K (2019) Resilience in dogs? lessons from other species. VetY Med: Res Rep 10:159–168. https://doi.org/10.2147/vmrr.s189566

Troisi CA, Mills DS, Wilkinson A, Zulch HE (2019) Behavioral and cognitive factors that affect the success of scent detection dogs. Comp Cogn & Behav Rev 14:51–76. https://doi.org/10.3819/ccbr.2019.140007

Udell, M.A.R., Ewald, M., Dorey, N.R., Wynne CDL (2014) Exploring breed differences in dogs (Canis familiaris): does exaggeration or inhibition of predatory response predict performance on human-guided tasks? In: Animal Behaviour. Accessed 2 May 2014

Wilsson E, Sinn DL (2012) Are there differences between behavioral measurement methods? A comparison of the predictive validity of two ratings methods in a working dog program. Appl Anim Behav Sci 141:158–172. https://doi.org/10.1016/j.applanim.2012.08.012

Wilsson E, Sundgren PE (1997) The use of a behaviour test for selection of dogs for service and breeding. II. Heritability for tested parameters and effect of selection based on service dog characteristics. Appl Anim Behav Sci 54:235–241. https://doi.org/10.1016/S0168-1591(96)01175-6

Canine Olfactometry: Tools, Techniques, and Procedures

Edgar O. Aviles-Rosa, Vidia Gokool, Nathaniel Hall, and Lauryn DeGreeff

Abstract

Canine olfaction has been leveraged across the globe for a wide variety of detection tasks, including medical, explosives, narcotic, and wildlife. The applications and usages of detection canines have grown substantially since the 1970s; however, technology to improve canine training and testing has largely lagged. Despite nearly 50 years of detection canine advancement, there have been few advancements in tools to present a controlled odorant to the canine for training. As such, wood containers, plastic boxes, and a wide variety of commercially available home storage containers remain popular odor delivery vessels. However, evidence suggests these methods may be non-ideal for odor presentation as there is no mechanism to provide standardized odor delivery, controlled odor concentration, or reproducibly create odor mixtures as might be required for canine olfactory testing or training. Nonetheless, based on advancements in human and small mammal olfactory testing, a small, but growing body of research on canine olfactory detection over the last 20 years has developed more advanced tools to provide standardized and controlled odorant delivery for the purposes of canine training and olfactory testing. These tools can largely be categorized based on use (for training of detection canines or research understanding canine olfaction) and technique (manual passive/diffusion odor delivery or olfactometer active delivery). As the tools and practice of detection

E. O. Aviles-Rosa · N. Hall (✉)
Department of Animal and Food Science, Texas Tech University, 1308 Indiana Ave., Lubbock, TX 79409, USA
e-mail: Nathaniel.j.hall@ttu.edu

V. Gokool · L. DeGreeff (✉)
Department of Chemistry and Biochemistry, Global Forensic and Justice Center, Florida International University, 11200 SW 8th St., Miami, FL 33146, USA
e-mail: ldegreef@fiu.edu

L. Lazarowski (ed.), *Olfactory Research in Dogs*,
https://doi.org/10.1007/978-3-031-39370-9_5

canine training advances, there is increasing overlap between the tools leveraged in the laboratory and those used for training in the field, but this review will highlight the range of odor delivery vessels that are utilized in both the field and laboratory with a focus on tools that provide a controlled and measurable odor to canines on demand, namely olfactometers. We will also discuss the various training and assessment paradigms that can be used in conjunction with odor delivery tools and the benefits and limitations of each paradigm.

Keywords

Olfactometry • Canine detection • Odor delivery • Olfactory testing

1 Odor Delivery Tools

Canine olfaction has been leveraged across the globe for a wide variety of detection tasks, including medical, explosives, narcotic, and wildlife. The applications and usages of detection canines have grown substantially since the 1970s; however, technology to improve canine training and testing has largely lagged. Despite nearly 50 years of detection canine advancement, there have been few advancements in tools to present a controlled odorant to the canine for training. As such, wood containers, plastic boxes, and a wide variety of commercially available home storage containers remain popular odor delivery vessels. However, evidence suggests these methods may be non-ideal for odor presentation as there is no mechanism to provide a standardized odor delivery, controlled odor concentration, or reproducibly create odor mixtures as might be required for canine olfactory testing or training.

Nonetheless, based on advancements in human and small mammal olfactory testing, a small, but growing body of research on canine olfactory detection over the last 20 years has developed more advanced tools to provide standardized and controlled odorant delivery for the purposes of canine training and olfactory testing. These tools can largely be categorized based on use (for training of detection canines or research understanding canine olfaction) and technique (manual passive/diffusion odor delivery or olfactometer active delivery). As the tools and practice of detection canine training advances, there is increasing overlap between the tools leveraged in the laboratory and those used for training in the field, but this review will highlight the range of odor delivery vessels that are utilized in both the field and laboratory with a focus on tools that provide a controlled and measurable odor to canines on demand, namely olfactometers. We will also discuss the various training and assessment paradigms that can be used in conjunction with odor delivery tools and the benefits and limitations of each paradigm.

2 Manual Passive Odor Presentation

2a. Manual devices with minimal odor control

Odor presentation can be as simple as a plastic squeeze bottle filled with solutions of odorants (Acree 1997) or a known concentration of an odorant in a glass jar for a participant to sniff. These examples are simple, have no moving or complicated parts, and can effectively present an odor as needed. Training materials for detection canines are commonly presented in boxes, jars, or other containment canisters. These options are used in training for simplicity of presentation and ease of integration into a training program. In cases of canines being trained to perform parcel checks or search large storage areas, the use of a box as a containment vessel may act as a good approximation of final duties.

There are a number of commercially available metal boxes of varying forms that are marketed to the canine detection community and are designed to minimize canine interaction with the target material itself while maximizing odor availability. However, with all these simple presentation devices the sample is at risk of contamination during the act of sniffing, exhaling, or licking at the sample or the container. Thus, a new/clean sample and container maybe necessary for every trial and every participant or there is a risk of sample contamination influencing results. Further, in the case of an open jar, the sample is subject to exposure to the environment which may contain other odorants and turbulent air flows in the room, making it difficult to ensure an equilibrium state is reached prior to the participant sampling the odorant. Thus, a simple presentation container may provide simplicity and ease of use but lacks odor control and protection of the sample from contamination sources. In a case where fresh samples can be prepared every trial; such devices may provide the right balance of simplicity for the user.

2b. Manual devices with odor containment

More recent containment devices are designed to provide security and protection of a training odorant from the environment (and can protect the environment from the training odorant). The SciK9 Training Aid Delivery Device (TADD) was patented by members of the U.S. Army Combat Capabilities Development Command Chemical Biological Center and allows restricted permeation of vapor into the surrounding environment (Mach et al. 2021). The device (Fig. 1) contains a hydrophobic and oleophobic odor-permeable membrane that retains sampling materials within the device, reducing loss of sample and introduction of physical contaminants. The lid which closes the device for storage can be removed to create an unrestricted opening, and there are also perforated lids available that allow for the restriction of delivered odor. This device varies from a design such as a sniffer tin by incorporating the additional membrane layer the vapor must pass through and by utilizing a Viton gasket that reduces odor loss and cross-contamination during long-term storage (Mach et al. 2021).

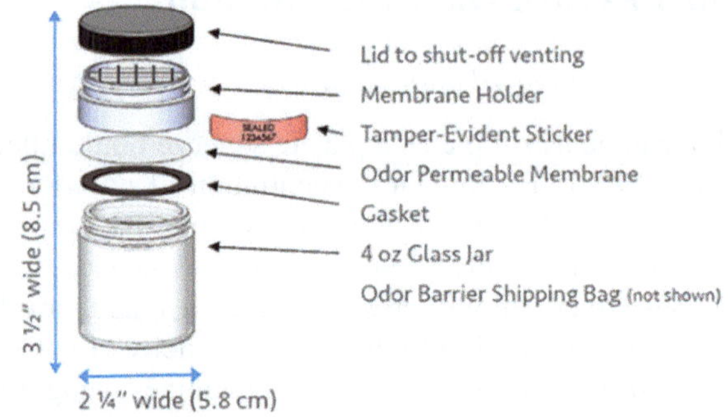

Fig. 1 SciK9 Training Aid Delivery Device (TADD) diagram, published by Sharpes et al., "Evaluation of the SciK9 Training Aid Delivery Device for Containment of Powders", U.S. Army Combat Capabilities Development Command Chemical Biological Center, DEVCOM CBC-TR-1705; this document is in the public domain (Mach et al. 2021)

2c. *Manual devices with controlled odor concentration*

Controlling concentration of an odorant can be a challenge with manual odor presentation devices, but there have been several engineered approaches. Among the first published studies in the field of olfactometry, Valentin's work in the mid-1800s referenced the encapsulation of a specified amount of odorant inside of a corked glass tube which allowed for the preparation of known quantities of odorous materials (Wenzel 1948). A similar approach with capillary tubes was followed shortly thereafter to assess canine detection thresholds (Neuhaus 1953). However, this technique requires precise glass instruments and can be difficult to replicate and control.

The Controlled Odor Mimic Permeation System (COMPS) (Fig. 2) patented by Furton and Harper (2017) is a simple device that allows odor to be delivered in known, reproducible amounts. COMPS are composed of a permeable polymer container housed inside nonpermeable packaging when not in use. This construction allows for multiple uses of the product; an example of COMPS is displayed in Fig. 14. The amount of odor delivered over time may be adjusted by changing the surface area of the COMPS bag or the thickness of the polymer used (Simon et al. 2019).

The COMPS approach is similar to a standard approach of utilizing permeable membranes for controlled odor diffusion. Permeation tubes can be used to emit stable odor concentrations under continuous flow rates. This approach, however, has rarely been used with canines with one notable exception (Walker et al. 2006). Outside of this, there are a few other approaches to manipulating concentration manually in a standardized manner.

2d. *Manual devices to present odor mixtures*

Each of the vessels that have been mentioned thus far has contained one central housing area for the deposition of a single substance. However, when presenting

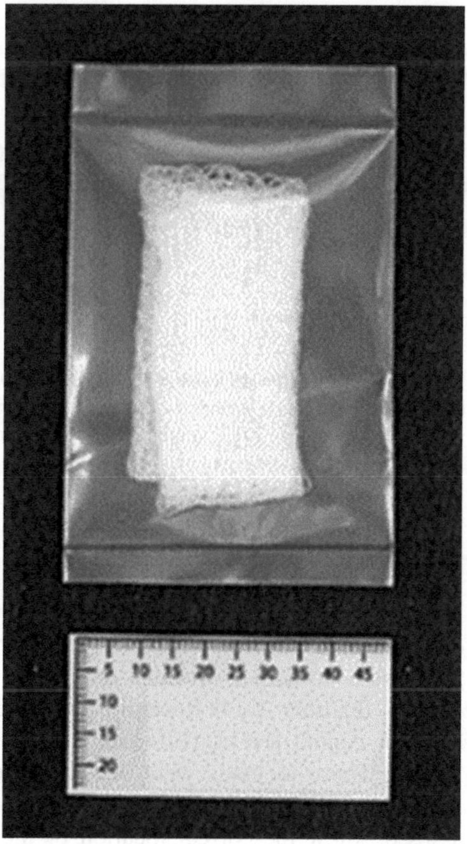

Fig. 2 COMPS interior permeable packaging, published by Simon et al., Chemical Senses, vol. 44, no. 6, 2019; this document is in the public domain (Simon et al. 2019)

materials such as explosive mixtures (e.g., homemade explosive mixtures) or drug mixtures (e.g., adulterated drug mixtures), physically mixing or adulterating the target compound may be limited due to safety or security concerns. In these circumstances, canines are trained on solely the parent compound with only limited access to the more operationally-relevant mixtures. This can deteriorate detection proficiency for mixed materials (DeGreeff and Peranich 2021). As such, the Mixed Odor Delivery Device (MODD) was developed and patented by the U.S. Naval Research Laboratory (NRL) for the presentation of odor mixtures without preparing the actual mixed product and is used to improve generalization across mixtures containing a common target DeGreeff and Peranich 2021). The MODD (Fig. 3) allows the user to insert up to four substances in separate wells. The odors of the separated components mix within the device as they diffuse from the wells to the outlet at the top of the device, allowing the canines to intake mixed odor from separated components. As seen with the TADD, the MODD includes a gasket to prevent vapor escape through the sides of the container (DeGreeff et al. 2017); however, unlike the TADD, the MODD is not meant for long-term storage.

Fig. 3 The Mixed Odor Delivery Device (MODD) used to contain separated components within the device and release a mixed odor at the sniffing outlet, published by DeGreeff et al., Forensic Chemistry, vol. 4, 2017; licensed by Elsevier and Copyright Clearance Center (DeGreeff et al. 2017)

3 Olfactometry

In contrast to manually presenting odorants in vials or containers that present an odor through diffusion principles, olfactometry (use of olfactometers) allows the investigator to actively present an odor on demand through a controlled air flow system (Buettner 2017) Recently, automated odor delivery systems have been created for canine testing (Edwards 2019; Jendrny et al. 2021; Aviles-Rosa et al. 2021a). These methods were developed to ensure precise odor delivery, as well as double-blind conditions, where neither the canine/handler team nor the test assessor know the correct location of the target. However, the development and use of olfactometers for canines is relatively novel and rare, although it has a long-established history in human and laboratory animal research.

Standardizations for olfactometry for measuring air quality by human assessors are already well established. The European Standard for "Air Quality—Determination of Odour Concentration by Dynamic Olfactometry" (EN13725) refers to an olfactometer as a dilution system that delivers an odor sample diluted or administered within a neutral carrier gas to an assessor. The standard specifies the methods and procedures to be used in the determination of the odor concentration of gaseous samples using dynamic olfactometry and human assessors and defines performance parameters to be met by the instrument for its accuracy of dilution, repeatability, and precision (Verhulst et al. 2011). EN13725 also puts forth a set of requirements for the testing environment and the human assessors used. The standards establish participant protocols such as avoiding perfumes/fragrances and refraining from eating, drinking, or smoking for at least 30 min before serving as an assessor. The sensitivity of the assessor is determined using n-butanol; the sample group must be within a detection range of 0.020 and 0.080 μmol/mol(Buettner 2017).

3a. Olfactometer construction

Figure 4 illustrates the basic components of an olfactometer. There are two paths of travel for the carrier gas to follow. Path 1 flows through the dilution unit where pure air is mixed with odorous sample air and diluted to the target concentration; this diluted air flows onward to the switch valve. The second path leads directly from the carrier gas source to the switch valve. The switch valve controls the presentation of flow paths to the assessor switching from Path 1 (odorous, sample air) to Path 2 (clean air). The recommended carrier gas is purified air. Multiple odor ports can be connected to the same olfactometer allowing multiple assessors to sample from the same device. The dilution unit, switch valve, and panelist responses are operated by a microcontroller (Buettner 2017). The unit is controlled by a computer and automatically dilutes sample concentrations, provides intermittent clean air flows to assessors, and controls which odor ports are supplied with air flows.

To prevent carry-over between subsequent runs of the device, all odorant-carrying components of the olfactometer must be constructed from inert, odorless, materials such as glass, stainless steel, or polytetrafluoroethylene (PTFE) that minimize absorption and are readily cleaned by heat or solvent (Buettner 2017). It is also suggested that all wetted surfaces be heated to at least the boiling point of the analyte of interest and the width of all tubing is maximized within the allowances of system requirements (Mullen et al. 2021).

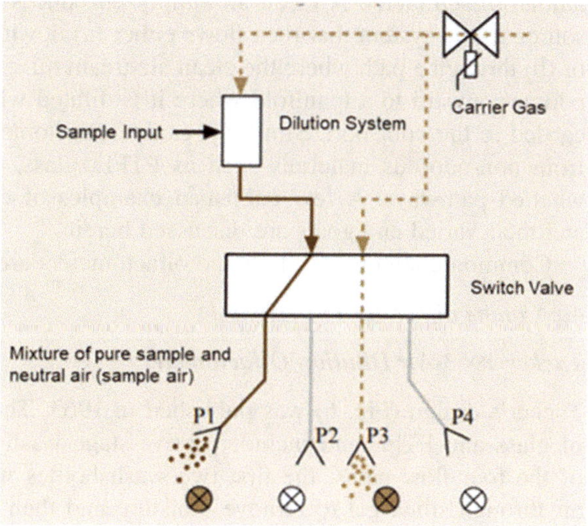

Fig. 4 Working principle of an olfactometer adapted from Buettner, Springer Handbook of Odor, chapter 24, 2017; reproduced with permission from Springer Nature

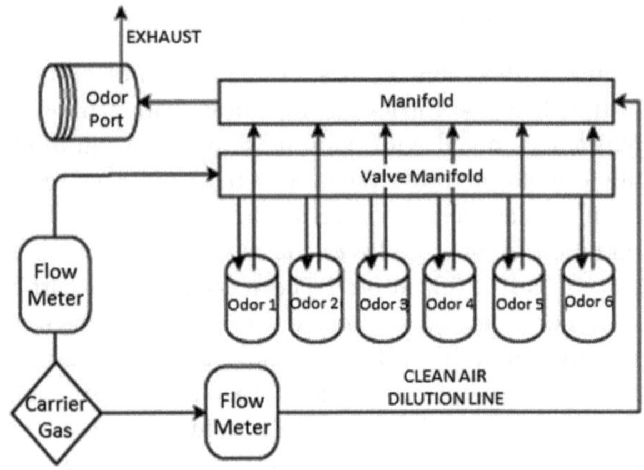

Fig. 5 Basic canine olfactometer, adapted from Aviles-Rosa et al., Front. Vet. Sci., vol. 8, 2021; licensed under a Creative Commons Attribution (CC BY) license (Aviles-Rosa et al. 2021a)

4 Olfactometers and Canine Assessors

Olfactometers are traditionally used to investigate aspects of human olfaction and are thus ergonomically designed for the human subject. In its application to canines, olfactometers are constructed to suit the canine end-user allowing for the delivery of odorous compounds via airstream to a canine assessor instead of a human. While there are many variations of the canine olfactometer, they all share the core components expressed by the human olfactometer. As shown in the example in Fig. 5, the base design of a canine olfactometer appears quite similar to a human olfactometer. A clean air supply introduces a non-odorous or purified air source to the system that then flows either (a) down a direct path to the odor port or (b) through a path where the clean airstream mixes with and transports a sample odorant onward to a manifold where it is diluted with clean air and subsequently carried to the odor port. Similarly, canine olfactometers are generally constructed from non-odorous materials such as PTFE, glass, or stainless steel along odor-whetted pathways. A few published examples of canine olfactometer structures and their varied end goals are discussed herein.

Common components of canine olfactometers are included in Table 1:

4a. Canine olfactometer designs

Tucker **1963-*Air Dilution Olfactometer***

Tucker's design (Fig. 6) was published in 1963. The olfactometer was composed of glass and PTFE and included a five-stage wash bottle construction. For each of the four flow paths, the first two wash bottles were used to flow compressed air through silica gel to remove moisture and then through activated charcoal to achieve odor removal from the air source. The third and fourth bottles were used to saturate the clean air; this is the point of odor introduction for the Odor A & Odor B paths. The last wash bottle served as an aerosol trap, retaining the saturated air until needed further downline. This system used the first flow path as a "clean"

Table 1 Common components of olfactometers

Part	Function
Mass Flow Controller (MFC)	A device that measures and regulates the flow of gases or liquids; designed and calibrated to function for a specific arrangement of input and flow rates
Electronic Flow Meter (EFM)/flow meter	A gauge used to measure the volume or mass of flow rates in a system
Manifold	A junction for multiple flow inputs; allows the combination of flow streams and redirection of path
Static mixer	A device whose physical design allows continuous mixing of flowed material without the need for moving parts

air system washing flow path; thus, this line did not receive odorant. The next two paths were able to contain separate odorants that could be released and mixed with the fourth flow path ("clean" air) to dilute said odorants. It was noted that the odorant could be excluded from the "Odor B" ports to create a second non-odorous pathway and allow two dilution events to occur before the presentation of odor occurs downstream (Tucker 1963; Krestel et al. 1984).

Hallowell **1994-*Air Dilution Olfactometer***

The olfactometer utilized a five-channel system with a series of mass flow controllers (MFC) used to dilute the air stream by controlling the amount of odorant gas that continued to the next phases (Fig. 7). It also allowed for the influx of additional non-odorous gas to the vapor stream for dilution. Sample odor was incorporated into the vapor stream by flowing clean air through a water bath-heated glass vessel containing the sample of choice. The vapor stream passed through the vessel and was incorporated with the sample air and continued through the next four channels of the system where it was subsequently diluted to the target range and delivered at the odor port. The net range of dilution factors reported for this construction is 10^{-2}–10^{-15} (Hallowell et al. 1994).

Dechant **2021-*Air Dilution Olfactometer***

The Dechant air dilution olfactometer (Fig. 8) used a "zero" air purified air source as its carrier gas. For sample introduction, air flows from the source, through a regulator, and into the sample containment unit. This setup utilized clean air to displace and flow the gaseous headspace of an odorant held in a glass vial within a temperature-controlled water bath. After the odorant was incorporated with the clean air, the vapor stream passed through a series of static mixers and MFCs. Each junction posed an opportunity to exhaust a portion of the vapor stream or incorporate additional clean diluent air to bring the initial vapor concentration to the targeted concentration level. After passing through three MFCsstatic mixers, and exhaust stages, the air-diluted sample arrived at the canine assessor's odor port (DeChant et al. 2021).

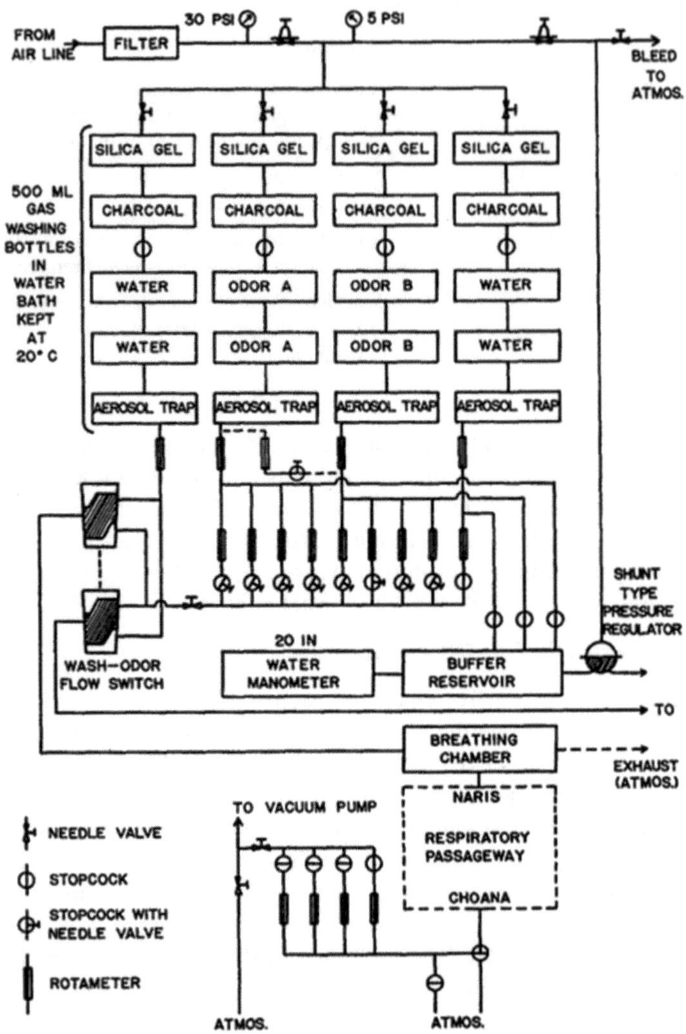

Fig. 6 Air dilution olfactometer, published by Tucker, Journal of General Physiology, vol. 46, no. 3, 1963; licensed under a Creative Commons Attribution (CC BY) license (Tucker 1963)

Hall et al. 2018-*Odor Mixture Olfactometer*

Unlike the above-described olfactometers, Hall and coworkers (2018) produced an olfactometer that allowed for the creation and delivery of odor *mixtures* to the canines (Fig. 9). The instrument was constructed of an automated 12-channel dynamic-dilution computer-controlled system. Each of the 12 channels contained an electronic flow meter and a saturation jar supplying a specific odorant. The flow of filtered air through each channel was controlled by the computer, the flows of multiple channels could be set to create a mixture of odors downstream in the odor

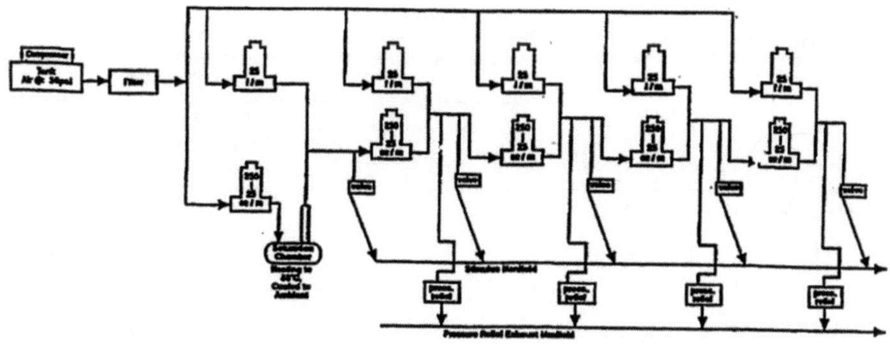

Fig. 7 Air dilution olfactometer, published by Hallowell et al., "Qualitative/semiquantitative chemical characterization of the Auburn Olfactometer", Proc. SPIE 2276, 1994; use of image granted by the author (Hallowell et al. 1994)

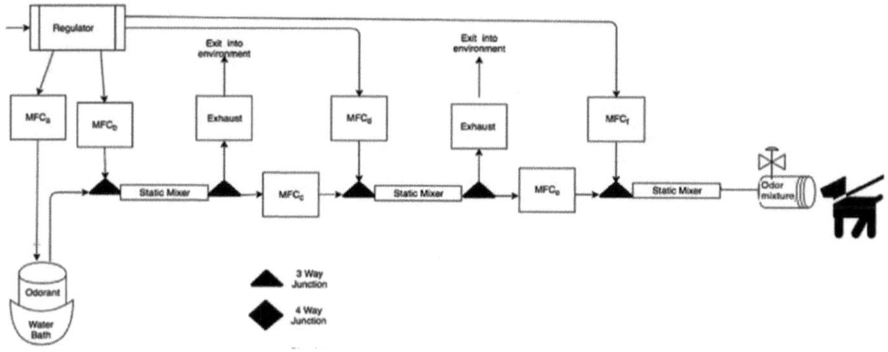

Fig. 8 Air dilution olfactometer, adapted from Dechant et al., Animals, vol. 11, no. 2, 2021; licensed under a Creative Commons Attribution (CC BY) license (DeChant et al. 2021)

manifold. The odor mixtures were created using proportional valves between the combined odorants (Hall and Wynne 2018).

Giordano et al. 2020 Modified Vapor Delivery Source

Trace Vapor Generator for Explosives and Narcotics (TV-Gen)

The U.S. Naval Research Laboratory created the Trace Vapor Generator for Explosives and Narcotics (TV-Gen) (Fig. 10). Originally designed to be used in the testing and validation of synthetic sensors and field vapor detection equipment, the device was adapted to be used as a canine olfactometer. The TV-Gen allowed for the introduction of liquid samples by two separate pathways; the first for sample introduction and a second for blank/control samples. Carrier gas from a zero-air generator was used to transport an aqueous solution of the analyte/odorant to a nebulizer, where the solution and analyte were vaporized and introduced to the

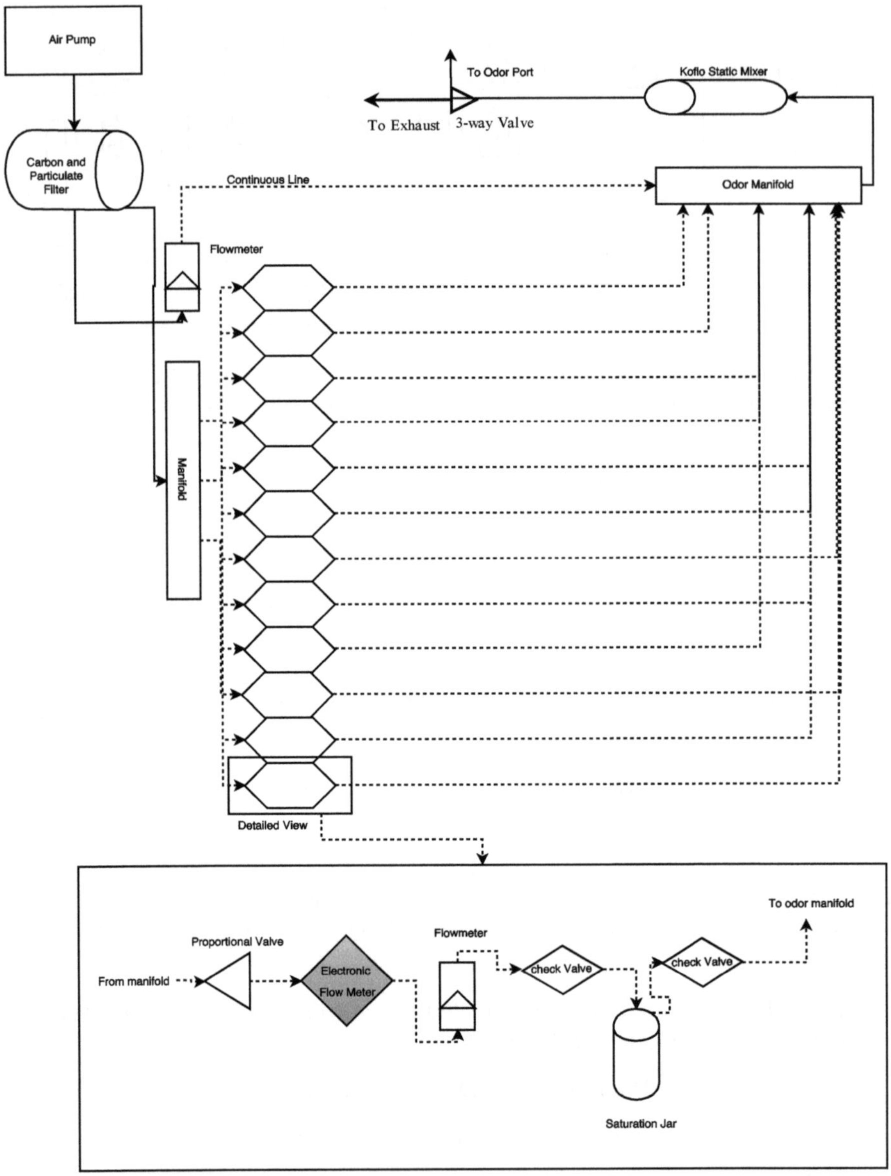

Fig. 9 Odor mixture olfactometer, published by Hall et al., Heliyon, vol. 4, no. 12, 2018; licensed under a Creative Commons Attribution (CC BY) license (Hall and Wynne 2018)

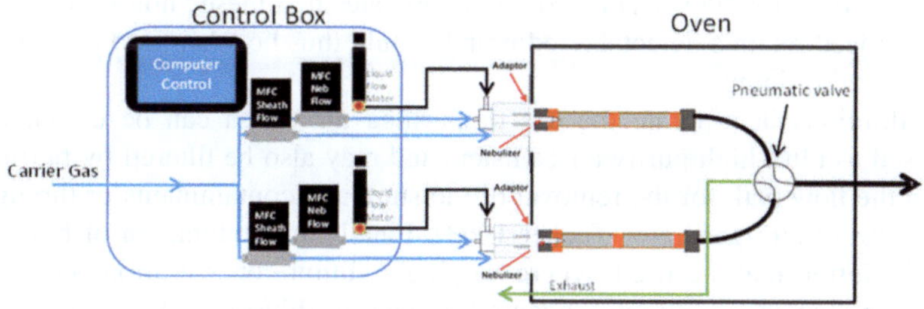

Fig. 10 TV-Gen, published by Giordano et al., Rev. Sci. Instrum., vol. 91, no. 8, 2020; licensed under a Creative Commons Attribution (CC BY) license (Giordano et al. 2020)

air flow path modulated by the MFC. The vapor stream was then diluted by additional air flow as needed. All vapor transport took place in a heated manifold (up to 130 °C). For the presentation of odor to the canine, an odor sampling port was developed. To ensure even distribution of the vapor in the sampling point, a vapor diffuser was designed to encourage the spread of analyte vapor throughout the cone instead of a narrow stream of analyte vapor down the middle of the port (Giordano et al. 2020; DeGreeff et al. 2021).

4*b. Considerations and Limitations in Odor Delivery and Olfactometer Design*

Odor Dilution

Odor dilution is an important concern in experimental design. Most odorant diluents are air, water, alcohol, or oil (Gamble and Smith 2009). The choice of a diluent is important not only for incorporating the stimulus but because a poor choice could reduce vapor phase availability and alter the sample's perception. In mixing an odorant with a diluent, both components become part of the presented stimulus. Due to the need to minimize the introduction of unwanted odorants to the sample, ultrapure diluents are preferred. Amongst the described methods of dilution, liquid (aqueous or oil) dilutions are the most common. There is greater control achieved when the stimulus can be diluted to the target concentration prior to its introduction in the olfactometer. In comparison, air dilution systems require extensive, and often expensive, mechanical components to perform sample dilution and delivery within the unit itself.

Water dilutions are often performed using distilled or highly filtered water, while ethanol is frequently used for alcohol dilutions (Le Berre et al. 2007). For oil dilutions, mineral oil or paraffin oil are most commonly used to dilute liquid odorant samples (DeChant et al. 2021; DeChant and Hall 2021; Concha et al. 2019) as they are assumed to be odorless for the practical purposes of experimentation (Gamble and Smith 2009); however, truly non-odorous material is difficult to achieve; (Koelega 1996; Pierce et al. 1996; Wysocki et al. 1997). Gamble et al. (2009) demonstrated that trained mice were able to discriminate between mineral oil vs. filtered air and between mineral oils sourced from different distributors

(Gamble and Smith 2009). This work demonstrates that these "non-odorous" diluents do indeed retain a detectable odor and should thus be taken into consideration in behavioral training.

Air dilutions incorporate the use of purified air which can be commercially purchased as ultra-high purity air canisters and may also be filtered by purifiers or traps in the flow path for the removal of moisture and contaminants or through the use of a "zero air" generator Tucker 1963). Finally, a combination of both liquid and air dilution may be used, where varying solution concentration accounts for large-scale changes in output concentration and air-dilution can be used to make finer changes in output concentration (Giordano et al. 2020).

4*c. Canine Olfactometer Design Considerations*

Flow Path

When designing and operating a canine olfactometer it is important to acknowledge that odorants transported through the system will not be delivered with 100% efficiency. There are losses of odor that occur due to adsorption and absorption to surfaces such as tubing, valves, and manifolds and loss due to inefficient delivery at the odor port (Mullen et al. 2021). Also, unintended dilution of odor can occur at multiple steps in the process, predominantly at the vapor outlet. Both instances decrease the intended vapor concentration (DeGreeff et al. 2021).

The presence of unattended compounds, due to insufficient clearing or cleaning of the vapor stream may cause contamination of the apparatus and analyte carryover. As a result of carryover, the assessor may believe the canine subject is detecting the intended analyte but is actually sampling the contaminating odorants (Hallowell et al. 1994). In one example, when an ion mobility spectrometer (IMS) was used to test the output of an olfactometer used to generate explosives vapor, significant contamination from the dinitrotoluene (DNT), an explosives-related compound, was detected. The presence of the DNT contamination was thought to be from the carryover of smokeless powder vapor used in testing the previous day. Though the olfactometer was purged overnight at 70 °C, the DNT contamination could not be entirely removed without more extensive system cleaning. As such, it is vital to consider the many points of possible carryover contamination in the olfactometer's design and carry out efforts to mitigate such effects.

An easily overlooked aspect of olfactometer design lies in the differences between sample and blank/control flow paths. As seen in Fig. 3, control samples are delivered through dedicated pathways reducing the opportunity for contaminant introduction and ensuring that blanks are being reproducibly delivered under the same conditions. However, sustaining a separate flow path for control sample introduction may result in differences between the sample and control paths due to differences in construction and operation, and minimizing such differences should be taken into account in instrument design (Collins et al. 2017). The delivery of an odorous stimulus requires more points of interaction and a more complex flow path than is needed in the delivery of a non-odorous control flow. At a minimum, sample lines must flow clean air through a sample flow meter, through a series of valves, into an odor incorporation stage, then, on toward an odor delivery port. The inclusion of air dilution steps adds to the complexity of this design, but as is, this path incorporates specialized equipment that accounts for space in the instrument's design and cost in its manufacturing. Logically, components that are not necessary for the delivery of a control sample would not be included to reduce costs and eliminate unnecessary redundancy, but this can lead to unintentional differences between the target and control paths (e.g. flow, pressure, sound). Canines are highly intelligent animals that use sensory cues to discern differences between stimuli. When conducting scenting experiments, the experimenter ideally wants the canine to rely on its olfactory ability to discern differences between samples, but olfactometers may yield non-olfactory, unintentional cues. A device that has an innate "tell" will not yield credible results and could adversely affect the training and performance of canines that use the device. Canines may sense the subtle differences in flow rates and the discrepancy in temperature caused by air flowing through a different path; or the difference in pressure caused by using a different flow meter on one line but not the other. They may pick-up on the delayed delivery of a sample flow that travels a longer path than the control; and they may feel differences in the vibrational frequency of operating the device in a specific mode or hear valves switching between flow paths denoting a change in sample type. All these considerations should be taken into account when designing and operating an olfactometer. The continual development of the field has seen active work to remove the influence caused by these differences resulting in constant improvement of canine olfactometers.

Delivery Interface (Odor Port)

The conditioning of the sample odor stream in terms of temperature, humidity, and flow rate can affect the canine's assessment of the odor. There are considerations to be made in terms of physical comfort in the sampling procedure, where samples should be monitored for temperature and humidity to ensure appropriate levels are established. Equally important, the delivery interface should be constructed with canine comfort, as well as odor diffusion and dilution, in mind (DeGreeff et al. 2021).

The type and shape of the odor port used with an olfactometer can most notably affect the concentration of the delivered stimulus. As a sample flow exits the olfactometer into the odor port it begins to mix with the ambient air surrounding the device. Particularly for ports with direct, single-stream introduction, a canine positioned farther away from the port's opening will experience a more diluted sample stimulus than one which is positioned more closely (DeGreeff et al. 2021). Below are a few basic designs and approaches to deliver stimulus to canines through a canine odor port.

Krestel et al. (1984) published an olfactometer design that describes an odor port that flowed sample odorant across the canine's sampling space. In Fig. 11, a canine can be seen placing its nose inside the rectangular opening of the odor chamber. The chamber included an input from the olfactometer system as well as an exhaust system to pull odor through the chamber, across the canine's snout, and

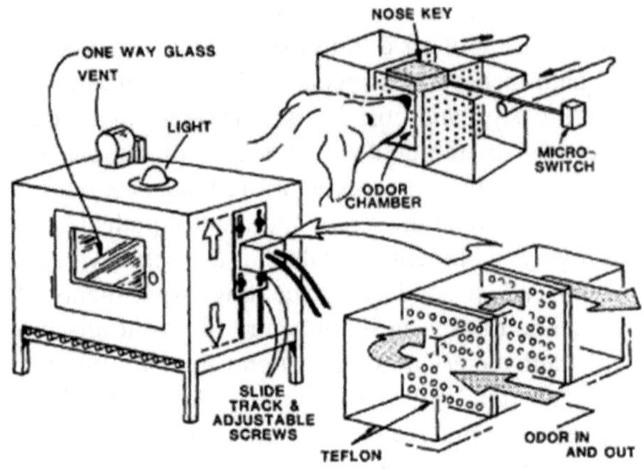

Fig. 11 Canine odor chamber published by Krestel et al., Neuroscience & Biobehavioral Reviews, vol. 8, no. 2, 1984; licensed by Elsevier and Copyright Clearance Center (Krestel et al. 1984)

out of the system allowing for circulation of sample air and exhausting odorous air prior to the presentation of a new sample. The design also incorporated a Teflon lever which was affixed inside the breathing chamber on its ceiling. This lever was used for canine reporting. The shape of the odor port promotes active dissipation of odor into the larger area and dilution of odor prior to its presentation to the canine (Krestel et al. 1984). Such a design was sufficient for a variety of types of olfactory testing but may not be appropriate for threshold measurements as the influx of air into the chamber and subsequent dilution of odor at the port are not taken into account.

Later, Johnston et al. (1994) performed olfactory threshold experiments using an experimental chamber including an interface panel with an odor port (Fig. 12). The odor port consisted of a 9 cm diameter aperture linking it to the olfactometer input. A vacuum pump system was used to remove odorous air from inside of the chamber as experiments were conducted. There were two levers contained within this chamber corresponding to "clean air" or "scented air" which the participating canines used to report their interpreted stimulus. Food reinforcement could be delivered directly to the chamber. This design of the chamber and its use of a vacuum pump to remove air from the enclosure requires that the odorous air travel from the olfactometer, through the odor port and into the larger chamber before it can be removed by the vacuum system. This design creates ample opportunity for odor mixing to occur due to incomplete removal of odorous air (Johnston et al. 1994).

Most modern odor ports resemble a blend between Krestel et al. and Johnston et al.'s designs. In their 2021 work, Aviles-Rosa et al. described the use of an odor port that resembled an aperture on a panel. Within the port, the odorant was fed in from a tube at the bottom and exhausted out using a fan at the top of the port. This design also incorporated a continuous flow of clean air through the port that allowed for continual clearing of odor (Krestel et al. 1984). This created odor circulation and exhausting odorous air after stimulus presentation, creating a clean

Fig. 12 Canine experimental sampling chamber published by Johnston et al., "Determination of canine olfactory thresholds using operant laboratory methods", Proc. SPIE 2092, 1994; use of image granted by the author (Johnston et al. 1994)

sample presentation field. These actions mitigate the likelihood of odor carryover or stagnation in the odor port.

The TV-Gen design utilized a specially devised canine odor port modeled from a human olfactometry port and crafted using the same materials, but with an extended conical shape to fit a canine muzzle (Fig. 13). Odor was presented through a multi-channel diffuser at the center of the cone allowing the odor stream to diffuse in multiple directions ensuring circulation of vapor throughout the port. Computational fluid dynamic modeling of the vapor distribution from a single vapor stream confirmed that, indeed, without the diffuser, the analyte vapor remained in a tight stream through the middle of the port, never reaching the walls. Furthermore, the air outside of the port was entrained by the jet stream and pulled into the cone, further diluting the air in the port. The port design reflected an active effort to improve stimulant delivery to the canine, increasing the efficacy of olfactometer use as a whole. Additionally, analytical measurements confirmed the absence of carryover contamination in this design (DeGreeff et al. 2021).

5 Future Olfactometry: Gas Chromatography-Olfactometry (GC-O)

Gas chromatography-olfactometry (GC-O) is the term used for the experimental set-up where samples are chromatographically separated and presented, in real time, to a human assessor for detection and evaluation of odorous compounds eluting from the GC separation. The technique has never been used for canine olfactory assessments in the published literature but possesses significant opportunity for future research. This section will primarily discuss the use of GC-O in human olfactory research, followed by a brief discussion of the potential of canine GC-O research.

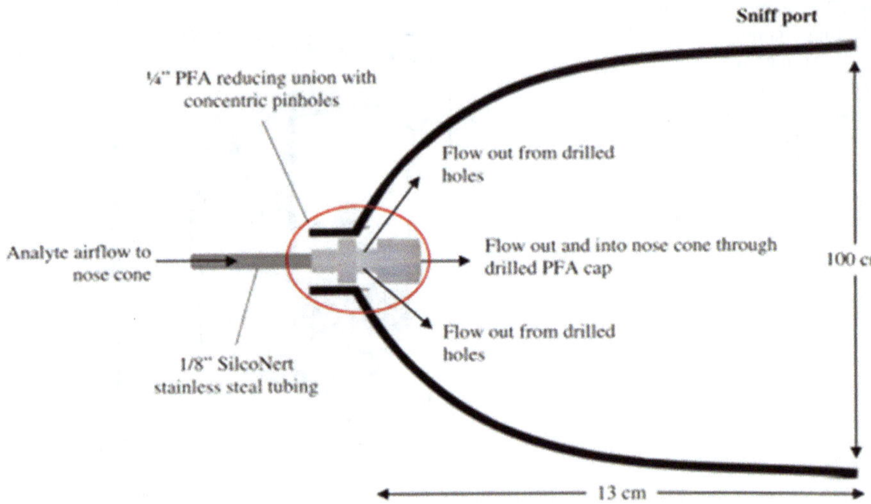

Fig. 13 Canine odor port for use with the TV-Gen, published by DeGreeff et al., Anal Bioanal. Chem, vol. 413, no. 3, 2021; licensed by Anal Bioanal. Chem and Copyright Clearance Center (DeGreeff et al. 2021)

5a. History of GC-O

The first application of GC-O is attributed to Fuller and colleagues published in 1964 (Fuller et al. 1964). Fuller et al. reported the construction and use of a GC-O comprised of a gas chromatograph flowing GC-separated vapor through an attached, heated transfer line, delivering the vapors to the nose of a professional perfumer (Fuller et al. 1964). The first iteration of the device fed the transfer line into a plastic head covering with a vacuum system removing air from above. The second iteration of the device fed the GC output into a booth where the perfumer was able to comfortably hold their head above the output of GC eluate. The experiments carried out by Fuller et al. demonstrated the novel use of human assessors in conjunction with gas chromatographic separation. The professional perfumer who participated in the study detected and assessed the odorous composition of more than 150 aromatic compounds (Fuller et al. 1964).

The initial invention posited by Fuller and colleagues was expanded upon in 1976. Acree et al. noted that the odor delivery system used in designs such as that of Fuller et al. delivered hot, dry air to the human assessor. This air would be uncomfortable for the human detector and cause inaccuracies due to the irritating effects of the dry air on the nasal cavities (Acree et al. 1976). The devised solution was the "sniffer" which incorporated an additional, post-GC, air stream that flowed through an in-line activated charcoal filter, mixed with the GC eluate, and flowed through a large volume of rapidly moving air. The resulting vapor composition was reported to be a moistened air flow of diluted concentration when compared to the direct GC eluate (Acree et al. 1976).

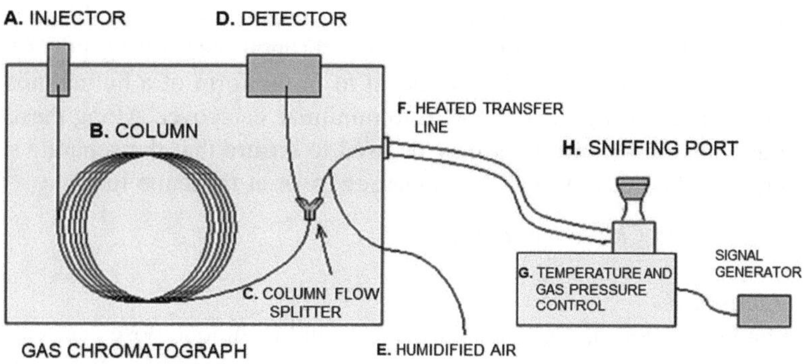

Fig. 14 Gas chromatography—olfactometry (GC-O) diagram adapted from Plutowska et al., Food Chemistry, vol. 107, no. 1, 2008; licensed by Elsevier and Copyright Clearance Center (Plutowska and Wardencki 2008)

5b. Principles of GC-O

Samples of interest can be investigated and prepared in numerous ways to allow for volatile compound extraction and transfer onto the GC. Depending on the composition of the sample, preparation procedures may require physical homogenization or centrifugation, other samples may be directly extracted and/or concentrated using methods such as steam distillation, solvent extraction, supercritical fluid extraction, solid phase extraction, and a number of headspace techniques including the use of sorptive traps and solid phase microextraction (SPME). There are many additional sampling methods; regardless of which sampling method is chosen, the next step is the introduction of the sample to the GC.

Samples are introduced to the GC through the instrument's inlet (Fig. 14a). The GC inlet is a heated entry port that will volatilize the sample to the vapor phase. The vapor phase sample is then passed, via carrier gas, to the GC column (B). Once deposited onto the GC column, a programmed cycle of column heating allows the compounds in the deposited sample to be separated by polarity and boiling point. The composition of the stationary phase coating the inside of the GC column influences boiling point and polarity-based separations. This separation process separates a larger, more complex sample into individual compounds making up its composition.

Once the sample has traveled through the GC column the entire sample may go to the human assessor, or, more commonly today, the sample exiting the GC column may be split and simultaneously routed to the human assessor and an instrumental detector. In the latter case, the gas phase sample passes from the GC column to a column flow splitter (C) where it is divided and sent along two paths. The first stream of gas is directed to the instrumental detector (D); a variety of instrumental detectors can be used in this scenario including thermal conductivity, photoionization, flame ionization, and mass spectrometers (Delahunty et al. 2006). The second split is mixed with humidified air (E) before passing through a heated transfer line (F) and into the sniffing port control modulator (G). The temperature and pressure of the gas are controlled by this modulator, fine-tuning the parameters of sample introduction to the human assessor through the sniffing port (H). The sniffing port is a conical-shaped port, meant to fit the form of a human nose; it is usually constructed from glass or PTFE to minimize carryover. Along these paths, pressure and gas flow controllers may be used to ensure that the separate streams of gas arrive at the detector and the human assessor at the same time.

5*c. Limitations and Challenges of GC-O*

System Limitations

The use and operation of GC-O as an instrumental technique poses issues, limitations, and inconveniences to the participant and researcher alike. Beginning with the introduction of samples into the device, there are restrictions placed upon what compounds can be deposited into a GC. Gas chromatographs cannot be used to analyze aqueous phase samples; additionally, samples that are phase compatible can still be a poor pairing for GC-O. Thermally labile compounds can decompose during the analysis process, the heating of samples can cause the breakdown of targeted analyte and the appearance of background artifacts, and the necessary sample preparation steps may preferentially trap certain types of compounds over others in the sample.

The information gathered from GC-O experiments is dependent on the combination of many factors including sample preparation, column choice, flow capacity, system resolution, and detector sensitivity. For instance, insufficient GC conditions for separating complex mixtures often result in poor selectivity, resulting in overlapping and unresolved peaks in the chromatogram. A response to this complaint is the adoption of Multidimensional GC-O (MDGC-O). MDGC-O incorporates the use of an additional GC column allowing for a multidimensional separation of compounds. However, like GC-O, MDGC-O has its limitations; it has been noted that while sensitivity and selectivity can improve, the additional technical components create opportunities for sample loss and increase the operating costs of the technique (Delahunty et al. 2006).

The Participant Factor

In the established use of humans as the GC-O assessor, the assessor-based challenges of its use are apparent. A challenge arises from the differences between analyte interpretation at the instrumental detector versus its interpretation by the assessor and discrepancies between responses acquired by different assessors. There is an inherent offset in signal processing that occurs with using a living being as a detector; the assessor must complete analyte uptake, sensory perception, and cognitive processing before providing a verbal or physical response (i.e., clicker, written, actuator slide bar) regarding the presented sample. Additionally, within these tasks, there are concerns over compounds missed during the assessor's completion of complex reporting procedures causing researchers to opt for simple reporting tasks such as noting one or two characteristics of the odor such as onset time and perceived profile. However, simplifying the task the assessor completes diminishes the amount of information that is gained from the session. Even when working within these simplified duties, there are some tasks such as reporting when odors end that are harder to attain reproducible results for; this outcome is viewed even when the same assessor provides repeated responses (Delahunty et al. 2006). When variations in the sensory abilities of differing assessors are incorporated, the many levels of variations due to human involvement become apparent.

5d. Canine Application of GC-O

GC-O is currently used as a manner of preparing and presenting odors to human assessors. However, this technology has the potential to be adapted for use with canines. The adaptation of this technology would allow the canine assessor to be presented with a flow of separated compounds. The canine would be tasked with acting as a detector for odorous compounds present in the GC eluate. This adaptation of technology would allow researchers to investigate such questions as (1) the presence or absence of known odor, (2) the beginning and end point of odorous compounds in a complex mixture, or (3) the intensity of odor. While the use of GC-O in this manner would require participating canine assessors to be trained to perform new, discrete trained responses, it is believed within the realm of canine research to incorporate GC-O in this manner. Recent advancements in the quality of portable field GC-based instrumentation may make this application more possible; however, the difficulty in GC-O for canine assessment would be in training a canine to wait, potentially for many minutes or even tens of minutes to detect the target, as a GC run can be from several minutes to as long as 30 min.

In a step towards GC-O testing using canines, researchers have collected fractions of the GC eluent onto sorbent materials and then delivered these materials to the canine in an odor recognition test. In research by Hudson (2009) and Vaughan t al. (2022), a GC-fractionation technique was used to probe the odorants of interest to trained detection canines (Vaughan et al. 2022) (Hudson 2009). Researchers studied the canine detection olfactory targets for human scent and crude oil, respectively; both highly complex mixtures of volatile compounds. In order to pinpoint, or narrow, the odors of olfactory interest, sections of the chromatograms were delivered to the canines, not through an olfactory port, but by collecting the fraction on sorbent materials and then presenting the fractions, as well as positive and negative controls prepared in the same manner, to canines trained to detect human scent or crude oil, respectively, in a series of controlled trials.

6 Animal Training Technologies

Sensory evaluation in animals is not as simple as asking a participant if two different stimuli are perceived as the same stimulus or to what degree they are perceptually different or similar to each other. Thus, different methods have been developed to conduct sensory evaluations in animals; each leverages animal behavior and operant conditioning to teach an animal to respond differently to different stimuli. By evaluating the animal's behavior and response to different stimuli,

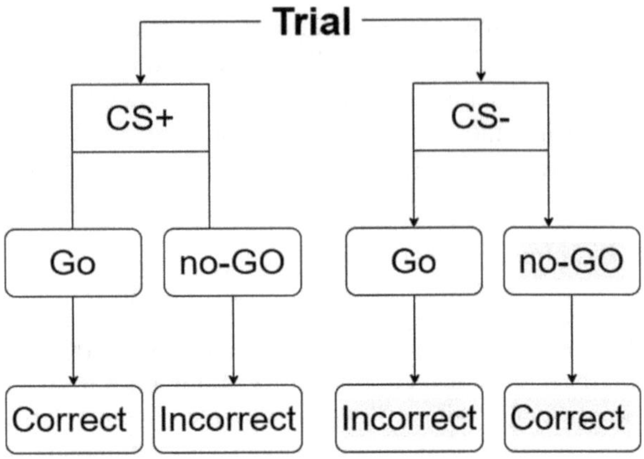

Fig. 15 A representation of the Go/no-Go paradigm. During a trial, the experimenter presents either the target odor (CS+) or other stimuli (CS−). The canine is trained to show the trained response (Go response) during a trial containing CS+ and to not show the trained response (No-Go response) during a trial containing CS−

researchers can assess how an animal perceives an olfactory stimulus in relation to another. Although we will focus our discussion on olfaction, it is important to note that the methods described below can be utilized to study any sensory modality. Nonetheless, by incorporating any of the odor-generating/presenting tools above and one of the behavior paradigms below, a wide range of animal sensory perception questions can be answered.

6a. Go/no-Go Paradigm

The Go/no-Go paradigm (GNG) is one of the most common methods used to study olfactory learning, discrimination, and generalization in animals, and a frequent choice of paradigm for use with olfactometers. A GNG paradigm consists of training the animal to show a behavioral response when a conditioned stimulus (CS+) is presented (Go response) and not showing the trained response when other stimuli (CS−) are presented (No-Go response) (Fig. 15). During a GNG, only one stimulus is presented to the animal during a trial.

GNG task has been widely used in different species to study cognitive and sensory processes (Bodyak 1999; Friedrich 2006; Kay et al. 2006; Frederick et al. 2011; Berditchevskaia et al. 2016; Carlson et al. 2016; Meule 2017; Hall and Wynne 2018; DeChant et al. 2021; Nakamura et al. 1987), and the behavior trained as the Go response varied among studies. One common trained response during a GNG task is to press a lever when the CS+ is presented and to not press the lever in the presence of a CS− (Kay et al. 2006; Nakamura et al. 1987). In rodents, it is also common to train the mice or the rat to hold their nose in an odor port until they perceive the CS+ and then go to another port where the reward (e.g., water) is delivered (Bodyak 1999; Carlson et al. 2016). The most common method used in rodents is to deliver water as a reinforcer within the port in the presence of the CS+ and not deliver water when a CS− is presented. The behavioral "Go" response is then to measure licks to the water delivery system (Otto et al. 1991; Abraham et al. 2012).

DeChant et al. (2021) used a GNG task with an air dilution olfactometer to evaluate canines' generalization to different concentrations of an odorant. In this study, canines were trained to hold their nose for 4 s in the odor port or to press a lever when the target odorant was delivered (both "Go" responses were trained) and not press the lever or hold their nose out of the port if the target was absent (DeChant et al. 2021). A similar procedure was used by Hall and Wynne (2018) to evaluate whether canines trained to detect a target odor when presented in odor mixtures showed improved generalization to novel mixtures containing the same target odor. In this study, researchers trained canines to hold the nose in the odor port when they perceived the target in a mixture and to remove the nose from the port when the target odor was absent (Hall and Wynne 2018).

GNG paradigms are easy to implement because they only require the presentation of an odor to a single sample port for a trial. This reduces the need for multiple odor ports to simultaneously present target and non-target odors. The main difficulty training canines with a GNG is to train the No-GO response, particularly to impulsive canines (Lazarowski et al. 2020). During initial training, canines may tend to respond to all stimuli presented, but by simply not reinforcing incorrect responses they should quickly learn to alert only to the target odor. One way to potentially overcome this and accelerate training is to reinforce correct Go and No-Go responses. This might reduce bias to the Go response, facilitating training, although reinforcing No-Go responses is not necessary and is frequently used in rodents (Slotnick and Restrepo 2001).

From a cognitive perspective, the GNG paradigm involves response inhibition (Helton 2009; Chikazoe et al. 2009). For instance, response inhibition control is needed to not alert to novel stimuli as the prepotent response in a GNG paradigm is the Go response (Helton 2009; Chikazoe et al. 2009). The frequency at which the CS+ and the CS− are presented can have an influence on the rate of false alerts (e.g., responding to CS−) and misses (e.g., not responding to the CS+). Higher rates of CS+ trials lead to higher false alerts and higher rates of CS− trials lead to more misses (Helton 2009; Chikazoe et al. 2009). To prevent the development of bias toward one response, CS+, and CS− trials should be randomized within a session and presented at equal rates (Chikazoe et al. 2009).

6b. Alternative Forced Choice (AFC)

The alternative forced choice (AFC) paradigm is another method used for sensory and cognitive analysis in animals (Shenoy and Yu 2012). Different from the GNG, AFC paradigms consist of presenting one or multiple CS− stimuli and a single CS+ stimulus simultaneously to the animal in the same trial. The animal is then trained to search the different samples presented and give the trained response only to the CS+ stimulus. As in the GNG paradigm, CS− are usually blank matrices (e.g., clean cotton gauze), the diluent of the CS+ (background), or just an empty vial. However, the use of distractor odors as CS− is frequently used, particularly when studying discrimination (e.g. Cleland et al. 2002). A distractor odor is an odor different from the target odor (CS+) used to ensure the animal is responding exclusively to the CS+ and not novel odors. Most of the studies in laboratory settings present the animal with only two (2-AFC) (Gomez et al. 2007; Frederick et al. 2011; Hall et al. 2016a; Shenoy and Yu) or three (3-AFC) (Fig. 16) (Reeve et al. 2018; Aviles-Rosa et al. 2021b; DeChant and Hall 2021) samples simultaneously, but an experimenter can adjust the method to present *n* number of samples during a trial. Independent of the number of samples presented in a trial, the CS+ or a testing odor (e.g., when studying generalization) must be always present in a trial and the animal must respond to one of the stimuli presented (hence forced choice). Failure to respond to a stimulus in a predetermined period of time results in the termination of the trial and the initiation of the next trial. Within these paradigms, how best to handle trials with no response can be challenging, but generally leads to either scoring the response as incorrect, repeating until a response is made, or removing the trial from analysis.

AFCs are commonly used in studying canine olfaction. Lazarowski et al. (2015) and Dorman et al. (2021) used a 2-AFC to evaluate generalization to untrained variations of Ammonium Nitrate (AN) compounds. In these studies, canines had to move an object containing the CS+. Moving the CS+ provided canines access

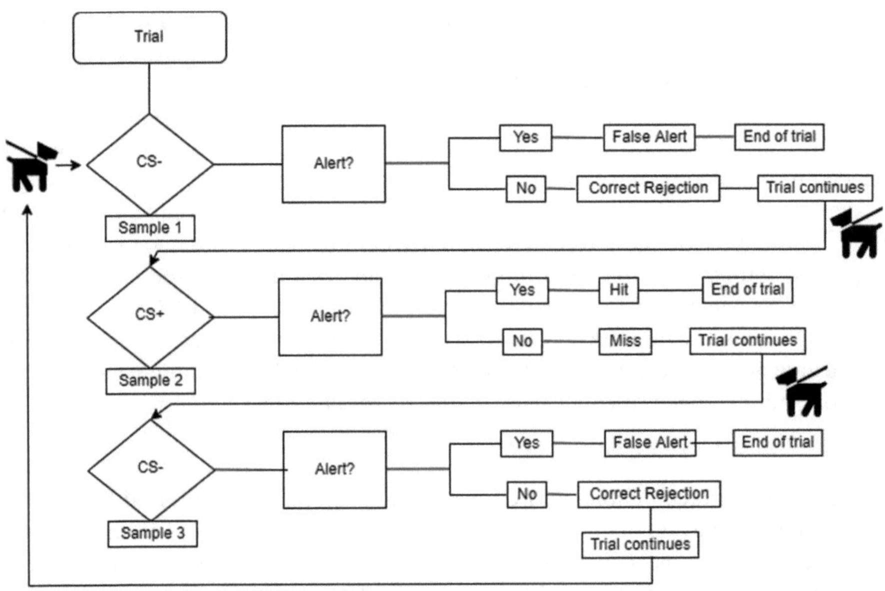

Fig. 16 A representation of a three alternative forced choice (3-AFC). Within a trial, there are two CS− and one CS+. The canine has to search all samples and alert to one of them (forced choice) before the end of the trial duration. Trial duration is established by the experimenter, and it is usually 30 s to 2 min. If the canine gives the trained final response for the CS+, it is recorded as a Hit or a correct alert. If the canine gives the trained response for a CS− sample, the trial response is recorded as a false alert or false positive. If the canine does not alert to any of the samples, the trial response is recorded as a miss or false negative. The experimenter might decide to repeat the trial or to terminate the trial if the canine does not alert to any of the samples at the end of the trial

to a treat. Hall et al. (2015, 2016b, a) used a 2-AFC where canines were trained to root in a bin containing the CS+ to evaluate alcohol discrimination and the effect of Pavlovian conditioning on odor acquisition and resistance to extinction. Reeve et al. (2018) and Martini et al. (2018) used AFC paradigms to evaluate canines' ability to detect breath samples and bar magnets, respectively. DeChant and Hall (2021) used a 3-AFC to determine the olfactory threshold of canines to isoamyl acetate; in this paradigm, isoamyl acetate concentration was gradually reduced (serial dilution) until canines were not able to accurately discriminate isoamyl acetate (CS+) from mineral oil (CS−). A 3-AFC and 5-AFC were used by Aviles-Rosa et al. (2021b) to evaluate generalization to different quantities of explosives (Aviles-Rosa et al. 2021b). Notably, AFCs are more frequently used with manual odor delivery devices and less frequently with olfactometry. This is in part because AFCs provide greater efficiency per trial statistically (probability of a correct response due to chance is related to the number of alternatives), allowing for fewer trails to be conducted. Thus, when manual labor is required, a more efficient test procedure is selected. Further, AFCs are less frequently used with olfactometry due to increased cost from replication of components and difficultly for olfactometry to present multiple stimuli at different locations simultaneously (Aviles-Rosa et al. 2021a).

Both the GNG and the AFC have been used to study different cognitive and olfactory phenomena in animals. Studies in rodents that compared their performance doing the same task (e.g., discrimination between odorants) with a GNG or 2-AFC, have found certain biases and differences between paradigms (Gomez et al. 2007; Frederick et al. 2011; Shenoy and Yu 2012). For instance, in a GNG paradigm, studies have found that rodents show a shorter response time and more false alerts relative to a 2-AFC (Shenoy and Yu 2012). This is thought to be a strategy to maximize the reinforcement rate and reduce the cost of the task (Gomez et al. 2007; Shenoy and Yu 2012). For instance, in a GNG paradigm, usually only Go trials are reinforced while in a 2-AFC every trial has a reinforceable response option. Reinforcement differences between methods can result in a bias toward the overt response in the GNG leading to more false alerts in the GNG relative to the AFC when performing the same task (Frederick et al. 2011). Frederick et al. (2011) were able to standardize both tasks by modifying certain parameters such as intertrial interval. This suggests that the difference between tasks may be due to the task parameters rather than the tasks differing in the cognitive processes they measure. Adjusting parameters, such as reducing intertrial interval and reinforcing correct No-Go responses, can make the GNG and AFC tasks more similar. However, from an olfactory/olfactometer perspective, the intertrial interval length may be a physical requirement to allow sufficient odor clearance. Furthermore, researchers might select one method over the other based on the aims of the study. For instance, if the aim of the study is to find whether canines are able to discriminate the CS+ from *n* different CS− then an AFC paradigm will be better suited as it allows the presentation of multiple CS − in a single trial. On the other hand, if the question is to determine the lowest concentration an animal can detect the

CS + from background, then a GNG might be better to leverage more precise olfactometry at lower cost than an AFC.

6c. GNG and AFC hybrid methods

Frequently, a hybrid between the GNG and the AFC is used in canine testing. Herein we consider procedures where multiple samples are presented to the animal in a trial, but the CS+ is not presented in every trial (Gazit et al. 2005; Porritt et al. 2015; Concha et al. 2019; Lazarowski et al. 2021c, b; Essler et al. 2021; Aviles-Rosa et al. 2021a; Waggoner et al. 2022), as hybrid between a GNG and an AFC because these procedures do not meet the methodological definition of one or the other. However, it is common that people to refer to what we call a hybrid as either GNG or AFC.

Hybrid methods or even AFC paradigms are sometimes erroneously described as GNG. For instance, a paradigm where five different samples are presented to the animal simultaneously is erroneously visualized as five independent GNG trials where the canine has to alert (Go response) or not alert (No-Go response) to each sample (Lazarowski et al. 2020). Analyzing the data in this way increases data collection (e.g., five data points per trial instead of only one data point) but this violates the assumption of statistical independence between trials. Statistical independence means that the occurrence of an event does not influence the occurrence of another event (Veech and Crist 2010). For instance, if each sample within a hybrid or AFC trial is independent from each other and each sample has an equal probability of being either CS+ or CS− , then there should be trials where the CS+ is presented more than once (e.g., 3 out of the five samples contain the CS+). Furthermore, if samples within a trial are independent from each other, canines' response to a sample should not result in the termination of the trial and the experimenter should allow the canine to investigate and respond (or not) to each sample independently of their response to other samples. This is not frequently the reported procedure in canine testing where an incorrect or correct response will result in the termination of a trial. Instead of being independent, samples within a trial are pseudorandomized. This means that within a trial only one sample is predetermined to contain the CS+ and the rest CS− . Thus, if sample one is the CS+ then the other samples have to be CS−. Because of this, the assumption of independence between samples is not correct. Furthermore, canines can easily learn this contingency and alert only to one sample and ignore the remaining samples after finding the CS+ (e.g., if the canine alerts to Sample 2, the canine will not even sniff the subsequent samples). Assuming independence will also influence how an experimenter determines performance levels above or below statistical chance. For instance, in a GNG paradigm where the CS+ and CS− are presented equally (e.g., 50% of the trials), chance performance is 50% (e.g., Go or No-Go response is a binomial outcome). However, chance performance in a hybrid method depends on the number of samples presented. In an AFC method where three samples are presented to the animal, chance performance is 33% instead of 50% (e.g., by chance the canine can alert to one of the three ports and be correct). Because of this, each sample within a trial should not be considered independent from each other. Evaluating canines' responses within a trial and not to each individual sample is the most appropriate way to collect and analyze the data in AFC and hybrid methods.

Similarly, sometimes hybrid methods are described as AFC. For instance, the example discussed above where five samples are presented simultaneously to the animal, can be erroneously visualized as a 6-AFC if we assume the canine is "forced" to alert to one of the 5 samples or to show an alternative "all clear" response indicating the CS+ is not present. The all-clear response can be a trained behavior (different from the alert to the CS+) or just not responding for a set period of time. The latter sometimes is mistakenly called a No-Go response. As mentioned above in an AFC paradigm, the animals are trained to expect that the CS+ is always present. Thus, because in a hybrid method, not all the trials contain the CS+, it is wrong to describe a hybrid method as an AFC because it does not meet the methodological definition of an AFC. The conceptualization of hybrid methods as AFC is less problematic than its conceptualization as GNG because it does not assume independence between samples but, it makes difficult the estimation of chance performance because this measurement will also need to take into consideration the number of target and blank trials withing a session.

Hybrid methods, where multiple samples are presented and the CS+ is not always present, are commonly used in detection canine olfactory testing because hybrid methods better resemble the detection canines' working environment than AFC. For example, working canines do not find a CS+ in every search or trial but during a search, they may encounter multiple CS-. This procedure has been recently integrated into an automated olfactometry paradigm for canines (Aviles-Rosa et al. 2021a).

6d. Matching-to-sample (MTS)

Matching-to-sample (MTS) paradigms consist of presenting a stimulus to the animal and subsequently the animal has to identify the same stimulus in a set of two or more different stimuli (Peña et al. 2006). In rodents, MTS paradigms are commonly performed by modifying GNG tasks. In an MTS-GNG task the animal is presented with an odorant and subsequently presented with the same or different odorant. The rodent has to show the Go response when both odorants presented are the same and not show the GO response when the odorants are different (Lu et al. 1993; Peña et al. 2006; Roddick et al. 2014). This allows for the paradigm to be conducted with a single odor port, and is readily amenable to automation with olfactometry. In canines, MTS are conducted using multiple simultaneously available options. First the canine samples or sniffs the sample stimulus. After sampling, the canine searches an array of samples in a line, circle, or carousel to identify the matching stimulus (Brisbin and Austad 1991; Marchal et al. 2016; Hale 2017; Lazarowski et al. 2021a; Schoon 1996). MTS paradigms require higher cognitive processing than regular discrimination tasks, and therefore are commonly used to evaluate working memory (Hartman et al. 2001; Krichbaum et al. 2021), and the ability of an animal to learn abstract concepts (e.g., the concept of same and different) (April et al. 2011; Lazarowski et al. 2021a).

MTS are frequently used to test canine olfaction for forensic purposes where canines have to match the human scent in an object linked to a crime with the scent of a suspect (Schoon and De Bruin 1994; Marchal et al. 2016; Schoon 1996). Importantly, MTS tasks require significantly more training time than simple discrimination tasks. Marchal et al. (2016) reported that initial training required to train canines five days a week for at least 18 months and that continued training was necessary throughout life. During training, samples and distractor odors should be changed daily or at least regularly to ensure canines learn the matching tasks and are not learning to avoid learned non-matches (e.g., see Hale 2017). If the canines learn the matching task, changing samples should not result in a significant performance decrement. A performance decrement when new samples are introduced during training may suggest that the canines are leveraging additional cues, such as familiarity with comparison stimuli rather than direct matching to the sample (Hale 2017).

7 Consideration and Limitations

The methods described above are validated methods to assess olfaction in canines. When selecting a method, a researcher must take into consideration the pros and cons of each method and select the one that fits best based on the purpose or the aim of the research, the labor, and the participant canines. For instance, GNG, AFC, and hybrid methods are excellent for proof-of-concept studies in laboratory settings where the aim is to investigate canines' olfactory capability to detect or discriminate different odors, but free searches may be better to evaluate different parameters in operational scenarios. Further, when controlling odor presentation and concentration are paramount, GNG paradigms maybe more convenient for use with expensive and sophisticated olfactometry equipment (e.g. Hall and Wynne 2018; DeChant et al. 2021), but when odor control can be adequately accomplished with more simplistic olfactometry, a hybrid approach may strike an ideal balance between odor control and operational relevance (e.g. Aviles-Rosa et al. 2021a).

Independent of the method or task used for canine olfactory assessment, double-blind conditions during testing and training should always be ensured to prevent canines from learning unintended cues from the handler, experimenter, or trainer. The learning of unintentional cues (e.g. a "Clever Hans Effect"; (Pfungst 1911; Samhita and Gross 2013) can be a significant concern in study designs. For instance, a handler's belief of a target odor presence can increase false alert rates. The iconic study in this topic by Lit et al. (2011) found that certified canines gave trained responses to an area with no target if the handler was told that a target was in the area (Lit et al. 2011); however, in a follow-up study DeChant et al. (2020) showed that when handlers were given details of the number of target odors or ambiguous information, but not intentionally mislead, handler knowledge changed search behaviors, but did not necessarily affect false alert rates (DeChant et al. 2020). Both studies suggest that while the effects of handler bias could occur in detection canines, it can be prevented. The use of automated olfactometry is ideal to ensure double-blind testing (Lit et al. 2011; DeChant et al. 2020).

Testing parameters can also produce some unintended biases. To prevent canines from showing bias to a position within the apparatus it is important that the experimenter randomize the order and place that the samples are presented within a session and that the sample appears in each position the same number of times (Lazarowski et al. 2020). This will reduce the possibility of a canine developing a side bias. In the case of GNG paradigms, the number of CS+ and CS− trials should be balanced and randomized to prevent canines from developing a bias toward the Go or No-Go responses. Further, response bias for Go or No-Go responses can be manipulated by changing the response requirements. For example, Edwards et al. (2022) found that canines' response bias for Go responses was reduced, leading to increased accuracy, when the response effort was increased (e.g., longer nose hold) (Edwards et al. 2022). Similarly, the topography of the canine alert (e.g. sit or nose hold) can have a potential impact on performance (e.g. Essler et al. 2020).

7a. The Importance of control tests

Even when the researcher designs testing in such a way to prevent bias or to prevent canines from learning unintended cues, the only way to be sure canines are responding exclusively to the intended olfactory stimulus is by conducting positive and negative controls. A negative control consists of a normal or abbreviated training session where the target odor is removed, but all other variables remain the same. This can be done by removing all odors from the apparatus (in the case of olfactometry) and replacing with clean non-target samples. This session then explicitly tests if canines can identify the "target" location by reinforcing canine indications to the target odor in the absence of the odor. Thus, if canines perform at a rate greater than chance during a control test, this indicates that canines can identify the "target" using a cue *other than* the intended odor. These other stimuli could be visual, auditory or another unintended odor stimulus from the apparatus (in the case of automated equipment), or contamination with materials used during odor preparation (Lazarowski et al. 2020). Thus, further investigation is required to identify the source of the unintended cue, but the procedure provides a direct and explicitly reinforced test of whether canines *can* make correct responses in the *absence* of the intended target.

Alternatively, in a positive control, the canine is presented with a new sample of the target odor to ensure canines correctly respond to the anticipated trained target (Lazarowski et al. 2020). The performance of canines during a positive control test is expected to be identical to its performance during training or testing. A significant performance decrement during a positive control test suggests that the canine was responding to unintended cues from the sample (e.g., contamination) rather than the specific desired target odor (Lazarowski et al. 2020). Conducting control tests are an important way a researcher or a handler can be sure that participant canines are conducting the task utilizing only the intended olfactory stimulus.

8 Conclusions

In this chapter, we have reviewed the various tools and technologies to (1) present odors to research participants in a controlled manner and (2) experimental paradigms to train and evaluate canine odor perception using these tools. The final application for the reader is to merge the desired odor control methods with the desired training paradigm.

Frequently, the choice made on either the odor delivery system or training paradigm leads to little freedom in another. For example, the selection of an expensive air dilution olfactometer may likely lead to only Go No-Go paradigms due to the expense of having multiple identical sample ports for alternative choice testing. An experimenter could manipulate the olfactometer to generate several different outputs from one device using simpler clean flow paths to reduce costs, but canines can frequently identify minimal differences in odor flow, pressure, temperature, etc., and leverage these cues to identify the target path.

Because of canines' remarkable ability to identify minor differences between targets and non-targets that may be unrelated to odor, the importance of controls that manipulate a stimulus thought to control performance (e.g., the target odor) is paramount by either removing it for a control session (negative control) or replacing the target sample (positive control).

As shown in the preceding text, there are a vast array of methods to present an odor to a canine and just as many behavioral paradigms to assess performance. There are even further potential combinations of odor delivery and behavioral testing preparations. Thus, researchers have ample room for creativity to develop and use a method that best suits their experimental needs.

References

Abraham NM, Guerin D, Bhaukaurally K, Carleton A (2012) Similar odor discrimination behavior in head-restrained and freely moving mice. PLoS ONE 7:e51789. https://doi.org/10.1371/journal.pone.0051789

Acree TE, Butts RM, Nelson RR, Lee CY (1976) Sniffer to determine the odor of gas chromatographic effluents. Anal Chem 48:1821–1822. https://doi.org/10.1021/ac50006a050

Acree TE (1997) GC/ olfactometry. Anal Chem News Featur 170–175

April LB, Bruce K, Galizio M (2011) Matching-and nonmatching-to-sample concept learning in rats using olfactory stimuli. J Exp Anal Behav 96:139–154. https://doi.org/10.1901/jeab.2011.96-139

Aviles-Rosa EO, Gallegos SF, Prada-Tiedemann PA, Hall NJ (2021a) An automated canine line-up for detection dog research. Front Vet Sci 8:775381. https://doi.org/10.3389/fvets.2021.775381

Aviles-Rosa EO, McGuinness G, Hall NJ (2021b) Case study: an evaluation of detection dog generalization to a large quantity of an unknown explosive in the field. Animals 11:1341. https://doi.org/10.3390/ani11051341

Berditchevskaia A, Cazé RD, Schultz SR (2016) Performance in a GO/NOGO perceptual task reflects a balance between impulsive and instrumental components of behaviour. Sci Rep 6:27389. https://doi.org/10.1038/srep27389

Bodyak N (1999) Performance of mice in an automated olfactometer: odor detection, discrimination and odor memory. Chem Senses 24:637–645. https://doi.org/10.1093/chemse/24.6.637

Brisbin IL, Austad SN (1991) Testing the individual odour theory of canine olfaction. Anim Behav 42:63–69. https://doi.org/10.1016/S0003-3472(05)80606-2
Buettner A (2017) Springer handbook of odor. Springer Cham
Carlson KS, Whitney MS, Gadziola MA et al (2016) Preservation of essential odor-guided behaviors and odor-based reversal learning after targeting adult brain serotonin synthesis. eneuro 3:ENEURO.0257–16.2016. https://doi.org/10.1523/ENEURO.0257-16.2016
Chikazoe J, Jimura K, Asari T et al (2009) Functional dissociation in right inferior frontal cortex during performance of go/no-go task. Cereb Cortex 19:146–152. https://doi.org/10.1093/cercor/bhn065
Cleland TA, Morse A, Yue EL, Linster C (2002) Behavioral models of odor similarity. Behav Neurosci 116:222–231. https://doi.org/10.1037//0735-7044.116.2.222
Collins GE, Malito MP, Tamanaha CR et al (2017) Trace explosives sensor testbed (TESTbed). Rev Sci Instrum 88:034104. https://doi.org/10.1063/1.4978963
Concha AR, Guest CM, Harris R et al (2019) Canine olfactory thresholds to amyl acetate in a biomedical detection scenario. Front Vet Sci 5:345. https://doi.org/10.3389/fvets.2018.00345
DeChant MT, Hall NJ (2021) Training with varying odor concentrations: implications for odor detection thresholds in canines. Anim Cogn 24:889–896. https://doi.org/10.1007/s10071-021-01484-6
DeChant MT, Ford C, Hall NJ (2020) Effect of handler knowledge of the detection task on canine search behavior and performance. Front Vet Sci 7:250. https://doi.org/10.3389/fvets.2020.00250
DeChant MT, Bunker PC, Hall NJ (2021) Stimulus control of odorant concentration: pilot study of generalization and discrimination of odor concentration in canines. Animals 11:326. https://doi.org/10.3390/ani11020326
DeGreeff LE, Peranich K (2021) Canine olfactory detection of trained explosive and narcotic odors in mixtures using a Mixed Odor Delivery Device. Forensic Sci Int 329:111059. https://doi.org/10.1016/j.forsciint.2021.111059
DeGreeff LE, Malito M, Katilie CJ et al (2017) Passive delivery of mixed explosives vapor from separated components. Forensic Chem 4:19–31. https://doi.org/10.1016/j.forc.2017.02.010
DeGreeff L, Katilie CJ, Johnson RF, Vaughan S (2021) Quantitative vapor delivery for improved canine threshold testing. Anal Bioanal Chem 413:955–966. https://doi.org/10.1007/s00216-020-03052-2
Delahunty CM, Eyres G, Dufour J-P (2006) Gas chromatography-olfactometry. J Sep Sci 29:2107–2125. https://doi.org/10.1002/jssc.200500509
Dorman DC, Foster ML, Lazarowski L (2021) Training with multiple structurally related odorants fails to improve generalization of ammonium nitrate detection in domesticated dogs (Canis familiaris). Animals 11:213. https://doi.org/10.3390/ani11010213
Edwards TL (2019) Automated canine scent-detection apparatus: technical description and training outcomes. Chem Senses 44:449–455. https://doi.org/10.1093/chemse/bjz039
Edwards TL, Giezen C, Browne CM (2022) Influences of indication response requirement and target prevalence on dogs' performance in a scent-detection task. Appl Anim Behav Sci 253:105657. https://doi.org/10.1016/j.applanim.2022.105657
Essler JL, Kane SA, Collins A et al (2021) Egg masses as training aids for spotted lanternfly Lycorma delicatula detection dogs. PLoS ONE 16:e0250945. https://doi.org/10.1371/journal.pone.0250945
Essler JL, Wilson C, Verta AC et al (2020) Differences in the search behavior of cancer detection dogs trained to have either a sit or stand-stare final response. Front Vet Sci 7. https://doi.org/10.3389/fvets.2020.00118
Frederick DE, Rojas-Líbano D, Scott M, Kay LM (2011) Rat behavior in go/no-go and two-alternative choice odor discrimination: differences and similarities. Behav Neurosci 125:588–603. https://doi.org/10.1037/a0024371
Friedrich RW (2006) Mechanisms of odor discrimination: neurophysiological and behavioral approaches. Trends Neurosci 29:40–47. https://doi.org/10.1016/j.tins.2005.10.004

Fuller GH, Steltenkamp R, Tisserand GA (1964) The gas chromatograph with human sensors: perfumer model. Ann N Y Acad Sci 116:711–724

Furton KG, Harper RJ (2017) Patent and Trademark Office. US Patent No. 9,706,755. Washington, DC, US

Gamble KR, Smith DW (2009) Discrimination of "Odorless" mineral oils alone and as diluents by behaviorally trained mice. Chem Senses 34:559–563. https://doi.org/10.1093/chemse/bjp036

Gazit I, Goldblatt A, Terkel J (2005) The role of context specificity in learning: the effects of training context on explosives detection in dogs. Anim Cogn 8:143–150. https://doi.org/10.1007/s10071-004-0236-9

Giordano BC, DeGreeff LE, Malito M et al (2020) Trace vapor generator for explosives and narcotics (TV-Gen). Rev Sci Instrum 91:085112. https://doi.org/10.1063/1.5142385

Gomez P, Ratcliff R, Perea M (2007) A model of the go/no-go task. J Exp Psychol Gen 136:389–413. https://doi.org/10.1037/0096-3445.136.3.389

Hale E (2017) Canine human-scent-matching: the limitations of systematic pseudo matching-to-sample procedures. Forensic Sci Int 279:177–186. https://doi.org/10.1016/j.forsciint.2017.08.014

Hall NJ, Wynne CDL (2018) Odor mixture training enhances dogs' olfactory detection of home-made explosive precursors. Heliyon 4:e00947. https://doi.org/10.1016/j.heliyon.2018.e00947

Hall NJ, Smith DW, Wynne CDL (2015) Pavlovian conditioning enhances resistance to disruption of dogs performing an odor discrimination: CONDITIONING ENHANCED DOGS' RESISTANCE. J Exp Anal Behav 103:484–497. https://doi.org/10.1002/jeab.151

Hall NJ, Collada A, Smith DW, Wynne CDL (2016a) Performance of domestic dogs on an olfactory discrimination of a homologous series of alcohols. Appl Anim Behav Sci 178:1–6. https://doi.org/10.1016/j.applanim.2016.03.016

Hall NJ, Smith DW, Wynne CDL (2016b) Effect of odorant pre-exposure on domestic dogs' sensitivity on an odorant detection task. Appl Anim Behav Sci 178:80–87. https://doi.org/10.1016/j.applanim.2016.02.003

Hallowell SF, Davies JP, Gresham GL (1994) Qualitative/semiquantitative chemical characterization of the Auburn olfactometer. In: Lawrence AH (ed) San Diego, CA, pp 437–448

Hartman M, Dumas J, Nielsen C (2001) Age differences in updating working memory: evidence from the delayed-matching-to-sample test. Aging Neuropsychol Cogn 8:14–35. https://doi.org/10.1076/anec.8.1.14.847

Helton WS (ed) (2009) Canine ergonomics: the science of working dogs. CRC Press/Taylor & Francis, Boca Raton

Hudson DT (2009) Variables affecting the collection and preservation of human scent components through instrumental and biological evaluations. Florida International University, Chemistry

Jendrny P, Twele F, Meller S et al (2021) Scent dog identification of SARS-CoV-2 infections in different body fluids. BMC Infect Dis 21:707. https://doi.org/10.1186/s12879-021-06411-1

Johnston JM, Myers LJ, Waggoner LP, Williams M (1994) Determination of canine olfactory thesholds using operant laboratory methods. In: Lanza RC, Myers LJ, Young PA (eds) Harding GL. Innsbruck, Austria, pp 238–243

Kay L, Krysiak M, Barlas L, Edgerton G (2006) Grading odor similarities in a Go/no-Go task. Physiol Behav 88:339–346. https://doi.org/10.1016/j.physbeh.2006.04.007

Koelega HS (1996) Sex differences in olfactory sensitivity and the problem of the generality of smell acuity. Percept Mot Skills 78:451–458

Krestel D, Passe D, Smith JC, Jonsson L (1984) Behavioral determination of olfactory thresholds to amyl acetate in dogs. Neurosci Biobehav Rev 8:169–174. https://doi.org/10.1016/0149-7634(84)90037-X

Krichbaum S, Lazarowski L, Davila A et al (2021) Dissociating the effects of delay and interference on dog (Canis familiaris) working memory. Anim Cogn. https://doi.org/10.1007/s10071-021-01509-0

Lazarowski L, Foster ML, Gruen ME et al (2015) Olfactory discrimination and generalization of ammonium nitrate and structurally related odorants in Labrador retrievers. Anim Cogn 18:1255–1265. https://doi.org/10.1007/s10071-015-0894-9

Lazarowski L, Krichbaum S, DeGreeff LE et al (2020) Methodological considerations in Canine olfactory detection research. Front Vet Sci 7:408. https://doi.org/10.3389/fvets.2020.00408
Lazarowski L, Davila A, Krichbaum S et al (2021a) Matching-to-sample abstract-concept learning by dogs (Canis familiaris). J Exp Psychol Anim Learn Cogn 47:393–400. https://doi.org/10.1037/xan0000281
Lazarowski L, Simon A, Krichbaum S et al (2021b) Generalization across acetone peroxide homemade explosives by detection dogs. Front Anal Sci 1:797520. https://doi.org/10.3389/frans.2021.797520
Lazarowski L, Waggoner P, Hutchings B et al (2021c) Maintaining long-term odor memory and detection performance in dogs. Appl Anim Behav Sci 238:105301. https://doi.org/10.1016/j.applanim.2021.105301
Le Berre E, Atanasova B, Langlois D et al (2007) Impact of ethanol on the perception of wine odorant mixtures. Food Qual Prefer 18:901–908. https://doi.org/10.1016/j.foodqual.2007.02.004
Lit L, Schweitzer JB, Oberbauer AM (2011) Handler beliefs affect scent detection dog outcomes. Anim Cogn 14:387–394. https://doi.org/10.1007/s10071-010-0373-2
Lu X-CM, Slotnick BM, Silberberg AM (1993) Odor matching and odor memory in the rat. Physiol Behav 53:795–804. https://doi.org/10.1016/0031-9384(93)90191-H
Mach P, Sharpes C, Carmany D (2021) Evaluation of the SciK9 Training Aid Delivery Device for containment of powders. U.S. Army Combat Capabilities Development Command Chemical Biological Center, Aberdeen Proving Ground, MD
Marchal S, Bregeras O, Puaux D et al (2016) Rigorous training of dogs leads to high accuracy in human scent matching-to-sample performance. PLoS ONE 11:e0146963. https://doi.org/10.1371/journal.pone.0146963
Martini S, Begall S, Findeklee T et al (2018) Dogs can be trained to find a bar magnet. PeerJ 6:e6117. https://doi.org/10.7717/peerj.6117
Meule A (2017) Reporting and interpreting task performance in go/no-go affective shifting tasks. Front Psychol 8:701. https://doi.org/10.3389/fpsyg.2017.00701
Mullen M, Katilie C, Collins GE, Giordano BC (2021) Empirical determination of explosive vapor transport efficiencies, vol 11
Nakamura RK, Myslobodsky MS, Coppola R, Mirsky AF (1987) Effects of 7-hydroxybutyrate on the performance of monkeys in a Go/no-go visual discrimination task, vol 9
Neuhaus W (1953) Über die riechschärfe des hundes für fettsäuren. Z Vgl Physiol 35:527–552
Otto T, Schottler F, Staubli U et al (1991) Hippocampus and olfactory discrimination learning: effects of entorhinal cortex lesions on olfactory learning and memory in a successive-cue, go–no-go task. Behav Neurosci 105:111–119. https://doi.org/10.1037/0735-7044.105.1.111
Peña T, Pitts RC, Galizio M (2006) Identity matching-to-sample with olfactory stimuli in rats. J Exp Anal Behav 85:203–221. https://doi.org/10.1901/jeab.2006.111-04
Pfungst O (1911) Clever Hans:(the horse of Mr. Von Osten.) a contribution to experimental animal and human psychology. Holt, Rinehart and Winston
Pierce JD, Doty RL, Amoore JE (1996) Analysis of position of trial sequence and type of diluent on the detection threshold for phenyl ethyl alcohol using a single staircase method. Percept Mot Skills 82:451–458. https://doi.org/10.2466/pms.1996.82.2.451
Plutowska B, Wardencki W (2008) Application of gas chromatography–olfactometry (GC–O) in analysis and quality assessment of alcoholic beverages–a review. Food Chem 107:449–463. https://doi.org/10.1016/j.foodchem.2007.08.058
Porritt F, Shapiro M, Waggoner P et al (2015) Performance decline by search dogs in repetitive tasks, and mitigation strategies. Appl Anim Behav Sci 166:112–122. https://doi.org/10.1016/j.applanim.2015.02.013
Reeve C, Wentzell P, Wielens B et al (2018) Assessing individual performance and maintaining breath sample integrity in biomedical detection dogs. Behav Proc 155:8–18. https://doi.org/10.1016/j.beproc.2017.08.008
Roddick KM, Schellinck HM, Brown RE (2014) Olfactory delayed matching to sample performance in mice: sex differences in the 5XFAD mouse model of Alzheimer's disease. Behav Brain Res 270:165–170. https://doi.org/10.1016/j.bbr.2014.04.038

Samhita L, Gross HJ (2013) The "Clever Hans Phenomenon" revisited. Commun Integr Biol 6:e27122. https://doi.org/10.4161/cib.27122

Schoon GAA, De Bruin JC (1994) The ability of dogs to recognize and cross-match human odours. Forensic Sci Int 69:111–118. https://doi.org/10.1016/0379-0738(94)90247-X

Schoon GAA Scent (1996) Identification lineups by dogs (Canis familiaris): experimental design and forensic application, vol 11

Shenoy P, Yu AJ (2012) Strategic impatience in Go/no-Go versus forced-choice decision-making, vol 9

Simon AG, DeGreeff LE, Frank K et al (2019) A method for controlled odor delivery in olfactory field-testing. Chem Senses 44:399–408. https://doi.org/10.1093/chemse/bjz031

Slotnick B, Restrepo D (2001) Olfactometry with mice. In: Current protocols in neuroscience. Wiley & Sons, Inc.

Tucker D (1963) Physical variables in the olfactory stimulation process. J Gen Physiol 46:453–489. https://doi.org/10.1085/jgp.46.3.453

Vaughan SR, Gokool V, Karpinsky M et al (2022) A preliminary study of the odorants of interest in native crude oils to oil detection canines. In: 44th Arctic and marine oilspill program techical seminar on environmental contamination and response. Environment and Climate Change Canada, Canada

Veech JA, Crist TO (2010) Diversity partitioning without statistical independence of alpha and beta. Ecology 91:1964–1969. https://doi.org/10.1890/08-1727.1

Verhulst NO, Qiu YT, Beijleveld H et al (2011) Composition of human skin microbiota affects attractiveness to malaria mosquitoes. PLoS ONE 6:e28991. https://doi.org/10.1371/journal.pone.0028991

Waggoner P, Lazarowski L, Hutchings B et al (2022) Effects of learning an increasing number of odors on olfactory learning, memory and generalization in detection dogs. Appl Anim Behav Sci 247:105568. https://doi.org/10.1016/j.applanim.2022.105568

Walker DB, Walker JC, Cavnar PJ et al (2006) Naturalistic quantification of canine olfactory sensitivity. Appl Anim Behav Sci 97:241–254

Wenzel BM (1948) Techniques in olfactometry: a critical review of the last one hundred years. Psychol Bull 45:231–247

Wysocki CJ, Dalton P, Brody MJ, Lawley HJ (1997) Acetone odor and irritation thresholds obtained from acetone-exposed factory workers and from control (occupationally unexposed) subjects. Am Ind Hyg Assoc J 58:704–712

Sources of Human Bias in Canine Olfactory Research

Sarah Krichbaum, Jordan G. Smith, Craig Angle, Paul Waggoner, and Lucia Lazarowski

Abstract

Canine olfactory research is susceptible to human bias that influences the reliability and validity of results. In this chapter, we provide case-study examples of handler, evaluator, and observer bias common to olfactory detection work. Dogs are socially apt and readily pick up on cues from their handler regarding the study parameters such as the location or presence of their target. Additionally, dog-handler team evaluators could unintentionally relay information to the handler or dog; therefore, double-blind testing, in which the handler and evaluator are unaware of the study parameters, is the gold standard of canine olfactory research. This chapter suggests blinding and other experimental controls for reducing human effects.

Keywords

Single-blind • Double-blind • Observer bias • Canine • Olfaction

1 Introduction

The study of canine olfaction has become increasingly important as dogs are used for several tasks that require the detection of harmful and illicit substances. Dogs provide national and personal security through the detection of explosives, narcotics, medical diseases, and biological warfare agents. However, to determine dogs' capabilities and form decisions regarding efficient implementation and evaluation of results, it is imperative that the methods for examining detection performance are reliable and valid. Human bias is one of the major threats to the

S. Krichbaum (✉) · J. G. Smith · C. Angle · P. Waggoner · L. Lazarowski
Canine Performance Sciences, Auburn University College of Veterinary Medicine, Auburn, AL, USA
e-mail: szk0138@auburn.edu

L. Lazarowski (ed.), *Olfactory Research in Dogs*,
https://doi.org/10.1007/978-3-031-39370-9_6

soundness of canine olfactory research, and therefore, it should be systematically examined. The aim of this chapter is to outline types of human bias, their effect on canine olfactory research, and provide strategies to reduce or eliminate them.

2 Handler Effects

For nearly a century it has been understood that animal behavior research is susceptible to human biases that can influence the results. This phenomenon is particularly apparent when the experimenter or observer has a vested interest in the performance of the animal and is present when the data is being collected. Critically, these effects often occur unknowingly or unintentionally. The powerful effect of unintentional human cueing on animal behavior was first widely recognized by the classic *Clever Hans* horse, who was claimed to have the ability to perform arithmetic by stomping his hoof a certain number of times in response to mathematical questions. In truth, Hans had learned to respond to unintentional cues by his handler, who exhibited subtle changes in body language and facial expressions as the horse approached the correct answer.

Dogs are extremely social and adept at producing strong bonds with people making them an ideal candidate for working roles involving cooperation with a handler (Zubedat et al. 2014). However, this sociability makes them proficient at picking up on conscious and unconscious cues given by their handlers which can create confounds in detection research. This effect, known as the *experimenter expectancy effect*, occurs when the experimenter's or observer's expectations of the study conditions or results inadvertently influence the subjects' performance. In the case of detection dog-handler teams, dogs are more likely to alert if their handler believes a target odor is present (Lit et al. 2011), will search longer if the handler knows that there are unfound target odors in the search, and spend less time in blank areas (DeChant et al. 2020).

It is important to reduce experimenter and handler influence that affects the validity of the results as it is necessary to know the true capabilities of the dog-handler team in an operational scenario in which they would be unaware of the parameters of their search (e.g., number of targets, length of search, etc.). One suggestion is to allow dogs to work off-leash to increase the dogs' independence and decision-making and reduce handler influence over the dog's movement; however, this method does not remove all potential cues dogs may utilize. Dogs can easily learn that certain hand signals, body orientation, and emotional content of their handler's speech are associated with the presence and location of a target (Edwards et al. 2017). For example, common handler errors include walking slower when in the presence of a target, faster when in a known blank area, and reaching for the reward in their pocket in anticipation of a correct alert.

3 Single-Blind Testing

The most efficient method to remove the potential of any handler cues is by testing dog-handler teams in situations in which the handler is uninformed of the testing parameters (e.g., presence or location of targets), known as single-blind testing. Single-blind testing is now widely accepted as standard practice in canine olfactory research. In fact, Johnen et al. (2017) suggest that the results gleaned from any study in which the handler is not blind should be carefully examined for potential handler effects. A recent study specifically evaluated how handler knowledge might affect search behavior and performance on a detection task (DeChant et al. 2020). In the study, the experimenter told the handlers in the Known Group how many hides were present in the search but did not tell the handlers in the Unknown Group. Teams in the Unknown Group spent significantly more time searching the blank areas than teams in the Known Group. Though an obvious effect of handler knowledge, it may not seem like a problem at face value. However, given that teams in the Known Group spent less time in the blank area, they were less likely to false alert and less likely to fatigue. These influences would not exist in an operational scenario; therefore, the performance of the Known Group is not a valid representation of performance in the field. In addition, dogs in the Unknown Group looked back at their handler more frequently than dogs in the Known Group, suggesting that they were looking for cues that they may otherwise receive.

Two studies from our group support and elaborate on the findings from DeChant et al. (2020) by directly comparing canine team search performance in non-blind and single-blind searches. In Lazarowski et al. (2021) detection dogs were tested for their ability to recall odors not experienced in 12 months as a test of long-term memory. Due to logistical constraints (i.e., two handlers running nine dogs in multiple searches), only a portion of the searches were run single-blind. Thus, whether the handler was blind or not on a given search was considered as a factor in the analyses to determine potential effects. The results indicated that hits (i.e., the number of responses to the target odor) were significantly higher in non-blind compared to single-blind searches. This finding suggested that the dogs were attuned to, likely unintentional, cuing by the handler as to the location or at least the presence of the target in the non-blind searches. Therefore, to address this, the authors excluded non-blind searches from the analyses and restricted results to only single blind searches. However, in a similar study, there was no difference in performance on single-blind and non-blind searches, likely due to the very high accuracy in performance leading to ceiling effects (Waggoner et al. 2022). Therefore, the influence of handler cues may vary based on the difficulty of the task.

We have further examined handler effects in ongoing (unpublished) work to analyze the effect of non-blind testing not only on dogs' responses to targets but also their false alerts, both of which are important for detection success. In this study, dogs were trained to detect 12 target odors across training steps culminating in a final criteria phase before advancing. A portion of the searches (26.78%) in

Table 1 Average (plus or minus standard error) hit rate and proportion of searches with a false alert across odors tested in the final step of criteria for single blind and non-blind searches

Condition	Hit rate	Proportion of searches with a false alert
Single blind	78.73 (3.27)	25.31 (4.70)
Non-blind	91.11 (1.66)	6.00 (1.55)

this phase were conducted single blind, so we were able to directly compare performance on single blind and non-blind runs. Table 1 shows the average hit rate (i.e., number of responses to the target odor divided by the total number of targets) and proportion of searches with a false alert (i.e., number of responses to a non-target odor divided by the total number of searches) for single blind and non-blind runs. Using independent-sample *t*-tests we found a significantly lower hit rate ($t(22) = -3.37$, $p = 0.003$) and a higher proportion of searches with a false alert ($t(10) = 3.90$, $p = 0.003$) in the single-blind than non-blind runs. Further examination of performance indicated that these effects were more pronounced when the handler was less experienced compared to a more experienced handler, and may vary due to target difficulty; however, these effects require further examination.

Together, these findings suggest that when olfactory detection research is conducted non-blind there is a significant influence of human bias on performance that threatens the internal validity of the study. The effects seem to inflate performance which could lead evaluators to believe that a dog-handler team is more proficient than they are. Due to this, single blind testing is critical in order to accurately assess performance. Specifically, this form of testing minimizes handler bias while allowing the evaluator to deliver timely feedback to the handler regarding the team's performance (e.g., when they have identified a target odor). However, though often preferred by the handler for these reasons, single blind testing does not account for any sources of bias from other parties present, such as the evaluator.

4 Double-Blind Testing

While single blind scenarios minimize handler effects, there is the potential influence of the non-blind evaluator on the team's performance. For example, the evaluator could cue the dog or the handler to the presence or location of a target (Centre for the Protection of National Infrastructure 2018). As above, this situation is not realistic to operational scenarios in which no one present would know where the target was located. Therefore, the gold standard for olfactory detection research is double-blind testing. In double-blind testing none of the individuals present, including the handler, evaluator, or any other participating observers are aware of the test conditions, thereby directly reflecting an operational scenario (Scientific Working Group on Dog and Orthogonal Detector Guidelines 2011). To demonstrate the influence of a non-blind evaluator on detection performance, DeChant et al. (2020) compared detection performance on single-blind and double-blind searches. Overall, no differences in hit rate were observed, demonstrating that an impartial observer can be present without influencing detection performance. However, it is important to note that the evaluator in this experiment was a trained researcher. A less impartial evaluator, especially one with a vested interest in the outcome such as the lead researcher or the supervising trainer, could impact team performance with intentional or unintentional cues.

To further understand the effects of an external evaluator, our group completed a study (unpublished) similar to DeChant et al. (2020) in which the evaluator was not a member of the experimental team but rather was an experienced canine training supervisor that routinely conducted team evaluations. In this study, we assessed 15 professional detection dog-handler teams on single- and double-blind operational search scenarios. Each team completed two searches consisting of ten small rooms, five of which contained a target and five contained a distractor. The handler was blind in all searches, but the information given to the evaluator was manipulated across conditions. Specifically, in one search the evaluator was blind to the location of targets (double blind condition), and in the other search, the same evaluator was not blind to the location of targets (single blind condition). In both conditions, the evaluator carried a clipboard with a camera attached and was asked to keep the camera pointed in the direction of the detection team. A proctor, remaining outside of the test area (out of view of the evaluator and team) monitored the video feed from the camera via a wireless connection between the camera and monitoring device. In the double-blind condition, the evaluator communicated when the team made an alert to the proctor through the wireless connection, and the proctor would reply "target" or "no" to signal if the response was correct. In the single-blind condition, the evaluator was given a map to ensure they understood the location of the targets.

On average there were more hits in the single blind ($M = 4.47$, $SEM = 0.17$) than the double blind ($M = 4.27$, $SEM = 0.18$; $t(14) = -1.38$, $p = 0.189$) condition and fewer false alerts in the single-blind ($M = 1.20$, $SEM = 0.30$) than the double-blind ($M = 1.80$, SEM = 0.31; $t(14) = 1.79$, $p = 0.095$) condition. Though the differences are not significant, the difference in total number of false alerts between the two conditions (double-blind: 27; single-blind: 18) is noteworthy. This data is visually represented in Fig. 1.

In addition to hits and false alerts, we compared the percentage of rooms in which communication occurred between the handler and evaluator (either verbal or non-verbal, scored from video by an independent observer) between the two conditions. We found levels of communication (in seconds) to be low regardless of conditions (double blind: $M = 9.76$, $SEM = 5.00$; single blind: $M = 7.86$, $SEM = 3.95$) and no significant difference between them ($t(13) = -0.46$, $p = 0.653$). However, anecdotally, we observed a difference in the type of verbal communication that occurred during some searches such that in specific situations, more direct forms of verbal communication were given by the evaluator on single blind searches when the handler was unsure of his dog's behavior. In one instance, a handler mentioned while searching a target room during a single blind search that he thought a target might be present as the dog was showing interest in certain

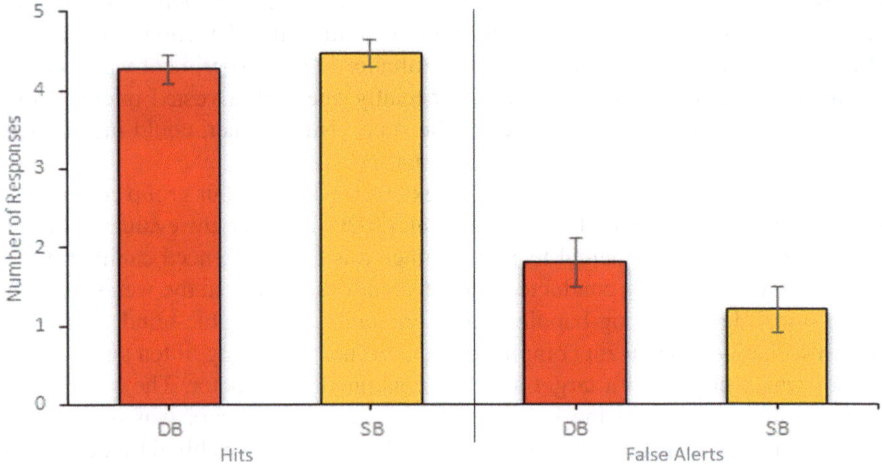

Fig. 1 Average number of hits and false alerts for each condition (SB: single blind; DB: double blind)

areas but had not given a final response. In response, the evaluator told the handler "good job" and allowed the team to continue searching until a final response was given. Situations like this, in which the handler is made privy to information prior to the dog performing an alert response, can potentially influence the behavior of the handler. Specifically, the handler may choose to remain in a certain area longer than he/she would have if blind to the presence of a target odor in that location, which could increase the dog's probability of detecting and responding to the odor.

We also evaluated the difference in average duration (in seconds) to search a room between the two conditions. We found an insignificant trend suggesting that average duration to search a room was longer in the double ($M = 56.99$, $SEM = 2.50$) than single blind ($M = 52.27$, $SEM = 3.29$; $t(13) = -1.28$, $p = 0.095$) condition (Fig. 2).

Together, these results suggest that non-blind evaluators can influence detection team performance. Though the effects shown here are not significant, there are trends suggesting that dogs are more likely to alert to targets and less likely to false alert when the evaluator is not blind compared to when the evaluator is blind. In addition, it seems that search duration is influenced by the blinding of the evaluator suggesting that teams do not search as long when the evaluator is not blind, which, as discussed above, affects the probability of a false alert as well as the team's endurance and fatigue. However, the above results may lack the number of observations (number of teams) to have the effect size needed to detect a significant effect. Therefore, further examination of these variables with a larger sample size is needed to further elucidate these findings.

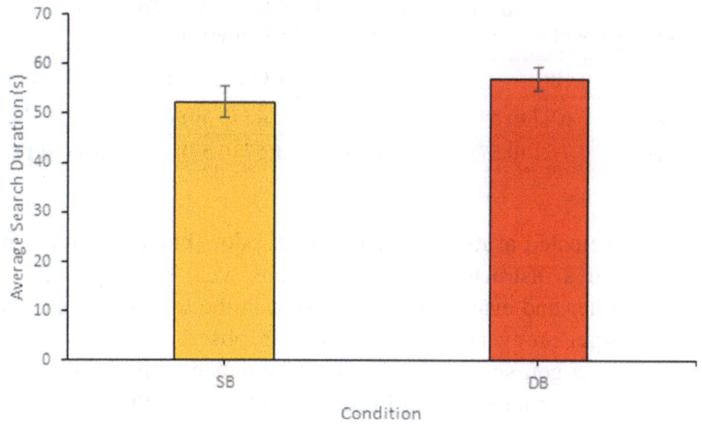

Fig. 2 Average duration to search a room in each condition (SB: single blind; DB: double blind)

5 Observer Bias

The types of influence discussed above describe situations in which the knowledge or beliefs of individuals present during testing directly or indirectly influence the team's performance. However, even when care is taken to minimize such influence over the behavior of participants, biases held by observers involved in the data collection can influence the interpretation of the dog's behavior and therefore the results. Observers often selectively attend to information that confirms hypotheses or certain beliefs based on prior knowledge. For example, Tuyttens et al. (2014) showed that providing observers with false information prior to scoring an animal's behavior influenced how they scored the behavior, though the behavior was not actually influenced. Thus, there is risk of observer bias in canine olfactory detection research when experimenters have expectations based on hypotheses or have a vested interest in the dog or team's success. Moreover, observer bias is more likely to occur when the behavior being observed is subtle, ambiguous, or subjective in nature (Tuyttens et al. 2014; van Wilgenburg and Elgar 2013). This scenario is especially likely to occur in canine olfactory research given that the alert response of the dog (e.g., sitting, lying down, freezing) is inherently variable and requires a certain degree of subjective interpretation. Further, individuals may differ in how conservative their interpretation of a dog's response is, leading to variability in observations (Edwards 2019).

Post-hoc examination of data from a previous odor detection study from our group (unpublished) sought to evaluate the effects of observer bias on several detection metrics. In this study 14 dogs were tested in an odor recognition test to determine their ability to alert to their trained targets as well as generalization to chemically similar odors. The test was conducted using a fixed sampling array in which ten discrete sampling positions were arranged in a circle. Test sessions consisted of twelve trials, with each containing either a target odor placed in one of

Table 2 Average (plus standard error) number of hits, false alerts, COB on targets, and COB on distractors scored by the blind evaluator and the non-blind observer

Scorer	Hit	False alert	COB on target	COB on distractor
Blind evaluator	6.93 (0.54)	1.36 (0.39)	1.43 (0.50)	1.93 (0.74)
Non-blind observer	7.21 (0.57)	1.57 (0.52)	1.57 (0.57)	0.43 (0.17)

the ten positions selected at random or no target odor (blank trial). All nine other positions contained a distractor. All test sessions were conducted double blind; therefore, the handler and evaluator, both present in the test area, were blind to the position of the target on every trial. However, an observer, who was not blind to the position of the target, scored the dog's performance behind a two-way mirror. Both the blind evaluator and the non-blind observer scored the dogs' responses on each trial. Responses recorded included an alert, defined as sitting in front of one of the positions in the circular array, or a "change of behavior" (COB), a distinctive pattern of behavior characterized by an alteration in ongoing behavior that occurs when a dog detects a trained odor (Furton et al. 2010). Change of behavior may include a head snap, change in direction, or other alterations in body posture and movement distinctive from normal searching behavior in the absence of a target, and are considered a valuable response by the dog especially when tested in a challenging scenario when it may be expected that the dog will not perform its trained final response (e.g., in a generalization test).

In order to determine the effects of observer bias on the interpretation of dogs' responses, we compared the scoring of the blind evaluator and the non-blind observer (see Table 2). Independent samples t-tests showed no significant difference in the average number of hits, false alerts, or COB on targets ($ps > 0.71$). However, there was an insignificant trend suggesting that the non-blind observer was less likely to record a COB on a distractor than the blind evaluator ($t(26) = 1.98$, $p = 0.06$). This could be explained by the non-blind observer being less attentive to the dog's behavior toward non-targets, or under-interpreting the dog's behavior toward non-targets compared to targets. Indeed, the non-blind evaluator was much more likely to record a COB to a target than a distractor (1.57 vs. 0.43), with only a slight discrepancy for the blind evaluator. This finding suggests that knowledge of the location of the targets and distractors can not only directly influence a participant's behavior, but how an observer interprets behavior. Thus, it is critical to operationally define a response (e.g., form and/or duration of the behavior) to reduce subjectivity and allow consistency across observers (Edwards 2019; Lazarowski et al. 2020).

6 Conclusion

Human bias, as in all research, is an important consideration when designing tests of canine olfactory abilities to ensure that the abilities are assessed without cues that influence the validity of the results. At a minimum, the handler, and when possible, the evaluator, should be blind to the testing conditions and target locations during assessments to prevent any intentional or unintentional cuing to the dog or handler regarding the presence or absence of target or other test parameters. Further, any observers involved in recording and scoring the dogs' behavior should also be blind to the study conditions to minimize biases during interpretation. With those parameters in place, the test results more closely represent operational performance and provide unbiased estimates of canine detection team performance and the factors affecting it.

References

Centre for the Protection of National Infrastructure (2018) Why you should train and test detection dogs 'double-blind.' Detection dog guidance notes. https://www.cpni.gov.uk/canine-detection guidance-notes53

DeChant MT, Ford C, Hall NJ (2020) Effect of handler knowledge of the detection task on canine search behavior and performance. Front Vet Sci 7. https://doi.org/10.3389/fvets.2020.00250

Edwards TL (2019) Automated canine scent-detection apparatus: technical description and training outcomes. Chem Senses 44(7):449–455. https://doi.org/10.1093/chemse/bjz039

Edwards TL, Browne CM, Schoon A, Cox C, Poling A (2017) Animal olfactory detection of human diseases: guidelines and systematic review. J Vet Behav 20:59–73. https://doi.org/10.1016/j.jveb.2017.05.002

Furton K, Greb J, Holness H (2010) The scientific working group on dog and orthogonal detector guidelines (SWGDOG). National Criminal Justice Reference Service

Johnen D, Heuwieser W, Fischer-Tenhagen C (2017) An approach to identify bias in scent detection dog testing. Appl Anim Behav Sci. https://doi.org/10.1016/j.applanim.2017.01.001

Lazarowski L, Krichbaum S, DeGreeff LE, Simon A, Singletary M, Angle C, Waggoner LP (2020) Methodological considerations in canine olfactory detection research. Front Vet Sci 7. https://doi.org/10.3389/fvets.2020.00408

Lazarowski L, Waggoner P, Hutchings B, Angle C, Porritt F (2021) Maintaining long-term odor memory and detection performance in dogs. Appl Anim Behav Sci 238:105301. https://doi.org/10.1016/j.applanim.2021.105301

Lit L, Schweitzer JB, Oberbauer AM (2011) Handler beliefs affect scent detection dog outcomes. Anim Cogn 14(3):387–394. https://doi.org/10.1007/s10071-010-0373-2

Tuyttens FAM, de Graaf S, Heerkens JLT, Jacobs L, Nalon E, Ott S, Stadig L, Van Laer E, Ampe B (2014) Observer bias in animal behaviour research: can we believe what we score, if we score what we believe? Anim Behav 90:273–280. https://doi.org/10.1016/j.anbehav.2014.02.007

Waggoner P, Lazarowski L, Hutchings B, Angle C, Porritt F (2022) Effects of learning an increasing number of odors on olfactory learning, memory and generalization in detection dogs. Appl Anim Behav Sci 247:105568. https://doi.org/10.1016/j.applanim.2022.105568

van Wilgenburg E, Elgar MA (2013) Confirmation bias in studies of nestmate recognition: a cautionary note for research into the behaviour of animals. PLoS ONE 8(1):e53548. https://doi.org/10.1371/journal.pone.0053548

Zubedat S, Aga-Mizrachi S, Cymerblit-Sabba A, Shwartz J, Leon JF, Rozen S, Varkovitzky I, Eshed Y, Grinstein D, Avital A (2014) Human–animal interface: the effects of handler's stress on the performance of canines in an explosive detection task. Appl Anim Behav Sci 158:69–75. https://doi.org/10.1016/j.applanim.2014.05.004

Human Scent Dynamics—Combining Theory and Practice in Locating People

G. A. A. Schoon and P. A. Moore

Abstract

The search and rescue and forensic communities train dogs to find, and sometimes identify, missing people and fugitives. In order to be able to do this effectively, knowledge on human scent and how it spreads is essential. Here we integrate knowledge from research into human volatile organic components (VOCs), aerosols, and skin rafts with current models of flow dynamics and gravitational effects, and include the effect of degradation into the dispersion model. Odorant availability is described for lost humans, articles they have left behind, and the path they have walked based on this model. Knowing what odors are available allows trainers to set up exercises to focus the dog effectively on the desired odor cue and to find odor sources more efficiently.

Keywords

Detection dogs • Search and rescue dogs • Human scent • Odor plume • Odor availability • Volatile organic components • Skin rafts • Flow dynamics

1 Introduction

When looking for people, dogs are trained for "tracking," "trailing," and/or "air scenting," based on what their operational deployment or specific sport requires. Although trainers are conscientious about trying to teach the dog team their task, it

G. A. A. Schoon (✉)
Animal Detection Consultancy, Vorchten, The Netherlands
e-mail: animaldetectionconsultacy@gmail.com

P. A. Moore
Department of Biological Sciences, Bowling Green State University, Bowling Green, OH 43403, USA
e-mail: pmoore@bgsu.edu

L. Lazarowski (ed.), *Olfactory Research in Dogs*,
https://doi.org/10.1007/978-3-031-39370-9_7

is sometimes difficult to see what the dog has actually learned in terms of odorants they respond to. Here we endeavor to clarify what odorants are available for the dog in different scenarios, in time, and in space. This should help in setting up efficient training exercises to focus the attention of the dog on the desired odor profile. We will attempt this by examining how diverse types of odor sources create plumes including live humans, articles they leave behind, and tracks they leave when moving away from where they were.

The physical processes that are involved in the dispersion of human-based scents can be grouped into four distinct areas: chemistry and volatility of the chemicals at the source, movement of chemicals downwind by flow and turbulence, gravitational movement of larger particles, and chemical interactions during dispersion (Conchou et al. 2019; Vickers 2000). The role that each of these processes plays in modulating the chemical dynamics of odors is fairly well understood on their own, but the odor plumes that dogs use to locate humans is a rich and complex mixture of all of these processes occurring simultaneously (Gu et al. 2022; Kowadlo and Russell 2006). Thus, the chemical composition of an odor that is detected by a dog, say one hundred meters downwind of a human, is different than the chemical composition of those odors emanating from the source person (Eckenrode et al. 2022). Understanding how each of these individual processes alters both the concentration and relative concentration of the individual components in an odor plume is critical to understanding how to train dogs. These processes will be discussed first in more general terms.

In addition, these four processes led us to conclude that diverse types of odor sources will have different plume dynamics. For example, a living source, such as a human or animal, will continually produce a relatively constant ratio of chemicals over the lifetime of any individual odor plume (Young et al. 2020) and can be considered to produce what we have termed an infinite odor plume. This infinite plume stands in contrast to a finite plume that is produced by a non-living source such as any article a person has left behind, such as a piece of clothing or something else they have dropped or discarded or scent they have left behind when walking. These odor sources contain a limited number of chemicals and the four processes described above interact with this source which results in a change in the concentration and composition of the odor source over time and space. This distinction will be covered in detail later (Packzkowski and Schütz 2011) for each of the three odor sources we will be discussing—live humans, scented articles, and tracks.

Understanding the significant differences in odor availability, which depend on the type of odor source present and its relationship to space and time, leads to a better understanding of common training scenarios and what a dog can learn under these circumstances. Alas, this is not as simple or straightforward as one would like, but we believe that understanding the processes involved will lead to more efficient and effective training.

2 Processes Involved in the Creation and Spreading of Odor Plumes

2.1 Odor Plumes

For the purposes of this chapter and ease of communication, we will label all scent trails, point marks, or other odor signatures that may be used by dogs to locate chemical sources as odor plumes. This singular label will include both the aerial borne odors that dogs use as well as tracks, locations, or sources on the ground or in building structures. While one could argue that the physical location (air vs solid substrate) is different, the physical and theoretical processes by which scents are dispersed from their source and eventually arrive at the dog's nostril are identical (Moore and Crimaldi 2004). By labeling all of these sources as odor plumes, we can treat the mechanics of dispersion under a single larger theory.

Before we describe the physical and chemical processes of odor plume creation we need to cover a small note on scales. Odor plumes can be considered to have two interrelated scales: spatial and temporal scales. An example of spatial scales includes the area, or space, over which an odor plume spreads downwind. Plumes can extend for meters or even kilometers downwind and can spread laterally in those same dimensions. Temporal scales can be associated with the age of the odor source. For example, an inanimate object sitting in a forest will age over time and, as such, the chemical composition of the source will be altered by the chemical processes covered below. Yet, these spatial and temporal scales can also be intertwined (Moore and Crimaldi 2004). In one situation, a dog that is downwind of the source of an odor plume will sample the odor plume temporally in a series of sniffs, but each sniff is actually a different spatial location within the plume. So, the dog samples the spatial distribution of a plume as a temporal distribution across its nose. The same phenomenon occurs as the dog moves upwind (spatially) but is guided by the temporally distinct odor sniffs. Finally, as a puff of odors moves downwind (spatially), the chemicals contained in the puff will interact with the environment over time (temporally) and become aged (Weissburg et al. 2002).

2.1.1 Chemistry and Volatility at the Source

The olfactory abilities of dogs are primarily tuned to volatile organic compounds (VOCs). Volatile organic compounds are characterized by high vapor pressure and low water solubility. Because of these two properties, these compounds are readily emitted into aerial environments as gases. In the terrestrial environment, classes of VOCs are responsible for odors produced by flowers, plants, and perfumes as well as aerial pollutants. These compounds are quite diverse in terms of chemical structure and function, but the majority of compounds produced by living organisms can be grouped as terpenoids, alcohols, and carbonyls. In regard to the creation of human odor plumes, VOCs are readily liberated into the air and move downwind with air currents. Such VOCs are too small to be impacted by gravitational forces and they spread following the physics of flow dynamics described below.

A second source of potential compounds for odor plumes includes aerosols produced by humans. Aerosols are larger than VOCs and are subject to different dispersion mechanics (covered below). Aerosols are a suspension of either solid or liquid particles that are released into the air. In regard to human sources, droplets from exhalation from either the mouth or nostrils can form aerosols that are part of the human odor plume. Sweat and coughing can be other sources of aerosols that eventually mix with air and are transported downwind. Typically, aerosols and particles are defined and categorized by the diameter of the object in question. The dividing point between aerosols and particles is typically set around 0.5 to 5 microns where aerosols are smaller than this and particles are larger, and their size is linked to their rate of settling.

A typical kind of particle in human odor plumes is called skin rafts or corneocytes. Rafts are considered naturally discarded or sluffed skin, skin cells, and patches of skin. These rafts contain VOCs, chemicals attached to the dermis, as well as bacterial communities associated with the skin. Rafts are larger and heavier than both VOCs and aerosols (30–50 microns) and are heavily influenced by gravitational forces. Given the size and speed at which rafts settle off an odor source, these sources are unlikely to play any significant role in the production of downwind odor sources. Each raft could be considered a new and singular point source for a small-scale odor plume.

2.1.2 Movement of Molecules as a Result of Flow and Turbulence

In air, odor plumes are transported from source to nose by advection or dispersion and these two processes work at different temporal and spatial scales. Advection is the larger scale, bulk movement of odor molecules by air flow. In all but the smallest microscopic situations, advection or air flow is the dominant transport mechanism. Dispersion can be attributed to smaller and slower process like stirring (turbulence interweaving of air parcels), molecular diffusion, and spreading due to shear effects in the flow. Shear can be imagined as two 'sheets' of air moving across each other at different velocities. Molecular diffusion (VOCs in air ~5×10^{-4} cm^2/s) only has an effect in 20 mm above an odor source over the course of 10 h and is unimportant for the formation of odor plumes at the spatial and temporal scales of dog searching. There is a great deal of confusion in the dog literature that attributes a significant role of diffusion in dog searches (particularly within 'sealed' containers). Even in these situations, convection or air flow redistributes odor molecules more than diffusion does. A physical and mathematical derivation to prove this last statement can be found elsewhere (Moore 2016).

Because dispersion and diffusion are slow and small-scale phenomenon, the majority of the spatial and temporal structure that appears in odor plumes is due to the diverse types of flow that occur in different situations (Elkinton and Cardé 1984). This flow influences all molecules (independent of size or shape) in the same manner. Here, the mechanisms of turbulence (related to the fluctuation of velocity) and convection (air movement due to differential heating and cooling) determine the concentrations and fluctuations of odor molecules downwind of a source (Pannunzi and Nowotny 2019). These two processes create the typical heterogeneity seen in images of smokestacks or smoke plumes. This same heterogeneity is present within odor plumes. As odorants move down wind, puffs or filaments of odorants are broken apart and stirred. Here turbulent stirring coupled with molecular mixing (diffusion at the corners of these filaments and puffs) redistributes the odorants and creates the characteristic fluctuations that appear within odor plumes.

The conclusion of all of this for dog searches is that flow creates the sensory landscape for dogs and that sensory landscape contains VOC signals that are patchy in space and fluctuate in time. Flow also affects the distribution of aerosols and rafts, but contrary to VOCs, these two odor sources are more strongly influenced by gravitational forces discussed below.

One final aspect to consider for odor plume searches and dogs is the interaction between air flow and the ground. Air flow across any solid surface (ground, vehicles, roads) forms a boundary layer (Schlichting and Kestin 1961; Jackson et al. 2007). The boundary layer is a gradient of decreasing air velocity as the surface is approached. Thus, even on windy days, the flow at or in grass and vegetation, or along buildings and other obstructions, is significantly slower than higher off the ground termed free space. Within this slower flow, odorants can be trapped and no longer transported downwind. Odorants trapped within the boundary layer will move more slowly and appear to be a higher concentration than those higher up off the ground. The spatial and temporal heterogeneity of odors is also significantly less in this boundary layer. This is likely why dogs will often spend a significant amount of time sniffing and investigating vegetation patches.

2.1.3 Gravitational Movements of Aerosols and Particles

Air flow and the associated turbulent dynamics will move odorants around in three dimensions, while gravitational forces will work solely on larger particles and aerosols to cause these odor sources to settle to the ground. The physics of settling of particles in both flow and stationary fields is well developed and modeled by Navier–Stokes equations and the Stokes-Cunningham law (Concha Arcil 2009; Tedeschi et al. 1999).

The important reference point for the consideration of gravitational forces in the generation of odor plumes is the relationship between the overall size of the particle and the velocity or rate at which those particles settle. It is important to note here that size is continuous from the smallest molecules to the largest particles. Gravity impacts all of these sources of odorants but can be ignored for molecules. Even small particles (less than 1 micron in diameter) will take hours to days to settle from a distance of five feet. In the same vein, air flow and turbulent dynamics also determine the dispersion of both molecules and aerosols. Yet, given the quick settling time of larger aerosols, air flow has extremely small impacts on the movement of larger particles. Recent work, based on COVID-19 infections, has produced estimates of settling velocities for aerosols of different diameters (Gu et al. 2022). For small aerosols (1 microns), settling velocities are quite slow and are centered less than 0.1 mm/s. At this velocity, an aerosol of this size would take over 200 h to settle to the ground if released at a height of 1.5 m. This time frame is long enough that flow dynamics will likely be important in dispersing these smaller aerosols within an odor plume. Conversely, an aerosol/particle around 100 microns in diameter would have a settling velocity of 1 to 2 mm/s. This droplet would hit the ground from the same height as above in just over 2 min.

In context of odor plume dynamics, these settling times allow us to return to the intertwined concept of space and time. If we return to our imagined odor source of a stationary human that is living, breathing, and maybe even speaking, then the odor plume downwind from this source is the combination of VOCs, aerosols, and skin rafts. In regard to aerosols, larger droplets and rafts will fall more quickly in time and thus, will be located closer to the human source in space. Conversely, smaller droplets with slower settling speeds will be moved farther downwind before they settle onto the ground. The differences in settling time will produce spatial differences in aerosol sizes. A gradient of aerosol size will be produced downwind from the human starting with larger aerosols and their associated odorants progressing to smaller and smaller aerosols.

2.1.4 Chemical Interactions During Movement Downwind

As molecules are liberated into the air and move downwind, they are subject to a number of chemical interactions that have the potential to change their structure and overall concentration within a plume as well as the relative ratios of chemicals within the plume. As a broad class of chemicals, VOCs vary in their functional groups, lengths, and degree of saturation. Because of this variation, a summary of the degree and nature of these interactions is difficult, but some general trends can be developed. There are a small number of gas-phase oxidants that are responsible directly for most gas-phase chemical transformations. Hydroxyl radical (OH) and ozone (O_3) are the most important. These oxidants rapidly combine and remove compounds from odor plumes (Kim et al. 2011). In addition, temperature, humidity, and light can impact the effects of oxidants as well as provide additional interactions to remove VOCs from odor plumes. Furthermore, the rates of oxidant reactions increase with humidity and are accentuated during sunlight hours as compared to nighttime reaction rates. Finally, those compounds that are saturated (hydrogen bonds at all of the sites) are less reactive with atmospheric oxidants than compounds with functional groups such as double bonds, aldehydes, and alcohols. Thus, functionalized compounds are more likely to react with oxidants and be removed from odor plumes as they move downwind (Atkinson 2000). Most odor compounds are highly functionalized making them even more susceptible to chemical reactions within a plume.

Interestingly, environmental conditions can play a significant role in altering the concentrations and composition of odor plumes. Humidity can affect VOCs, aerosols, and skin rafts in differential fashion. The presence of increased humidity will also increase the presence of free OH^- ions contained within water droplets, rain, or humid vapors (Tobias et al. 2000). VOCs can react with elevated ozone concentrations as well as OH^- in water vapor to form products that are lower in volatility and can be incorporated into aerosol particles. In addition, humidity can serve to increase the rate at which aerosol particles grow in size which, in turn, increases the settling velocity of the particles. Increased light (as mentioned above and through increased UV radiation) as well as increased temperature will increase decay rates of gas phase compounds.

2.2 Categorizing Plume Types

Keeping in mind that the purpose of these plume descriptions is for dog searches, we can summarize the preceding theory to generate diverse types of odor plumes. Each of these plume types will have distinctive characteristics and mixtures of odorants downwind which will be important for both the training of dogs to track odors as well as for the overall understanding of plume dynamics for handlers.

2.2.1 Source Concentrations

There can be two diverse types of odor plumes based on the types of sources that may exist. Living sources (primarily humans in these cases) can be considered infinite sources. The term infinite refers to the fact that biological processes within the living organism will provide a constant source of chemicals that will be liberated downwind. As odorants move from the source to the air, they are quickly replaced by the metabolic processes involved in biological functioning. These sources can be contrasted with finite sources. These sources are associated with non-living or once-living material. Examples include worn clothing, human-scented articles, biological stains, and residual odor. The critical distinction between infinite and finite revolve around the source concentration of compounds. In a finite source, the concentration of chemicals at the source gets depleted over time as chemicals are dispersed downwind. Since there are no active biological processes to replace the chemicals, it is possible that both the concentration and the mixture of compounds in finite sources change over time.

For example, consider a finite odor source that consists of two compounds, A and B. Furthermore, consider that the volatile nature of compound A is three times greater than compound B. In this simple case, the concentration of compound A will be depleted faster over time than the concentration of compound B. Thus, the mixture of the two compounds in the odor plume will also change as the source ages. As long as compounds remain in the original odor source, there will be an odor plume downwind, but the relative ratios of compound A and B will change over time. This is not the case for infinite odor sources.

Thus, the first axis of odor plumes can be those with finite odor sources or those with infinite odor sources.

2.2.2 Source Composition

The second axis of odor plumes revolves around the source composition. As noted above, humans are complex sources of odors consisting of VOC molecules, droplets or aerosols, and particles. For the purpose of this categorization, particles are large enough and gravity strong enough that particles like skin rafts are unlikely to play a significant role in the creation of odor plumes. This leaves two diverse types of odor plumes and, unlike the finite and infinite, a possible combination of both. With molecular plumes comprised primarily of VOCs, the dominant dispersal mechanics associated with this plume will be bulk flow and turbulence. Plumes composed of primarily aerosols will be dispersed by a combination of flow for smaller aerosols and droplets and gravity for larger ones. Finally, plumes consisting of both VOCs and aerosols will be dispersed downwind by both flow and gravitational forces.

2.2.3 A Plume System for Training Dogs

Finally, we can combine these two different plume axes to create a matrix of plume types. The importance of these plume types resides in the changes in chemical concentration and composition as the plume ages (either downwind or through time) and as the plume is dispersed.

Finite molecular plumes: These are plumes primarily composed of VOCs emanating from a non-living source. These plumes are generated by biological stains, inanimate objects like clothing and articles a person has dropped or thrown away. The concentration and relative composition of these plumes are based on the original sources size and the types of chemicals being volatilized. Small sources, such as skin rafts and blood spots will lose compounds rather quickly (minutes to hours) and the relative composition of the odor plume will also change given the small source concentrations and differences in rates of volatilization. Larger sources such as clothing will last longer given the large concentration of chemicals in the original source of the odors. Both sources of these plumes are also subject to photooxidation (breakdown of chemicals exposed to sunlight), oxidation, and other mechanics of decay that impact chemicals at the source. Still, as long as the source exists, there will be an odor plume moving downwind of the source.

Infinite molecular plumes: These plumes arise from either living sources or sources so large that they can be considered infinite compared to the rates of volatilization of chemicals in the source. Because the concentration of chemicals in the source is constantly being replaced through metabolic processes, the relative concentration of chemicals being liberated into odor plumes remains largely unchanged. Thus, the relative ratios of chemicals in the odor plume as it moves downwind or even as the age of the plume changes remain constant. Within searching contexts, these plumes will have a stable ratio of odors contained in the plume compared to finite molecular plumes. It is important to note that the concentration of source chemicals can change through time as the quality of human scent changes with diet, age, health, and even with stress. Although infinite molecular plumes move as a singular plume downwind, differences in the chemical properties of odorants may cause some molecules to get entrapped in the environment due to adsorption onto surfaces and structures in the environment or some odorants may have different rates of decay as they move downwind. These processes are relatively slow compared to the transport mechanics of turbulence and flow.

Infinite aerosol plumes: Droplets, particles, and aerosols given off by living organisms can be dispersed by flow mechanics but are subject to gravitational forces because of their size. As in the previous two plumes, the composition of chemicals in the source may change due to diet, health, age, or seasonality. Unlike the previous two plumes, these particles do not move together as they are dispersed downwind and by gravitational forces. These plumes consist of droplets and particles ranging widely in size and shape and as such, gravity will act differentially on the particles. The particles will be deposited in a size-sort manner with larger particles being closer to the source of the droplets and smaller particles being transported longer distances away before settling out from the plume. Because of these gravitational mechanics, the relative ratio and concentration of odorants will vary greatly as a function of distance (or age) from the odor source. In this manner, finite aerosol plumes are similar to finite molecular plumes in regard to the change in concentration and ratios within the plume.

Finite aerosol plumes: These plumes would consist of a set concentration of droplets being released from a non-living source. It is possible that plumes like this exist as in raindrops that form on leaves or a person walking through a forested area. In these situations, the odor sources are finite. As the rain ends, the formation of droplets on leaves or grasses also ends. For a person walking through soil settings, the aerosols liberated by the interaction between the footfall, water or dew on plants, and the crushed plant elements form miniature finite aerosol plumes. As the person moves on, the particular source also moves in space and time. Given the nature of settling for these particles, it is possible that finite aerosol plumes are important for very small droplets, but as far as dog searches over larger distances are concerned, these plumes are unimportant.

3 Humans as Scent Source

When looking for a person, dogs use the erratic scent plume from a person on the breeze. They "air scent": holding their heads up in the air, taking extremely long inhalations whilst breathing out through their mouths, they work upwind toward their scent source, a human being—human scent.

Although recent chemical analysis of human scent has focused on VOCs produced by people as a primary scent source (Curran et al. 2007), human scent had been characterized as a combination of volatiles and skin rafts since Syrotuck published his "Scent and the scenting dog" in 1972. Results obtained from the study of infectious diseases, focusing on microbial clouds and aerosols, have recently spiked, and can now be added to form a more complete picture. Let us look at these different groups more closely.

The skin is a major source of volatiles emitted by people. Stoddart described this in his book titled "The Scented Ape—the biology and culture of human odor" (1990). People have several types of glands that are unevenly distributed over the body: eccrine glands for thermoregulation all over the body, apocrine glands conveying sexually interesting information mainly in the armpit and groin regions, and sebaceous glands conveying individuality everywhere a person has hair (so not on the palms of the hand or soles of the feet). The main products of these glands (water, cholesterol, and sebum, respectively) are basically "odorless" but skin bacteria break them down into smaller, more volatile molecules that contribute to our odor signature. Recently, it has also been shown that higher ozone concentrations lead to increased VOC production (Gao et al. 2015).

Odor signatures are studied by several research groups in different countries, and with technology developing further, a clearer picture is developing. Different body parts emit different odor signatures, which fits in with the uneven distribution of the glands over our bodies: hands only have eccrine glands, whilst armpits have all three. Although odor signatures consist of the breakdown products of bacteria, people have quite stable skin bacteria populations (that are difficult to change even if you wanted to) and the resulting signature is very different between people (Schoon et al. 2009). Based on such profiles, people can be reliably differentiated. The odor signatures seem to be quite stable (Prada et al. 2014), but a recent study (Gokool 2022) showed that the longer apart (up to 35 days was tested) samples are collected, the more different they become. Studies have also shown that there are systematic differences in VOC composition between men and women and between ethnic groups (Colon-Crespo et al. 2017), and since gland productivity is also linked to age, age differences are also apparent in odor signatures (Haze et al. 2001). Airborne chemicals also differ with the emotional state of people (Williams et al. 2016), perhaps serving as a "danger" signal for other people.

In summary: the human skin can be categorized as an infinite source of VOCs and people have uniquely different odor signatures, although they slowly change in time and there are some common factors between groups of people. Since people usually have a temperature of around 37 °C, these volatiles are emitted in a steady stream.

Besides these skin volatiles, volatiles are also emitted by the lungs. Exhaled breath is a mixture of nitrogen, oxygen, carbon dioxide, water vapor, inert gasses, and hundreds of volatile compounds (de Lacy Costello et al. 2014). In a study on metabolite plumes emitted by trapped people (Huo et al. 2011), carbon dioxide, ammonia, and acetone were found to be reliable indicators of life. The spread of carbon dioxide was influenced by water in the debris. Huo found higher levels of ammonia than previously reported in breath, and the conclusion was that it was also emitted by the skin. The amount decreased during sleep. The concentration of acetone in breath varies widely: between people, but also within a person with diurnal and dietary effects (Španěl et al. 2011).

From an odor availability point of view, it is important to conclude that although a stable indicator, carbon dioxide is said to be odorless at low concentrations, and that the levels of ammonia and acetone were very variable, on top of a very variable VOC composition that differed between people and in time.

The skin VOCs and breath volatiles spread following the rules described in the theoretical section above. The source of the odors is infinite as long as the person remains alive and in place, and the released volatiles follow the rules of flow dynamics. This would imply that the odorant signature travels, as a whole, roughly downwind in a heterogeneous plume, gradually diluting as the plume widens. However, there are several factors that lead to changes in odorant signature with increasing distance:

- Local conditions may lead to VOCs and other volatiles being selectively adsorbed to the environment, which could lead to a change in the relative proportions of the components in the air, and thus change the odor signature with increasing distance.
- Differences in solubility between VOCs can have an effect in moist environments when some VOCs dissolve and others do not.
- Another factor is the degradation of VOCs through a reaction with ozone in the air. Different VOCs do this in different degrees (Carter 1994). This process is stimulated by UV radiation, so this will occur more on a sunny day than on a shady one and is also influenced by rain and humidity. This degradation also leads to differences both in quality and quantity of the odor signature with time/distance from the source.

Besides volatiles produced by bacterial breakdown of skin gland products and breath, skin rafts also spread VOCs. Syrotuck (1972) described rafts as cornflake-shaped flakes that were consumed by resident bacteria producing such VOCs. Eckenrode et al. (2022) described rafts (which they more correctly called "corneocytes") in much more detail, showing how VOCs contained within the keratin structure could be released at a later moment in time when the keratin structures degraded as a result of bacterial activity or environmental processes. Besides this release of VOCs that originated from the human body, new VOCs are thought to be produced through bacterial metabolism as a result of bacteria consuming the rafts.

The rafts themselves are 30–50 micron in diameter and 1 micron thick (Piérard et al., 2015), making them particles so they do not spread following the laws of flow dynamics. Aerodynamic qualities of their shape and gravity determine their spreading. In essence, they do not move very far before they settle, and they can settle in uneven concentrations, similar to leaves in the fall. Each raft in itself is a finite source of VOCs. Within its lifetime, the resulting odorant signature will vary, depending on local conditions for the raft and bacteria on it. At a higher temperature, the VOCs emitted directly from the raft will be depleted more quickly than at a lower temperature because of increased bacterial activity. As temperatures increase and conditions begin to dry, bacterial activity may become limited changing the odorant signature. In another scenario, a temporary drop in temperature will decrease both the volatility of the already present VOCs and decrease bacterial activity, but both may revive when the conditions improve (higher temperature and humidity), leading to a temporary lull in the production of odorants. Once the bacteria have consumed the raft, they no longer produce volatiles but go into a resting phase. Rafts are therefore finite sources of VOCs and since they do not travel far they are present close to the person (or deposited along a path they have travelled; this will be discussed later).

Another odor source people emit is aerosols. Humans emit aerosols directly through coughing and sneezing, but also by simply talking. Studies have shown that there are major interpersonal differences and that the volume of our speech impacts the number of aerosols we emit (Asadi et al. 2019). These tiny (1 micron) aerosols can carry viruses and bacteria and contain proteins and non-volatile metabolites. Smaller particles remain buoyant for a longer period of time and thus aerosols travel farther downwind than the larger rafts. Still, these aerosols are subject to gravity and degradation. They can be breathed in easily (which is how diseases get transmitted), and they may release odorants outside and inside the nose.

The total result of VOCs, skin rafts, and aerosols emitted from the human body and spreading has consequences for the availability of odorants for a searching dog. In essence, it follows that the amount of odorant information increases as the dog gets closer. At a great distance, there will be some VOC information, but this may be selectively diluted due to the adsorption of volatiles to the environment and may consist in part of degradation products of the original volatiles. As the dog nears an odor source, the VOC odorant profile will become less diluted, fresher, and more complete. Closer by, information from aerosols is added, and even closer the information from skin rafts is added as well, completing the full odorant signature of the person. A visual analogy would be that far away, you see something moving, closer by you see it is a happy person dancing, and when you get even closer you can see it is a middle-aged man wearing a raincoat and boots. Environmental conditions also impact the availability of the odorants. The visual counterpart of this would be wisps of mist clouding the view.

4 Articles as a Scent Source

In the course of locating people, dogs can come across articles that people have dropped. These articles can provide valuable information in search and rescue or tracking operations—the person came along this path, so the dog is on the right track. These articles have been handled by a person, so their scent is on them. What does this mean in terms of odor availability for the dog?

Whenever you touch something, you leave skin residue. This consists of water, lipids, amino acids, and inorganic salts, as separate chemical components but also as rafts. There is a limited amount of this residue left on the article, and therefore as an odor source, it is finite. This is in contrast to a human being who continually produces volatiles, skin rafts, and aerosols.

Besides being finite, the skin residue interacts with the article material. As a result of this interaction, some components are adsorbed more strongly than others, leading to differences in the odor signature in the vapor phase. For example, several studies have been done as part of optimizing human scent sampling and significant differences have been found between different textiles (Prada et al. 2014). Similar studies using blood residue have shown differences in odorant profiles through interaction with porous and non-porous materials (Rust et al. 2016).

Since the source is finite, the more volatile components will evaporate first, leaving the less volatile ones to disperse more slowly. Aside from this, microbial activity may serve to degrade the residue. This has a significant effect on the odor signature in time, as was shown by Filetti et al. (2019): whole classes of components disappeared before the articles were 10 days old, leaving only aromatic hydrocarbons, fatty acids, and amines being released from day 10–15.

Environmental circumstances, like temperature and humidity, have a significant effect on chemical alteration of an odor source. This effect can be direct, for example, higher temperatures and wind lead to faster evaporation, or indirect, for example, influencing moisture and temperature conditions for bacterial activity. The impact of raindrops may lead to aerosol formation, releasing volatiles into the air when it is light rain, but heavy rain may wash away residue, depleting the article as a source of scent.

The total result of this process is similar to that of human beings in that the closer the dog gets to the article, the more complete odorant information becomes available. However, the scale of this change is much more limited as a result of non-proportional availability of the odor signature due to interaction with the article material, differences in vapor pressure of the different odorants, and differences in degradation by microbial activity. This also implies that even at close range, the odor profile coming off from such an article will differ from that of the live human who touched it.

5 Track as a Scent source

When following the path a person has walked, people use the words "tracking" and "trailing" to describe what the dog is doing and what scents it is following. The precise definitions and interpretations of these terms vary.

Syrotuck (1972) described these terms based on the dog's observable behavior. A tracking dog works in a very characteristic head-down posture, indicating almost each of the subjects' footsteps, not varying more than one or two feet away from them. He defined a trailing dog as being able to work some distance from the track, overshooting some corners and cutting others, and assumed the dogs were more oriented to the rafts fallen beside the track. He also defined air-scenting dogs as working with a characteristic head held high, searching for scent in air currents, and completely ignoring ground deposits or airborne scent from the tracks.

Jeff Schettler's (2013) definition of tracking seems in line with this, but he links available odorants to his definitions. When tracking, the dog's nose is in the tracks made by a human on a soft surface. By this definition, tracking is impossible on the hard surface of urban environments since there is no ground disturbance there. His definition of trailing combines Syrotucks trailing and air scenting into one. He describes trailing as a dog following a particular human scent pattern wherever it might lie—on the ground or in the air—so "scent specific," following that one particular person based on their unique scent profile.

In this chapter, we are examining tracks from the viewpoint of the availability of odorants—depending on where these tracks were laid, environmental conditions at the time, and the effect of aging. Taking this approach, the model Syrotuck set up in a time when understanding of human scent was in its infancy, is still very valid. He described three main groups of odorants: "crushed plant vapor" as odorants coming from physically broken bits and pieces a person has stepped on (think of grass being mown); "vegetative scent" as odorants arising from the surface as a result of bacterial life in a changed biological environment striving toward a new equilibrium (footsteps releasing nutrients—changing water/air availability etc.); and "human scent" as residue the person leaves behind (skin rafts).

Combining this with current knowledge, the presence of crushed plant odorants is clearly apparent in areas where there is vegetation (dead or alive). However, physical abrasion of lichen on hard surfaces may also produce an odorant peak (García-Plazaola et al. 2017), as may the shifting of pebbles or seashells on a path. Some research has been done on the release of odorants from mowed grass (Harvey et al. 2014). Mown grass clippings emit a class of VOCs called Green Leaf Volatiles (GLV), and some of these react with ozone to produce aerosols. These GLVs changed over the 60 min Harvey studied them in response to different conditions. The short peak of crushed plant odorants Syrotuck described is therefore dependent on the degree of abrasion (i.e., weight and shoe profile of the tracklayer), it may be more widely available than only on grass/soft surfaces, it may last longer than Syrotuck thought depending on what was broken, and the signal may change with time depending on the availability of ozone and other environmental conditions.

Vegetative scent caused by ground disturbance leads to all kinds of changes. Syrotuck described this in terms of bacteria that multiply as a consequence of ground disturbance. Ground disturbance caused by a footstep causes the release of nutrients, changes the availability of oxygen, and may push water to new places. As a result, the footstep leads to changed circumstances for bacteria that can begin to multiply, and thus to a change in VOCs being released. However, the disturbance has more effects. For example on insects; when disturbed, a particular species of ground beetles (*Anchomenus dorsalis*) emit specific volatiles presumably as a defense mechanism or some other kind of chemical communication signal (Bonacci et al. 2011). Multiple ground organisms emit particular alcohols that are characteristic of an "earthy" smell, and Conrady et al. (2021) found a significant increase of these volatiles after soil disturbance. They measured for up to 60 min in their study without finding a systematic decrease of VOCs during this time. Their soil disturbance was pretty massive compared to a footstep (they disturbed 3.5 L of soil) but they conducted this study in the field and not in laboratory circumstances which makes this study very relevant for our topic. Fungi have been shown to emit VOCs that can be measured directly (Hung et al. 2015). These VOCs also mediate interactions between other organisms in their surroundings, which may in turn have an effect on odorants being released. In conclusion, Syrotuck's idea of vegetative scent caused by ground disturbance stands. It is however caused by a wider group of organisms than only bacteria, and since some of these organisms are not limited to living in soft soil, "vegetative scent" is a wider phenomenon. These odorants are released during a longer period of time until a new equilibrium is reached, but the available odorants will decrease and may also change during that time.

The third main group, the available human scent, is minute compared to what is present close to a person or on an article the person has had in his possession given the short contact time with the soil and the brief presence of a walking person at a particular spot. Syrotuck described this only in terms of rafts that served as food for the resident skin bacteria. However, we now understand that rafts also carry volatiles within them that may be released when the raft physically degrades, we understand that environmental bacteria also feed on rafts, we know that VOCs directly emitted from people may adsorb into the surroundings and linger instead of being blown away and that rafts and aerosols will be deposited more closely to the track.

Little is known about the direct disposal of odorant residue through the soles of shoes aside from Neuhaus (1953) who experimented with butyric acid (one of the smaller human VOCs) and calculated that this should permeate through shoes. Hepper and Wells (2005) found that sealing off shoes prevented dogs from picking up directional cues in a track on carpet squares, leading them to conclude that dogs used odorants that had leaked through the shoes of the person who walked the track, but to our knowledge, this has not been validated by chemical analysis. Of course, if a person stands somewhere for a longer period of time, a "scent pool" will develop in the vicinity of the location. More volatiles seep through their shoes, volatiles a person releases directly may get adsorbed to the surroundings, and the larger components such as rafts and aerosols will obey the laws of gravity and be deposited on the ground. When the person continues on his way, VOCs may be desorbed, and rafts/aerosols resuspended depending on environmental circumstances such as wind and physical stimulation. And similarly to what happens with articles containing human scent, the most volatile components will disappear first since this track is a "finite source," so the odorant profile being released in time changes.

When walking, all three groups of components contribute to the scent track but depending on what one is wearing and where one is walking, the relative contribution to the total picture will differ. At one extreme, walking on vegetation on soft soil in a pleasantly warm, moist environment will enhance the availability of crushed plant odorants in the short term and the vegetative odorants caused by the ground disturbance in the longer term, overpowering the available human scent.

At the other extreme, walking on a dry hard surface will lead to relatively more human odorants being available but in hot conditions, these may be depleted relatively quickly. Intermediate conditions will produce other mixtures. And walking across multiple surface types, or even moving from a shady area to a sunny area on the same type of surface, can lead to changes in the available odorants, both in terms of amount and quality. Even along a track on a single surface, the total scent picture may vary since volatiles, aerosols, and particles may "heap" together in pockets as a result of local conditions. Think of leaves heaping up in particular spots in the fall as a result of very local wind turbulence. In consequence, the signal along the track may vary from "huge" to "absent" and may vary in quality depending on the local surface, available microorganisms, speed the person walked, type of shoes they wore, and many other factors described earlier in the topic "Humans as scent source."

Similarly to the articles as a scent source described above, environmental factors play a key role in how the scent picture on a track develops in time. Wind may stimulate evaporation by blowing away volatiles changing the local source concentrations. Temperature will have a direct effect by increasing the volatility and an indirect one by influencing the metabolism of organisms. Moisture is necessary for microorganisms to interact with the environment and the residue, and oxygen (or lack thereof) also affects organic life. Light rain impacting the soil generates aerosols with the characteristic smell associated with light rain (Joung and Buie 2015), but heavy rain washes out residuals.

Changing environmental factors after the track has been laid may lead to a temporary dip in available odorants. For example, temporary lower temperatures (at night), or a temporary lack of moisture (extremely hot midday) may temporarily decrease organic life activity and therefore decrease the production of volatiles. But when circumstances change—a higher temperature when the sun comes out after a cold night; an increase in moisture when the dew sets in after a dry day—the system comes alive again and more volatiles are produced.

Whatever the track consists of, it is a finite source. After some time, the crushed plants will no longer emit volatiles or aerosols. The vegetative scent that is the result of ground disturbance will cease when a new equilibrium has been reached. The human VOCs will have dispersed or broken down, and the aerosols and rafts will have been degraded or consumed.

The total result of these processes is that when tracking, the dog is confronted with a continually changing scent picture that it has to follow. In an otherwise stable, undisturbed, and uncontaminated background, this is relatively easy since there is nothing else and dogs are very good at "novelty detection," or finding things that stick out. But in more disturbed and contaminated environments the dog has to follow a particular shade of grey, whilst this shade of grey may become a dotted line instead of a solid one and may change hue along the way. There is a lot of information available on tracks and we cannot say what components the dog is paying attention to just by watching—careful testing is needed to figure that out. The debate on how long a dog can follow a track is never-ending.

6 Practical Conclusions

Live humans can be considered infinite odor sources, and their odors can be considered a complex source comprised of three different major sources of odors. As outlined above, volatiles are likely the largest component. Volatiles will remain a relatively stable (although particular environmental circumstances cause degradation) and continuous source of odorants that are transported downwind by the dynamics of air flow in a gradually widening plume with patches of higher concentrations. Given that the molecules are small, their movement is unaffected by gravitational forces. Aerosols are tiny particles given off by living sources from specific areas on the body, e.g., mouth, sweat glands. It is possible that aerosols may also contain volatile organic compounds as potential sources of odorants. Aerosols will be spread both by the flow dynamics of air movement as well as settling dictated by the interaction between the size of the aerosols and gravitational forces. Finally, skin rafts are larger particles of shed skin that contain VOCs as well as the bacterial fauna present on the skin. These sources are subject to strong gravitational forces and will be deposited relatively close to the source compared to VOCs. These particles are dispersed mainly by the movement of the source and not by air flow dynamics. Skin rafts are also subject to decomposition which will liberate additional odorants after having been shed.

A person standing still will, in principle, produce these three odor sources infinitely, VOCs spreading far, aerosols traveling less far and rafts falling close by. An object a person drops is a finite odor source that will mostly produce VOCs. And besides being finite, the relative concentration of different volatiles will change as the source ages. This will happen because of differences in volatility, but also due to degradation of the human skin residue left on the object. The VOCs from this finite odor source will spread by flow dynamics but will gradually fade away as the source depletes.

A person walking leaves behind a finite odor source after they have left. The VOCs will travel away in a moving plume; the aerosols deposited slightly downwind are finite sources in themselves and will deplete, and the rafts settling closer to the path are also a finite source of human scent that will deplete. Aerosols and rafts are likely to be distributed patchily along the path. Added to this are the VOCs that are produced by the abrasion of plants and other organic material by the person stepping on them, VOCs that are produced by microorganisms in response to this disturbance, and rainfall may release these odorants in the air by generating aerosols.

Taken together, the complexity of a human odor source will increase as a dog moves from a distant location to the source, revealing a more complete scent profile the closer the dog gets. Far downwind from a stationary source, the human odor plume will be composed of only VOCs. As a dog approaches the odor source, aerosols along with VOCs will be contained within the mixture of odorants. Finally, close to the odor source, the odor plume will become complete: consisting of VOCs, aerosols, and skin rafts.

How much information does a dog need to be able to identify a person? Dogs are sometimes trained to be "scent specific," meaning they have to use the characteristic individual scent of the person to follow their track, point out objects these people have handled, and identify this person standing somewhere. Usually, this is done by giving the dog a "smeller" as a starting scent, and the dog is expected to match the scent in this example to whatever we want him to point out: the track, another article, or the person himself; a typical "match to sample" paradigm. But let us look at this problem from the point of scent availability.

Matching the scent of a "smeller" to a live human standing close by seems the simplest, especially if the smeller is made of inert material and has been in the possession of the person for long enough to have collected enough skin residue. An important variable is the age of the smeller—the "fresher" it is, the closer it will resemble the person standing there, thus making it easier.

Matching the scent of a "smeller" to that of the same person on another article is the next step in complexity. Important variables are the degree in which the materials of "smeller" and "article" differ (the more similar the easier), and difference in age between the "smeller" and the article (the closer in time the two were scented, the easier).

The third step in complexity is matching a smeller, or the scent of a person on any article, or a person directly, to a track, since the track combines human scent with other odorants. Here it follows that tracks with less additional odorants should be easier (meaning typical "hard surface tracks"), and the closer in age the two are the easier it is as well.

Finding a human track is easier in areas with little disturbance than in busy areas. Locating a specific human track is much more difficult. Human scent along a slightly aged track is, at best, patchy. It resembles a grainy picture of the person who walked along the path, possibly obscured by other odorants generated by disturbances and abrasions in the environment. And like us, dogs find it easier to recognize familiar people in such a grainy picture than unknown people. But the true measure of what a dog is capable of may not be finding his "friend." It may be finding and identifying a complete stranger. Which is not easy based on a very grainy, black-and-white, faded picture. Be aware of this! Determining what dogs actually learn to do requires careful testing (Schoon 2022).

This also holds for identifying articles left along a path as having been dropped there by the person whose track the dog is following. Dogs are great novelty detectors and will pay attention to anything "new" in the environment; things that have not blended into the general background odor. Dogs are not usually explicitly trained to be scent specific and ignore items left by someone else during tracking exercises—people just expect them to be able to do that. They probably could, but did they learn to do that? Confirmation can only be had by (correctly) evaluating the individual dog.

Looking at training exercises from the point of view of what odorants are available for a dog can help focus the dogs on what we want them to focus on. By using unfamiliar people, we are focusing the dog to get as much odorant information as possible. By training in varied locations and varying environmental conditions, we are focusing the dog on how varied in time and space human scent profiles and plumes can be. By training on hard surfaces, we are focusing the dog on the human scent in the track but may have to live with this only being possible for a limited track age. By training on vegetated soft soils, we are focusing the dog on vegetative odorants. This may not be very human scent specific, but especially in remote areas where people tend to get lost the dog may be capable of doing very aged tracks. By using live people at the end of a track, we are stimulating the use of air scenting. There is no right or wrong here, it just depends on what you want to teach the dog. So think, and then do.

References

Asadi S, Wexler AS, Cappa CD, Barreda S, Bouvier NM, Ristenpart WD (2019) Aerosol emission and super emission during human speech increase with voice loudness. Sci Rep 9(1):1–10

Atkinson R (2000) Atmospheric chemistry of VOCs and NOx. Atmos Environ 34(12–14):2063–2101

Bonacci T, Brandmayr P, Zetto T, Perrotta ID, Guarino S, Peri E, Colazza S (2011) Volatile compounds released by disturbed and undisturbed adults of Anchomenus dorsalis (Coleoptera, Carabidae, Platynini) and structure of the pygidial gland. ZooKeys 81:13

Carter WP (1994) Development of ozone reactivity scales for volatile organic compounds. Air Waste 44(7):881–899

Colón-Crespo LJ, Herrera-Hernández D, Holness H, Furton KG (2017) Determination of VOC marker combinations for the classification of individuals by gender and race/ethnicity. Forensic Sci Int 270:193–199

Concha Arcil F (2009) Settling velocities of particulate systems. Kona Powder Part J 27:18–37

Conchou L, Lucas P, Meslin C, Proffit M, Staudt M, Renou M (2019) Insect odorscapes: from plant volatiles to natural olfactory scenes. Front Physiol 972

Conrady MW, Bauer M, Jo KD, Cropek DM, Busby RR (2021) Solid-phase microextraction (SPME) for determination of geosmin and 2-methylisoborneol in volatile emissions from soil disturbance. Chemosphere 284:131333

Curran AM, Ramirez CF, Schoon AA, Furton KG (2007) The frequency of occurrence and discriminatory power of compounds found in human scent across a population determined by SPME-GC/MS. J Chromatogr B 846(1–2):86–97

Eckenrode B, Riley P, Dailey A, Couch R (2022) Towards the development of a human scent model. In: Canines, pp 275–315. Jenny Stanford Publishing

Elkinton JS, Cardé RT (1984) Odor dispersion. In: Chemical ecology of insects, pp. 73–91. Springer, Boston, MA

Filetti V, Di Mizio G, Rendine M, Fortarezza P, Ricci P, Pomara C, Messina G, Riezzo I, Zammit C, Messina A, Salerno M (2019) Volatile organic compounds: instrumental and canine detections link an individual to the crime scene. Egypt J Forensic Sci 9(1):1–11

Gao K, Xie J, Yang X (2015) Estimation of the contribution of human skin and ozone reaction to volatile organic compounds (VOC) concentration in aircraft cabins. Build Environ 94:12–20

García-Plazaola JI, Portillo-Estrada M, Fernández-Marín B, Kännaste A, Niinemets Ü (2017) Emissions of carotenoid cleavage products upon heat shock and mechanical wounding from a foliose lichen. Environ Exp Bot 133:87–97

Gokool VA, Holness HK, Furton KG (2022) The influence of intra-personal variations in human hand odor on the determination of sample donor. Forensic Sci Int 334:111235

Gu AY, Zhu Y, Li J, Hoffmann MR (2022) Speech-generated aerosol settling times and viral viability can improve COVID-19 transmission prediction. Environ Sci Atmos 2(1):34–45

Harvey RM, Zahardis J, Petrucci GA (2014) Establishing the contribution of lawn mowing to atmospheric aerosol levels in American suburbs. Atmos Chem Phys 14(2):797–812

Haze S, Gozu Y, Nakamura S, Kohno Y, Sawano K, Ohta H, Yamazaki K (2001) 2-Nonenal newly found in human body odor tends to increase with aging. J Investig Dermatol 116(4):520–524

Hepper PG, Wells DL (2005) How many footsteps do dogs need to determine the direction of an odour trail? Chem Senses 30(4):291–298

Hung R, Lee S, Bennett JW (2015) Fungal volatile organic compounds and their role in ecosystems. Appl Microbiol Biotechnol 99(8):3395–3405

Huo R, Agapiou A, Bocos-Bintintan V, Brown LJ, Burns C, Creaser CS, Devenport NA, Gao-Lau B, Guallar-Hoyas C, Hildebrand L, Malkar A (2011) The trapped human experiment. J Breath Res 5(4):046006

Jackson JL, Webster DR, Rahman S, Weissburg MJ (2007) Bed roughness effects on boundary-layer turbulence and consequences for odor-tracking behavior of blue crabs (Callinectes sapidus). Limnol Oceanogr 52(5):1883–1897

Joung YS, Buie CR (2015) Aerosol generation by raindrop impact on soil. Nat Commun 6(1):1–9

Kim S, Guenther A, Karl T, Greenberg J (2011) Contributions of primary and secondary biogenic VOC tototal OH reactivity during the CABINEX (Community Atmosphere-Biosphere INteractions Experiments)-09 field campaign. Atmos Chem Phys 11(16):8613–8623

Kowadlo G, Russell RA (2006) Using naïve physics for odor localization in a cluttered indoor environment. Auton Robot 20(3):215–230

de Lacy Costello B, Amann A, Al-Kateb H, Flynn C, Filipiak W, Khalid T, Ratcliffe NM et al (2014) A review of the volatiles from the healthy human body. J Breath Res 8(1):014001

Moore P, Crimaldi J (2004) Odor landscapes and animal behavior: tracking odor plumes in different physical worlds. J Mar Syst 49(1–4):55–64

Moore PA (2016) Aerodynamics of odor plumes and plume structures in different habitats. In: Jezierski T, Ensminger J, Papet LE (eds) Canine olfaction science and law: Advances in Forensics, medicine, conservation, and environmental remediation. Taylor and Francis

Neuhaus W (1953) Über die riechschärfe des hundes für fettsäuren. Z Vgl Physiol 35(6):527–552

Paczkowski S, Schütz S (2011) Post-mortem volatiles of vertebrate tissue. Appl Microbiol Biotechnol 91(4):917–935

Pannunzi M, Nowotny T (2019) Odor stimuli: not just chemical identity. Front Physiol 10:1428

Piérard GE, Courtois J, Ritacco C, Humbert P, Fanian F, Piérard-Franchimont C (2015) From observational to analytical morphology of the stratum corneum: progress avoiding hazardous animal and human testings. Clin Cosmet Invest Dermatol 113–125

Prada PA, Curran AM, Furton KG (2014) Human scent evidence. CRC Press

Rust L, Nizio KD, Forbes SL (2016) The influence of ageing and surface type on the odour profile of blood-detection dog training aids. Anal Bioanal Chem 408(23):6349–6360

Schettler J (2013) Trailing versus tracking: The keys to success. https://www.police1.com/police-products/k9/k9-training/articles/trailing-versus-tracking-the-keys-to-success-bjtllYuJ6GIHxvAS/

Schlichting H, Kestin J (1961) Boundary layer theory, vol 121. McGraw-Hill, New York

Schoon GAA (2022) The use of tracking/man trailing dog results as evidence in courts. J Vet Behav 52–53:14–20

Schoon GAA, Curran AM, Furton KG (2009) Odor and biometrics. In: Li SZ, Jain AK (eds) Encyclopedia on biometric recognition. Springer-Verlag

Španěl P, Dryahina K, Rejšková A, Chippendale TW, Smith D (2011) Breath acetone concentration; biological variability and the influence of diet. Physiol Meas 32(8):N23

Stoddart DM (1990) The scented ape: the biology and culture of human odour. Cambridge University Press

Syrotuck WG (1972) Scent and the scenting dog. Arner publications, New York

Tedeschi G, Gouin H, Elena M (1999) Motion of tracer particles in supersonic flows. Exp Fluids 26(4):288–296

Tobias HJ, Docherty KS, Beving DE, Ziemann PJ (2000) Effect of relative humidity on the chemical composition of secondary organic aerosol formed from reactions of 1-tetradecene and O3. Environ Sci Technol 34(11):2116–2125

Vickers NJ (2000) Mechanisms of animal navigation in odor plumes. Biol Bull 198(2):203–212

Weissburg MJ, Dusenbery DB, Ishida H, Janata J, Keller T, Roberts PJW, Webster DR (2002) A multidisciplinary study of spatial and temporal scales containing information in turbulent chemical plume tracking. Environ Fluid Mech 2(1):65–94

Williams J, Stönner C, Wicker J, Krauter N, Derstroff B, Bourtsoukidis E, Klüpfel T, Kramer S (2016) Cinema audiences reproducibly vary the chemical composition of air during films, by broadcasting scene specific emissions on breath. Sci Rep 6(1):1–10

Young BD, Escalon JA, Mathew D (2020) Odors: from chemical structures to gaseous plumes. Neurosci Biobehav Rev 111:19–29

Olfaction and Behavior in Dogs: Applications to Cognition, Training, and Welfare

Dog Olfactory Cognition

Adam Davila, Jordan G. Smith, Emma Cox, Lane Montgomery, Sarah Krichbaum, Lucia Lazarowski, and Jeffrey S. Katz

Abstract

Davila and colleagues cover methods and review findings regarding dog olfactory cognition. The chapter begins by introducing the topic of dog cognition. Specific aspects of dog olfactory cognition are explored. The authors review methods and procedures that investigate different aspects of olfaction and memory, including episodic memory, working memory, and long-term memory. Next, olfactory category and concept learning in dogs is reviewed followed by a discussion of the nature of olfactory representations in dogs, as well as studies that involve an olfactory search image. Finally, the last section reviews olfactory quantity judgments. The authors conclude by emphasizing the importance of dog cognition and suggesting a converging operation approach for future research.

Keywords

Dogs • Olfaction • Memory • Concepts • Cognition

The field of dog cognition seeks to understand the cognitive processes of dogs. This includes aspects of how cognitive processes evolved as well as the function of these processes, from sensation, perception, and discrimination to dogs' ability

A. Davila · J. G. Smith · E. Cox · L. Montgomery · J. S. Katz (✉)
Department of Psychological Sciences, Auburn University, Auburn, AL, USA
e-mail: katzjef@auburn.edu

A. Davila
Animal Studies, Eckerd College, St. Petersburg, FL, USA

J. G. Smith · S. Krichbaum · L. Lazarowski
Canine Performance Sciences, Auburn University College of Veterinary Medicine, Auburn, AL, USA

L. Lazarowski (ed.), *Olfactory Research in Dogs*,
https://doi.org/10.1007/978-3-031-39370-9_8

to learn complex procedures utilizing multiple memory systems. Dogs may not be unique in their cognitive abilities (Lea & Osthaus, 2018), yet when considering their roles in human society, dogs enjoy a privileged and important position. Dogs serve as companions, models of human cognition and aging, and emotional support animals. It is their highly developed sense of smell combined with their ability to work well with human handlers that have given rise to their use in roles that require them to detect a range of olfactory and chemical stimuli, as well as attend to human needs.

Utilizing their primary sense, olfaction, dogs have proved reliable in a variety of scent detection roles, including detecting pathologies such as SARS-CoV-2 (Lippi et al., 2021) or prostate cancer in urine samples (Taverna et al., 2015), conservation work (Beebe et al., 2016), detection and location of human remains (Riezzo et al., 2014), and explosives detection (Furton & Myers, 2001). Relative to humans, dogs have a larger repertoire of genes that encode olfactory receptors (Quignon et al., 2003). The dog's *umwelt*, their unique perception of the world (von Uexküll, 1957) depends heavily on their sense of olfaction. Dogs use smell to identify other dogs, other animals, and humans. There is some evidence that dogs may use their sense of smell as a self-recognition cue (Horowitz, 2017), and evidence of olfactory dominance over other senses. In one study (Gazit & Terkel, 2003), dogs made use of olfactory cues over visual cues when searching for explosives. Dogs and other macrosmatics (i.e., keen-smelling species) have hundreds more active olfactory receptor (OR) genes than humans. It is hypothesized that olfactory information is encoded by specific receptors that respond to the chemical nature of each odorant. The "one neuron-one receptor" rule suggests that each olfactory gene is responsible for the expression of one olfactory receptor neuron (Bystrova & Kolesnikov, 2021; Mombaerts, 2004), which in turn is receptive to a variety of odorants.

Just as it can be difficult to imagine life as a pentachromat (having five color receptors, like pigeons), it is difficult to imagine a dog's olfactory experience, both in terms of the intensity of olfactory inputs and the range of smells. Therefore, it is important to understand the cognitive processes of dogs as they pertain to olfaction and to not treat dogs as merely better scent detectors. The best way to do this is through careful behavioral experimentation that allows researchers to examine cognitive capacities of dogs as they pertain to olfaction. Understanding the processes underlying canine olfaction also has important applied implications. First, because dogs are heavily relied upon in applied settings such as detection and service animals, training and performance expectations can be updated to ensure efficient training practices and humane treatment of all working dogs (Cobb et al., 2015). Second, dogs have also been identified as potential models for progressive human diseases such as Alzheimer's disease (AD) and dementia. Both AD and dementia and their precursor, mild cognitive impairment, are associated with declines in olfactory ability (Doty & Kamath, 2014; Jung et al., 2019; Windon et al., 2020) and cognitive functioning (Murman, 2015). Cognitive dysfunction syndrome is a similar progressive aging disease in dogs (as well as cats; Landsberg et al., 2012) that may serve as a model for human neurodegenerative diseases (Chapagain et al.,

2018). Understanding the extent to which dogs can serve as a valuable translational model of age-related decline in humans depends on fully understanding dog cognition, as does maximizing their effect as service animals. This chapter will review recent methodological advances in the study of dog cognition as it pertains to olfaction. Specifically, we focus on the areas of memory, concept learning and categorization, odor representation, and quantity discrimination.

1 Odor Memory

Memory research in dogs has provided valuable translational benefits for the study of aging and neurogenerative disease. As interest in canine cognition has grown, the study of dog memory in its own right has expanded to examine factors that influence dogs' memory processes (Foraita et al., 2021), as well as how individual differences relate to dogs' ability to learn and perform various tasks (MacLean et al., 2017).

To date, the majority of studies have used visual or visuospatial tasks to assess dog memory (e.g., Fiset et al., 2003; Kaminski et al., 2008; Milgram et al., 1994). However, given the importance of olfaction in dogs' behavior and their highly developed sense of smell (Hayes et al., 2018), as well as its relevance to detection roles that dogs are commonly utilized for, examining dog memory through olfaction may provide valuable insights. Researchers have examined various memory systems to address basic and comparative questions about olfactory memory in dogs as well as applied research regarding the extents and limits of detection dogs' memory for trained odors. As declines in olfactory functioning and cognition are associated with aging in humans, assessing odor memory in aging dogs could have translational benefits for the study of neurogenerative diseases in humans. In this section we will review various types of memory in dogs that have been examined using the olfactory sense, highlighting differences in comparison to non-olfactory modalities.

1.1 Working Memory

Working memory (WM) is a cognitive system used to hold and process a limited and temporary amount of information and is critical for executing cognitive functions, including learning and problem-solving. In non-human animals, WM is defined as short-term memory of a stimulus, object, or location within a single experimental session (Dudchenko, 2004), and is measured in terms of duration and capacity (for a theoretical review of WM in humans, see Cowan, 2017). Working memory duration refers to the amount of time that a previously encountered stimulus is remembered in a single session. The delayed matching-to-sample task (dMTS) is commonly used to assess working memory duration in animals (Lind et al., 2015). In this task, a subject is presented with a sample stimulus (e.g., red circle), after which it is removed for a delay period. After the delay, the subject is presented with two comparison stimuli, one that matches the sample (red circle) and another that does not (e.g., blue square); a response to the matching comparison is rewarded. By varying the delay interval between the sample presentation and the comparison stimuli, WM duration can be probed. A subject's WM duration is defined as the longest delay at which the subject maintains above-chance performance in accurately choosing the matching comparison (Lind et al., 2015).

Studies involving the dMTS task with dogs have demonstrated differences in WM durations for auditory (60 s; Kuœmierek & Kowalska, 2002) and visuospatial (110 s; Chan et al., 2002) stimuli. Recently, Krichbaum et al. (2021) developed the first olfactory dMTS for dogs to assess WM duration using a potentially more dominant stimulus modality for the species. In this task, dogs were presented with a sample odor followed by a choice between two comparison odors, one matching the sample odor (S+) and one non-matching odor (S−); dogs were rewarded for choosing the S+. Once each dog learned to perform the MTS task with no delay and trial-unique stimuli (i.e., each of 48 odors only appeared in one trial per session), WM duration was probed by inserting variable delays (0-, 30-, 60-, and 90-s) between the sample and comparisons. Dogs demonstrated a typical memory function with performance decreasing across delays and indicated a WM duration between 60 and 90 s. To further challenge the limits of WM duration, proactive interference was manipulated in subsequent tests by repeating stimuli throughout the session. Proactive interference occurs when memory for earlier events in time influences memory for later ones (Wright et al., 2012), which is minimized in experimental settings when stimuli are trial unique (i.e., they appear once per session) and increases when they are reused (i.e., there is repetition within a session). When the number of odors used in each dMTS session was reduced to sets of 2 and 6 (stimuli repeated on each trial with a 2-odor set, and 8 times per session in the 6-odor set, thus increasing proactive interference to different degrees), similar delay functions were observed across all odor sets, and accuracy on the 48- and 6-odor sets was comparable. However, dogs only performed significantly above chance on 0-s delays for the 2-odor set, suggesting that the combined effect of delay and proactive interference diminished dogs' olfactory working memory duration (Krichbaum et al., 2021). Figure 1 illustrates these forgetting functions. Overall, dogs appear to demonstrate a similar WM duration for olfactory and auditory stimuli, while studies using visuospatial stimuli report longer durations (however, see Krichbaum et al., 2021 for an explanation of potential methodological factors contributing to this difference).

Another aspect of WM is its *capacity*, or the number of items that can be remembered in a single session. Krichbaum et al. (2020) adapted the odor span task (OST), used in rats (Dudchenko et al., 2000) and humans (Levy et al., 2003), to evaluate dog olfactory WM capacity. The OST is an incrementing non-matching-to-sample procedure in which on every trial the animal has a choice between a novel odor (S+) and an odor presented on a previous trial (S−), thus requiring memory for all previously encountered odors in the session. WM capacity is reflected by the accuracy in correctly identifying the novel odor as the number of odors increases across the session. Krichbaum et al. (2020) initially trained dogs

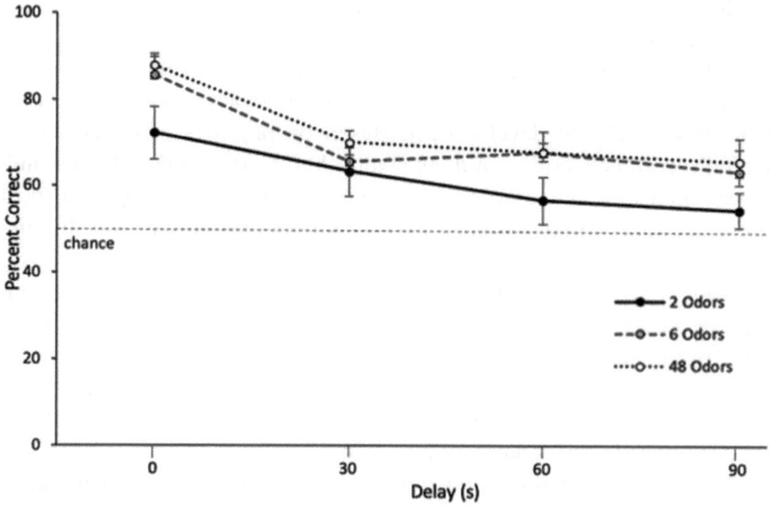

Fig. 1 From Krichbaum et al. (2021) showing the forgetting functions for the 2, 6, and 48-odor sets. As delay increases, dogs perform worse. This decrease is strongest when PI is at its highest (i.e., 2-odor set)

in sessions consisting of 24 trials (i.e., 24 odors to remember). Working memory capacity was then assessed in sessions where the number of odors in the session was expanded to 36, 48, and 72 trials, increasing the length of the span of odors that dogs needed to remember. There was no difference in overall session accuracy across the 36, 48, and 72 trial testing conditions, and dogs maintained above-chance accuracy on all set sizes. The study demonstrated that dogs performed similarly on the OST to rats and were able to maintain high accuracy for up to 72 odors (April et al., 2013). Krichbaum et al. (2020) conducted an additional analysis to determine if accuracy of the OST in dogs was affected by the number of intervening trials (i.e., the number of trials since the S– was last encountered, an analysis similar to the n-back task used to measure WM capacity in humans; cf., Kirchner, 1958). Krichbaum et al. (2020) found that accuracy decreased as the number of intervening odors increased, and above-chance performance was maintained up to 7–8 trials since last encountered. The results of this study indicate that the OST may be a valuable method of measuring WM capacity in dogs.

1.2 Long-Term Memory

Long-term memory refers to memory for prior events that is maintained over extended periods of time (e.g., Cowan, 2008). In dogs, the salience of olfactory stimuli may lead to the long-term retention of species-relevant odors that can ultimately influence interactions later in a dog's life. Understanding the extent and limits of long-term odor memory in dogs also has important implications for the training of detection dogs.

Dogs' olfactory system develops within the first few weeks post-birth (Lord, 2013). Therefore, memories regarding species-relevant odors may be formed within the first weeks of development. Based on previous evidence demonstrating prenatal olfactory learning in dogs (Wells & Hepper, 2006), Hepper and Wells (2006) evaluated how prenatal, postnatal, and perinatal (both pre- and postnatal) exposure to an odor would influence long-term retention of that odor. In this study, puppies were exposed to aniseed through the mother's diet, either through the amniotic fluid during prenatal development, the mother's milk during postnatal development, or a combination of exposure through the amniotic fluid and breast-milk (i.e., perinatally). Puppies in the prenatal group were exposed to aniseed during the last 20 days of gestation while puppies in the postnatal group were exposed to aniseed during the first 20 days after birth. The perinatal exposure group experienced both prenatal and postnatal exposure to aniseed. At 10 weeks of age, all puppies were presented with food flavored with and without aniseed to assess their preference. Overall, puppies with perinatal odor exposure demonstrated a stronger preference for aniseed relative to the other groups for at least 5 weeks after the last exposure, suggesting that both pre- and postnatal experiences may influence preferences for extended periods of time and result in long-term changes in behavior.

Another study demonstrated long-term memory for odors experienced during early development by evaluating kinship recognition in dogs (Hepper, 1994). At 4–5 weeks of age, puppies spent more time investigating cloths with the odor of their siblings and their mothers relative to unfamiliar dogs, illustrating a greater preference for the scent of both their siblings and their mother. Mothers also demonstrated a preference for their offspring based on olfactory cues alone. When retested after 2 years of separation, the odor preferences between the mothers and the offspring remained, although sibling recognition appeared to be modulated by social experiences after infancy. The extent to which kin recognition through olfaction is based on long-term odor memory due to experience and associations with individuals or whether it is influenced by the detection of genetic cues of relatedness is unclear. However, research overall suggests that exposure to odors during early developmental periods can result in long-term memory for those odors.

Long-term memory can also be assessed by measuring responses to repeated stimuli (e.g., habituation or sensitization; Squire & Zola, 1996). This type of test is based on the spontaneous tendency to explore items less as they become familiar across repeated exposures and to show more interest in exploring novel stimuli. Thus, decreased exploration of a familiar stimulus indicates recognition of that stimulus. In dogs, one study evaluated odor recognition by presenting familiar and novel odors to dogs and measuring the time spent investigating each odor (Salvin et al., 2012). The results showed that dogs habituated to a novel odor (male conspecific urine sample) after two presentations (i.e., investigation time significantly decreased between the first and second presentation of the odor). The authors also observed a trend indicating that older dogs demonstrated less habituation relative to young and middle-aged dogs, possibly demonstrating age-related memory decline. However, when Salvin et al. (2012) presented dogs with a novel odor to measure dishabituation effects, only half of the dogs showed a novelty response to the unfamiliar odor. These results are potentially due to the sample of dogs tested being detection dogs, which are often discouraged from investigating urine in their environments. Therefore, this study presents evidence that memory for odors may be assessed in dogs through habituation to previously presented odors, but more research with different populations of dogs is needed to demonstrate this effect more clearly.

Other studies measuring long-term olfactory memory in dogs have examined the potential limits of long-term memory for trained odors. Studies have used a variety of retention intervals to measure long-term olfactory memory for previously trained odor discriminations, reporting high performance on memory tests after periods of 6 weeks (Wright et al., 2017), 69 days (Lubow et al., 1973), 4 months (Johnston, 1999), 233 days (Lo et al., 2020), and 1 year (Lazarowski et al., 2021a, b; Waggoner et al., 2022). Results from these studies have also found that long-term memory for odors is not influenced by the number of intervening odor discriminations learned as additional discriminations are trained, with dogs demonstrating equivalent performance for discriminations learned early and later in training (Lo et al., 2020; Williams & Johnston, 2002; Waggoner et al., 2022). Thus, dogs' long-term odor memory does not seem to be susceptible to retroactive interference in these cases (i.e., the memory of an event affected by memories of more recent items or events). Comparative tests between species have also shown that dogs outperformed both rats and humans on the same odor memory test (Lo et al., 2020).

Such research also has practical applications to the training of detection dogs. Detection canine training is often challenged by constraints on time and access to training locations and materials. Therefore, how long a detection dog can remember odors is important for informing training practices and allocating resources towards maintenance training. Lazarowski et al. (2021a, b) assessed whether detection dogs could accurately locate a set of explosive odors after a period of 1 year without exposure to those odors. More importantly, the study examined the effectiveness of minimal maintenance training that utilized a non-hazardous odor that could safely and easily be used in a variety of training settings. In this study, two groups of detection dogs were trained to detect 10 different target odors followed by a 12-month period in which one group received once-a-month maintenance training with only one of the odors from the initially trained set, specifically a non-hazardous (i.e., non-explosive) odor. The other group did not participate in any odor-detection activity during the maintenance period. All dogs were then retested on the previously trained target odors not experienced in 12 months. Dogs that received no odor training across the 12 months recognized the odors with moderate accuracy (85%), whereas dogs that received maintenance training with the single odor demonstrated nearly perfect recognition for the other 9 target odors not experienced in 12 months. The results of this study indicate that dogs are able to remember odors fairly well after extended periods and that minimal maintenance training with a single odor can sustain long-term memory at high levels for other odors not used in maintenance training. Importantly, the single maintenance training odor was a non-hazardous odor unrelated to the training explosives, suggesting that safe "surrogate" odors can be used for the maintenance training of detection dogs when access to hazardous materials may be limited.

Recently, our laboratory assessed long-term retention for previously learned *rules* rather than specific stimuli. Using a convenience sample of dogs previously trained for dMTS and OST procedures described above, we re-tested 9 of the original dogs in the dMTS ($n = 3$) and OST ($n = 6$) studies after an average of two years (1–3 years) since their last session. Dogs had engaged in routine odor detection tasks but had not participated in dMTS or OST procedures since the conclusion of the previous studies. To test long-term memory for the matching/non-matching rules, we replicated the last test session that each dog had performed 1–3 years prior. Our results (unpublished) demonstrated that two out of three dogs reassessed on MTS performed significantly above chance, while only one dog that was reassessed on the OST scored significantly above chance. Together these findings suggest that dogs are able to remember task procedures and associated rules for extended periods with no practice.

Overall, research indicates that long-term odor memory in dogs is particularly robust and resilient to extended periods of time and interference, possibly due to the sophisticated nature of the dog olfactory system and the enhanced salience of odors for the species. Dogs not only remember odors experienced during critical periods of early development for extended periods of time, but long-term memory for trained odors has also been demonstrated in a number of studies. These results, along with findings indicating dogs' long-term memory for task procedures involving olfactory stimuli, could help inform decisions regarding training practices for specific detection tasks.

1.3 Episodic Memory

Episodic memory is a form of long-term memory that includes encoding information related to specific events that an individual experiences. Because these memories are associated with an event in the past, individuals must not only remember *what* happened during the event but also *where* and *when* it occurred. These pieces of information (what, where, when) must also be bound to memories of a specific event, and not derived from other sources.

Episodic memory research in dogs is only in its initial stages, however it seems dogs can remember the "what" and "where" of specific events (Kaminski et al., 2008; Fujita et al., 2012). One study has attempted to demonstrate episodic memory using olfactory stimuli in dogs. Lo and Roberts (2019) used a what, where, and when (WWW) task that utilized four different boxes with various odors (i.e., what) that were placed in separate locations (i.e., where) and visited at different times (i.e., when). In the first experiment, dogs were presented with all four boxes at specific locations and times. The dogs were then shown the first and last box simultaneously and only rewarded for choosing the first box. Even when controlling for handler cues, dogs demonstrated high accuracy on the WWW task. However, because dogs were only ever presented with the first and last box, they may have only remembered one stimulus (i.e., the odor from the first box) and ignored all other events and cues. As a result, the authors ran a second experiment in which any of the four boxes could be presented during the test phase, and dogs were always rewarded for picking the box visited earlier in the sequence. Dogs were able to meet criteria on average in six test sessions, demonstrating high performance overall. The authors conducted two additional experiments to determine what strategies dogs could be using to solve this task by running tests that forced dogs to use specific components of episodic memory (i.e., a what-when test that eliminated the use of spatial information). Dogs continued to perform well on the WWW task even when tested on various components separately, indicating dogs are likely using all three components of episodic memory to solve the task. While this study provides evidence that dogs may have episodic memory for olfactory cues, additional research is needed to determine if memory for the "what," "where," and "when" of an event is encoded into one single memory or if dogs remember all three components separately. Outside of olfaction, Fugazza et al. (2020) report that dogs seem to exhibit episodic memory for their own spontaneous actions. Dogs were trained to repeat a set of behaviors when given a command ("repeat"). Not only could dogs learn to repeat the trained set of behaviors, but most could also repeat untrained but recognizable and discrete behaviors that had been untrained when commanded to repeat. It has been suggested that these sorts of "surprise" tests represent episodic memory, as they are more likely to rely on incidentally encoded memories that would have to be remembered in an episodic-like manner (Zentall et al., 2008).

1.4 Context-Dependent Memory

While the majority of odor memory studies in dogs have focused on assessing memory processes in dogs using odor stimuli, another approach is to evaluate how odors can serve as contextual cues for memory retrieval. Human research has demonstrated the powerful impact that odors can have on the formation of our memories, but little research has investigated these effects in dogs. Quaranta et al. (2020) used a spatial learning task in which dogs were required to remember the location of five rewards they had observed as they were hidden in different locations. During the memory encoding phase, the odor of vanilla was dispersed in the environment and dogs were allowed to investigate until they located all of the hidden rewards. Dogs were then brought back for the memory retrieval test phase 24 hours later and were tested in the presence of the vanilla odor or a control odor (apple). Overall, the dogs demonstrated better memory for the reward locations when vanilla was present in the environment during the encoding and retrieval phases, providing evidence that, like humans, odors can enhance context-dependent memory for dogs.

Animals developed memory stores and processes to retain information for different durations. These processes vary in terms of the amount of information or duration that the information is held for, but they all serve the purpose of allowing animals to retain information that might be useful in the future. Aspects of dog memory, including capacity and duration as well as the effects of interference and context on memory, especially regarding olfaction, have been emerging topics of research. As the domestic dog has adapted to a human-centered environment, they provide researchers with a unique opportunity to not only identify environmental risk factors influencing neurogenerative diseases in humans, but also the neural and genetic factors associated with these diseases (Ruple et al., 2022; Topál et al., 2019) and how they affect memory. Research on dog memory specifically has provided significant support for the use of dogs as translational models for neurodegenerative diseases (Tapp, 2003).

2 Categories and Concepts

Animals inhabit spaces that are full of constantly changing sensory information. Making sense of such an environment would be extremely challenging if not for the fact that often there is considerable overlap in the environment (Smith & Medin, 1981), and animals have specific cognitive processes to interpret their environment. The ability to form categories, representations of classes of stimuli, is based on the groupings of similar stimuli and provides stability in a variable environment by taking disparate sensory input and creating coherent perceptual experiences in the form of internal representations (Medin & Smith, 1984). Such categorization is the basis of forming concepts (e.g., birds, cars, colors, etc.). Concepts are the internal, mental representations that are functionally used to determine whether a particular stimulus is a member of a one or more categories (Smith & Medin, 1981). Categories, then, are the groupings of stimuli based on shared features and traits while concepts provide unifying representations of these categories. The value of categorization is being able to generalize quickly and efficiently what is learned about one member of a category to all novel members, even when these items are not exact replicas of the original members of the category. Stimuli that share some number of the features with the concept of a category will be added to that category and treated as a member of that category. A key endeavor in comparative cognition, including canine cognition, is to understand the neural, behavioral, and evolutionary bases of concept formation. The ability to form concepts is an important and general process, one that disparate species seem to share (Soto & Wasserman, 2012).

Dogs have emerged as subjects in concepts and category research recently, using a variety of stimulus modalities. Range et al. (2008) trained dogs to touch their nose to a touchscreen that displayed images of dogs with a landscape background and images of landscapes without a dog. Nose touches to the stimuli with dogs on them were reinforced while those without a dog were not, creating dog/non-dog categories. All dogs learned to pick out the images with dogs in them, even when they were against a familiar background, demonstrating visual category learning.

Understanding of olfactory categorization in all animals lags far behind that of visual categorization (Batty, 2014), and to a lesser extent audition. To some extent, this is due to relying on vision as a framework for perception broadly (Barwich, 2019). However, there are recent studies that look at olfactory categorization in dogs. Wright et al. (2017) split dogs into two groups: a category group and a pseudo-category control group. Both groups were trained to make a go/no-go response, depending on their group assignment. The category group made a go response (e.g., sit) when presented with a combination of a substrate odor that had been mixed with an accelerant (burnt or unburnt), and a no-go response (e.g., inhibit sitting/remain standing) when the substrate was presented without the accelerant. Therefore, they were trained to categorize the odors based on the presence or absence of accelerants. A go response could be either rewarded or nonrewarded for each substrate odor, depending on whether they were mixed with the accelerants. The control group in this experiment received substrates that in some cases had accelerants (burnt or unburnt) but were not exclusively rewarded for selecting the stimulus mixed with the accelerants. The control group received the same stimuli as the experimental group but without the rule. After reaching a performance criterion, dogs were given a transfer test (referred to as a generalization test), which consisted of two probe trials with novel substrate and accelerant combinations inserted into normal sessions. Four of the initial 11 dogs learned to categorize the training stimuli and reached transfer testing, and three of these four successfully transferred while one failed to finish testing.

This study is important because it suggests that dogs may learn to categorize odors based on common chemical characteristics that they can detect by smell. This could help in training dogs to generalize one common characteristic to all members of a class without necessarily needing to be trained on each member of that class. For example, a dog trained to detect all floral essential oils, as opposed to citrus-based essential oils, might only need to be trained on a small subset of floral essential oils (e.g., rose and tulip) in order to generalize the response to all floral essential oils. In a live detection setting, this would be a significant advantage as there may be small variations in the exact components used in an explosive compared to ones from training. A dog that can detect improvised explosive devices (IEDs) that are similar to ones it was trained on but still unfamiliar would be far more effective than one that cannot. However, other attempts to train dogs to generalize have produced mixed results. Recently, Dorman et al. (2021) trained dogs to detect either two or six variants of ammonium nitrate. Dogs were then tested for their ability to generalize to a novel AN compound (i.e., AN mixed with aluminum shavings). All dogs, whether trained on two or six variants of AN, failed to generalize to the mixture. DeGreeff et al. (2020) attempted to find chemical factors that could be related to categorization among odors. The olfactory system can discriminate odors in two ways: either in terms of the length of the carbon chain associated with that odor or in the functional group (the specific atoms attached to a molecule that determine how it reacts) associated with the odor. DeGreeff et al. were able to control one aspect of the odor while varying the other (i.e., changing carbon chain length for molecules of the same functional group and vice versa) to determine which encouraged dogs to generalize and which increased selectivity (e.g., caused greater discrimination and less generalization). By way of visual analogy, this procedure is akin to determining whether an animal is more or less likely to generalize a visual category rule by either expanding the diameter of an object or changing its color. Dogs, after training, were more likely to generalize between different compounds of a similar carbon chain length rather than similar functional groups.

3 Abstract Concepts

Thus far, concepts have been discussed based on rote memorization. That is, the categorization is based on perceptual similarity between the to-be-categorized item and existing representations of the item. This type of concept is often referred to as a natural concept. An alternative form of concept learning is the abstract concept, so-called because abstract concepts transcend the perceptual qualities of stimuli and instead involve categorization based on relational characteristics between two or more stimuli (Katz & Wright, 2021). Abstract concepts were once considered "the very keel and backbone" of human thought (James, 1890), but research has demonstrated that a variety of species are capable of such learning (Wright et al., 2021).

The matching-to-sample task (MTS), described above, is a common method for assessing abstract-concept learning. In terms of abstract concepts, a generalized matching rule would require subjects to select matching pairs even among never before encountered stimuli (e.g., transfer the matching rule to novel stimuli). The MTS procedure and the dMTS variant have a long history in comparative cognition, as a wide variety of animals have learned to successfully complete the task with little to no procedural differences, including different species of birds (Wright et al., 1988), monkeys (Overman & Doty, 1980), great apes (Matsuno et al., 2004), rodents (Lazarowski et al., 2019; Peña et al., 2006), fish (Aellen et al., 2022), and bees (Giurfa et al., 2001). Using the same procedure with different species presents an opportunity to explore similarities and differences across species, as well as provide insight into how these processes may have evolved.

In terms of dog olfaction, Lazarowski et al. (2021a, b) trained six dogs on an olfactory variant of the MTS task to test for evidence of abstract-concept learning. Dogs were trained to investigate a sample odor followed by matching (S+) and non-matching odors (S−). Each session consisted of 24 trials, each of which was trial-unique (i.e., each odor appeared once per session). Once dogs reached two consecutive sessions of at least 20 out of 24 correct trials, they moved on to the transfer tests, which were identical to baseline training sessions except that six of the baseline trials were replaced with six transfer trials. Transfer trials contained completely novel odors as the S+ and S−, which never repeated across sessions or trials. All dogs learned the initial MTS task as well as transferred to novel stimuli, providing evidence of abstract-concept learning according to the established criteria: performance on transfer tests is equivalent to baseline, the analysis is restricted to the first presentation of the stimulus, and the procedures are identical on baseline and transfer trials (Katz & Wright, 2021). Similar evidence comes from dogs trained to detect human scents in forensics cases. Marchal et al. (2016) found that extensively training dogs on a scent detection line, where they must match human scents in lineup to a sample scent, led to a high degree of accuracy (90%). Dogs were also tested on odors collected from crime scenes and potential subjects, where they continued to perform well. The data is not reported in terms of a transfer test and is not experimentally controlled as in Lazarowski et al. (2021a, b), yet still provides similar evidence and an example of a practical application of MTS learning in dogs. These results demonstrate dogs' ability to form abstract concepts and further support a general process account of abstract-concept learning; that is, the ability to form abstract concepts is shared amongst a variety of animal species.

The ability to categorize objects and to represent abstract concepts is a core cognitive process that is shared across disparate species (Soto & Wasserman, 2012), including primates, birds, and others. The development of key procedures such as pseudo-category training and controlled transfer testing elucidated the general nature of these abilities. Including dogs in the species tested for category and concept learning further expands the knowledge of cognitive evolution. Further, testing with olfactory stimuli represents an important expansion of these procedures to sensory modalities other than vision and audition. Properly training dogs to categorize related odors can help train detection dogs more efficiently. Incorporating the findings outlined in this section could help improve training and selection of service dogs.

4 Odor Representations

Of recent interest in canine cognition is the relationship between dogs' olfactory perception and cognition—specifically how dogs mentally represent olfactory information. While dogs have long been acknowledged for their remarkable olfactory capabilities and have been employed in a variety of olfactory detection settings, including explosives, drugs, and biomedical agents (Kokocinska-Kusiak et al., 2021), dogs' perception of individual odors and understanding of what these odors represent is underexplored.

The first examination of olfactory object perception in dogs assessed dogs' ability to form what is referred to as a search image (Gazit et al., 2005). A search image refers to an animal's enhanced ability to detect stimuli based on prior experience with that stimuli. For example, an animal showing an enhanced ability to find a specific type of cryptic prey amongst, and instead of, other equally rewarding alternatives as a result of prior repeated experience with that prey suggests that animal has a "search image" of that prey (Dukas & Kamil, 2001; Katz & Cook,

2000; Tinbergen, 1960). Here, Gazit et al. (2005) assessed dogs' ability to detect different odors following differential levels of exposure to each odor. They found that repeated exposure to a particular target odor prior to testing resulted in greater detection levels for that target as compared to baseline levels of detection, suggesting that a search image was formed for that particular odor and dogs' ability to detect that odor was enhanced by increased exposure. However, when the probability of encountering that target odor was reduced such that it was encountered less than another odor during training, detection performance for the target odor during testing declined, suggesting that search image is maintained only as long as the probability of encountering a particular target remains high.

The findings of Gazit et al. (2005) suggest that dogs can preferentially attend to specific odors. An important extension of these findings is understanding how dogs mentally represent target odors. In other words, do they incorporate specific odors into their representations of specific objects? Brauer and Belger (2018) explored this question through an olfactory violation-of-expectation paradigm. Dogs were required to follow an odor trail for a specific toy ("A") to a target location. In the baseline condition, the dogs would always find Toy A at the target location, while in the experimental condition, the dogs would always find Toy B. In the experimental condition, in which the unexpected Toy B was present, dogs were significantly more hesitant to fetch the toy than in the baseline condition, suggesting that dogs had an expectation of what they would find at the end of the odor trail.

Dogs' understanding of olfactory information as representing specific objects has been extended along social dimensions to include knowledge of odors as representing specific human individuals, as well as the self. In a similar violation-of-expectation paradigm as described above, Brauer and Blasi (2021) explored dogs' understanding of odors as representing specific humans. Dogs in this experiment were able to track the odor trail of Person A (the dog's owner). At the target location, dogs found either the expected Person A, or another unexpected Person B (another owner or person that the dog was very familiar with). Dogs demonstrated more behavioral excitement when the unexpected person was present at the final location than the expected person, providing evidence of surprise in this condition and suggesting that dogs had been searching specifically for Person A.

Not only can dogs recognize odors from individual humans, but it is suggested that they can also recognize their own. Horowitz (2017) developed an olfactory self-recognition test for dogs as an analogy of the classic mirror self-recognition test (cf. Gallup, 1970), which may not be an appropriate test of assessing self-recognition in some species. Horowitz (2017) found that dogs spent a significantly longer time investigating a modified sample of their own odor than samples of their own odor without the modifier or the modifier in isolation, suggesting that dogs recognized the odor as being of themselves but different. Further, Horowitz explored dogs' investigations of their own versus conspecifics' odors and found that consistent with observations in natural environments, dogs spent significantly more time investigating the odors of others as compared to themselves. It should be noted that the interpretation of these results as evidence of self-recognition is controversial due to lack of proper controls. Gallup and Anderson (2018) suggest that a more parsimonious explanation of Horowitz's results might be that dogs are habituated to their own odor and that dishabituation to this odor occurs when it is modified. Therefore, dogs' olfactory self-recognition remains in need of further exploration.

Of note, these experiments are treating olfactory objects as qualities used to represent their sources. In other words, the object is a visual thing with a smell. As an example, consider the difference between a representation of the smell of a banana as a characteristic of the visual object banana, as opposed to the representation of the smell existing as its own olfactory object. This would suggest, among other things, that one could separate the smell of a banana from a background of other smells, without visual cues. The extent to which the mammalian olfactory system represents olfactory objects is far from settled (cf. Barwich, 2019; Millar, 2019; Wilson & Stevenson, 2003). Taken together, these studies suggest that dogs are able to associate specific odors with specific targets and can use this information to search for a certain target preferentially. Further, dogs are able to use olfactory information to represent specific multisensory objects and individuals.

5 Olfactory Quantity Discrimination

Animals use quantity discrimination to recognize the difference between two or more amounts (Jackson et al., 2021), allowing them to make informed decisions regarding food, conflicts, and other environmental situations. For example, scavenging wild dogs may decide between two food piles based on the amount of food each contains (Banerjee & Bhadra, 2019). In animals, quantity discrimination studies often focus on the visual modality, viewing olfaction as a confound to be controlled for instead of examined (Baker et al., 2011; Petrazzini & Wynne, 2016). Visual quantity discrimination is important, but for animals that rely heavily upon their sense of smell, quantity discrimination should not be viewed from one modality alone.

Recently, issues of the importance of olfaction in olfactory quantity discrimination have been explored in dogs (Ferrando & Dahl, 2022; Horowitz et al., 2013; Jackson et al., 2021). By understanding olfactory quantity discrimination in dogs, we may better prepare them for jobs in working fields, as well as understand how their cognitive processes differ from other species. Studies on dogs' olfactory quantity discrimination abilities have primarily relied upon experimental designs that have been similar to those used in visual tasks, but altered for olfaction (Ferrando & Dahl, 2022; Horowitz et al., 2013; Jackson et al., 2021). While there are differences between the two experimental designs, some aspects remain the same. These designs rely upon a two-alternative forced choice (2-AFC) paradigm with two experimenters, one acting as the handler and one controlling the food placement (Ferrando & Dahl, 2022). Two plates containing differing amounts of food are placed in the middle of the floor, 0.5–1 m apart. Both the dog handler and the food handler are positioned 1.5 m away from the plates, on opposite sides. These

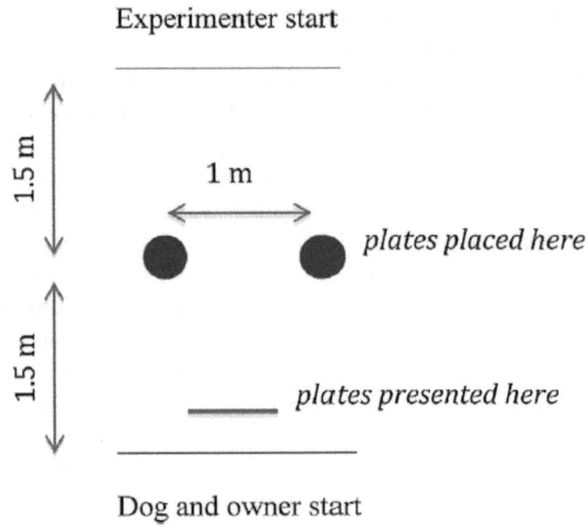

Fig. 2 Take from Horowitz et al. (2013) showing a common setup for these procedures. This schematic shows locations of the dog and owner (bottom) and experimenter (top) at the beginning of each trial. Plates were placed 1 m apart from each other after they were originally presented to the dog. The plates were 1.5 m from both the experimenter and the dog-owner pair

measurements vary some across experiments, but this general setup is used for both olfactory and visual quantity discrimination tasks (Baker et al., 2011; Banerjee & Bhadra, 2019; Horowitz et al., 2013; Jackson et al., 2021; Petrazzini & Wynne, 2016; Petrazzini et al., 2020). An example of this design, taken from Horowitz et al. (2013), can be seen in Fig. 2.

In the Horowitz et al. (2013) adaptation of the original visual experiment, pieces of hot dog were placed on paper plates, with one plate containing one hot dog wedge while the other contained five hot dog wedges. The paper plates were then each taped shut with another paper plate on top—in order to remove any visual cues—and were presented to the dogs either sequentially or simultaneously, before placement in the middle position. From there, the dogs were able to choose one of the plates and allowed to eat whatever was under that plate. Up to six trials were conducted in this primary condition. The results obtained from this study were not statistically significant (Horowitz et al. 2013), suggesting that dogs do not use olfaction when discriminating between food amounts. However, Jackson et al. (2021) found different results following a similar setup to Horowitz et al. (2013) but with some important differences. As in the Horowitz et al. (2013) study, hot dog slices were used, with one plate containing one hot dog slice and the other containing five hot dog slices, both of which were covered to avoid the potential use of visual cues. However, instead of taping shut the plates, Jackson et al. (2021) covered the plates with slitted plastic containers, allowing scent molecules to disperse while still preventing visual quantity discrimination. Rather than presenting the plates and then putting them down, the dogs were able to move forward and sniff each of the containers before being pulled back to the starting position. From there, after a three-second delay, the dogs were released and allowed to choose between the two plates, consuming the contents of the plate chosen. There were twenty trials with a 30-s intertrial interval, and the plate presentation of food was counterbalanced (so that each side contained each food quantity for an equal number of trials). Under these conditions, the difference between choosing the larger and smaller quantity of food was statistically significant.

Considering the overall similarities in experimental design, the different results are worth further examination. There are multiple factors that may impact the results obtained in these experiments. There is the buildup of odor within the experimental apparatus. Numerosity (an ability to distinguish the actual numbers involved between the two amounts) is a relevant factor in visual quantity discriminations, but it may not apply to the olfactory setting (Petrazzini & Wynne, 2016). However, odor buildup within the plates could impact task performance just as much as numerosity.

Another factor that must be considered is Weber's Law, which is the concept that ratios that are closer together (1:2) are harder to discriminate than ratios that are farther apart (1:6) (Jackson et al., 2021; Petrazzini et al., 2020). In visual quantity discrimination tasks, Weber's Law has been shown to generally hold in studies of adult dogs (Petrazzini & Wynne, 2016), coyotes (Baker et al., 2011), and—in a more limited form—puppies as young as 2 months of age (Petrazzini et al., 2020). In studies of olfactory quantity discrimination, however, Weber's Law has not been examined (Horowitz et al., 2013) or has not held (Jackson et al., 2021). While this is important for understanding the cognitive processes of dogs, it is especially important for the setup of the task. If Weber's Law impacts how dogs perceive the task, then the results that are obtained will be impacted by it as well. While the food quantities were the same across studies, it remains that there could be differences in the amount of scent molecules available. The roles of Weber's Law and odor buildup should be considered together to better understand the true nature of dogs' olfactory quantity discrimination abilities.

Ferrando and Dahl (2022) developed another version of the experiment, addressing both of these issues. They used a 2-AFC paradigm, with some slight differences in measurements from those seen with Horowitz et al. (2013) and Jackson et al. (2021). Importantly, they used glass plates and upside-down plastic cups, rather than paper plates. These were, respectively, cleaned and replaced between trials. Furthermore, dogs did not examine the plates beforehand. The plates were placed 1 m away from the dog and from there, after a ten second delay, the dog made their decision (however, the criteria for a choice was more restrictive, requiring the dog to be within 30 cm of the cup and to stay there for three seconds). Eighteen trials were conducted, spread out over three days in three different sessions. Cervelat sausage was used, but the quantities tested diverged from the previous two experiments in that three different quantity comparisons were used (1:2, 1:3, 1:4) in order to explicitly examine the impact of Weber's Law. Differences in performance based on quantity comparisons were not statistically significant but were close to it (7%) (Ferrando & Dahl, 2022). Regardless, the results were similar to Jackson et al. (2021) in terms of Weber's Law. Regarding odor buildup, Ferrando and Dahl (2022) aimed to address this by refreshing the experimental apparatus, however this still remains a factor. Ferrando and Dahl (2022) recognized this drawback but preferred to examine other task factors (i.e., breed differences) that required an emphasis on task difficulty over a control of odor molecules. Overall, they found that dogs performed around chance level on the olfactory quantity discrimination task (Ferrando & Dahl, 2022).

The studies described above have yielded mixed findings on dogs' ability to discriminate between olfactory quantities. Horowitz et al. (2013) and Ferrando and Dahl (2022) found that dogs could not consistently discriminate food quantities. Of the dogs tested, 61% preferred the plate with more food, but the difference in choosing the larger amount of food over the smaller amount of food was not significant (Horowitz et al. 2013). Meanwhile, Jackson et al. (2021) found that dogs chose the greater food quantity 70.5% of the time. When the experiment was repeated with pieces of kibble instead of hot dog, however, the dogs performed around chance level (Jackson et al., 2021). Lastly, when Ferrando and Dahl (2022) performed the experiment with differing numbers of Cervelat sausage, dogs performed at around 50% accuracy, no better than chance level. In light of these mixed results, there are several factors to be considered. This is a spontaneous choice task (Banerjee & Bhadra, 2019), which may result in the task not only examining dogs' ability to smell a difference in quantity, but whether they have learned to discern the rewards attached to this discrimination as well. There were multiple trials carried out in all experiments, giving the dogs the opportunity to learn, but the dogs were rewarded regardless of how they performed (Ferrando & Dahl, 2022; Horowitz et al., 2013; Jackson et al., 2021). Further, these studies operated on the assumption that the dogs would naturally prefer and choose the greater quantity of food, but there was no requirement for dogs to do so. Dogs may have the ability to discriminate quantities but may not be able to perform at their full capacity in a spontaneous condition. Horowitz et al. (2013) even noted that dogs attended to the larger quantity of food significantly more than to the lower quantity of food regardless of their choice, which may point to an ability to discriminate quantity, but that the dogs did not learn what that means for the task and rewards. This setup is also not an everyday experience for companion dogs. As Banerjee and Bhadra (2019) note, performance in quantity discrimination tasks could be dependent on the context for the dog. Since the dogs in two of these experiments were companion dogs (Horowitz et al., 2013; Jackson et al., 2021), and in the other were shelter dogs (Ferrando & Dahl, 2022), they were not usually asked to use their noses in this capacity (Horowitz et al., 2013). If performed with working dogs, or with dogs trained on the task, results may differ. Also, while many breeds were involved in these experiments, there could be differences—both at a breed level and at an individual level—in dogs' olfactory abilities. Ferrando and Dahl (2022) explicitly examined this, finding differences in detection performance based on cephalic index that could impact discrimination as well (e.g., breeds with brachycephalic skulls were worse at detection tasks). The impact of type of scent (certain odors may be more attractive) and saliency of scent (certain scents may emit odor molecules more rapidly), while touched on by Jackson et al. (2021) in their comparison of kibble versus hot dog, could also be investigated further.

Further research is needed in order to understand dogs' olfactory discrimination abilities. Future studies could be performed with different populations, different trial and reward setups, and with a greater focus on the influence of Weber's Law. Replicating such experiments with working dogs may show increases in performance based on selective breeding and training history for enhanced olfactory capabilities. If dogs were rewarded for specific choices—picking the larger quantity of food or the smaller quantity of food—then the impacts of cognitive reward evaluations and physical scent capabilities could be separated. Only when further replication is performed will there be a more comprehensive view of dogs' ability to perform olfactory quantity discriminations.

6 Conclusion and Future Directions

Dogs are currently used around the world as a means to detect chemical compounds through olfaction. While dogs may not be unique in their ability to detect these compounds to a high degree of accuracy, they are the only species that serve a variety of roles in addition to chemical detections, as service animals and companion animals as well. Understanding cognitive processes in dogs as they relate to olfaction, such as memory, category and abstract-concept learning capabilities, the nature of their representations of olfactory objects, and their ability to extract measures of quantity from olfactory discrimination is vital to ensuring dogs are employed efficiently in these jobs while also ensuring their welfare. When training dogs for these roles, it is important to consider their capabilities and deficiencies; operating within these limits will minimize costs and maximize their utility. Further, considering the link between olfaction and cognition in humans, understanding this link in dogs and its underlying mechanisms has implications for canine aging and neurogenerative diseases.

Future directions for studies of olfactory dog cognition include furthering our understanding of the processes underlying dog cognition and olfaction. Connecting the brain processes (e.g., Thompkins et al., 2021) and genetics (Eyre et al., 2022) of olfaction with cognitive and behavioral measures of olfaction in dogs, both in the laboratory and real-world scenarios, is vital to not only understanding dog cognition but also to train detection dogs. In the future, a converging operations approach integrating neuroscience, cognitive, and behavioral methods can help answer basic and applied questions about olfaction, from sensation and perception to complex cognitive processes.

References

Aellen M, Siebeck UE, Bshary R (2022) Cleaner wrasse Labroides dimidiatus perform above chance in a "matching-to-sample" experiment. PLoS One 17(1):e0262351

April LB, Bruce K, Galizio M (2013) The magic number 70 (plus or minus 20): variables determining performance in the rodent odor span task. Learn Motiv 44(3):143–158. https://doi.org/10.1016/j.lmot.2013.03.001

Baker JM, Shivik J, Jordan KE (2011) Tracking of food quantity by coyotes (Canis latrans). Behav Process 88(2):72–75. https://doi.org/10.1016/j.beproc.2011.08.006

Banerjee A, Bhadra A (2019) The more the merrier: dogs can assess quantities in food-choice tasks. Curr Sci 117(6):1095–1100. https://doi.org/10.18520/cs/v117/i6/1095-1100

Barwich AS (2019) A critique of olfactory objects. Front Psychol 10:1337

Batty C (2014) Olfactory objects. In: Perception and its modalities, pp 222–224

Beebe SC, Howell TJ, Bennett PC (2016) Using scent detection dogs in conservation settings: a review of scientific literature regarding their selection. Front Vet Sci 3:96

Brauer J, Belger J (2018) A ball is not a Kong: odor representation and search behavior in domestic dogs (Canis familiaris) of different education. J Compar Cogn 132(2):189–199

Brauer J, Blasi D (2021) Dogs display owner-specific expectations based on olfaction. Sci Rep 11:3291

Bystrova MF, Kolesnikov SS (2021) The "One Neuron-One Receptor" rule in the physiology and genetics of olfaction. Neurosci Behav Physiol 51(7):1008–1017

Chan ADF, Nippak PMD, Murphey H, Ikeda-Douglas CJ, Muggenburg B, Head E, Cotman CW, Milgram NW (2002) Visuospatial impairments in aged canines (*Canis familiaris*): the role of cognitive-behavioral flexibility. Behav Neurosci 116(3):443–454

Chapagain D, Range F, Huber L, Virányi Z (2018) Cognitive aging in dogs. Gerontology 64(2):165–171. https://doi.org/10/1159/000481621

Cobb M, Branson N, McGreevy P, Lill A, Bennett P (2015) The advent of canine performance science: offering a sustainable future for working dogs. Behav Proc 110:96–104

Cowan N (2008) What are the differences between long-term, short-term, and working memory? Prog Brain Res 169:323–338. https://doi.org/10.1016/S0079-6123(07)00020-9

Cowan N (2017) The many faces of working memory and short-term storage. Psychon Bull Rev 24(4):1158–1170

DeGreeff LE, Simon AG, Peranich K, Holness HK, Frank K, Furton KG (2020) Generalization and discrimination of molecularly similar odorants in detection canines and the influence of training. Behav Proc 177:104148

Dorman DC, Foster ML, Lazarowski L (2021) Training with multiple structurally related odorants fails to improve generalization of ammonium nitrate detection in domesticated dogs (Canis familiaris). Animals 11(1):213

Doty RL, Kamath V (2014) The influences of age on olfaction: a review. Front Psychol 5:20. https://doi.org/10.3389/fpsyg.2014.00020

Dudchenko PA (2004) An overview of the tasks used to test working memory in rodents. Neurosci Biobehav Rev 28(7):699–709. https://doi.org/10.1016/j.neubiorev.2004.09.002

Dudchenko PA, Wood ER, Eichenbaum H (2000) Neurotoxic hippocampal lesions have no effect on odor span and little effect on odor recognition memory but produce significant impairments on spatial span, recognition, and alternation. J Neurosci 20(8):2964–2977. https://doi.org/10.1523/JNEUROSCI.20-08-02964.2000

Dukas R, Kamil AC (2001) Limited attention: the constraint underlying search image. Behav Ecol 12(2):192–199

Eyre AW, Zapata I, Hare E, Lee K, Bellis C, Essler JL et al (2022) Genome scanning of behavioral selection in a canine olfactory detection breeding cohort. Sci Rep 12(1):1–12

Ferrando E, Dahl C (2022) An investigation on the olfactory capabilities of domestic dogs (Canis lupus familiaris). Anim Cogn. https://doi.org/10.1007/s10071-022-01640-6

Fiset S, Beaulieu C, Landry F (2003) Duration of dogs' (Canis familiaris) working memory in search for disappearing objects. Anim Cogn 6(1):1–10. https://doi.org/10.1007/s10071-002-0157-4

Foraita M, Howell T, Bennett P (2021) Environmental influences on development of executive functions in dogs. Anim Cogn 24(4):655–675. https://doi.org/10.1007/s10071-021-01489-1

Fugazza C, Pongrácz P, Pogány Á, Lenkei R, Miklósi Á (2020) Mental representation and episodic-like memory of own actions in dogs. Sci Rep 10(1):1–8

Fujita K, Morisaki A, Takaoka A, Maeda T, Hori Y (2012) Incidental memory in dogs (Canis familiaris): adaptive behavioral solution at an unexpected memory test. Anim Cogn 15(6):1055–1063. https://doi.org/10.1007/s10071-012-0529-3

Furton KG, Myers LJ (2001) The scientific foundation and efficacy of the use of canines as chemical detectors for explosives. Talanta 54(3):487–500

Gallup GG Jr (1970) Chimpanzees: self recognition. Science 167(3914):86–87

Gallup GG Jr, Anderson JR (2018) The "olfactory mirror" and other recent attempts to demonstrate self-recognition in non-primate species. Behav Proc 148:16–19

Gazit I, Terkel J (2003) Domination of olfaction over vision in explosives detection by dogs. Appl Anim Behav Sci 82(1):65–73

Gazit I, Goldblatt A, Terkel J (2005) Formation of an olfactory search image for explosives odours in sniffer dogs. Ethology 111:669–680

Giurfa M, Zhang S, Jenett A, Menzel R, Srinivasan MV (2001) The concepts of 'sameness' and 'difference' in an insect. Nature 410:930–933. https://doi.org/10.1038/35073582

Hayes JE, McGreevy PD, Forbes SL, Laing G, Stuetz RM (2018) Critical review of dog detection and the influences of physiology, training, and analytical methodologies. Talanta 185:499–512. https://doi.org/10.1016/j.talanta.2018.04.010

Hepper PG (1994) Long-term retention of kinship recognition established during infancy in the domestic dog. Behav Proc 33(1–2):3–14. https://doi.org/10.1016/0376-6357(94)90056-6

Hepper PG, Wells DL (2006) Perinatal olfactory learning in the domestic dog. Chem Senses 31(3):207–212. https://doi.org/10.1093/chemse/bjj020

Horowitz A (2017) Smelling themselves: dogs investigate their own odours longer when modified in an "olfactory mirror" test. Behav Proc 143:17–24

Horowitz A, Hecht J, Dedrick A (2013) Smelling more or less: investigating the olfactory experience of the domestic dog. Learn Motiv 44(4):207–217. https://doi.org/10.1016/j.lmot.2013.02.002

Jackson SM, Martin GK, Roberts WA (2021) The olfactory capability of dogs to discriminate between different quantities of food. Learn Behav 49(3):321–329. https://doi.org/10.3758/s13420-021-00463-8

James W (1890) Principles of psychology, vol 1. Holt, New York

Johnston JM (1999) Canine detection capabilities: operational implications of recent R&D findings. Institute for Biological Detection Systems, Auburn University, Auburn, Alabama, USA

Jung HJ, Shin I, Lee SJ (2019) Olfactory function in mild cognitive impairment and Alzheimer's disease: a meta-analysis. The Laryngoscope 129(2):362–369. https://doi.org/10.1002/lary.27399

Kaminski J, Fischer J, Call J (2008) Prospective object search in dogs: mixed evidence for knowledge of what and where. Anim Cogn 11(2):367–371. https://doi.org/10.1007/s10071-007-0124-1

Katz JS, Cook RG (2000) Stimulus repetition effects on texture-based visual search by pigeons. J Exp Psychol: Anim Behav Process 26:220–236. https://doi.org/10.1037/0097-7403.26.2.220

Katz JS, Wright AA (2021) Issues in the comparative cognition of same/different abstract-concept learning. Curr Opin Behav Sci 37:29–34

Kirchner WK (1958) Age differences in short-term retention of rapidly changing information. J Exp Psychol 55(4):352–358. https://doi.org/10.1037/h0043688

Kokocinska-Kusiak A, Woszczylo M, Zybala M, Mociocha J, Barlowska K, Dziecol M (2021) Canine olfaction: physiology, behavior, and possibilities for practical applications. Animals 11:2463

Krichbaum S, Rogers B, Cox E, Waggoner LP, Katz JS (2020) Odor span task in dogs (Canis familiaris). Anim Cogn 23(3):571–580. https://doi.org/10.1007/s10071-020-01362-7
Krichbaum S, Lazarowski L, Davila A, Cox E, Smith JG, Katz JS (2021) Dissociating the effects of delay and interference on dog (Canis familiaris) working memory. Anim Cogn 24(6):1259–1265. https://doi.org/10.1007/s10071-021-01509-0
Kuœmierek P, Kowalska D (2002) Effect of sound source position on learning and performance of auditory delayed matching-to-sample task in dogs. Acta Neurobiol Exp 62:251–262
Landsberg GM, Nichol J, Araujo JA (2012) Cognitive dysfunction syndrome: a disease of canine and feline brain aging. Vet Clin Small Anim Pract 42(4):749–768.z
Lazarowski L, Goodman A, Galizio M, Bruce K (2019) Effects of set size on identity and oddity abstract-concept learning in rats. Anim Cogn 22:7331–74210
Lazarowski L, Davila A, Krichbaum S, Cox E, Smith JG, Waggoner LP, Katz JS (2021a) Matching-to-sample abstract-concept learning by dogs (Canis familiaris). J Exp Psychol Anim Learn Cogn 47(3):393–400. https://doi.org/10.1037/xan0000281
Lazarowski L, Waggoner P, Hutchings B, Angle C, Porritt F (2021b) Maintaining long-term odor memory and detection performance in dogs. Appl Anim Behav Sci 238:105301. https://doi.org/10.1016/j.applanim.2021.105301
Lea SE, Osthaus B (2018) In what sense are dogs special? Canine cognition in comparative context. Learn Behav 46(4):335–363
Levy DA, Manns JR, Hopkins RO, Gold JJ, Broadbent NJ, Squire LR (2003) Impaired visual and odor recognition memory span in patients with hippocampal lesions. Learn Mem 10(6):531–536. https://doi.org/10.1101/lm.66703
Lind J, Enquist M, Ghirlanda S (2015) Animal memory: a review of delayed matching-to-sample data. Behav Proc 117:52–58. https://doi.org/10.1016/j.beproc.2014.11.019
Lippi G, Mattiuzzi C, Henry BM (2021) Are sniffer dogs a reliable approach for diagnosing SARS-CoV-2 infection? Diagnosis 8(4):446–449
Lo KH, Roberts WA (2019) Dogs (Canis familiaris) use odor cues to show episodic-like memory for what, where, and when. J Comp Psychol 133(4):428–441. https://doi.org/10.1037/com0000174
Lo GKH, Macpherson K, MacDonald H, Roberts WA (2020) A comparative study of memory for olfactory discriminations: dogs (Canis familiaris), rats (Rattus norvegicus), and humans (Homo sapiens). J Comp Psychol 134(2):170–179. https://doi.org/10.1037/com0000205
Lord K (2013) A comparison of the sensory development of wolves (Canis lupus lupus) and dogs (Canis lupus familiaris). Ethology 119(2):110–120. https://doi.org/10.1111/eth.12044
Lubow RE, Kahn M, Frommer R (1973) Information processing of olfactory stimuli by the dog: I. The acquisition and retention of four odor-pair discriminations. Bull Psychon Soc 1(2):143–145. https://doi.org/10.3758/BF03334324
MacLean EL, Herrmann E, Suchindran S, Hare B (2017) Individual differences in cooperative communicative skills are more similar between dogs and humans than chimpanzees. Anim Behav 126:41–51. https://doi.org/10.1016/j.anbehav.2017.01.005
Marchal S, Bregeras O, Puaux D, Gervais R, Ferry B (2016) Rigorous training of dogs leads to high accuracy in human scent matching-to-sample performance. PLoS One 11(2):e0146963
Matsuno T, Kawai N, Matsuzawa T (2004) Color classification by chimpanzees (Pan troglodytes) in a matching-to-sample task. Behav Brain Res 148(1–2):157–165. https://doi.org/10/1016/S0166-4328(03)00185-2
Medin DL, Smith EE (1984) Concepts and concept formation. Annu Rev Psychol 35(1):113–138
Milgram NW, Head E, Weiner E, Thomas E (1994) Cognitive functions and aging in the dog: acquisition of nonspatial visual tasks. Behav Neurosci 108(1):57–68
Millar B (2019) Smelling objects. Synthese 196(10):4279–4303
Mombaerts P (2004) Odorant receptor gene choice in olfactory sensory neurons: the one receptor–one neuron hypothesis revisited. Curr Opin Neurobiol 14(1):31–36
Murman DL (2015) The impact of age on cognition. Semin Hear 36(3):111–121. https://doi.org/10.1055/s-0035-1555115

Overman WH, Doty RW (1980) Prolonged visual memory in macaques and man. Neuroscience 5(11):1825–1831. https://doi.org/10/1016/0306-4522(80)90032-9
Peña T, Pitts RC, Galizio M (2006) Identity matching-to-sample with olfactory stimuli in rats. J Exp Anal Behav 85(2):203–221
Petrazzini MEM, Wynne CDL (2016) What counts for dogs (Canis lupus familiaris) in a quantity discrimination task? Behav Proc 122:90–97. https://doi.org/10.1016/j.beproc.2015.11.013
Petrazzini MEM, Mantese F, Prato-Previde E (2020) Food quantity discrimination in puppies (Canis lupus familiaris). Anim Cogn 23(4):703–710. https://doi.org/10.1007/s10071-020-01378-z
Quaranta A, d'Ingeo S, Siniscalchi M (2020) Odour-evoked memory in dogs: do odours help to retrieve memories of food location? Animals 10(8):1249. https://doi.org/10.3390/ani10081249
Quignon P, Kirkness E, Cadieu E, Touleimat N, Guyon R, Renier C et al (2003) Comparison of the canine and human olfactory receptor gene repertoires. Genome Biol 4(12):1–9
Range F, Aust U, Steurer M, Huber L (2008) Visual categorization of natural stimuli by domestic dogs. Anim Cogn 11(2):339–347
Riezzo I, Neri M, Rendine M, Bellifemina A, Cantatore S, Fiore C, Turillazzi E (2014) Cadaver dogs: unscientific myth or reliable biological devices? Forensic Sci Int 244:213–221
Ruple A, MacLean E, Snyder-Mackler N, Creevy KE, Promislow D (2022) Dog models of aging. Annu Rev Anim Biosci 10(1). https://doi.org/10.1146/annurev-animal-051021-080937
Salvin HE, McGrath C, McGreevy PD, Valenzuela MJ (2012) Development of a novel paradigm for the measurement of olfactory discrimination in dogs (Canis familiaris): a pilot study. J Vet Behav 7(1):3–10. https://doi.org/10.1016/j.jveb.2011.04.005
Smith E, Medin D (1981) Categories and concepts. Harvard University Press, Cambridge, MA
Soto FA, Wasserman EA (2012) Visual object categorization in birds and primates: integrating behavioral, neurobiological, and computational evidence within a "general process" framework. Cogn Affect Behav Neurosci 12(1):220–240
Squire LR, Zola SM (1996) Structure and function of declarative and nondeclarative memory systems. Proc Natl Acad Sci 93(24):13515–13522. https://doi.org/10.1073/pnas.93.24.13515
Tapp PD, Siwak CT, Estrada J, Holowachuk D, Milgram NW (2003) Effects of age on measures of complex working memory span in the Beagle dog (Canis familiaris) using two versions of a spatial list learning paradigm. Learn Mem 10(2):148–160. https://doi.org/10.1101/lm.56503
Taverna G, Tidu L, Grizzi F, Torri V, Mandressi A, Sardella P et al (2015) Olfactory system of highly trained dogs detects prostate cancer in urine samples. J Urol 193(4):1382–1387
Thompkins AM, Lazarowski L, Ramaiahgari B, Gotoor SSR, Waggoner P, Denney TS et al (2021) Dog–human social relationship: representation of human face familiarity and emotions in the dog brain. Anim Cogn 24(2):251–266
Tinbergen L (1960) The natural control of insects in pine woods. 1. Factors influencing the intensity of predation by songbirds. Arch Neerlandaises de zoologie 13:265–343
Topál J, Román V, Turcsán B (2019) The dog (Canis familiaris) as a translational model of autism: it is high time we move from promise to reality. Wiley Interdiscip Rev Cogn Sci 10(4):e1495. https://doi.org/10.1002/wcs.1495
von Uexküll J (1957) A stroll through the worlds of animal and men. In: Schiller CH (ed) Instinctive behavior; the development of a modern concept. International University Press
Waggoner P, Lazarowski L, Hutchings B, Angle C, Porritt F (2022) Effects of learning an increasing number of odors on olfactory learning, memory and generalization in detection dogs. Appl Anim Behav Sci 247:105568. https://doi.org/10.1016/j.applanim.2022.105568
Wells DL, Hepper PG (2006) Prenatal olfactory learning in the domestic dog. Anim Behav 72(3):681–686. https://doi.org/10.1016/j.anbehav.2005.12.008
Williams M, Johnston JM (2002) Training and maintaining the performance of dogs (Canis familiaris) on an increasing number of odor discriminations in a controlled setting. Appl Anim Behav Sci 78(1):55–65. https://doi.org/10.1016/S0168-1591(02)00081-3
Wilson DA, Stevenson RJ (2003) The fundamental role of memory in olfactory perception. Trends Neurosci 26(5):243–247

Windon MJ, Kim SJ, Oh ES, Lin SY (2020) Predictive value of olfactory impairment for cognitive decline among cognitively normal adults. The Laryngoscope 130(4):840–847. https://doi.org/10.1002/lary.28166

Wright AA, Cook RG, Rivera JJ, Sands SF, Delius JD (1988) Concept learning by pigeons: matching to sample with trial-unique video picture stimuli. Anim Learn Behav 16:436–444

Wright AA, Katz JS, Ma WJ (2012) How to be proactive about interference: lessons from animal memory. Psychol Sci 23(5):453–458. https://doi.org/10.1177/0956797611430096

Wright AA, Kelly DM, Katz JS (2021) Same/different concept learning by primates and birds. Learn Behav 49:76–84. https://doi.org/10.3758/s13420-020-00456-z

Wright HF, Wilkinson A, Croxton RS, Graham DK, Harding RC, Hodkinson HL, Keep B, Cracknell NR, Zulch HE (2017) Animals can assign novel odours to a known category. Sci Rep 7(1):9019. https://doi.org/10.1038/s41598-017-09454-0

Zentall TR, Singer RA, Stagner JP (2008) Episodic-like memory: pigeons can report location pecked when unexpectedly asked. Behav Process 79:93–98. https://doi.org/10/1016/j.beproc.2008.05.003

Olfactory Learning and Training Methods

Lyn Caldicott, Helen E. Zulch, Thomas W. Pike, and Anna Wilkinson

Abstract

Collaboration between researchers and detection dog practitioners is key to optimizing training practices for operational detection dogs, yet published research is often difficult to obtain and not always easy to apply to practice. This chapter sets out the most common detection dog training practices and explains the science currently available in each area of the detection task: odor learning, odor discrimination, indication/alert and finally the search. Odor generalization and factors that impact it are clearly set out alongside the potential for recently researched new training paradigms to be incorporated into practice. Factors that affect operational performance and potential causes for incorrect final responses are discussed including; reinforcement schedules, training session structure, motivation and arousal, and the impact of the human half of the detection team. Throughout the chapter knowledge gaps in odor learning and training have been highlighted and suggestions of key areas for future research are made. An easy-to-read glossary provides the reader with clear definitions of the key terms used within the literature relating to operational detection dogs. This chapter aims to start to bridge the gap between laboratory findings and their field application by presenting the science to practitioners in an accessible way.

Keywords

Odour learning • Odour generalization • Detection training • Olfaction • Detection dogs

L. Caldicott · H. E. Zulch · T. W. Pike · A. Wilkinson (✉)
School of Life and Environmental Sciences, University of Lincoln, Lincoln, UK
e-mail: awilkinson@lincoln.ac.uk

L. Lazarowski (ed.), *Olfactory Research in Dogs*,
https://doi.org/10.1007/978-3-031-39370-9_9

Training dogs to detect an odor is far from a new concept. The tracking of animal and human scent by dogs can be traced back to Roman times; however, the training of detection dogs as we know them today began to surface in the mid-twentieth century (for a brief review of the history of working dogs see Hall et al. 2021). Training methods used in the field for dog detection work are rooted in tradition, handed down from trainer to trainer and little has changed in training approaches in nearly a century (Hall et al. 2021; Hayes et al. 2018; Troisi et al. 2019; Ramirez 2020). In contrast, our scientific understanding of how animals learn has increased substantially over the same period (Ramirez 2020). Knowledge is slow to filter through from the laboratory to the field, frequently hampered by the fact that scientific literature and the language used is not easily accessible to the average person in the field. Commonly, wider dissemination of the research is left to the writers of easily digestible web and social media posts which are often abridged and missing crucial points, or quoted with a confirmation bias to the writers' own beliefs.

Key to improving the performance of dogs in the field is the ability of trainers and handlers to apply up to date science to their work. This chapter will set out today's most common detection training practices and consider the research currently available in each area of the detection task: odor learning, discrimination training, the alert and finally the search. It will examine the effectiveness of training methods, based on current knowledge informed by research findings, as well as identify where knowledge is limited and investigation into alternative approaches may be warranted based on current understanding of learning theory. Implications for operational settings and factors that affect performance will be identified and recommendations or potential solutions provided. In addition, we will highlight some of the knowledge gaps remaining and give suggestions for key areas for future research. This chapter aims to start to bridge the gap between laboratory findings and their field application by presenting the science to practitioners in an accessible way.

1 Odor Learning

The first stage of detection training requires the dog to learn the relevance of the target odor. In practice, this is often referred to as "*imprinting*," although it should be noted that this term is adapted from ethology referring to any rapid, time-sensitive learning, typically in infancy (for example when a young animal becomes socially attached to an object, which is usually, but not necessarily, its parent; Hess 1959). It would never be used scientifically in the context of learning a target odor and so, for clarity, we will refer to it as "*odor learning*."

Exposure to the odor without pairing it with an *unconditioned stimulus* can have differing impact on odor learning, having been shown to be positive (Mandairon 2006; Mandairon et al. 2006; Yee and Wysocki 2001), negative (Dalton and Wysocki 1996) and neutral (Hall 2014; Fletcher and Wilson 2002; Cunzeman and Slotnick 1984). This is likely to be due to a variety of factors that modify salience;

see some of the discussion below. However, all studies that, during initial exposure, paired the odor exposure with a positive (or negative) outcome yielded improved discrimination ability (Fletcher and Wilson 2002; Mandairin et al. 2006) and therefore with our current understanding we would not recommend starting training with exposure to the odor alone.

1.1 Classically Conditioning the Odor

To learn the relevance of the target odor, training set-ups that facilitate *associative learning* (see glossary) are usually utilized to enable the dog to associate the presence of the target odor with a reinforcer (reward). It is widely accepted that *classical conditioning* (a type of associative learning) has an important role to play in all learning; including odor learning (Hall et al. 2021). However, how classical conditioning is implemented impacts the effectiveness of the learning. Simultaneous conditioning (reward placed alongside the odor) has been shown to be ineffective for facilitating odor learning, most likely due to blocking and/or overshadowing (Hall 2013). **Overshadowing** occurs when two stimuli are presented at the same time and one is more salient than the other (Jezierski 2016; Hall and Wynne 2018); for example, if the odor and food are placed in the same container and presented to the dog (e.g. Hall 2013) it is likely that the odor will not be salient to the dog as the food odor is intrinsically relevant. **Blocking** results in the neutral stimulus (target odor) failing to become a *conditioned stimulus* as a result of a second stimulus having been previously associated with the outcome (Smith 1994; Jezierski 2016). For example, if one odor is learned by the dog, and a second placed with it, it is unlikely the second odor will achieve the required association.

In contrast to simultaneous conditioning, *delayed conditioning* results in improved learning speed (Hall 2014), a reduction of the effect of disruptors (distractions) (Hall 2015) and improved low odor detection levels (Hall 2016). Delayed conditioning is a process where the target odor and unconditioned stimulus (the stimulus that will later become the reinforcer) overlap in their presentation, with the odor being presented first. For example, the target odor is presented to the dog and food rewards are delivered by the handler shortly after, but still in the presence of the odor (at source). This ensures a clear association is made between the odor being presented and the delivery of the reinforcer. We therefore recommend that delayed conditioning is used for odor learning wherever possible.

2 Discrimination

Discrimination is the ability of the dog to identify the target odor as different from all other environmental stimuli. In the field, the dog encounters an enormous number of potentially distracting stimuli (Hall and Wynne 2018). These might be visual, auditory or olfactory and the dog must learn to ignore everything apart from the trained odor(s). Training usually involves placing the target stimulus in a container with small holes to allow the odor to escape (e.g., a flour dredger). The target odor is presented alongside multiple identical containers that are empty or contain distractor odors (irrelevant odors that may or may not be of interest to the dog, or odors that could commonly be found in operational searches which the dog needs to learn to ignore). The dog is required to sniff each of the containers and reinforcement is given when they show an interest in the container containing the target odor. Initially, the dog may only have two or three choices; the number of containers is gradually increased as the dog learns the task and the range of distractor odors is adapted to facilitate learning to required *sensitivity* and *specificity* (Johnen 2013).

For stepwise instructions on the placement of target, blanks, and distraction containers see "Using blanks and interferents to ensure effective detection dog training" (DSTL 2018c).

3 The Final Response (Indication/Alert)

When a dog detects the target odor, they are required to perform a trained behavior to communicate to the handler the precise location of the odor. The "indication" or "alert" required may differ depending on the task at hand; some may require a solid sit/lie down and stare, whereas others may require the dog to be more specific and freeze with the nose close to the source of the odor (Mancini et al. 2015; Jezierski 2016). Proximity to the odor source will vary depending on the odor being trained, and in some instances may require a significant distance, particularly when the target is potentially harmful to the dog or handler (e.g. explosives or live animal detection; Jezierski 2016). The alert can be trained as part of the conditioning process often through *shaping* (Johnen 2013; Mancini et al.2015); or prior to odor learning as an operant behavior, which is then transferred via *higher order conditioning*, resulting in the odor becoming a *discriminative stimulus* (cue) for the behavior (Hayes et al. 2018).

When the dog is searching there are several different discreet outcomes:

- **True Positive**: Target odor is present and the dog alerts at its location.
- **True Negative**: Target odor is not present and the dog does not perform the alert behavior.
- **False Negative**: Target odor is present and the dog does not perform the alert behavior.
- **False Positive**: Target odor is not present and the dog performs an alert behavior.

In addition, there are two other potential responses classified by those who work in the field, but rarely reported on scientifically:

- **Interest**: The dog shows overt interest in an area but does not perform the alert behavior (DSTL, 2021b).

- **Nuisance Alert**: The dog alerts on a substance that contains elements of the target odor signature but the substance itself is not of operational interest (DSTL, 2021b). For example, plasticizers which are contained in many plastics are also added to explosives to reduce sensitivity (DSTL, 2021b). Nuisance alerts are a necessary side effect of effective generalization as the dog is technically correct at detecting what he perceives as the target odor. Attempting to train the dog not to alert on a nuisance substance can result in a narrowing of the generalization curve and an increase in false negatives (DSTL, 2021b).

4 The Search

There is little literature available on the process of training a dog to search; the general assumption is that dogs are proficient hunters and the skill comes naturally to them. However, well-planned training can assist the dog to learn to follow odor trails in a range of situations and under different conditions more effectively and improve the development of systematic search behavior. In addition, training handlers to read situations effectively can ensure that they best assist their dog by knowing when to direct their dog's search versus allowing a free search. Better understanding of the different roles of the partnership during the search phase and how to optimize this is likely to lead to more effective and efficient detection.

4.1 Target Odor Placement

It is important to ensure training provides adequate learning opportunities for the dog to understand how odor moves and thus how to search for an odor source. This is key to creating detailed and systematic search behavior. Careful consideration of the placement of the target odor in training can aid this learning. For example:

- To encourage a dog to systematically search the edges of a room or building, multiple targets can be placed repeatedly around the perimeter.
- Tables, chairs, and other obstacles can be used to create voids and corners, where odor may be pool and become trapped (Wasser et al. 2004).
- Hides near heat sources can result in the odor rising, traveling, and falling elsewhere in the room (Wasser et al. 2004).
- A hide placed on a vehicle facing the wind can result in odor being trapped in the crevices of the under carriage of the vehicle away from the target odor location.
- A hide that has been in place for several hours/days (aged) will have a different odor plume to one that had only been placed a few moments before (Wasser et al.2004).

4.2 Handler Assessment of the Search Area

Assessing whether the dog has effectively searched an area is an important handler skill and one that is not straightforward to achieve. There is evidence that handler perception of the likelihood of a target being present impacts the dog's propensity to locate the target (DeChant et al. 2020; Lit et al. 2011). As such, it is important to ensure the handler can use rules to ensure the search is systematic. One potentially helpful strategy is the use of "vigilance points," imaginary lines drawn at 45 degrees to corners within the area, creating zones which would not be crossed unless the dog was intentionally searching into the corners (Porrit et al. 2015). This could be a useful measure of the dog's effectiveness at covering an area during a free search as a lack of vigilance point crossings has been found to be correlated to increased number of false negatives (Porrit et al. 2015).

5 Generalization

While the dog is required to discriminate the substance they are searching for from other non-target odors, they must also allow for a level of variation within the item's odor profile and generalize their learning to account for this.

In the field, the substances dogs are required to detect will have some degree of variation from sample to sample (Jezierski 2016; Marshall and Oxley 2009; Hall and Wynne 2014). Factors such as the ambient temperature (Thiesan et al. 2005; Lotspeich et al. 2012; Lazarowski et al. 2020, manufacturing process, country of origin (Moser 2019), and the age of the sample can all affect the odor emitted. Furthermore, extrinsic factors may impact the dog's ability to detect the odor: for example, contraband is often concealed in, or mixed with substances that themselves have a strong odor (Ferton et al. 2002; Lazarowski and Dorman 2014; Hall and Wynne 2018); explosives are typically found alongside a wide variety of accelerants such as icing sugar or fuel oil (Hall and Wynne 2018); organic matter (plants/scat) will differ depending on the growing location or food consumed by the animal; or substantially different quantities from the samples used in training can hinder the dog's ability to detect a substance (Aviles-Rosa et al. 2021; DeChant and Hall 2021).

The required breadth of generalization (*sensitivity*), or discrimination (*specificity*), to a target odor will depend on the substances the dog is required to detect (DeChant and Hall 2021); too much generalization results in an unacceptable number of *false positive* indications; too little risks increasing the number of false negatives (DeChant and Hall 2021). Clearly, there is a degree of "trade off" which must be considered to determine the required level of generalization vs. discrimination, dependent on the needs of the task and the inherent risks associated with false positives or false negatives (Marshall and Oxley 2009; Hayes et al. 2018).

There are several factors that can be incorporated into the training method to assist in aiding the dog to generalize effectively (Mandairon et al. 2006; Marshall and Oxley 2009; Wright et al. 2017; Hall et al. 2021):

1. If the same odor sample is used consistently in training, the generalization curve will narrow (Moser 2019; Aviles-Rose et al. 2022). Ideally, a range of samples from different origins should be used to enhance generalization to account for variations within the composition of a substance (DSTL, 2018b).
2. The quantity of available odor can affect the dog's ability to respond to it (Aviles-Rosa 2021; DeChant and Hall 2021). Training on a variety of sample sizes is important for the dog to be able to generalize to both small and larger amounts of odor (DSTL 2018b). Odor chemistry is outside the remit of this chapter; however, it should be noted that it is not only the physical quantity that affects the amount of odor emitted, the surface area of the substance will also have an impact. For example, if odor is placed in a container with a small hole the amount of available odor emitted will be similar regardless of the physical quantity inside (DSTL 2018b).
3. Incorporating a variety of distractor odors, both separate from the target odor, as well as mixed in with the target odor will help the dog to differentiate the target odor alongside interferents that they may encounter in the field and still recognize them as target odor (Keep et al. 2021; Wright et al.2017; Hall 2017; DSTL 2018b).

In addition to the above suggestions, different training paradigms have recently reported an effect on generalization learning:

5.1 Sequential Training

Historically, detection dogs were trained using a sequential, or single odor, method (Hayes et al. 2018). This is where the dog is trained to detect a target odor in isolation, and additional target odors are learned individually and consecutively (Hayes et al. 2018). Efficacy of sequential learning has been supported in terms of learning individual odors (Williams and Johnston 2002). However, more recent research demonstrates that sequential training results in a reduced generalization when compared with other training methods (Lazarowski and Dorman 2014; DeChant and Hall 2021; Keep et al. 2021; Caldicott et al., in prep).

5.2 Compound Training

Compound training is the process of presenting the target odor as a mixture of target odors and/or target plus non-target odors. Recent research suggests that dogs trained in this manner demonstrated higher levels of generalization to a target mixed with a novel interferent than those trained sequentially (Lazarowski and Dorman 2014; Hall and Wynne 2018; Hall et al. 2021). Despite the improved efficacy, the compound training method is susceptible to the possibility of *blocking* or *overshadowing* (Moser 2019; Keep et al. 2021, Caldicott et al., in prep), and a mixture of two or more odors may result in a completely different odor profile (Lazarowski and Dorman 2014). Depending on the characteristics of the substances involved, this can impact the perception of the target odor, something which is often difficult to predict during training (Hayes et al. 2018).

5.3 Intermixed Training

Keep et al. (2021), investigated the efficacy of an "Intermixed" training method in which rats were trained to detect multiple individual target odors concurrently, together with the presentation of distractors. The results revealed a significant improvement in odor generalization, with animals on intermixed training outperforming both compound and sequential training (Keep et al. 2021). A follow-on study with dogs has also revealed similar results (Caldicott et al., in prep). More research with an increased number of target odors would be beneficial, however these findings suggest that concurrent training of multiple target odors provides improved generalization and odor learning, compared to sequential and compound methods.

5.4 Categorization

Odor generalization can be described as the process of assigning similar odors to "categories" that are likely to result in the same outcome, and thus produce the same behavioral response. Visual categorization is well understood in a number of animal species (for a review see Huber and Aust 2006). However, until recently this process had not been explored in the odor domain. Wright et al. (2017) translated these well-understood approaches to investigate *olfactory categorization* in dogs. Dogs were presented with accelerants on a variety of substrates that were either burned or unburned, such that the only consistent aspect of the target odors was the presence of the accelerant. Four of the six dogs in the experimental group succeeded in learning the task and succeeded in generalizing to new odors within their category, whereas all dogs in the control group, which did not have a categorical rule and had to rote learn the stimuli entirely failed the task. These findings suggest that dogs are able to categorize olfactory information in a similar manner to how many species categorize visual information and the use of alternative cognitive approaches could substantially enhance our approach to detection dog training. However, there are more questions that need to be asked: How do number and variation in training stimuli impact category formation? How does a categorical approach compare to a standard sequential learning approach or an intermixed approach and under what conditions, if any, might it be superior? Using well-understood cognitive approaches such as this has the potential to transform the process of detection training.

6 Schedules of Reinforcement and Extinction

Work requirements in the field impact performance. One key area for consideration in this context is reinforcement schedules as these can quickly affect motivation. The frequency of delivery of reinforcers (schedule of reinforcement) can affect the acquisition and maintenance of the trained behavior (Hall 2015). In most training situations, during the initial learning phase, the dog receives their reward every time they successfully encounter and alert to the target odor (continuous reinforcement), however once the search and locate behavior is effectively learned, maintaining it may require a different approach (Hall 2015). Continuous schedules of reinforcement allow for rapid conditioning of the target odor due to the *contingency and contiguity* between the reinforcer and odor. However, an intermittent reinforcement schedule is generally believed to strengthen the performance of a trained behavior (*see* Hall 2017) and provide more resistance to the effect of *extinction* (the reduction of a behavioral response when reinforcement provided in learning is withdrawn) (Thrailkill et al. 2016).

Traditionally an intermittent schedule of reinforcement consists of primary reinforcement only being given for a proportion of correct responses, resulting in some correct responses not being reinforced. The aim of this practice is to increase motivation and performance as the dog "works harder" in an attempt to gain reinforcement. Which responses will and which will not be reinforced should be unpredictable, as if the dog is able to predict when reinforcement is likely to be forthcoming, performance will decrease during the intervening repetitions (Thrailkill et al. 2016; Ramirez 1999). An intermittent schedule of reinforcement has been shown to improve persistence in searching when no reinforcement is given (*Partial Reinforcement Effect on Extinction*) (Thrailkill et al. 2016). In the field, continuous reinforcement presents substantial logistical challenges as it is not always practical to reward the dog in operational situations. For example, there may be safety concerns, or an inability to confirm what the substance is—risking reinforcing a false positive to a distraction odor. In addition, the target odor may be infrequently present which means that the search portion of the task receives no reinforcement and once again performance may reduce (Porritt et al. 2015; Hall et al. 2017). This lack of reinforcement (operational extinction) in the field is compounded by the fact that dogs are excellent contextual learners (Gazit 2004); should operational situations differ substantially from training, they soon learn when reinforcement is unlikely and performance may then reduce in a context-specific manner (Hall 2017; Porritt et al. 2015). Due to these constraints, when a dog is working operationally, they are effectively working under *extinction* conditions, which if not effectively addressed through appropriate training can result in reduced performance and a greater risk of false responses (Hall 2017; Porritt et al. 2015).

To prepare a dog for operational extinction, changing the reinforcement schedule after the initial odor learning is complete, from continuous to intermittent, is likely to benefit detection dog training programs (DSTL 2019). However, deciding on the optimum ratio for intermittent reinforcement and how best to transition a dog from a continuous rate of reinforcement to an intermittent schedule requires further research. If incorrectly applied it can result in a decrease in performance (via extinction) and motivation, as well as increased frustration (Ramirez 1999). There is little research into this area, but Aviles-Rosa et al. (2022) reported a successful transition from a continuous schedule to a 60% random rate of reinforcement in two increments; continuous to 80 to 60%. However, they do not report the time taken to reach criterion at any stage.

6.1 Secondary Reinforcers

Incorporating *secondary reinforcers* alongside *primary reinforcers* within the randomization schedule (rather than a void) may potentially be helpful (Ramirez 1999). Praise and social interactions are commonly used as secondary reinforcers and evidence suggests they are effective (Dudley et al. 2019; Ramirez 1999; Venniro et al. 2019). The value of praise, or any other meaningless stimuli, can be created/increased and maintained if it is consistently and proactively classically conditioned in the same way as the odor—paired with a primary reinforcer. In addition to reducing the frequency with which primary reinforcers are delivered, which may have benefits in some situations, the use of a range of reinforcers has been reported to increase motivation and attention (Cameron and Pearce, 1994; Ramirez 1999), presumably through the unpredictability of the reward received. It should be noted that for optimum effect the conditioning process of secondary reinforcers must be maintained (Ramirez 1999). This approach warrants investigation for use in detection training—the potential to choose a secondary reinforcer to suit the situation creates a level of flexibility that could be optimized for individual dogs; combat some of the concerns regarding the inability to reward with a primary reinforcer in the field; as well as avoiding lean intermittent schedules which may risk decreasing performance (Hall et al. 2021; Troisi et al. 2019).

6.2 Use of Non-Hazardous Odors

After the initial task has been learnt, performance needs to be maintained over time, especially if the dog is required to be able to locate and alert to targets that are not frequently encountered in everyday work. As mentioned above, there are times, for example, in airport security and explosive detection, where dogs are often tasked to repeatedly search large areas for prolonged periods where they do not (thankfully) make a find at the end of the search. This is demotivating for the dog, particularly if the area is often re-searched (e.g. airports/stadiums) (Gazit et al. 2004), and search behavior is seen to decrease (during free search) with the risk of false negatives increasing (Gazit et al. 2004; Porrit et al. 2015).

One suggestion for overcoming this is the use of a "non-contraband" odor as part of the dog's target odor bank, providing the opportunity for the dog to make a "safe" find in operational scenarios (Porritt et al. 2015). In this study, after initial training, dogs were split into three groups. For one group search training was maintained on three target explosives, another group was maintained on a single non-contraband "safe" find, and the third group was exposed to the same search area but without any targets present. Target odors were then reintroduced to the area and performance assessed. The results provide clear evidence of the impact of repeated blank searches in the same location, with detection rates falling to 25% when target odors were reintroduced into the area. Interestingly, the dogs exposed to repeated blank searches retained a similar search vigilance score to the other groups in free searches (where the dog chooses the search pattern); that is, they achieved similar proximity to the target odors as the other two groups, despite not alerting to the target. This suggests that dogs exposed to repeated blank searches in a specific area may cease alerting to targets, regardless of their perceived search behavior *and* successful target detection during concurrent maintenance training in a different location. However, during directed searches (where the handler directs the dog to search specific areas) search vigilance is reduced; implying handler behavior may also have changed when there was little expectation of a target being present.

Maintenance training on the non-contraband odor resulted in significantly improved detection rates when compared to dogs exposed to repeated blank searches. However, it should be noted that when compared to dogs maintained on all the target explosives, results revealed significantly lower detection rates and significantly higher false alerts. A further study found maintenance training over a 12-month period which was carried out on one odor out of a bank of 10 initially trained, increased discrimination accuracy on all 10 odors (Lazarowski et al. 2021). Interestingly though, in this study the dog's performance in operational searches was not maintained, suggesting that while odor memory is long-lasting, learning regarding the complexities of searching and alerting in a working environment, where odor accessibility and movement will be variable, is accomplished through training repetition and is an important component for successful search outcomes.

Currently, there is no more effective method proposed for maintaining detection rates in areas where the target odor is unlikely to be found than the use of a "co-trained" non-contraband odor (Porritt et al. 2015). This study generates many more research questions that need to be fully explored to appreciate the impact of this method on detection training. For example, Porritt et al. (2015) tested the reintroduction of target odors after six weeks, and so is not able to predict the long-term impact and whether the "non-contraband effect" will be maintained indefinitely. Also, the total number of trained odors learned, as well as the ratio of non-contraband versus target odors need to be fully explored as, for example, increasing the variability of detected odors in areas where true targets are rare, may have the potential to positively affect overall detection rates.

6.3 Unknown Substances

For dogs that have a high rate of finds in operational searches (e.g., drugs/tobacco) the lack of field reinforcement can be a particularly challenging problem. The presence of drugs often has to wait for laboratory confirmation (Jezierski, 2014), resulting in a high rate of un-rewarded true positives. We have already discussed the use of an intermittent schedule of reinforcement to combat this problem (Thrailkill et al. 2016), however, for these dogs in particular, there is a high contingency between the non-rewarding of target odors and an operational setting, increasing the opportunity for the dog to make the association over time (Porritt et al. 2015). If the handler is proficient at understanding the dogs' search and alert behavior, and the training plan is robust with generalization and discrimination effectively accounted for, it could be reasonably proposed that an operational dog would be unlikely to give false positive, or at least will do so at a rate that is understood. It may be that a "trade-off" between the risk of rewarding a false positive or nuisance alert, and that of reduced performance over time resulting from non-rewarded true positives, needs to be considered. Understanding the effect of sporadic rewards, both primary and secondary, on false positives, and the ability to correct the error through training, is an area that has not been investigated, however systematic and detailed research is essential to elucidate this, reduce the risk of drop out and assist in developing optimal operational guidelines.

7 False Positive and False Negative Alerts

Potential causes of incorrect final responses relating to training shortfalls could be attributed to factors such as:

- Similarities between target and distractor odors that require additional decision-making time, may result in interest being shown prior to, or instead of the final alert. Increasing time spent on discrimination training through adapting interferents and distractors may help mitigate this (DSTL 2018c).
- Air flow trapping drifted odor in foliage, crevices or corners, can result in interest being shown in an area where the target odor is not located. This can result in an inexperienced dog having difficulty determining the precise location of the target odor. Increased training in search to source and alerting at source may help mitigate this.
- Contamination of training samples during learning will hinder the correct associative learning taking place. For example, if the dog is rewarded on samples that are contaminated with human scent or other substances, blocking or overshadowing may result in odor/reward association being misdirected. The importance of selection and management of training samples in the widest terms is beyond the scope of this chapter, but must always be considered when performance drops (*see* DSTL 2018d).

- A lack of generalization to the target odor (section above) will result in the dog not alerting on variations of the target odor. This can be adapted by considering the distractors and novel interferents used during training and variation in the samples of used of the target odor.
- The dog has learned to indicate based on the handler's body language; this can often occur if the handler frequently knows the location of the target odor during training and provides the dog with 'clues' through their positioning or timing of the reward (*section below).*
- If the required associative learning is not made effectively at the start of training, or the training set ups do not sufficiently account for discrimination learning, the dog may learn to indicate on any anomaly (i.e., novel or "out of place") odors in the environment, rather than the intended target odor. Again, the importance of a systematic approach to the discrimination phase of training is essential (DSTL 2021b).

Such shortfalls can be rectified by putting in place a systematic review of the training procedures. Additionally, it should be noted that maintaining good records during training is important to provide early evidence to inform any adaptations that may be necessary (see: DSTL 2022).

Particular attention should be paid to minimizing false negative indications; missing a target usually has the potential to result in the most serious consequences, for example explosives, drugs or live victims remaining undetected. False negatives are usually reported to be the result of poor discrimination/generalization, lack of systematic searching or a problem with the dog's focus and/or motivation for the task. However, there is some evidence to suggest that the alert behavior itself may also impact the likelihood of both false positives and false negatives.

Preliminary evidence suggests a commonly trained sit/stare indication may increase the risk of false positives when compared to a stand/stare, at least in medical detection scenarios. A stand/stare position is a natural 'pointing' behavior for the dog; whereas a sit is an operantly trained behavior that potentially requires higher cognitive processing to perform (Mancini et al. 2015). The increased challenge of performing the sit alert has the potential to shift the focus from the detection task to anticipation of performing the final behavior, increasing the risk of a false positive (Mancini et al. 2015). More recently, Essler et al. (2020) investigated the effect of a sit and stare versus a stand and stare, in medical detection dogs and although the sample size was small it revealed interesting results. Dogs performing a sit alert spent less time sniffing the samples compared to dogs performing a stand indication (Essler et al. 2020). While a sit indication moves the dog's nose away from the sample, a stand position allows the dog to maintain its nose close to the source, allowing for prolonged information gathering prior to, and following the initial decision making and alert behavior, giving the dog an opportunity to "change their mind" (Essler et al. 2020). While more investigation is required, and not withstanding safety aspects in some detection scenarios, these findings suggest a traditional sit may not be the best alert for the dog to attain optimal accuracy.

7.1 The Go/No-Go Paradigm

In contrast to the final response required to the presence of the target odor, a blank search, where no target is found, expects no response (DSTL 2021b; Aviles-Rose et al. 2022; Mancini et al. 2015). This scenario is known as a Go/No-Go paradigm, where a negative response (no target present) is identified by the *absence* of a behavior; i.e., the dog is required to search until the handler decides the area has been sufficiently covered, if the dog has not alerted in that time, no target present is assumed. The end of the search in this scenario is not contingent on the dog's behavior but occurs when the handler decides that the area has been covered, which in turn will be influenced by the handler's ability to interpret their dog's search behavior effectively (see section below).

This Go/No-Go paradigm is known to have a bias to a 'Go' response (Schoon, 1996; Huber et al. 2005; Shenoy and Yu 2012; Ramirez 2020) which is of particular interest in relation to detection dogs (who are specifically selected for their high drive) given the level of impulse control required to inhibit a response (refrain from alerting) (Fadel et al. 2016). In addition, evidence suggests, the presence or absence of the target is potentially determined by the dog much sooner during the search than perhaps handlers realize. Time spent sniffing a sample without target odor (true negative), is significantly less than when target odor is present (true positive), or perceived as present (false positive) (Mancini et al. 2015; Concha et al. 2014; Essler et al. 2020). Requiring the dog to continue searching once they have assessed it as "clear" could increase the likelihood of another behavior (i.e., the alert) being offered. An alternative approach to the go/no-go paradigm is a two-choice paradigm; where an 'all clear' alert is trained in addition to the target present alert. This has been reported as successful in detection dog training (Ramirez 2020) and has been used in some papers relating to detection dogs (Johnen et al. 2017; Edwards 2019; Murarka et al. 2019). To date the benefits and/or limitations of adopting this paradigm are not yet understood; however, given the possible challenges of a go/no-go approach, it is essential that this is investigated thoroughly as it may have the potential to improve detection accuracy.

Training of the final alert response is an area that has been neglected by current research and key questions remain. Investigations into the effect of using operantly trained vs. natural alert behaviors and the potential for a two-choice training paradigm need to be explored. Further, research into the communicative signals of dogs while searching and the potential for artificial intelligence and machine learning to assist in human interpretation of dogs' behavior in the presence vs absence of targets, are also areas that have the potential to uncover practices that will improve the efficacy of detection in the real world (Mancini et al. 2015).

8 Training Session Structure

There has been little research on the length and regularity of training sessions in detection dogs and most current recommendations are based on traditional practice. Some evidence suggests that fewer regular sessions (e.g., weekly rather than daily) assist in maintaining interest and improving arousal and motivation resulting in fewer sessions required to learn the task (Troisi et al. 2019; Meyer and Ladewig 2008); however this has to date only been tested in the learning of simple behaviors. Demant et al. (2011) similarly found that 1–2 short training sessions per week significantly improved task acquisition when compared to longer daily or weekly sessions in an obedience task. This is an area that requires further investigation in circumstances specific to detection training. Operational detection dogs are often trained daily and for long periods of time (Alexander et al. 2011), especially in the early stages; should this finding translate to detection training less frequent training may be beneficial for optimal outcomes.

8.1 Length of Training Searches

Evidence suggests that search duration is negatively correlated with successful outcomes. Jezierski et al. (2008), found that the longer the dog spent in the search the more likely they were to give false negative or false positive alerts. DSTL (2019b) recommends an optimal training session should include one long search and several short searches. Theoretically, this provides an opportunity to improve search fitness and prolonged motivation on the longer searches, while finding the target quickly in the shorter searches encourages intense search behavior from the beginning of the task. While this appears a logical recommendation there is no evidence as yet that supports any particular training for searching on operational performance in searches of differing length.

8.2 Fitness

Searching is a physically demanding activity for the dog and fitness should be built up gradually (Gazit and Terkel 2003). Environmental factors that affect the dog's body temperature (e.g. hot/cold weather, stress levels) need to be carefully assessed during each search as the dog must pant to regulate their body temperature under hot conditions. The dog is physically unable to sniff and pant at the same time, and so will be unable to search effectively when panting (Gazit and Terkel 2003; Jezierski et al. 2016; Troisi et al. 2019).

9 Motivation, Arousal, and Performance

Behaviors carried out based on the receipt of rewards or punishers given following the behavior are extrinsically motivated, whereas behaviors carried out simply for their own sake because the individual enjoys the activity in its own right, are intrinsically motivated (Deci, 1971; Ryan and Deci 2020). Intrinsically motivated behaviors are more resistant to disruption from distractions and extinction, generally resulting in improved levels of performance when compared to extrinsically motivated behaviors (Ryan and Deci 2020). Extrinsically motivated behaviors are also influenced by "establishing operations"; factors that temporarily affect the value of a reinforcer. For example, being hungry prior to training will increase the desire for food rewards, but being full will decrease it (Cameron and Pierce 1994).

Intrinsically motivated behaviors will differ between individuals; some breeds have been selectively bred for certain characteristics and those who have been selected for search/hunt behavior may find detection work more intrinsically rewarding than others (Lazarowski et al. 2020; Hall et al. 2021). While often it is believed that the increased "drive" to search that this may create is a pre-requisite for many operational dogs, it should be noted that it may also result in an increased need for significant levels of "impulse control" in order to cease the search behavior and perform an operantly conditioned "sit"—an extrinsically motivated behavior (Mancini et al. 2015; Fadel et al. 2016). The tradeoff between different temperament traits and intrinsic motivations and their relationship to detection dog training and performance is an area that requires further investigation.

9.1 Motivation and Deprivation

It is not uncommon for deprivation to be used within detection training to increase motivation; often the dog is only allowed access to a favorite toy during training sessions, or the meal prior to a training session may be skipped (Hall et al. 2020). Increasing motivation in extrinsically motivated search dogs through the deprivation of reinforcers (or infrequent opportunities to work for intrinsically motivated dogs), must be carried out with care, not only due to potential welfare concerns but also to ensure the arousal and resulting motivation levels do not pass the optimum level for performance (Hall et al. 2020). The commonly cited Yerkes-Dodson law outlines the link between arousal and performance; arousal must be maintained at an intermediate level (specific to the individual and the task) to obtain peak performance (Yerkes and Dodson, 1908; Bray et al. 2015). If the reinforcer is of particularly high value, and deprivation results in over arousal, the dog may not be able to focus on the task at hand, increasing the risk of performance decrement such as the potential for a spontaneous increase in generalization, resulting in a false positive alert (*see* Hall et al. 2020).

10 Human Handler Influence

Dogs have an incredible ability to read human body language and very quickly pick up on any cues handlers unknowingly give (Jezierski et al. 2008; Lit et al. 2011; DeChant et al. 2020). During training, handlers must be aware of their body language, ensuring they don't inadvertently indicate to the dog the location of the hide (DSTL 2021; Wasser et al. 2004). For example, consistently standing next to the hide, or at the other end of the room to the hide, can act as a cue to the dog and the association will soon be learned; the most miniscule of changes can be perceived by the dog, for example even a deep breath as the dog gets near to the hide. This can not only result in the dog waiting for these cues before alerting, but also increase the risk of false positives should the handler produce any of the 'false cues' during the search (Jezierski et al. 2008; Lit et al. 2011).

It is also not uncommon for handler's body language to change during a 'blind' search (in which the handler does not know where the hide is), where the handler believes there will be no target odor, or in any situation where the handler is under increased stress (DSTL 2021b; Jezierski et al. 2008; DeChant et al. 2020; Lit et al. 2011). In these situations, the handler may crowd the dog in an attempt to see for themselves what the dog is doing or they may be unusually disengaged from the dog as he works (Lazarowski et al. 2020; Lit et al. 2011). Interestingly research has shown that a distracted handler is beneficial to the dog's search performance and handlers placed under stress resulted in a lower latency to detection, when compared to non-stressed handlers, implicating handler 'interference' in reduced performance (Zubedat et al. 2014).

Another impact of the human handler on the dog's performance is their competence in understanding their dog's behavior. It has been demonstrated that handlers and trainers are not always proficient at reading their dogs body language (Tami and Gallagher 2009), and the nuances of behavior during a search are often missed or misinterpreted by the handler resulting in a breakdown of communication (Trosi et al. 2019; Mancini et al. 2015). For example, Wasser et al. (2004) found incorrect false negatives were reported due to the handler moving the dog away and not recognizing the indication. Mancini et al. (2015) discusses the body language of dogs during the search at length; they highlight the importance of being able to recognize not only the final response but also interest, and the decision-making processes throughout the search. This area provides interesting potential for future research. Advances in our understanding of dogs' body language and the advent of modern technology for recording and analyzing search behavior and utilizing machine learning may in future facilitate more in-depth investigation and potentially enable us to work toward the development of systems that can support handlers' decision making.

To combat some of these concerns regarding the handler's effect on dog performance, incorporating scenarios akin to operational work into the training regime—potentially through regular formal assessments that may create similar pressure—is good preparation for the dog / handler team, reducing the risk of unwanted learning and habits developing in both. Trainers should ensure that any contextual cues which allow the dog to differentiate between operational and training scenarios are removed (Porrit et al. 2015). For example, are there always certain people/other dogs present in training that are absent in operational work? Does the daily routine change on training days compared to operational days? If training cannot take place in operational locations does the training always take place in the same or similar locations? Could an "on location" training session with an out-of-area team, to share knowledge and training samples be arranged? Is the equipment carried and used the same in both training and operational days (for example the same rewards carried/harness used/clothing worn)? This is an area that should be prioritized for further research; fully understanding the impact of the human handler and work vs training-related routines on the dog in detection work is key in identifying and overcoming such barriers that prevent optimal performance from the dog.

11 Extraneous Factors that Affect Performance

There are many reasons a dog's performance may deteriorate and commonly the first port of call is to increase or adapt the training; however, assuming the training regime is the cause may result in overlooking important factors. There are numerous external events that may affect the dog's ability to work (DSTL 2019; Hall et al. 2021): what has happened in the dog's life over the preceding few days/weeks? Has a conspecific been ill/away? Has the dog had more or less exercise than normal? Is there any potential pain/upset tummy? Have there been visitors (canine or human) to the home/kenneling? Such external stressors, good or bad, can affect the dog's ability to perform at normal levels for several days.

12 Conclusion

While this chapter has sought to present an overview of current training methods alongside the scientific evidence base, and understanding of learning theory, it should be noted that there are many more areas of behavioral, cognitive and social psychology that are of interest, both to odor learning and learning in a wider context. Areas such as social learning and errorless learning are emerging within the literature currently and may provide additional understanding that will expand the traditional approach taken in detection training (Troisi et al. 2019; Hall et al. 2021).

Our understanding of the processes involved in training detection dogs has progressed rapidly in recent times, and there is no doubt it will continue to do so, as it is clear that there remains much more research to be done. Each study not only informs effective changes to training methodology but also adds to our understanding of where knowledge is lacking, and so directs future research. Designing and carrying out scientific studies which stand up to scrutiny by peers is a challenging and complex task; if this effort is not to be wasted, effective transference of the results, to those directly involved in the training of operational dogs on the ground, is imperative (Hall et al. 2021; Troisi et al. 2019). Consideration must be given to accessible and appropriate dissemination of research results, and to how operational handlers and trainers can realistically incorporate findings into their training protocols. Operational handlers and trainers should be open to adapting their methods and trust that the knowledge science brings will enhance their outcomes (Hall et al. 2021). Conversely, the scientific community needs to be available and open to receiving feedback from those undertaking the training on the ground, as they hold a wealth of knowledge and experience that should not be underestimated (Hayes et al. 2018; Hall et al. 2021). We hope that this chapter goes some way to bridging the gap between science and practice.

13 Glossary of Learning Theory Terminology Commonly Applied in the Field.

13.1 Associative Learning

The majority of learning occurs through the animal making associations between events or objects in the environment (stimuli). There are two types of associative learning: Classical Conditioning (Pavlovian or Respondent Conditioning) and Operant Conditioning (Instrumental Conditioning). *For a detailed description of the principles detailed below see* Pearce (2013).

13.2 Neutral Stimulus

Something that attracts attention but produces no emotional or behavioral response from the dog.

13.3 Unconditioned Stimulus

Something that produces an involuntary or reflux response from the dog. (E.g., the smell of food resulting in salivation).

13.4 Conditioned Stimulus

The unconditioned stimulus becomes the conditioned stimulus after the conditioning process is complete.

13.5 Classical Conditioning

The pairing of a "neutral stimulus" with an "unconditioned stimulus." Over time the neutral stimulus becomes *'conditioned'* and presentation results in a similar reflex response to the unconditioned stimulus that it was paired with (Pearce 2013). *For example, if a dog is given food every time he hears the fridge door opening, he will salivate at the sound of the fridge door opening.*

Simultaneous Conditioning: A neutral stimulus (odor) is presented at the same time as the unconditioned stimulus (food-toy). This method produces a weak association as presenting both stimuli at the same time risks the potential for blocking or overshadowing (see main text).

Trace Conditioning: The neutral stimulus (odor) is presented and withdrawn prior to the presentation of the unconditioned stimulus (food/toy) (Pearce 2013). The gap between the withdrawal of first and presentation of the second stimuli is important, and must be relatively short for an effective association to be made; this is known as temporal contiguity (closeness in time). In addition, spatial contiguity (proximity) can affect the associative learning; if the unconditioned stimulus is presented too far away from the neutral stimulus the link between the two diminishes. For example, in a training scenario, this can be the conditioning principle at work when the dog locates an odor and moves away from it to receive the reward from the handler.

Delayed Conditioning: The neutral stimulus (odor) and the unconditioned stimulus (food/toy) overlap; with the neutral stimulus (odor) being presented first and the unconditioned stimulus being presented while the neutral stimuli is still present (Pearce 2013). This represents the most effective method of conditioning as the overlap in presentation eliminates any delay in temporal or spatial contiguity, ensuring the neutral stimuli is salient and entirely predictive of the unconditioned stimuli. The length of delay between the presentation of the first and second stimuli and the duration of both stimuli being present, giving the dog the opportunity to process the events, can improve the strength of the association. For example, the odor is presented for several seconds, food is then given at source (delivered at the same location as the odor), food continues to be given for several seconds until the odor and food are removed.

Backward Conditioning: This is usually a training error rather than an intended method of training where the unconditioned stimulus (food/toy) is presented before the neutral stimulus (odor). The reversal in the presentation of the neutral and unconditioned stimulus lacks any predictive nature making associative learning unlikely (Pearce, 2013). For example, a handler who is keen to reduce temporal contiguity pre-empts the dog's approach to the odor and produces the reward before the odor is salient to the dog.

13.6 Operant Conditioning

Learning that occurs due to the consequence of performing a behavior (Pearce 2013). A behavior is regularly followed by food (rewarded), the behavior is more likely to be repeated, if a behavior is regularly followed by a zap from a shock collar (punished) it is less likely to be repeated.

Note: *Although both classical and operant conditioning differ in their effect on learning, they are not mutually exclusive, both will occur simultaneously—regardless of the intent of the trainer (*Pearce 2013*).*

13.7 Contingency

The reliability of the reinforcement (or punishment) following a behavior. The more predictable the outcome of the behavior, the quicker learning will take place (Pearce, 2013).

13.8 Contiguity

In associative learning the time delay between the presentation of one stimulus and second stimulus (temporal contiguity), the distance between the presentation of the two stimuli (spatial contiguity) affects the speed and strength of the association (Pearce 2013).

13.9 Positive Reinforcement

The addition (positive) of a stimulus that the dog finds rewarding (reinforcing), e.g., food or toy (Pearce 2013). Evidence reports the use of positive reinforcement training improves learning and attentiveness to the handler, thereby improving overall performance (Haverbeke et al. 2008; Deldalle and Gaunet, 2014). If the handler and dog's relationship is built on an effective reward-based system, the dog is likely to be more motivated to engage in the tasks requested, and compliant to the handlers' directions (Deldalle and Gaunet, 2014; Troisi et al. 2014).

13.10 Primary Reinforcers

Primary reinforcers have an innate reinforcing effect and are usually linked to a biological need (e.g. food, water, sleep) (Ramirez 1999).

13.11 Secondary Reinforcers:

Reinforcers that are learned through association with a primary reinforcer (e.g. clicker, light, sound) (Ramirez 1999).

13.12 Differential Reinforcement

Odor learning via discrimination training requires differential reinforcement. This is the process of providing positive reinforcement when the dog makes the correct choice and withholding reinforcement for an incorrect choice (negative punishment) (Pearce 2013). The dog learns through operant conditioning that proximity to the odor results in reinforcement (Wasser et al. 2004). *For example, the dog is presented with two or more identical containers—one holding the target odor while the others are either empty or contain novel 'distractor' odors. The dog is reinforced when they investigate the container containing the target odor.*

13.13 Intermittent Reinforcement (see Pearce 2013)

Reinforcement of correct behavioral responses at varying intervals. Intermittent reinforcement can be provided on four schedules:

Variable ratio: Reinforcement is given on a schedule that is unpredictable (random) but within the constraints of a given ratio. For example, in a 1:5 variable ratio, reinforcement is provided once for every five correct responses; which of the five responses is reinforced each time is random.

Variable interval: Reinforcement is provided for one correct response regardless of how many correct responses are completed within a given timeframe. For example; in a five-minute variable interval schedule, reinforcement is given once at different points within the five minutes.

Fixed ratio: Reinforcement is given after a fixed number of correct responses. For example, in a 1:5 fixed ratio schedule, reinforcement is provided after every fifth correct response.

Fixed interval: Reinforcement is given after a pre-determined time period. For example, in a 5-min fixed interval schedule, reinforcement is provided after every fifth minute.

13.14 Shaping

Shaping is the process of rewarding spontaneous behaviors offered by the dog in ever closer approximations to the desired behavior, usually via the use of marker or clicker training (Ramirez 1999; Pearce 2013). In detection this can be applied by "marking" the dog at the precise moment the dog's nose is above the odor container, this is repeated while very gradually increasing the delay between the dog's nose arriving at the correct container and delivering the "mark" (DSTL 2021a; Wasser et al. 2004). This effectively allows the dog to learn the required behavior of holding the nose over the container containing the odor. Over time a skilled clicker trainer can wait for even more desired responses, such as a moment of stillness, and continue to shape the indication into a full 'nose hold freeze' (DSTL 2021a; Ramirez 1999).

13.15 Discriminative Stimulus—S^D

A stimulus, usually a verbal cue, that is a predictor of a behavior, e.g., "sit." For it to be a true discriminative stimulus the behavior should not happen in the absence of the cue (Ramirez 1999).

13.16 Higher Order Conditioning

The transference of an operantly conditioned behavior to finding the target odor (Ramirez 1999). A behavior is taught to a reliable verbal cue e.g. "sit"—the *discriminative stimulus*. The required new cue (the odor) is presented, immediately followed by the old cue ("sit"); eventually, the dog will associate the odor (new cue) with the "sit"; the old cue becomes redundant and the odor becomes the new discriminative stimulus.

13.17 Habituation

A decrease in a reflex response to an environmental stimulus through repeated exposure. An adaptive process that avoids unnecessary physiological reactions to prevent becoming overwhelmed and fatigued (Pearce 2013).

13.18 Latent Inhibition

A process that inhibits associative learning due to previous un-reinforced exposure to a stimulus (Pearce 2013). For an association to be made between the neutral stimulus (odor) and the unconditioned stimulus (toy/food), the neutral stimulus must be salient to the animal (Ramirez 1999). If the brain has already learned to disregard the odor through constant exposure resulting in no consequence, associative learning will, at best, be slow to manifest (Dalton and Wysocki, 1996).

13.19 Sensitivity

The ability of the dog *to correctly detect the target odor*. Focus on true positive responses and so requires more generalization than discrimination. High sensitivity increases the risk of false positives (alerting where there is no target odor).

13.20 Specificity

The ability of the dog to *correctly identify NO target odor*. Focus on true negative and so requires more discrimination than generalization. High specificity increases the risk of false negatives (not alerting where there is target odor).

References

Alexander MB, Friend T, Haug L (2011) Obedience training effects on search dog performance. Appl Anim Behav Sci 132:152–159. https://doi.org/10.1016/j.applanim.2011.04.008

Aviles-Rosa EO, Fernandez LS, Collins-Pisano C et al (2022) The use of an intermittent schedule of reinforcement to evaluate detection dogs' generalization from smokeless-powder. Anim Cogn. https://doi.org/10.1007/s10071-022-01648-y

Aviles-Rosa EO, McGuinness G, Hall NJ (2021) Case study: an evaluation of detection dog generalization to a large quantity of an unknown explosive in the field. Animals 11:1341. https://doi.org/10.3390/ani11051341

Axe JB, Laprime AP (2016) The effects of contingent pairing on establishing praise as a reinforcer with children with autism. J Dev Phys Disabil 29:325–340. https://doi.org/10.1007/s10882-016-9526-9

Bray EE, MacLean EL, Hare BA (2015) Increasing arousal enhances inhibitory control in calm but not excitable dogs. Anim Cogn 18:1317–1329. https://doi.org/10.1007/s10071-015-0901-1

Cameron J, Pierce WD (1994) Reinforcement, reward, and intrinsic motivation: a Meta-Analysis. Rev Educ Res 64:363–423. https://doi.org/10.3102/00346543064003363

Concha A, Mills DS, Feugier A et al (2014) Using sniffing behavior to differentiate true negative from false negative responses in trained Scent-Detection dogs. Chem Senses 39:749–754. https://doi.org/10.1093/chemse/bju045

Cunzeman PJ, Slotnick BM (1984) Prolonged exposure to odors in the rat: effects on odor detection and on mitral cells. Chem Senses 9:229–239. https://doi.org/10.1093/chemse/9.3.229

Dalton P, Wysocki CJ (1996) The nature and duration of adaptation following long-term odor exposure. Percept Psychophys 58:781–792. https://doi.org/10.3758/bf03213109

DeChant MT, Ford C, Hall NJ (2020) Effect of Handler Knowledge of the Detection Task on Canine Search Behavior and Performance. Front VetY Sci 7: https://doi.org/10.3389/fvets.2020.00250

DeChant MT, Hall NJ (2021) Training with varying odor concentrations: implications for odor detection thresholds in canines. Anim Cogn 24:889–896. https://doi.org/10.1007/s10071-021-01484-6

Deci EL (1971) Effects of externally mediated rewards on intrinsic motivation. J Pers Soc Psychol 18:105–115. https://doi.org/10.1037/h0030644

Defence Science and Technology Laboratory (DSTL) (2018a) Using blanks and interferents to ensure effective detection dog training. https://www.cpni.gov.uk/canine-detection-guidance-notes. Accessed 17 Jul 2022

Defence Science and Technology Laboratory (DSTL) (2018b) Why and how to control contamination. https://www.cpni.gov.uk/canine-detection-guidance-notes. Accessed 17 Jul 2022

Defence Science and Technology Laboratory (DSTL) (2018c) Canine odour discrimination test. https://www.cpni.gov.uk/canine-detection-guidance-notes. Accessed 17 Jul 2022
Defence Science and Technology Laboratory (DSTL) (2018d) How to train your dog to generalise across varying types and sizes of targets. https://www.cpni.gov.uk/canine-detection-guidance-notes. Accessed 17 Jul 2022
Defence Science and Technology Laboratory (DSTL) (2019) Maintaining the operational performance of detection dogs. https://www.cpni.gov.uk/canine-detection-guidance-notes. Accessed 17 Jul 2022
Defence Science and Technology Laboratory (DSTL) (2022) Using training records to optimise detection dog performance. https://www.cpni.gov.uk/canine-detection-guidance-notes. Accessed 17 Jul 2022
Defence Science and Technology Laboratory (DSTL) (2021a) Understanding detection dog indications ("hits", "misses", "false alarms" and "nuisance alarms"). https://www.cpni.gov.uk/canine-detection-guidance-notes. Accessed 17 Jul 2022
Defence Science and Technology Laboratory (DSTL) (2021b) An introduction to marker training. https://www.cpni.gov.uk/canine-detection-guidance-notes. Accessed 17 Jul 2022
Demant H, Ladewig J, Balsby TJS, Dabelsteen T (2011) The effect of frequency and duration of training sessions on acquisition and long-term memory in dogs. Appl Anim Behav Sci 133:228–234. https://doi.org/10.1016/j.applanim.2011.05.010
Deldalle S, Gaunet F (2014) Effects of 2 training methods on stress-related behaviors of the dog (Canis familiaris) and on the dog–owner relationship. J Vet Behav 9(2):58–65. https://doi.org/10.1016/j.jveb.2013.11.004
Dudley LL, Axe JB, Allen RF, Sweeney-Kerwin EJ (2019) Establishing praise as a conditioned reinforcer: Pairing with one versus multiple reinforcers. Behavioural Interventions 34:534–552. https://doi.org/10.1002/bin.1690
Edwards TL (2019) Automated canine Scent-Detection apparatus: Technical description and training outcomes. Chem Senses 44: https://doi.org/10.1093/chemse/bjz039
Essler JL, Wilson C, Verta AC, et al (2020) Differences in the search behavior of cancer detection dogs trained to have either a sit or stand-stare final response. Front VetY Sci 7: https://doi.org/10.3389/fvets.2020.00118
Fadel FR, Driscoll P, Pilot M, et al (2016) Differences in trait impulsivity indicate diversification of dog breeds into working and show lines. Sci Rep 6: https://doi.org/10.1038/srep22162
Ferton KG, Hong Y, Luo T, et al (2002) Identification of odor signature chemicals in cocaine using Solid-Phase Microextraction—Gas chromatography and Detector-Dog response to isolated compounds spiked on U.S. paper currency. J Chromatogr Sci 40:
Fletcher ML, Wilson DA (2002) Experience modifies olfactory acuity: Acetylcholine-Dependent learning decreases behavioral generalization between similar odorants. J Neurosci 22:RC201–RC201. https://doi.org/10.1523/jneurosci.22-02-j0005.2002
Gazit I, Goldblatt A, Terkel J (2004) The role of context specificity in learning: the effects of training context on explosives detection in dogs. Anim Cogn 8:143–150. https://doi.org/10.1007/s10071-004-0236-9
Gazit I, Terkel J (2003) Explosives detection by sniffer dogs following strenuous physical activity. Appl Anim Behav Sci 81:149–161. https://doi.org/10.1016/s0168-1591(02)00274-5
Hall NJ (2017) Persistence and resistance to extinction in the domestic dog: Basic research and applications to canine training. Behav Proc 141:67–74. https://doi.org/10.1016/j.beproc.2017.04.001
Hall NJ, Johnston AM, Bray EE, et al (2021) Working dog training for the Twenty-First century. Front VetY Sci 8: https://doi.org/10.3389/fvets.2021.646022
Hall NJ, Smith DW, Wynne CDL (2014) Effect of odor preexposure on acquisition of an odor discrimination in dogs. Learn Behav 42:144–152. https://doi.org/10.3758/s13420-013-0133-7
Hall NJ, Smith DW, Wynne CDL (2015) Pavlovian conditioning enhances resistance to disruption of dogs performing an odor discrimination. J Exp Anal Behav 103:484–497. https://doi.org/10.1002/jeab.151

Hall NJ, Smith DW, Wynne CDL (2016) Effect of odorant pre-exposure on domestic dogs' sensitivity on an odorant detection task. Appl Anim Behav Sci 178:80–87. https://doi.org/10.1016/j.applanim.2016.02.003

Hall NJ, Smith DW, Wynne CDL (2013) Training domestic dogs (Canis lupus familiaris) on a novel discrete trials odor-detection task. Learn Motiv 44:218–228. https://doi.org/10.1016/j.lmot.2013.02.004

Hall NJ, Wynne CDL (2018) Odor mixture training enhances dogs' olfactory detection of Home-Made Explosive precursors. Heliyon 4:e00947. https://doi.org/10.1016/j.heliyon.2018.e00947

Haverbeke A, Laporte B, Depiereux E et al (2008) Training methods of military dog handlers and their effects on the team's performances. Appl Anim Behav Sci 113:110–122. https://doi.org/10.1016/j.applanim.2007.11.010

Hayes JE, McGreevy PD, Forbes SL et al (2018) Critical review of dog detection and the influences of physiology, training, and analytical methodologies. Talanta 185:499–512. https://doi.org/10.1016/j.talanta.2018.04.010

Hess E (1959) Imprinting. New Series 130:133–141

Huber L, Apfalter W, Steurer M, Prossinger H (2005) A new learning paradigm elicits fast visual discrimination in pigeons. J Exp Psychol Anim Behav Process 31:237–246. https://doi.org/10.1037/0097-7403.31.2.237

Huber L, Aust U (2006) A Modified Feature Theory as an Account of Pigeon Categorization. In: Zentall TR (ed) Wasserman EA. Experimental explorations of animal intelligence. Oxford University Press, Comparative Cognition, pp 497–512

Jezierski T, Adamkiewicz E, Walczak M et al (2014) Efficacy of drug detection by fully-trained police dogs varies by breed, training level, type of drug and search environment. Forensic Sci Int 237:112–118. https://doi.org/10.1016/j.forsciint.2014.01.013

Jezierski T, Ensminger J, Papet LE (2016) Canine olfaction science and law: Advances in forensic science, medicine, conservation, and environmental remediation. Taylor & Francis

Jezierski T, Walczak M, Górecka A (2008) Information-seeking behaviour of sniffer dogs during match-to-sample training in the scent lineup. Pol Psychol Bull 39: https://doi.org/10.2478/v10059-008-0010-y

Johnen D, Heuwieser W, Fischer-Tenhagen C (2013) Canine scent detection—Fact or fiction? Appl Anim Behav Sci 148:201–208. https://doi.org/10.1016/j.applanim.2013.09.002

Johnen D, Heuwieser W, Fischer-Tenhagen C (2017) An approach to identify bias in scent detection dog testing. Appl Anim Behav Sci 189:1–12. https://doi.org/10.1016/j.applanim.2017.01.001

Keep B, Pike TW, Moszuti SA et al (2021) The impact of training method on odour learning and generalisation in detection animals. Appl Anim Behav Sci 236:105266. https://doi.org/10.1016/j.applanim.2021.105266

Lazarowski L, Dorman DC (2014) Explosives detection by military working dogs: Olfactory generalization from components to mixtures. Appl Anim Behav Sci 151:84–93. https://doi.org/10.1016/j.applanim.2013.11.010

Lazarowski L, Krichbaum S, DeGreeff LE, et al (2020) Methodological considerations in canine olfactory detection research. Front VetY Sci 7: https://doi.org/10.3389/fvets.2020.00408

Lazarowski L, Waggoner P, Hutchings B et al (2021) Maintaining long-term odor memory and detection performance in dogs. Appl Anim Behav Sci 238:105301. https://doi.org/10.1016/j.applanim.2021.105301

Linster C, Smith BH (1999) Generalization between binary odor mixtures and their components in the rat. Physiol Behav 66:701–707. https://doi.org/10.1016/s0031-9384(99)00007-4

Lit L, Schweitzer JB, Oberbauer AM (2011) Handler beliefs affect scent detection dog outcomes. Anim Cogn 14:387–394. https://doi.org/10.1007/s10071-010-0373-2

Lotspeich E, Kitts K, Goodpaster J (2012) Headspace concentrations of explosive vapors in containers designed for canine testing and training: Theory, experiment, and canine trials. Forensic Sci Int 220:130–134. https://doi.org/10.1016/j.forsciint.2012.02.009

Mancini C, Harris R, Aengenheister B, Guest C (2015) Re-Centering Multispecies practices. In: Proceedings of the 33rd Annual ACM Conference on Human Factors in Computing Systems. https://doi.org/10.1145/2702123.2702562

Mandairin N, Stack C, Linster C (2006) Olfactory enrichment improves the recognition of individual components in mixtures. Physiol Behav 89:379–384. https://doi.org/10.1016/j.physbeh.2006.07.013

Mandairon N, Stack C, Kiselycznyk C, Linster C (2006) Enrichment to odors improves olfactory discrimination in adult rats. Behav Neurosci 120:173–179. https://doi.org/10.1037/0735-7044.120.1.173

Marshall M, Oxley J (2009) Aspects of explosives detection. Elsevier, Amsterdam, Boston

Meyer I, Ladewig J (2008) The relationship between number of training sessions per week and learning in dogs. Appl Anim Behav Sci 111:311–320. https://doi.org/10.1016/j.applanim.2007.06.016

Moser AY, Bizo L, Brown WY (2019) Olfactory generalization in detector dogs. Animals 9:702. https://doi.org/10.3390/ani9090702

Murarka M, Vesley-Gross ZI, Essler JL et al (2019) Testing ovarian cancer cell lines to train dogs to detect ovarian cancer from blood plasma: A pilot study. J VetY Behav 32:42–48. https://doi.org/10.1016/j.jveb.2019.04.010

Pearce JM (2013) Animal learning & cognition : an introduction, 3rd edn. Psychology Press, Hove, New York

Porritt F, Shapiro M, Waggoner P et al (2015) Performance decline by search dogs in repetitive tasks, and mitigation strategies. Appl Anim Behav Sci 166:112–122. https://doi.org/10.1016/j.applanim.2015.02.013

Ramirez K (2020) The eye of the trainer : animal training, transformation, and trust. Sunshine Books Inc., Waltham

Ramirez K (1999) Animal training: successful animal management through positive reinforcement. Karen Pryor Clicker Training, Massachusetts USA

Ryan RM, Deci EL (2020) Intrinsic and extrinsic motivation from a self-determination theory perspective: Definitions, theory, practices, and future directions. Contemp Educ Psychol 61:101860. https://doi.org/10.1016/j.cedpsych.2020.101860

Schoon GAA (1996) Scent identification lineups by dogs (Cunis familiaris): experimental design and forensic application. Appl Anim Behav Sci 49:257–267

Shenoy P, Yu A (2012) Strategic impatience in Go/NoGo versus Forced-Choice Decision-Making. Adv Neural Inf Process Syst 25:2132–2140

Slotnick B (2007) Odor-Sampling Time of Mice under Different Conditions. Chem Senses 32:445–454. https://doi.org/10.1093/chemse/bjm013

Smith BH, Cobey S (1994) The olfactory memory of the honeybee Apis mellifera. II. Blocking between odorants in binary mixtures. J Exp Biol 195:91–108. https://doi.org/10.1242/jeb.195.1.91

Tami G, Gallagher A (2009) Description of the behaviour of domestic dog (Canis familiaris) by experienced and inexperienced people. Appl Anim Behav Sci 120:159–169. https://doi.org/10.1016/j.applanim.2009.06.009

Thiesan L, Hannum D, Murray DW, Parmeter JE (2005) Survey of commercially available explosives detection technologies and equipment 2004. U.S. Department of Justice

Thrailkill EA, Kacelnik A, Porritt F, Bouton ME (2016) Increasing the persistence of a heterogeneous behavior chain: Studies of extinction in a rat model of search behavior of working dogs. Behav Proc 129:44–53. https://doi.org/10.1016/j.beproc.2016.05.009

Troisi CA, Mills DS, Wilkinson A, Zulch HE (2019) Behavioral and cognitive factors that affect the success of scent detection dogs. Comp Cogn & Behav Rev 14:51–76. https://doi.org/10.3819/ccbr.2019.140007

Venniro M, Russell TI, Zhang M, Shaham Y (2019) Operant social reward decreases incubation of heroin craving in male and female rats. Biol Psychiat 86:848–856. https://doi.org/10.1016/j.biopsych.2019.05.018

Wasser SK, Davenport B, Ramage ER et al (2004) Scat detection dogs in wildlife research and management: application to grizzly and black bears in the Yellowhead Ecosystem, Alberta, Canada. Can J Zool 82:475–492. https://doi.org/10.1139/z04-020

Williams M, Johnston JM (2002) Training and maintaining the performance of dogs (Canis familiaris) on an increasing number of odor discriminations in a controlled setting. Appl Anim Behav Sci 78:55–65. https://doi.org/10.1016/s0168-1591(02)00081-3

Wright HF, Wilkinson A, Croxton RS, et al (2017) Animals can assign novel odours to a known category. Sci Rep 7: https://doi.org/10.1038/s41598-017-09454-0

Yee KK, Wysocki CJ (2001) Odorant exposure increases olfactory sensitivity: olfactory epithelium is implicated. Physiol Behav 72:705–711. https://doi.org/10.1016/s0031-9384(01)00428-0

Yerkes RM, Dodson JD (1908) The relation of strength of stimulus to rapidity of habit-formation. J Comp Neurol Psychol 18:459–482. https://doi.org/10.1002/cne.920180503

Zubedat S, Aga-Mizrachi S, Cymerblit-Sabba A et al (2014) Human–animal interface: The effects of handler's stress on the performance of canines in an explosive detection task. Appl Anim Behav Sci 158:69–75. https://doi.org/10.1016/j.applanim.2014.05.004

Olfaction and Dog Welfare

Nicola J. Rooney and Zoe Parr-Cortes

Abstract

Smell is a well-developed and predominant sense in dogs and so it is unsurprising that olfaction is tightly linked to dog welfare. The extent to which a dog can use their olfactory abilities and the types of odours they smell, have been shown to affect their behaviour, memory, and wellbeing, and similarly an animal's state of health and welfare impacts upon its olfactory ability. There are several natural scents whose effects upon dogs have been well studied. Lavender has been shown to induce calm and resting behaviour, whilst there is some evidence that peppermint, rosemary, and laurel may lead to increased activity in kennelled dogs. Several olfactory-administered products have sought to mimic these natural calming effects and have been marketed to mitigate fear and anxiety in dogs. Pet Remedy a Valerian-based product and synthetic Dog Appeasing Pheromone (DAP) are relatively well tested. Whilst some studies present promising findings of their potential benefits, particularly of DAP in puppies, the body of research remains unconvincing regarding the efficacy of both products especially for adult dogs. With the increasing array of scent-based tasks that dogs are trained to perform, the way they are reared, housed, trained, and cared for all have the capacity to affect their welfare. There is evidence of the value of a range of aspects of housing and husbandry that can improve welfare. Research on the interface between olfaction and welfare although growing, remains in its infancy, which we see as evidence of sensory biases in human approaches to dog behaviour and welfare.

Keywords

Welfare • Wellbeing • Enrichment • Calming • Kennel • Olfaction

N. J. Rooney (✉) · Z. Parr-Cortes
Animal Welfare and Behaviour Group, Bristol Veterinary School, University of Bristol, Bristol, UK
e-mail: Nicola.Rooney@bristol.ac.uk

L. Lazarowski (ed.), *Olfactory Research in Dogs*,
https://doi.org/10.1007/978-3-031-39370-9_10

1 Introduction

Since smell is such a well-developed and predominant sense in dogs (see Chap. XX), it is unsurprising that olfaction is tightly linked to dog welfare. The extent to which a dog is able to use their olfactory abilities and the types of odours they smell have been shown to affect their behaviour, cognition and wellbeing, and similarly an animal's state of health and welfare will impact upon its olfactory ability (Nielsen et al. 2015). What's more, with the increasing array of scent-based tasks that dogs are trained to perform, the way they are reared, housed, trained, and cared for, all have the capacity to affect their welfare. In this chapter, we review and discuss current evidence for each of these impacts. When referring to welfare, we describe a construct that varies from very poor to very good, and has a psychological as well as a physical component (i.e. to have good welfare one must be physically fit but also psychologically fulfilled). Welfare is not only a measure of the absence of negative states, but also the presence of positive ones (Boissy et al. 2007; Ogi et al. 2020). There is no one definitive measure of welfare, but indicators of welfare states include behavioural, physiological and cognitive measures, a range of which have been validated for dogs (Polgár et al. 2019).

2 The Value of Olfaction

Many dogs are trained to carry out scent-based tasks for working purposes, but companion dogs are also trained in scent detection for hobby or competition. Since most dogs are strongly motivated to smell, there is a belief that such scent work is rewarding, and indeed "find it" or "hide and seek" games are commonly recommended by behaviourists as a way to stimulate dogs (RSPCA 2020). This idea is supported by a study by Duranton and Horowitz (2019), which used cognitive bias testing, a technique that measures individual dogs' propensity to judge ambiguous stimuli as positive (suggesting optimistic tendencies) or negative (suggesting pessimistic tendencies) (Harding et al. 2004; Paul et al. 2005). Previous studies have shown links between decision-making under ambiguity and affective states in a range of species including dogs (Mendl et al. 2010; Titulaer et al. 2013). Duranton and Horowitz (2019) compared dogs that underwent two weeks of either "nose work" where they were encouraged to search for high value treats hidden under one of three boxes, or "heelwork" in which they were encouraged to follow their owner around the room. Dogs in both groups spent the same amount of time practising their respective task (five minutes a day, three task repetitions) and received the same volume of food rewards for completing the tasks. However, the dogs that engaged in daily search games showed more optimistic responses to ambiguous cues in the cognitive bias test post-treatment, compared to baseline. In contrast, dogs in the group that practised daily heelwork showed no change in cognitive bias test performance (Duranton and Horowitz 2019). This suggests that search training can change judgment bias, possibly improving dogs' emotional state compared to other rewarded activities. However, whether this difference is due to the actual act of sniffing, or for example, the difference in motivation during "fun" energetic search games as compared to more controlled obedience exercises requires further elucidation. Is it the act of searching that had the effect and would visual or auditory activities be similarly rewarding or is the olfactory element crucial?

What is clear is that dogs have a strong motivation to use their sense of smell. When dogs meet each other, sniffing facilitates direct olfactory communication. They engage in mutual sniffing of both the rear and head regions (Holcova et al. 2021). When on a walk, most dogs will also spend a large amount of time sniffing the ground and other aspects of the environment. This facilitates indirect inter-dog communication, allowing them to use the information gained from sniffing urine of conspecifics to derive information about social (Lisberg and Snowdon 2009) as well as reproductive status (Jezierski et al. 2019; Woszczylo et al. 2020). Dogs will also roll in strong scents such as carcasses and faeces. The motivation behind this behaviour is still debated and authors have posed olfactory camouflage and social enhancement theories, but whatever the cause the behaviour appears especially rewarding (Horowitz 2016; Allen et al. 2016). These aspects of olfactory communication are all very normal behaviour for dogs, and it could be argued that denying these opportunities denies the dog's ability to behave normally. This is one of the five basic welfare needs, corresponding to the Five Freedoms (FAWC 1979) that form the basis of much animal welfare legislation (Animal Welfare Act: DEFRA 2006). Therefore, if dogs are not given the opportunity to use olfaction, for example those that are prevented from sniffing, this may impact upon their wellbeing.

When on a walk, humans and dogs often have different interests and goals. Observations show that when walked on a lead down a straight path, dogs tend to investigate smells, follow scent trails, stop frequently and veer off path whilst humans tend to walk straight to reach the end of the walk (Aspling et al. 2015). The opportunity for sniffing is also affected by the type of walk (Budzinski and Budzinski 2019). On average dogs spent 2.8 times longer sniffing when on a long (5 m) rather than a short (1.5 m) lead and 3.2 times longer sniffing when off-lead than on a short lead. Dogs also showed a decrease in heart rate when sniffing especially when sniffing was considered "intense." It is possible that sniffing on a walk reduces stress and/or tension, however heart rate can increase at the start and end of odour search tasks (Brugarolas et al. 2019) and varies with the respiratory cycle (Tilley and Smith 2015). Hence whether changes in heart rate are due to a positive experience of olfactory searching or a reflex response in heart rhythm due to respiratory changes remains to be ascertained.

The extent to which the act of smelling is rewarding, is an inelastic behavioural need, and/or induces a positive affect are yet to be fully investigated. However, techniques in animal welfare science exist that could be used to investigate these effects further. One method explores the amount an animal will work in order to access the opportunity to engage in an activity. For example, using these techniques, it was demonstrated that mink (*Mustela vision*) will work hard to gain access to a swimming pool (Cooper and Mason 2000). To what extent will dogs work to gain access to opportunities to sniff? If dogs will work hard to have the opportunity to sniff, it follows that not having the opportunity to sniff would be detrimental. If so, then the loss of their sense of smell may also represent a welfare concern.

3 Loss of Sense of Smell

When a dog loses one of their five senses, it can make navigating everyday life challenging, both for the dog and its owner. When this occurs, actively engaging the use of their other senses can help them better cope with this loss. Dogs that lose their vision, can have difficultly navigating their environment and can injure themselves when bumping or stumbling over objects and furniture. The use of non-toxic scents to mark obstacles and objects in the household, such as table corners or chair legs, has been suggested as a navigation aid to improve quality of life by making use of their olfactory sense to compensate for their lack of vision (Hedges 2016). Indeed, products such as Tracerz® Scent Markers for Blind Dogs (Innovet Pet Products) are marketed specifically as navigation aids for blind dogs. However, the ability of scent markers to improve navigation and quality of life for blind dogs has not yet been empirically measured.

Loss of vision in dogs is often easily recognised by owners and can be conclusively diagnosed by a vet. However, the same is not true for olfaction. Clinical assessment of olfactory function in dogs is challenging due to the lack of objective testing methods for this neuronal pathway. Clinical evaluation is usually based on subjective assessment of a dog's response to an aromatic odour while blindfolded (e.g. sniffing, licking the nose or aversive head movements) (Platt et al. 2013). However, the absence of these behaviours does not necessarily mean the sense of smell is reduced (hyposmia) or absent (anosmia). Therefore, diagnosis of hyposmia or anosmia is often based on owner reports, for example of decreased appetite (since taste is highly dependent upon smell). But, even in conditions where significant damage to the olfactory epithelium is present (e.g. tumours of the nasal cavity or aspergillosis infections), these clinical signs are often not apparent (Platt et al. 2013). Knowledge from experiments and from the performance of detection dogs, for whom olfactory ability is closely monitored, show there are a number of medications and diseases that can negatively affect olfaction including steroids (Ezeh et al. 1992), metronidazole (Jenkins et al. 2016), Canine Distemper Virus (Myers et al. 1988a) and parainfluenza virus (Myers et al. 1988b). The welfare impact of such loss of olfactory ability is not fully explored. Extrapolating from humans; one in 140 people suffer from anosmia, and within these approximately 30% are clinically depressed (Croy et al. 2012). Those with anosmia had significantly poorer Quality of Life in several domains (Croy et al. 2014), compared to those with a functional sense of smell. Since sniffing is such an important and potentially rewarding activity for dogs, it is likely that loss of smell significantly negatively impacts their welfare, and more research is needed.

3.1 Has Selective Breeding Negatively Impacted "Olfactory" Welfare?

Whilst some breeds have been purposefully selected for olfactory work, there has also been a long tradition of selective breeding for physical appearance at the expense of health and welfare (Rooney and Sargan 2010). This has resulted in high levels of inherited diseases in specific breeds and exaggerated morphological features that can directly compromise welfare; issues that affect the majority of dog breeds. Several breeds have been selected for shortened muzzles or brachycephalia (McGreevy et al. 2013). This has resulted in changes to the respiratory system compared to mesocephalic and dolichocephalic breeds including structures important for olfaction, such as narrowed nostrils, a shortened nasal cavity and relatively enlarged nasal turbinates (Buzek et al. 2022). Brachycephalic breeds suffer from a suite of health issues (O'Neill et al. 2020) and the welfare implications of obstructive airway syndromes and respiratory distress are now well-evidenced (O'Neill et al. 2022). However, whether these extreme conformations impact olfactory ability is less known.

Polgár et al. (2016) used a single session Natural Detection Task to compare the olfactory ability of "scent dogs" (breeds typically selected for scent detection, (e.g. basset hound and beagle)), "non-scent dogs" (not typically selected for scent detection, e.g. Afghan hound and Siberian husky) and short-nosed brachycephalic breeds (e.g. English bulldog and pug). They found "scent dogs" were best at identifying which one of four containers had food hidden under it, with the short-nosed group showing the poorest performance (Polgár et al. 2016). However, at the highest difficulty level, neither the "non-scent dogs" nor short-nosed dogs performed above chance level, indicating a similar olfactory ability at this level. In contrast, in a multiday odour discrimination task, pugs performed better than German Shepherd Dogs; learning the task faster and making more correct choices at low odour concentrations (Hall et al. 2015).

Although short-nosed breeds performed worst in the Natural Detection Task (Polgár et al. 2016), their performance was above chance level in all but the highest difficulty, in line with other non-brachycephalic breeds not typically selected for scent detection. Hence, these two small studies suggest that the olfactory ability of brachycephalic breeds is not significantly compromised and is unlikely to negatively impact their welfare. Furthermore, when sniffing games as enrichment were compared between brachycephalic and mesocephalic dogs, no breed-specific differences were found in ability to find hidden food treats (Chvala-Mannsberger et al. 2021). This suggests that despite their conformational differences, brachycephalic breeds too may benefit from olfactory enrichment.

4 Specific Odours Can Have a Calming Effect

Beyond the act of sniffing, research has started to explore how specific odours can affect the behaviour and welfare of dogs. There are plant species that are known to have relaxing effects on a range of animal species when smelled. Their potential value for mitigating the stress of dogs in rehoming kennels and during transport has started to be systematically tested.

In one study, dogs were shown to respond to diffused essential oils (Graham et al. 2005). Singly housed dogs in a rehoming kennel were exposed to each of five odour treatments: Lavender, chamomile, rosemary, peppermint, and no odour (control) for four hours a day on five consecutive days. Dogs spent more time resting and less time moving and vocalising when exposed to lavender and chamomile than any of the other olfactory stimuli. The diffusion of rosemary and peppermint encouraged standing, moving and vocalising. Whilst increased sedentary behaviour is not always a sign of improved welfare (Fureix et al. 2022), the authors argued that dogs' welfare may be enhanced through exposure to appropriate forms of olfactory stimulation, especially lavender and chamomile, since the relaxing behaviours they promote are likely considered desirable by potential adopters. Wearing collars impregnated with lavender also seems to ameliorate the increase in stress behaviours experienced within a kennel environment (Stanghellini 2019).

Lavender has similarly been shown to reduce stress-related behaviours in dogs when travelling by car (Wells 2006). The behaviour of companion dogs referred to a behaviourist for over-excitement during car journeys was compared in both the presence and absence of the scent of lavender for three consecutive days. Dogs spent significantly less time moving and vocalising and more time resting and sitting during the lavender treatment, consistent with a calming effect (Wells 2006).

A more recent study utilised behavioural, physiological (cortisol) and cognitive measures to explore the effects of wearing cotton collars containing one of nine different essential oils (including lavender) and a blend of the nine oils together for three hours, in comparison to a control (no odour), on the behaviour of dogs in a rehoming kennel (Uccheddu et al. 2018). Since dogs in such environments generally struggle to concentrate, a shortened version of the cognitive bias test was used (Mendl et al. 2010). This study revealed a significant decrease in latency to approach the ambiguous probes, suggesting increased optimism, after exposure to the blend of nine essential oils, but none of the individual oils. Salivary cortisol (a stress indicator) significantly decreased after exposure to lavender supporting the calming effect, as has also been shown in humans (Sayorwan et al. 2012) where it also improves self-reported emotional state, and increases free radical scavenging activity important in reducing the effects of oxidative stress that can be harmful to the body (Atsumi and Tonosaki 2007). However, a similar decrease in cortisol was also observed in the control group, which could indicate the cognitive bias test alone was a stress-relieving activity. Behavioural analysis revealed an increase in "high posture" following exposure to *Laurus nobilis* (laurel) oil which may indicate an increase in self-confidence (Fatjó et al. 2007), or increased vigilance as has been seen in humans (Matsubara et al. 2011).

Other plant-based compounds (vanilla, Valerian, coconut, and ginger) have also been shown to have positive effects on the behaviour of dogs in rehoming shelters. When dogs were exposed to the essential oils for two hours on three consecutive days, stress-related vocalisations (barking, whining, and whimpering) and activity were significantly reduced by each of the four scents compared to control conditions. There was also a significant increase in resting behaviour in all four odour treatments and an increase in sleep with exposure to ginger and coconut (Binks et al. 2018). As in all the studies described above, observations were recorded in person, so blinding was not possible as the type of odour would have been detectable to the experimenter.

The smell of cut grass, also known as "green odour," has been found to have stress-relieving and hypothalamic–pituitary–adrenal (HPA) modulatory effects in a number of different species including rodents (Spiers et al. 2015; Nakashima et al. 2004), cattle (Sutoh et al. 2013) and humans (Oka et al. 2008). To test whether it was similarly calming for dogs, the behaviour and salivary cortisol of companion dogs was measured before, during and after a three minute separation from their owner (Carlone et al. 2018) with and without exposure to green odour. Dogs spent significantly less time in proximity and following their owners pre-separation, in the presence of green odour compared to no odour, but there was no significant difference in behaviour nor cortisol during separation between odour conditions. These findings suggest dogs may have felt more secure and independent in the presence of green odour and therefore did not seek as much reassurance from their owner. The absence of an increase in cortisol in either condition suggests the stressor may have been too mild, or too brief, to cause a measurable stress response, making any effect of green odour impossible to detect. However, the biological relevance of green odour for a predator species such as the dog is not obvious, and the possibility of secondary effects via green odour calming the owner cannot be ruled out.

5 Plant-Based Products

The body of evidence that plant-derived products can positively enhance dog welfare has inevitably led to the development and marketing of plant-based products as well as nutraceuticals and herbal products that can be administered through an olfactory route. With no requirement for systematic testing of alternative therapies, robust evidence for the value of such products remains sparse. However, in the case of one product: Pet Remedy, marketed as a fast-acting remedy for calming stressed or anxious pets (Unex Designs Ltd 2022), there is a growing body of research.

Pet Remedy is a blend of essential oils with the active ingredient valerian (*Valeriana officinalis*), with small amounts of vetiver, sweet basil, and clary sage essential oils. The calming effects of Valerian are attributed to its effects on the gamma-amino butyric acid (GABA) pathways. As an inhibitory neurotransmitter,

GABA is responsible for reducing neuronal excitability, with low levels of endogenous GABA being linked to anxiety disorders in humans (Lydiard 2003). Valerian has been used as an alternative therapy in humans for hundreds of years and has been shown to increase endogenous brain GABA levels (Awad et al. 2007; Trauner et al. 2008; Yuan et al. 2004).

The Pet Remedy website describes three unpublished clinical trials and reports positive effects for reducing stress in dogs (Unex Designs Ltd 2022). Dog groomers using a Pet Remedy spray or diffuser over five months reported moderate improvements in behaviour in 51.6% of 244 dogs and significant improvements in 32.8% of cases (Hale 2021). Two trials comparing behaviour modification programmes, accompanied with either Pet Remedy or placebo treatment, both reported improvements in excitement ratings in the treatment group only (Hale and Meaney 2021), but relied entirely on subjective ratings.

In contrast, the only published study that included a double-blind, placebo-controlled trial, saw no noticeable effect (Taylor and Madden 2016). Dogs reported by their owners to show anxious behaviour when in an unfamiliar environment visited a novel room for 30 min; once during Pet Remedy diffusion and once during a placebo odour diffusion (carrier agent). No significant difference in behaviour was detected between conditions, indicating Pet Remedy exposure did not reduce stress-related behaviours in this context (Taylor and Madden 2016). Therefore overall, in spite of Valerian alone being shown to have some calming effects comparable to other essential oils (Binks et al. 2018), there is limited evidence of the value of Pet Remedy for stress relief in dogs. This led Buckley (2019) to conclude in her review that using Pet Remedy is unlikely to cause any direct animal welfare harm and may have a positive effect, but more robust studies with better external validity are required (Buckley 2019). These studies should also test whether the levels of Valerian cause any sedation effect, as has been explored for rabbits (Unwin et al. 2020), since sedating fearful animals and rendering them unable to avoid the cause of fear can lead to sensitisation and hence can be detrimental to welfare.

6 Pheromonotherapy: Dog Appeasing Pheromone (DAP)

There is a growing body of research on a range of species on the effect of pheromonotherapies. Pheromones are chemicals that animals use to communicate within a species (Pirner and McGlone 2016a). Dog appeasing pheromone (DAP) is a synthetic compound based on fatty acids secreted by the mammary gland of bitches after parturition (Pageat and Gaultier 2003). It has been studied in puppies and adult dogs as a possible pheromone therapy for separation issues (i.e. reducing vocalisation, door scratching and house-soiling issues), but also for other behavioural problems and for reducing stress in a range of contexts.

Mothers were seen to be more tolerant of their puppies when exposed to DAP during weaning (Santos et al. 2020). Puppies exposed to DAP cried significantly less than those exposed to placebo, although there were nonsignificant effects on the number of nights that puppies soiled indoors (Taylor and Mills 2007a). A triple-blinded study of puppies adopted from a pet shop found that significantly fewer of the puppies with the DAP collars showed signs of fear when facing unfamiliar people at home and/or during outings compared to puppies without the DAP collars (Gaultier et al. 2009). Similarly, when DAP collars were used on puppies during training sessions they appeared to be less fearful and more sociable and showed improved learning (Denenberg and Landsberg 2008). Most other studies into the efficacy of DAP for behavioural and stress-related issues have concentrated on adult dogs in a range of contexts and show a range of results. One study showed improvement in travel-related problem behaviours after dogs wore impregnated DAP collars for six weeks. The greatest improvement was in those that had shown motion sickness signs (vomiting and salivating), and the least improvement in excitable dogs that had shown barking, jumping and whining (Estellés and Mills 2006).

For dogs kennelled in a rehoming centre, DAP diffused for seven days reduced barking amplitude and frequency when people walked by the kennels (Tod et al. 2005). Similarly, dogs in rehoming kennels exposed to DAP spray showed a reduction in barking intensity in the 30 min after the treatment compared to a control (no spray) condition. However, none of the other measured stress related-behaviours were affected (Hermiston et al. 2018). A small-scale study of eight dogs used for teaching purposes and housed in university kennels similarly saw no difference in kennel behaviour over 21 days between those wearing a DAP collar and those not (Grigg and Piehler 2015).

More recently Amaya et al. (2020a) compared three hours a day of DAP exposure for five days to two other enrichments (lavender and slow tempo instrumental music) and a control (no treatment) within a rehoming shelter. Dogs exposed to DAP lay down more, and those exposed to music lay down more with their head down, compared to controls. Those in the control group stood on their hind legs, panted and vocalised much more than dogs in the three enrichment treatments (Amaya et al. 2020a). The study suggests that DAP, like music and lavender, had some positive benefits for dogs in shelters, as well as being non-invasive and easy to apply. However, in a follow-up study, the authors measured the physiological effects, specifically Heart Rate Variability (HRV) (variation in time between heart beats that decreases in response to stress) of dogs during these four treatments, and found DAP and lavender led to nonsignificant changes in HRV whilst playing music led to a significant decrease in HRV (Amaya et al. 2020b).

In a veterinary clinic setting, DAP diffusers appeared to reduce anxiety signs, but there was no evidence of reduced aggression during the clinical exam after a single exposure (Mills et al. 2006). When DAP exposure was compared to acupuncture for efficacy at mitigating separation-related behaviours in hospitalised dogs over three days, treatment led to a decrease in destruction, licking, aggression and hyper-salivation, however there were no significant differences between treatments (Kim et al. 2006). In a follow-up placebo controlled, double-blind study, improvements were seen in a number of separation-related behaviours of hospitalised dogs exposed to a DAP diffuser for four days (Kim et al. 2010), including significant reductions in elimination (urinating/defecating), excessive licking and pacing, whilst placebo exposed dogs generally experienced increases in separation-related behaviours. However, a study following dogs undergoing ovariohysterectomy saw no significant differences in cortisol between placebo and treatment groups which had DAP sprayed into their intensive care unit cage (Siracusa et al. 2010); although they did see a smaller decrease in visual exploration, alertness and prolactin (a hypothesised biomarker of stress (Gutiérrez et al. 2019)) in DAP exposed dogs. A recent randomised, triple-blind, placebo-controlled study assessed the effect of a gel formulation of DAP applied to the hands of owners and veterinarians prior to physical examination of 27 dogs attending a veterinary clinic. DAP exposure decreased lip-licking behaviour but increased panting in dogs in the waiting room compared to the placebo treatment, but had no effect on behaviour during the veterinary examination. Heart rate, respiratory rate, blood pressure, rectal temperature and salivary cortisol were also measured, however none were found to differ significantly between DAP and placebo groups (Puglisi et al. 2022).

In the home environment, Sheppard and Mills (2003) saw improvements in 9 of 14 owner-reported behavioural signs of fear in dogs exposed to DAP diffusers for two weeks before firework night (Sheppard and Mills 2003). Landsberg et al. (2015) conducted a placebo-controlled and blinded study and found greater decreases in fear of simulated thunder in dogs wearing DAP collars compared to placebo collars. However, the DAP-exposed dogs also spent longer in a hide box during and after the sounds (Landsberg et al. 2015) making it difficult to separate the effect of DAP from the comfort provided by the hide. In contrast, an online survey saw that owner-reported effectiveness of pheromonotherpies for firework fear was no higher than would be expected of a placebo effect (Riemer 2020).

Taylor et al. (2020) examined separation-related problems in a laboratory environment, but saw no effect of DAP administration on orienting behaviour, eye or ear temperature (hence no evidence of stress-induced hyperthermia), heart rate or HRV compared to when the same dogs were exposed to placebo (Taylor et al. 2020). A trial reported positive outcomes of using both dog and cat-appeasing pheromones for increasing calmness and positivity of dog-cat interactions within a household (Prior and Mills, 2020). Only a single published study has tested DAP on working dogs. This study randomly assigned DAP and placebo collars to 51 Belgian Malinois on arrival at Lackland Airforce base kennels and saw no significant treatment effects on behaviour monitored over five weeks (Broach and Dunham 2016).

The results within these studies are very mixed and overall lack strong evidence of the value of DAP, likely partly due to different doses and routes of administration (McGlone et al. 2022). This is echoed in two review papers. Frank et al's. (2010) systematic review of the use of pheromones to treat undesirable behaviours in cats and dogs concluded that of the seven studies reviewed, only one on puppies (Denenberg and Landsberg 2008) provided sufficient evidence of its value (Frank et al. 2010). They also raised legitimate concerns about the internal and external validity of the studies, potential selection biases, poorly controlled variables, inconsistent methodologies, and subjective outcome measures. Similarly, in a review examining whether DAP reduces the frequency and/or severity of non-specific stress behaviours, Wong and Govendir (2021) concluded that the evidence for DAP as an effective treatment for anxiety in adult dogs remains weak. Some authors have posed that DAP shows promise for use in wild canids (Riddell et al. 2021) suggesting it can prevent androgen surges and may reduce contact dominance and active submission after stressful interventions in African wild dogs (Van den Berghe et al. 2019). However, for domestic dogs the evidence is not convincing and seems to be stronger for puppies than adults; unsurprising as DAP is a synthetic analogue of a pheromone dogs are exposed to when very young, so puppies are the pheromone's natural recipients. This lack of evidence is reflected in owner opinions, who when surveyed were less comfortable with the use of pheromone products to treat behavioural problems (58.1% "very" or "somewhat" comfortable) than nutritional or herbal supplements (71.9%), with their level of comfort around pheromonotherapies being similar to that of fast acting medications (62.3%) (van Haaften et al. 2020). Most owners regarded proven effectiveness as important when choosing between medical and alternative therapies and personal experience of taking herbal or nutritional supplements also increased acceptance. It follows that since humans are less able to relate to the olfactory experience of dogs and unlikely to have had olfactory-administered treatments, they must instead rely on objective scientific evidence of the effectiveness of such therapies, for which there are mixed results.

The pheromone studies described above have concentrated on signalling within the same species, but recent work has started to examine "intermone" pheromones from one species which may elicit a different effect in another species. For example, androsterone is a sex pheromone secreted in the saliva of male domestic pigs which promotes acceptance of mounting behaviour in females and also reduces aggression in group-housed swine. A small-scale case study tested the effect of spray-administered androsterone on dog behaviour, and saw a reduction in barking (McGlone et al. 2014). A second study suggested that it may be effective at reducing lead pulling but not jumping up (Pirner and McGlone 2016b). However, all treatments were accompanied by a burst of air spayed at the dogs' head designed to interrupt the behaviour. The results suggest that this air burst and the odour are aversive and hence even though they are described as "mild" they still constitute positive punishment and have the potential to adversely affect welfare as described below. One could argue that utilising pheromones or natural remedies, although not proven to be effective, could be beneficial and hence there is no harm done in their use. However, if their use delays the use of more effective proven interventions, this can be a welfare concern.

7 Olfactory Enrichment and Stimulation

Whilst most research has focussed on calming effects of scents it follows that some scents can be stimulating. A recent study explored the effect of adding scents to toys as a method of olfactory enrichment for dogs in a rehoming centre (Murtagh et al. 2020). Dogs underwent three test conditions: no toys, three unscented toys, and scented toys (one lavender scented, and one rabbit scented presented with an unscented toy). The presence of the two scented toys increased the likelihood of engagement with all three of the toys present in that condition, even the unscented one. Dogs spent significantly longer interacting with the scented toys than the unscented ones, but there was no difference between the two scents. There was also a significant decrease in the frequency of stress-related behaviours, and an increase in exploratory behaviours observed in dogs during the scented toy condition compared to the other two conditions. Although only two scents were used during this study, these findings suggest olfactory enrichment with numerous odours may improve engagement with toys and overall wellbeing of dogs in kennels. This idea has been expanded into developing "sensory gardens" for dogs within rehoming centres (RuffleSnuffle 2021), which amongst other sensory experiences provide a range of scents to smell and plant species to encourage zoopharmacognosy (self-medication). However, not all scents will exert positive effects.

8 Aversive Scents Can Impact Welfare Negatively

In comparison to humans, dogs can smell significantly lower concentrations of odorants due to a higher number and density of olfactory neurons, higher air flow and specificity of central processing. Therefore, levels of scent that would go undetected by people may be extremely aversive to dogs, and scents that may be pleasant in small amounts such as essential oils, in higher concentrations may be aversive.

Living in such close proximity to people means that we often share olfactory space. Household cleaning products, perfumes, diffusers, tobacco, as well as dog cosmetics (e.g. shampoos) are but a few examples of widely used aromas whose effects on dogs remain unknown. At the least, these strong odours may mask some of the odours on which the dogs rely, and at worst they may be unpleasant, aversive or even painful to their sensitive olfactory systems. There has been very little research examining preferences and aversions to scents, although Kokocinska et al. (2022) conducted scent preference tests by presenting 33 common scents to dogs. They found dogs showed use of the left nostril or both nostrils simultaneously when sniffing lavender, which could suggest that this odour is interesting, and pleasant. Dogs were significantly more likely to interact with orange oil and to avoid linalyl acetate and showed a tendency to prefer raspberry, rose, peppermint and strawberry. Overall, most of the odours presented did not seem unpleasant or aversive (Kokocinska et al. 2022), at least at the concentrations and durations presented.

Scent preferences should be considered when adding fragrances to dog products (e.g. shampoos and cosmetics), since forcing dog to wear a scent that is unpleasant to them could negatively affect their welfare. Conversely, choosing scents that dogs find pleasant or calming may be beneficial to their welfare. It is also important to remember that increasing the concentration or potency of an odour, even a pleasant one, can make it overwhelming or unpleasant.

Dogs also respond differently to the scent of predator and non-predator species, showing an increase in heart rate and decrease in time spent near the scent of faeces from bears and lynx compared to beavers and control odour (water) (Samuel et al. 2020). Therefore, living within olfactory reach of potentially threatening species can present a welfare concern.

In addition, kennel environments potentially present high levels of ammonia (released from urine and faeces) which have been shown to be aversive to other species (The Pig Site 2014). In fact, Codes of Practice that interpret the Animal Welfare Act (DEFRA 2020) for farmed species such as pigs and chicken stipulate maximum ammonia levels in living spaces 2006, whereas for the dog no such guidelines exist, and underlying research is yet to be conducted.

The aversiveness of specific scents however is evident from the fact that the odour of citronella and lemon are used in anti-bark collars, which release a spray of odour either automatically or when activated by the owner, when the dog barks. They have been found to reduce unwanted barking behaviour in pet dogs by 88.9% compared to 44.4% using electronic shock collars (Juarbe-Diaz and Houpt 1996). Wells (2001) compared dogs given intermittent and continuous exposure to a citronella collar over a three-week period. She found that the collar was effective at reducing some types of barking (e.g. in the car more than at the TV) and intermittent exposure was more effective (Wells 2001). A similar study found citronella collars effective at reducing coprophagia when compared to sound therapy (Wells 2003). In contrast, when used in a veterinary hospital setting, although bark-activated citronella collars resulted in a 77% reduction in barking in the subsequent five-minute period, dogs wearing a scentless spray collar which releases HFC134A tetra-fluoroethane (a pharmaceutical-grade, ozone-friendly propellant used in human asthma inhalers) also experienced a 59% reduction in barking (Moffat et al. 2003). There were no significant differences between the two collars, suggesting they were both effective punishers.

A home-based study saw that Aboistop™ citronella (lemon scented) spray collars (Dynavet®), applied to dogs by their owners over a 3-month period, were effective for three of the seven dogs, with vocalisations being completely eliminated for two dogs, and reduced for one (Dynavet 2022). However, the study observed distress reactions which in some dogs were marked (Sargisson et al. 2011). Citronella collars are perceived by owners to be more humane than electronic collars, for example for reducing excessive barking (Juarbe-Diaz and Houpt 1996), but does this result from our anthropogenic bias, in which we find it hard to perceive the strength of an olfactory punisher? Indeed, Steiss et al. (2007) saw no significant difference in mean plasma-cortisol values between dogs that wore shock and "lemon-spray" collars during six 30-min training sessions over a two-week period (Steiss et al. 2007), which they deemed to show that neither were stressful. However electronic collars utilised by companion dog owners have been demonstrated to present welfare concerns (Cooper et al. 2014). The use of citronella is similarly based on the principles of positive punishment, applying an aversive stimulus to reduce the frequency of an unwanted behaviour. It follows that if the effects are indiscernible, then the strength of aversion to the electric shock and the scent of citronella must be similar, and hence we would argue the use of citronella is likely also a welfare concern.

9 Olfaction and Memory Are Tightly Linked

We know that for humans, odours can evoke strong emotional memories; for example, memories of the smell of a grandparent's house or reminder of trauma for people suffering from posttraumatic stress disorder (PTSD) (Herz 2011). There is increasing evidence of the link between memory and olfaction for dogs also.

Odour is linked to recognition of objects. Dogs show surprise when there is a mismatch between the odour and the appearance of an object they had previously encountered. Bräuer and Belger (2018) used the violation-of-expectation theory to study olfactory representation of objects. Dogs were instructed to fetch a toy by being directed to the beginning of a scent trail created by the toy. Dogs showed hesitation when what they found at the end of the scent trail did not match the toy they had followed and hence they expected. However, this effect was only observed initially and not in subsequent trials. This may be because dogs were rewarded regardless of which toy they retrieved, so learned that whether the toy matched the scent trail or not was not important. Alternatively, the scent from previous trials may still have been detectable in subsequent trials, or dogs may have started using other learning strategies with subsequent trials that were not completely reliant on scent tracking (Brauer and Belger 2018). Similarly, dogs form an expectation of their owner, based on odour and are surprised when the person at the end of an odour trail is not the owner (Bräuer and Blasi 2021), and we may expect that this may affect their affective state.

Quaranta et al. (2020) showed that odour can improve dogs' memory during a spatial memory task. Dogs were exposed to the odour of vanilla during "memory encoding" when they observed the experimenter placing food rewards under plastic cups. In the "memory retrieval" phase, dogs were required to recall the reward locations in the presence of the vanilla odour, a control odour (apple) or no odour (blank control). Dogs exposed to the original vanilla odour during this phase spent less time searching and made fewer mistakes. Hence, odour acts as a memory trigger, in this case improving performance at a task. This indicates odours are good contextual memory cues and are important in context-dependent memory for dogs (Quaranta et al. 2020).

Olfaction, memory and emotions are all controlled and processed by the brain's limbic system (Rolls 2019; Campellone 2022), therefore smells have the power to trigger memories and elicit strong feelings since their neuronal pathways are closely linked. This means there is potential to use olfaction to to benefit welfare. Scents used as aromatherapy for managing dog behaviour could be classically conditioned and paired with a positive experience to improve effectivity of any positive or calming effects of the odours themselves (Hedges 2021). But similarly, associations can easily be formed between negative events and odours, such as the smell of a veterinary surgery which is an important welfare consideration. Hence scents can be used to reduce unwanted behaviours. Conditioned food aversion (CFA) is a learned avoidance of food that is previously associated with a negative experience (e.g. illness). CFA is a learned mechanism that animals use to prevent future poisoning when consuming food or prey in the wild (Gustavson et al. 1974). By adding an aversive chemical substance (e.g. levamisole, an anti-parasite treatment) that induces mild and short-acting gastrointestinal effects (e.g. vomiting and diarrhoea) to a food source, CFA can be used as a non-lethal intervention to deter predators such as dogs, wolves, foxes and coyotes from killing livestock and wildlife (Tobajas et al. 2019, 2020a, 2020b, 2021). The coupling of levamisole with odours such as vanilla has been shown to be effective in aiding CFA in wolves (Tobajas et al. 2020b) and foxes (Tobajas et al. 2021), where after initial exposure to both substances (levamisole and vanilla), baiting food with just vanilla was enough to deter predators from the food source. This demonstrates how even non-aversive olfactory cues can trigger memory recall of a prior negative experience and influence decision-making and aversive behaviour, and hence shows how olfactory triggers can impact dogs' wellbeing.

10 Value of Odour from Familiar People and Dogs

The scent of familiar individuals is widely believed to be comforting and hence conducive to improved welfare in dogs. It is common advice for puppies leaving their litter or dogs going into boarding kennels to be given a clothing item smelling of their mother or owner, respectively (DoggieDiva 2022). Scientific study has started to confirm this principle.

The presence of an owner's scent has been shown to reduce cortisol levels in stressful situations (Shin and Shin, 2016). A study of dogs reported to show separation-related behaviour measured salivary cortisol concentrations before, during and after a 20-min period of separation from their owner. Dogs were provided with the owner's odour (worn T-shirt), audio (voice recording) or neither (control). The relative increase in salivary cortisol in the first five minutes of separation was significantly reduced in dogs provided with the odour or sound of their owner compared to the control condition.

Owner odours are also processed and interpreted differently in the brain. Dogs presented with scents from a familiar human (owner), an unfamiliar human, a familiar dog, an unfamiliar dog, and their own scent were studied under conscious fMRI (Berns et al. 2015). Only the owners' scent led to significant increases in activation of the caudate nucleus. Since this region of the brain is associated with positive expectations (Knutson et al. 2001), social rewards (Rilling et al. 2002; Izuma et al. 2008) and motivation ("wanting" or "seeking") (Panksepp 2004), this suggests that the odour of the owner causes a positive expectation and/or social reward.

11 Odour of Individuals in Different Emotional States Affects the Recipient

It is, however, not just the identity of the odour that potentially impacts on welfare but also the emotional state of the donor. A recent study found that dogs can discriminate the odour of human sweat and breath collected after a 3-min mental arithmetic stress test from that taken from the same individuals before the test (baseline) (Wilson et al. 2022). Four dogs were trained to discriminate control blank cloth from stress odour. During a series of double-blind, forced-choice tests they were able to discriminate stress odour samples from the baseline (pre-test) samples. When tested with samples from 36 individual people dogs showed accuracies between 90% and 97%. This indicates that there is a difference in odour associated with acute stress in humans that is detectable by dogs. However the effect these odours had on the dogs' own emotional state was not investigated.

There is a growing body of evidence of emotional contagion both between dogs and between human and dogs. Dogs show increases in alertness, stress-related behaviours, cortisol and comfort offering after hearing recordings of distress vocalisations from familiar conspecifics compared to unfamiliar conspecifics or control vocalisations (Quervel-Chaumette et al. 2016). They show person-oriented behaviours such as looking, approaching and making contact when hearing a familiar or unfamiliar person crying compared to humming or talking (Custance and Mayer 2012). Dogs will work to release their owner from a box, and will do so quicker, with higher heart rate (Carballo et al. 2020) and showing more stress behaviours if the owner is distressed (Van Bourg et al. 2020). Dogs' performance in a spatial task is improved if their owners are moderately stressed (Sümegi et al. 2014), raising the possibility that an owners' state anxiety is contagious to dogs, that the anxiety experienced by the owner influences their dog's behaviour and that these effects are manifested in the cognitive domain.

Some studies show that dogs yawn more after a person yawns (Joly-Mascheroni et al. 2008; Harr et al. 2009), although the extent of this effect varies between studies (Harr et al. 2009) and individual dogs (Buttner and Strasser 2014) and is greater when the person is familiar (Silva et al. 2012). It has also been debated whether this effect is empathy-induced or a sign of the stress of the experimental situation (Kis et al. 2020). Research continues to investigate the extent to which these effects are due to empathic distress (feelings of discomfort and anxiety in response to the perceived physical or emotional pain of another; Kis et al. 2020). However, these studies primarily focus on visual and auditory triggers, such as the dog watching yawning or hearing a person crying. Owners often report that their dogs pick up on their mood, for example when they are sad the dog reflects this state in their own behaviour. Undoubtedly this involves dogs detecting subtle behavioural cues from their owner such as their posture and body language, but evidence from functional lateralisation studies suggests it may also involve an olfactory component.

Functional lateralisation of the brain occurs in a multitude of sensory and motor functions as well as cognitive processes and emotions. It has been demonstrated across animal species, with similar processing pathways implicated (Vallortigara et al. 1999, 2011). The right hemisphere is thought to be involved in the processing and control of emotions, avoidance/escape behaviours and fear across species (Rogers and Andrew 2002). The right forebrain in particular is associated with sympathetic activity and emotional responses such as arousal, aversion, danger-perception, survival and negative affect (Craig 2005). By examining asymmmetries in how animals approach stimuli, one can draw conclusions about how that stimuli is perceived. This has been applied to vision (Bourne 2010), hearing (Altenmüller et al. 2002) but also olfactory stimuli (Simon et al. 2022). For example, preferences for right nostril sniffing behaviour have been observed in horses presented with adrenaline and oestrus urine, but not other odours such as food and repellents (Siniscalchi et al. 2015).

When it comes to dogs, lateralisation in sniffing behaviour has also been seen in response to different odour stimuli (Siniscalchi et al. 2011). Researchers mounted a video camera at the level of the odour samples (presented on cotton swabs) to record which nostril was used first and whether dogs switched nostril in subsequent presentations of the same odour. Six different odours were presented multiple times; blank cotton swabs, food, lemon, vaginal secretion from a bitch in oestrus, and adrenaline and sweat from a famailiar veterinarian. Dogs showed preferential and sustained use of the right nostril for sniffing the arousing stimuli of adrenaline and sweat from a veterinarian. Because olfactory neuronal pathways ascend ipsilaterally (on the same side), from the right nostril to right hemisphere, these findings suggest involvement of the sympathetic hypothalamic-pituitary axis controlled by the right hemisphere, where alarming or threat stimuli are processed. When presented with the other odours there was an initial right nostril preference which then switched to the left nostril with repeated presentations, suggesting initial right hemisphere processing of novel stimuli followed by transfer to the left hemisphere once the odour was identified to be familiar and non-threatening (MacNeilage et al. 2009; Siniscalchi et al. 2011). Such behavioural exhibitions can help us to understand how an odour is perceived. A later study by Siniscalchi et al. (2016) using the same methodology found significant asymmetrical sniffing behaviour in dogs presented with stress odours (perianal, interdigital and salivary secretions) from conspecifics during an isolation condition and human sweat collected during fear-eliciting film and physical exercise conditions (Siniscalchi et al. 2016). Dog were also presented with odours collected from conspecifics and humans after pleasurable encounters such as play (dog) or watching a joyous film, and neutral conditons such as showering and sleeping. Similar to the previous findings, there was a right nostril bias when sniffing the dog isolation condition, but there was a left nostril bias for samples from humans in fear and physical stress (exercise) conditions. No significant nostril biases were found for the other odours presented. This potentially suggests different pathways and brain hemispheres are involved in processing these odours depending on the species from which they originated (Siniscalchi et al. 2016). It is not clear why the left hemisphere may be involved in processing odours from humans in this study, however studies into the function of the amygdala indicate the roles of the left and right sides may be distinct (Gläscher and Adolphs 2003). While the right side is involved in rapidly and non-discriminately detecting potential threats and is activated by easily identifiable threat stimuli, the left amygdala is involved in more fine-tuned assessment of perceived threats and determining whether a real threat is posed. So, it could be that dogs require further fine-tune processing of odour signals from humans before determining whether the signal is interpreted as a real threat. However this is a novel field and requires greater study.

Siniscalchi et al. (2016) also found higher maximum heart rates in dogs when presented with odours from stressed conditions (isolated and dogs disturbed from sleep, human fear and exercise) compared to the neutral and dog "play" odour (Siniscalchi et al. 2016). The duration for which the heart rate was above the baseline average was also significantly increased when presented with these stress odours than neutral odours. Although when taken alone, increases in heart rate cannot indicate whether the response is associated with a positive (anticipation or excitement), negative (anxiety or fear) or neutral (arousal or alertness) experience, when taken together with observed changes in stress-related behaviours (D'Aniello et al. 2018) and olfactory processing (Siniscalchi et al. 2016), it could suggest an association with stress or threat perception.

Other studies have also looked at responses to the odour of fear, by examining more general behaviour. Dogs showed an increase in the frequency and duration of owner-oriented and stress-related behaviours, and a decrease in stranger-oriented behaviours when presented with sweat samples from human donors collected during a "fear" condition (watching fear-inducing video) compared to a "happiness" condition and control samples (D'Aniello et al. 2018). The heart rate of recipient dogs remained higher following exposure to odour samples from the fear condition than the happiness condition or blank samples (D'Aniello et al. 2018). Exposure to "happiness odours" increased the frequency and duration of stranger-oriented behaviours. Together, these findings suggest modulation of social interests and interactions with humans when exposed to chemosignals associated with the different emotional states of the donor, showing a preference to interact with a familiar human during fear odour exposure and an increase in sociability with strangers during happiness odour exposure (D'Aniello et al. 2018). This makes biological sense given that animals have evolved to limit risk-taking behaviours when environments are unsafe, and for social species, responding to the fear of other group members by avoiding novel situations or animals when others are fearful is protective. But why are these distinguishable across species? Semin et al. (2019) poses two possibilities; that odours produced in emotional contexts (of fear and happiness) contain distinctive chemical compounds that invariably activate the same responses across the species and thus have a pheromone-type quality, or that the responses to the odour are acquired in the process of socialisation and the two species acquire a sensitivity to the specific emotion-induced odours (Semin et al. 2019). Either way, if dogs smell odour taken from individuals in different states with different nostrils and behave differently following exposure, it is plausible that the odour affects their own emotional state. If this is the case, then being housed next to a stressed dog, for example in rehoming or training kennels, or being handled by a stressed handler may have adverse effects on a dog's own welfare. This concept is yet to be fully explored but should be considered when housing and husbandry regimes are planned.

12 Welfare of Dogs Used for Olfactory Work

The use of dogs for olfactory detection appears limitless, but as use continues to expand, it is imperative that ethical consideration ensures that dog welfare within each role is protected. Some olfactory tasks have an inherent risk attached to them such as land mines (Phelan 2002), explosives (Rooney and Clark 2021), and even predator (e.g. grizzly bear (Wasser et al. 2004)) detection. For these, it is essential that dogs are selected and trained to maximise their performance and that operational protocols are devised to minimises risk to the dog. Other uses such as detecting contaminants like DDT (dichloro-diphenyl-trichloroethane) (Arner et al. 1986), and diseases such as COVID-19 (Mancilla-Tapia et al. 2022), require extensive risk assessment to minimise the risk to the dog's health. However, there may also be some tasks that, although not detrimental to health, are aversive or less rewarding to the dog. Very repetitive work such as vehicle searching and remote sample searching for diseases can require very high stamina and hence may require stricter selection, innovative training, and greater efforts to optimise welfare when not working to retain dogs' motivation. One may predict, given a dog's sensitivity to the scent of affect in others (see section 11), that smelling stress or disease in others may cause them distress, and hence using dogs for disease detection, or to assist people suffering from fear or anxiety may impact their welfare. The fact that dogs can be trained for these purposes with seeming success and longevity suggests this may not be the case or that effective classical conditioning can result in successful pairing of scent and reward, overriding any inherent aversion or negative affect. It is, however, important that welfare is monitored to ensure this is always the case, and dogs do not start to show any signs of stress during their work.

Search and Rescue (SAR) dogs involved in the 9/11 response have been monitored annually and compared to similar SAR dogs not involved in the event using the Canine Behavioral Assessment and Research Questionnaire (C-BARQ; a standardised, behavioural evaluation tool for dog owners/guardians, handlers and professionals) (Hare et al. 2021). Using this instrument, no significant behavioural differences have been detected, leading the authors to conclude the dogs were resilient. Similar welfare monitoring of long-term use of working dogs is recommended, and should use a variety of instruments, health and physiological measures as well as questionnaires.

The way the dog is trained has a major impact upon its welfare. Most working dogs worldwide are now trained predominantly using positive reinforcement (Martin et al. 2020), and this applies also to scent work where reward is usually paired with the target scent. However, positive punishment and negative reinforcement are still utilised and are common in specific disciplines and organisations (Haverbeke et al. 2009). Studies on companion animals have found links between the way dogs are trained and their welfare and also their task performance. Surveys and observations of companion dogs strongly suggest punishment-based training is linked to increased incidence of behaviour problems (Hiby et al. 2004), including fear, anxiety and aggression (Blackwell et al. 2008), wariness towards strangers, reduced playfulness (Rooney and Cowan 2011), and increased anxiety-related aggression and excitability (Arhant et al. 2010). Dogs regularly trained with electric shocks show fear and distress behaviour in the presence of their owner, even outside the training context (Schilder and van der Borg 2004). Dogs subjected to physical reprimands have been shown to score significantly higher for aggression (Hsu and Sun 2010) and dogs whose owners report using a higher proportion of punishment are less likely to interact with a stranger (Rooney and Cowan 2011), and more likely to show a negative cognitive bias (Casey et al. 2021). Similarly in working dogs, studies have reported better performance and fewer behaviour problems in police dogs trained using rewards (Fattah and El Abdel-Hamid 2020), and an association between better performance and less use of more aversive stimuli by handlers in military dogs (Haverbeke et al. 2008). Those military dogs suspected to have been handled roughly in the past were perceived to be more fearful (Lefebvre et al. 2007). During protection and obedience work, dogs which had received more punishment tended to show more fear behaviours (Haverbeke et al. 2009).

13 Aspects of Housing and Husbandry are also Vitally Important

Many working dogs used for olfactory roles are kennelled. Kennelling has been shown to result in considerable stress especially to dogs that are unaccustomed to it (Hiby et al. 2006; Rooney et al. 2007). Infrared thermography showed that working police dogs were more stressed in their home kennel than in a police car on the way to work (Bartolome et al. 2021). Dogs kept in a home environment and then brought into military training kennels experienced a significant increase in cortisol (Rooney et al. 2007). This stress can be reduced by gradual introduction to the kennel environment from an early age using a program involving positive rewards (Rooney et al. 2007).

There are numerous aspects of a kennel environment that can be stressful; limited space and exercise, excessive noise, limited social contact and separation from attached figures (Cobb et al. 2020). Many kennel environments do not provide for all that the animal needs; for example, the dog has little control over their environment and may thus have difficulty coping and may experience negative feelings. Research has shown the importance of a variety of factors within the kennel environment (Gaines et al. 2008; Rooney et al. 2009; Taylor and Mills 2007b; Cobb et al. 2022). Hence, there are numerous interventions which have been shown to be effective ways of improving welfare of kennelled dogs including those used for scent detection.

Intraspecific social contact (Gfrerer et al. 2018) can benefit working dogs as can social housing for dogs that are sociable, resulting in lower levels of hair cortisol (Grigg et al. 2017). Increasing time outside the kennels and exercise can reduce barking (Clark et al. 1997), repetitive behaviours and cortisol, as can the addition of in-kennel toys (Nogueira et al. 2021) or chewing devices such as Kong (Gaines et al. 2008). Interactions with humans can be calming for many dogs, and stroking leads to increases in oxytocin in working dogs suggesting a positive effect on emotional state (Ogi et al. 2020). Play sessions with other dogs (Coppola et al. 2006; Shiverdecker et al. 2013) or humans (Conley et al. 2014) can be especially beneficial. Allowing dogs vantage points can increase control over their visual environment (Hubrecht 1993) whilst providing bedding and/or heating increases thermal comfort (Hiby 2005). The olfactory environment is also vitally important. As seen above, being able to smell potential prey or predator species can be a cause of stress, as can smelling neighbouring conspecifics that cannot be reached nor avoided. It is especially frustrating for intact males to be able to smell females that are in oestrus, although the extent of arousal without physical contact was questioned by the single published study examining this (Jezierski et al. 2019). Olfactory calming agents may have some value as may olfactory enrichment (see section 4–7 above).

14 In Conclusion

It is clear that dog research, like that on other species, suffers from anthropocentric sensory biases, favouring investigating visual information and cues despite the dominance of other sensory cues in many nonhuman research subjects, and the excellent olfactory performance in dogs (Horowitz and Franks 2020). Welfare research has similarly concentrated upon our predominant senses of vision and hearing and neglected olfaction. For dogs, the impact of odour is potentially great (Nielsen et al. 2015). Odours have the capacity to calm, stimulate and cause harm. Our knowledge of this area is gradually growing. However, those training, handling and researching with dogs need to be aware of possible olfactory influences if they are to optimise dogs' potential use whilst also protecting their welfare.

References

Allen ML, Gunther MS, Wilmers CC (2016) The scent of your enemy is my friend? The acquisition of large carnivore scent by a smaller carnivore. J Ethol 1:13–19

Altenmüller E, Schürmann K, Lim VK, Parlitz D (2002) Hits to the left, flops to the right: different emotions during listening to music are reflected in cortical lateralisation patterns. Neuropsychologia 40:2242–2256

Amaya V, Paterson M B A, Phillips C J C (2020a) Effects of olfactory and auditory enrichment on the behaviour of shelter dogs. Animals 10

Amaya V, Paterson M B A, Descovich K, Phillips C J C (2020b) Effects of olfactory and auditory enrichment on heart rate variability in shelter dogs. Animals 10

Arhant C, Bubna-Littitz H, Bartels A, Futschik A, Troxler J (2010) Behaviour of smaller and larger dogs: effects of training methods, inconsistency of owner behaviour and level of engagement in activities with the dog. Appl Anim Behav Sci 123:131–142

Arner L, Johnson G, Skovronek H (1986) Delineating toxic areas by canine olfaction. J Hazard Mater 13:375–381

Aspling F, Juhlin O, Chiodo E (2015) Smelling, pulling, and looking: unpacking similarities and differences in dog and human city life. In: Proceedings of the 12th International Conference on Advances in Computer Entertainment Technology. Iskandar, Malaysia: Association for Computing Machinery

Atsumi T, Tonosaki K (2007) Smelling lavender and rosemary increases free radical scavenging activity and decreases cortisol level in saliva. Psychiatry Res 150:89–96

Awad R, Levac D, Cybulska P, Merali Z, Trudeau VL, Arnason JT (2007) Effects of traditionally used anxiolytic botanicals on enzymes of the γ-aminobutyric acid (GABA) systemThis article is one of a selection of papers published in this special issue (part 1 of 2) on the Safety and Efficacy of Natural Health Products. Can J Physiol Pharmacol 85:933–942

Bartolome E, Perdomo-González DI, Sánchez-Guerrero MJ, Valera M (2021) Stress at rest in working dogs assessed with infrared thermography. Dog Behavior 7:13–21

Berns GS, Brooks AM, Spivak M (2015) Scent of the familiar: An fMRI study of canine brain responses to familiar and unfamiliar human and dog odors. Behav Proc 110:37–46

Binks J, Taylor S, Wills A, Montrose VT (2018) The behavioural effects of olfactory stimulation on dogs at a rescue shelter. Appl Anim Behav Sci 202:69–76

Blackwell EJ, Twells C, Seawright A, Casey RA (2008) The relationship between training methods and the occurrence of behavior problems, as reported by owners, in a population of domestic dogs. J Vet Behav 3:207–217

Boissy A, Manteuffel G, Jensen MB, Moe RO, Spruijt B, Keeling LJ, Winckler C, Forkman B, Dimitrov I, Langbein J (2007) Assessment of positive emotions in animals to improve their welfare. Physiol Behav 92:375–397

Bourne VJ (2010) How are emotions lateralised in the brain? Contrasting existing hypotheses using the Chimeric Faces Test. Cogn Emot 24:903–911

Bräuer J, Blasi D (2021) Dogs display owner-specific expectations based on olfaction. Sci Rep 11

Brauer J, Belger J (2018) A ball is not a Kong: Odor representation and search behavior in domestic dogs (*Canis familiaris*) of different education. J Comp Psychol 132:189–199

Broach D, Dunham AE (2016) Evaluation of a pheromone collar on canine behaviors during transition from foster homes to a training kennel in juvenile Military Working Dogs. J Vet Behav-Clin Appl Res 14:41–51

Brugarolas R, Yuschak S, Adin D, Roberts DL, Sherman BL, Bozkurt A (2019) Simultaneous monitoring of canine heart rate and respiratory patterns during scent detection tasks. IEEE Sens J 19:1454–1462

Buckley LA (2019) Does pet remedy reduce stress in dogs? Vet Evidence 4

Budzinski A, Budzinski C (2019) Dog flied study – at the heart of the walk [Online]. Available: http://www.dogfieldstudy.com/en/pulse-study/at-the-heart-of-the-walk [Accessed 4 September 2022 2022]

Buttner AP, Strasser R (2014) Contagious yawning, social cognition, and arousal: an investigation of the processes underlying shelter dogs' responses to human yawns. Anim Cogn 17:95–104

Buzek A, Serwanska-Leja K, Zaworska-Zakrzewska A, Kasprowicz-Potocka M (2022) The shape of the nasal cavity and adaptations to sniffing in the dog (canis familiaris) compared to other domesticated mammals: a review article. Animals 12

Campellone JV (2022) Limbic system [Online]. A.D.A.M. Medical Encyclopedia. Available: https://medlineplus.gov/ency/imagepages/19244.htm [Accessed 28 September 2022]

Carballo F, Dzik V, Freidin E, Damián JP, Casanave EB, Bentosela M (2020) Do dogs rescue their owners from a stressful situation? A behavioral and physiological assessment. Anim Cogn 23:389–403

Carlone B, Gazzano A, Gutiérrez J, Sighieri C, Mariti C (2018) The effects of green odour on domestic dogs: A pilot study. Appl Anim Behav Sci 207:73–78

Casey RA, Naj-Oleari M, Campbell S, Mendl M, Blackwell EJ (2021) Dogs are more pessimistic if their owners use two or more aversive training methods. Sci Rep 11:1–8

Chvala-Mannsberger S, Amtmann M, Vogl C, Binder R (2021) The sniffing carpet – a suitable enrichment tool also for French Bulldogs: odour perception and search behaviour of dogs of a brachycephalic breed. Wiener Tierarztliche Monatsschrift 108:260–270

Clark JD, Rager DR, Crowell-Davis S, Evans DL (1997) Housing and exercise of dogs: effects on behavior, immune function, and cortisol concentration. Comp Med 47:500–510

Cobb ML, Lill A, Bennett PC (2020) Not all dogs are equal: perception of canine welfare varies with context. Anim Welf 29:27–35

Cobb M L, Carter A, Lill A, Bennett P C (2022) Perceived importance of specific kennel management practices for the provision of canine welfare. Appl Anim Behav Sci 249

Conley MJ, Fisher AD, Hemsworth PH (2014) Effects of human contact and toys on the fear responses to humans of shelter-housed dogs. Appl Anim Behav Sci 156:62–69

Cooper JJ, Mason GJ (2000) Increasing costs of access to resources cause re-scheduling of behaviour in American mink (Mustela vison): implications for the assessment of behavioural priorities. Appl Anim Behav Sci 66:135–151

Cooper JJ, Cracknell N, Hardiman J, Wright H, Mills D (2014) The welfare consequences and efficacy of training pet dogs with remote electronic training collars in comparison to reward based training. PLoS ONE 9:e102722

Coppola CL, Grandin T, Enns RM (2006) Human interaction and cortisol: can human contact reduce stress for shelter dogs? Physiol Behav 87:537–541

Craig AD (2005) Forebrain emotional asymmetry: a neuroanatomical basis? Trends Cogn Sci 9:566–571

Croy I, Negoias S, Novakova L, Landis BN, Hummel T (2012) Learning about the functions of the olfactory system from people without a sense of smell. PLoS ONE 7:e33365

Croy I, Nordin S, Hummel T (2014) Olfactory disorders and quality of life—an updated review. Chem Senses 39:185–194

Custance D, Mayer J (2012) Empathic-like responding by domestic dogs (*Canis familiaris*) to distress in humans: an exploratory study. Anim Cogn 15:851–859

D'Aniello B, Semin GR., Alterisio A, Aria M, Scandurra A (2018) Interspecies transmission of emotional information via chemosignals: from humans to dogs (Canis lupus familiaris) 21, 67

DEFRA (Department for Environment, Food and Rural Affairs) (2006) Animal Welfare Act. [online] Available at: https://www.legislation.gov.uk/ukpga/2006/45/contents

DEFRA (2020) Code of practice for the welfare of pigs. https://publishing.service.gov.uk

Denenberg S, Landsberg GM (2008) Effects of dog-appeasing pheromones on anxiety and fear in puppies during training and on long-term socialization. J Am Vet Med Assoc 233:1874–1882

DoggieDiva (2022) The scent of home and familiar surroundings is important to dogs [Online]. Available: https://doggie-diva.com/pages/the-scent-of-home-and-familiar-surroundings-is-important-to-dogs [Accessed 23 September 2022 2022]

Duranton C, Horowitz A (2019) Let me sniff! Nosework induces positive judgment bias in pet dogs. Appl Anim Behav Sci 211:61–66

Dynavet (2022) Dynavet® [Online]. Available: http://www.dynavet.com/ [Accessed 19 November 2022]

Estellés MG, Mills DS (2006) Signs of travel-related problems in dogs and their response to treatment with dog appeasing pheromone. Veter Record 159:143–148

Ezeh PI, Myers LJ, Hanrahan LA, Kemppainen RJ, Cummins KA (1992) Effects of steroids on the olfactory function of the dog. Physiol Behav 51:1183–1187

Fatjó J, Feddersen-Petersen D, Ruiz de la Torre JL, Amat M, Mets M, Braus B, Manteca X (2007) Ambivalent signals during agonistic interactions in a captive wolf pack. Appl Anim Behav Sci 105:274–283

Fattah AFA, El Abdel-Hamid S (2020) Influence of gender, neuter status, and training method on police dog narcotics olfaction performance, behavior and welfare. J Adv Veter Animal Res 7:655–662

FAWC (1979) Five Freedoms [Online]. Available: https://webarchive.nationalarchives.gov.uk/ukgwa/20110909181150/, http://www.fawc.org.uk/freedoms.htm [Accessed 26 September 2022]

Frank D, Beauchamp G, Palestrini C (2010) Systematic review of the use of pheromones for treatment of undesirable behavior in cats and dogs. J Am Vet Med Assoc 236:1308–1316

Fureix C, Trevarthen AC, Finnegan EM, Bučková K, Paul ES, Mendl MT (2022) Do greater levels of in-cage waking inactivity in laboratory mice reflect a spontaneous depression-like symptom? A pharmacological investigation. Pharmacol Biochem Behav 212:173311

Gaines SA, Rooney NJ, Bradshaw JW (2008) The effect of feeding enrichment upon reported working ability and behavior of kenneled working dogs. J Forensic Sci 53:1400–1404

Gaultier E, Bonnafous L, Vienet-Lague D, Falewee C, Bougrat L, Lafont-Lecuelle C, Pageat P (2009) Efficacy of dog-appeasing pheromone in reducing behaviours associated with fear of unfamiliar people and new surroundings in newly adopted puppies. Veterinary Record 164:708–714

Gfrerer N, Taborsky M, Würbel H (2018) Benefits of intraspecific social exposure in adult Swiss military dogs. Appl Anim Behav Sci 201:54–60

Gläscher J, Adolphs R (2003) Processing of the arousal of subliminal and supraliminal emotional stimuli by the human amygdala. J Neurosci 23:10274–10282

Graham L, Wells DL, Hepper PG (2005) The influence of olfactory stimulation on the behaviour of dogs housed in a rescue shelter. Appl Anim Behav Sci 91:143–153

Grigg E, Piehler M (2015) Influence of dog appeasing pheromone (DAP) on dogs housed in a long-term kennelling facility. Vet Rec Open 2:e000098

Grigg EK, Nibblett BM, Robinson JQ, Smits JE (2017) Evaluating pair versus solitary housing in kennelled domestic dogs (*Canis familiaris*) using behaviour and hair cortisol: a pilot study. Veterinary Record Open 4:e000193

Gustavson CR, Garcia J, Hankins WG, Rusiniak KW (1974) Coyote predation control by aversive conditioning. Science 184:581–583

Gutiérrez J, Gazzano A, Pirrone F, Sighieri C, Mariti C (2019) Investigating the role of prolactin as a potential biomarker of stress in castrated male domestic dogs. Animals 9:676

Hale A, Meaney C (2021) Study into the use and efficacy of Pet Remedy in support of training and behaviour modification in dogs

Hale A (2021) Study into efficacy of Pet Remedy as an aid to reducing stress in canines during grooming

Hall NJ, Glenn K, Smith DW, Wynne CD (2015) Performance of Pugs, German Shepherds, and Greyhounds (Canis lupus familiaris) on an odor-discrimination task. J Comp Psychol 129:237–246

Harding EJ, Paul ES, Mendl M (2004) Cognitive bias and affective state. Nature 427:312–312

Hare, E., Kelsey, K. M., Niedermeyer, G. M. & Otto, C. M. 2021. Long-term behavioral resilience in search-and-rescue dogs responding to the September 11, 2001 terrorist attacks. *Applied Animal Behaviour Science,* 234.

Harr AL, Gilbert VR, Phillips KA (2009) Do dogs (Canis familiaris) show contagious yawning? Anim Cogn 12:833–837

Haverbeke A, Laporte B, Depiereux E, Giffroy J-M, Diederich C (2008) Training methods of military dog handlers and their effects on the team's performances. Appl Anim Behav Sci 113:110–122

Haverbeke A, De Smet A, Depiereux E, Giffroy J-M, Diederich C (2009) Assessing undesired aggression in military working dogs. Appl Anim Behav Sci 117:55–62

Hedges S (2021) Practical Canine Behaviour : For Veterinary Nurses and Technicians, [S.l.], CAB International.

Hermiston C, Montrose VT, Taylor S (2018) The effects of dog-appeasing pheromone spray upon canine vocalizations and stress-related behaviors in a rescue shelter. J Vet Behav-Clin Appl Res 26:11–16

Herz R S (2011) Odor-evoked memory. The Oxford handbook of social neuroscience, 265–275

Hiby E, Rooney N, Bradshaw J (2004) Dog training methods: their use, effectiveness and interaction with behaviour and welfare. Anim Welf.13:63–70

Hiby EF, Rooney NJ, Bradshaw JW (2006) Behavioural and physiological responses of dogs entering re-homing kennels. Physiol Behav 89:385–391

Hiby E (2005) The welfare of kennelled domestic dogs. University of Bristol. PhD Thesis

Holcova K, Koru E, Havlicek Z, Rezac P (2021) Factors associated with sniffing behaviors between walking dogs in public places. Appl Anim Behav Sci 244:105464

Horowitz A (2016) Being a dog: Following the dog into a world of smell. Simon and Schuster

Horowitz A, Franks B (2020) What smells? Gauging attention to olfaction in canine cognition research. Anim Cogn 23:11–18

Hsu Y, Sun L (2010) Factors associated with aggressive responses in pet dogs. Appl Anim Behav Sci 123:108–123

Hubrecht RC (1993) A comparison of social and environmental enrichment methods for laboratory housed dogs. Appl Anim Behav Sci 37:345–361

Izuma K, Saito DN, Sadato N (2008) Processing of social and monetary rewards in the human striatum. Neuron 58:284–294

Jenkins EK, Lee-Fowler TM, Craigangle T, Behrend EN, Moore GE (2016) Effects of oral administration of metronidazole and doxycycline on olfactory capabilities of explosives detection dogs. Am J Vet Res 77:906–912

Jezierski T, Dzieciol M, Szumny A, Nizanski W, Woszczylo M, Pieczewska B, Godzinska EJ (2019) Discrimination of estrus odor in urine by male dogs in different experimental settings. J Vet Behav-Clin Appl Res 29:25–30

Joly-Mascheroni RM, Senju A, Shepherd AJ (2008) Dogs catch human yawns. Biol Let 4:446–448

Juarbe-Diaz SV, Houpt KA (1996) Comparison of two antibarking collars for treatment of nuisance barking. J Am Anim Hosp Assoc 32:231–235

Kim YM, Kang BT, Gu SH, Yoo JH, Park C, Jung DI, Jeon HW, Hahm DH, Park HM (2006) Comparison of acupuncture with dog-appeasing pheromone (DAP) Application in separation anxiety of hospitalized dogs. Exp Neurobiol 15:81–87

Kim YM, Lee JK, Abd El-aty AM, Hwang SH, Lee JH, Lee SM (2010) Efficacy of dog-appeasing pheromone (DAP) for ameliorating separation-related behavioral signs in hospitalized dogs. Can VetY J-Rev Vet Can 51:380–384

Kis A, Tóth K, Kanizsár O, Topál J (2020) The effect of oxytocin on yawning by dogs (*Canis familiaris*) exposed to human yawns. Appl Anim Behav Sci 223:104916

Knutson B, Adams CM, Fong GW, Hommer D (2001) Anticipation of Increasing Monetary Reward Selectively Recruits Nucleus Accumbens. J Neurosci 21, RC159-RC159

Kokocinska A, Woszczylo M, Sampino S, Dzieciol M, Zybala M, Szczuka A, Korczynska J, Rozempolska-Rucinska I (2022) Canine Smell Preferences-Do Dogs Have Their Favorite Scents? Animals (Basel) 12

Landsberg G M, Beck A, Lopez A, Deniaud M, Araujo J A, Milgram N W (2015) Dog-appeasing pheromone collars reduce sound-induced fear and anxiety in beagle dogs: a placebo-controlled study. Vet Record 177

Lefebvre D, Diederich C, Delcourt M, Giffroy J-M (2007) The quality of the relation between handler and military dogs influences efficiency and welfare of dogs. Appl Anim Behav Sci 104:49–60

Lisberg AE, Snowdon CT (2009) The effects of sex, gonadectomy and status on investigation patterns of unfamiliar conspecific urine in domestic dogs, Canis familiaris. Anim Behav 77:1147–1154

Pet Remedy'(2022) Pet Remedy Ltd, https://petremedy.co.uk Accessed 26 September 2022.

Lydiard RB (2003) The role of GABA in anxiety disorders. J Clin Psychiatry 64:21–27

MacNeilage PF, Rogers LJ, Vallortigara G (2009) Origins of the left and right brain. Sci Am 301:60–67

Mancilla-Tapia JM, Lozano-Esparza V, Orduna A, Osuna-Chavez RF, Robles-Zepeda RE, Maldonado-Cabrera B, Bejar-Cornejo JR, Ruiz-Leon I, Gonzalez-Becuar CG, Hielm-Bjorkman A, Novelo-Gonzalez A, Vidal-Martinez VM (2022) Dogs detecting COVID-19 from sweat and saliva of positive people: a field experience in Mexico. Front Med (lausanne) 9:837053

Martin C, Diederich C, Verheggen F (2020) Cadaver dogs and the deathly hallows—a survey and literature review on selection and training procedure. Animals 10:1219

Matsubara E, Fukagawa M, Okamoto T, Fukuda A, Hayashi C, Ohnuki K, Shimizu K, Kondo R (2011) Volatiles emitted from the leaves of *Laurus nobilis* L. improve vigilance performance in visual discrimination task. Biomed Res 32:19–28

McGlone JJ, Thompson W, Guay KA (2014) Case Study: The pig pheromone androstenone, acting as an interomone, stops dogs from barking. Prof Animal Scientist 30:105–108

McGlone JJ, Archer C, Henderson M (2022) Interpretive review: Semiochemicals in domestic pigs and dogs. Front Vet Sci 9:967980

McGreevy PD, Georgevsky D, Carrasco J, Valenzuela M, Duffy DL, Serpell JA (2013) Dog behavior co-varies with height, bodyweight and skull shape. PLoS ONE 8:e80529

Mendl M, Brooks J, Basse C, Burman O, Paul E, Blackwell E, Casey R (2010) Dogs showing separation-related behaviour exhibit a 'pessimistic' cognitive bias. Curr Biol 20:R839–R840

Mills DS, Ramos D, Estelles MG, Hargrave C (2006) A triple blind placebo-controlled investigation into the assessment of the effect of Dog Appeasing Pheromone (DAP) on anxiety related behaviour of problem dogs in the veterinary clinic. Appl Anim Behav Sci 98:114–126

Moffat KS, Landsberg GM, Beaudet R (2003) Effectiveness and comparison of citronella and scentless spray bark collars for the control of barking in a veterinary hospital setting. J Am Anim Hosp Assoc 39:343–348

Murtagh K, Farnworth M J, Brilot B O (2020) The scent of enrichment: Exploring the effect of odour and biological salience on behaviour during enrichment of kennelled dogs. Appl Anim Behav Sci 223

Myers LJ, Hanrahan LA, Swango LJ, Nusbaum KE (1988a) Anosmia associated with canine distemper. Am J Vet Res 49:1295–1297

Myers LJ, Nusbaum KE, Swango LJ, Hanrahan LN, Sartin E (1988b) Dysfunction of sense of smell caused by canine parainfluenza virus infection in dogs. Am J Vet Res 49:188–190

Nakashima T, Akamatsu M, Hatanaka A, Kiyohara T (2004) Attenuation of stress-induced elevations in plasma ACTH level and body temperature in rats by green odor. Physiol Behav 80:481–488

Nielsen BL, Jezierski T, Bolhuis JE, Amo L, Rosell F, Oostindjer M, Christensen JW, McKeegan D, Wells DL, Hepper P (2015) Olfaction: an overlooked sensory modality in applied ethology and animal welfare. Front Vet Sci 2:69

Nogueira LB, Palme R, Mendonça-Furtado O (2021) Give them a toy or increase time out of kennel at lawn areas: what is the influence of these interventions on police dogs' welfare? Animals 11:2264

O'Neill D, Pegram C, Crocker P, Brodbelt D, Church D, Packer R (2020) Unravelling the health status of brachycephalic dogs in the UK using multivariable analysis. Sci Rep 10:1–13

O'Neill D G, Skipper A, Packer R M A, Lacey C, Brodbelt DC, Church DB, Pegram C (2022) English Bulldogs in the UK: a VetCompass study of their disorder predispositions and protections. Canine Med Genet 9

Ogi A, Mariti C, Baragli P, Sergi V, Gazzano A (2020) Effects of stroking on salivary oxytocin and cortisol in guide dogs: preliminary results. Animals 10:708

Oka T, Hayashida S, Kaneda Y, Takenaga M, Tamagawa Y, Tsuji S, Hatanaka A (2008) Green odor attenuates a cold pressor test-induced cardiovascular response in healthy adults. BioPsychoSocial Medicine 2:2

Pageat P, Gaultier E (2003) Current research in canine and feline pheromones. Vet Clin North America-Small Anim Pract 33:187-+

Panksepp J (2004) Affective neuroscience: the foundations of human and animal emotions. Oxford University Press

Paul ES, Harding EJ, Mendl M (2005) Measuring emotional processes in animals: the utility of a cognitive approach. Neurosci Biobehav Rev 29:469–491

Phelan JM (2002) Chemical sensing for buried landmines-fundamental processes influencing trace chemical detection. Sandia National Lab.(SNL-NM), Albuquerque, NM (United States); Sandia

Pirner G, McGlone J (2016a) Rabbit maternal pheromone delivered in ointment decreases heart rate in domestic dogs during a simulated thunderstorm. J Anim Sci 94:206–206

Pirner G, McGlone J (2016b) Impact of androstenone on leash pulling and jumping up in dogs. Animals 6

Platt SR, Olby NJ (2013) BSAVA manual of canine and feline neurology. Fourth, edition. British Small Animal Veterinary Association, Cheltenham

Polgár Z, Kinnunen M, Újváry D, Miklósi Á, Gácsi M (2016) A Test of Canine Olfactory Capacity: Comparing Various Dog Breeds and Wolves in a Natural Detection Task. PLoS ONE 11:e0154087

Polgár Z, Blackwell EJ, Rooney NJ (2019) Assessing the welfare of kennelled dogs—A review of animal-based measures. Appl Anim Behav Sci 213:1–13

Prior MR, Mills DS (2020) Cats vs. Dogs: The efficacy of feliway friends (TM) and Adaptil (TM) products in multispecies homes. Front Vet Sci 7

Products IP (2022) *Tracerz®* scent markers for blind dogs [Online]. Available: https://www.innovetpet.com/products/blind-dog-scent-markers [Accessed 22 September 2022]

Puglisi I, Masucci M, Cozzi A, Teruel E, Navarra M, Cirmi S, Pennisi MG, Siracusa C (2022) Effects of a Novel Gel Formulation of Dog Appeasing Pheromone (DAP) on Behavioral and Physiological Stress Responses in Dogs Undergoing Clinical Examination. Animals 12:2472

Quaranta A, D'Ingeo S, Siniscalchi M (2020) Odour-Evoked Memory in Dogs: Do Odours Help to Retrieve Memories of Food Location? Animals 10:1249

Quervel-Chaumette M, Faerber V, Farago T, Marshall-Pescini S, Range F (2016) Investigating Empathy-Like Responding to Conspecifics' Distress in Pet Dogs. PLoS ONE 11:e0152920

Riddell P, Paris MCJ, Joone CJ, Pageat P, Paris D (2021) Appeasing pheromones for the management of stress and aggression during conservation of wild canids: Could the solution be right under our nose? Animals 11

Riemer S (2020) Effectiveness of treatments for firework fears in dogs. J Vet Behav-Clin Appl Res 37:61–70

Rilling JK, Gutman DA, Zeh TR, Pagnoni G, Berns GS, Kilts CD (2002) A Neural Basis for Social Cooperation. Neuron 35:395–405

Rogers LJ, Andrew RJ (2002) Comparative Vertebrate Lateralization. Cambridge University Press, New York

Rolls ET (2019) Chapter 2–The cingulate cortex and limbic systems for action, emotion, and memory. In: Vogt BA (ed.) Handbook of Clinical Neurology, Elsevier

Rooney NJ, Clark CCA (2021) Development of a Performance Monitoring Instrument for Rating Explosives Search Dog Performance. Front Vet Sci 8:545382

Rooney NJ, Cowan S (2011) Training methods and owner–dog interactions: Links with dog behaviour and learning ability. Appl Anim Behav Sci 132:169–177

Rooney NJ, Sargan DR (2010) Welfare concerns associated with pedigree dog breeding in the UK. Anim Welf 19:133–140

Rooney NJ, Gaines SA, Bradshaw JW (2007) Behavioural and glucocorticoid responses of dogs (Canis familiaris) to kennelling: Investigating mitigation of stress by prior habituation. Physiol Behav 92:847–854

Rooney N, Gaines S, Hiby E (2009) A practitioner's guide to working dog welfare. J Vet Behav-Clin Appl Res 4:127–134

RSPCA. 2020. *5 training games that will help build your dog's trust* [Online]. Available: https://www.rspca.org.uk/-/5-training-games-that-will-help-build-your-dog-s-trust [Accessed 09 September 2022]

RuffleSnuffle (2021) *Creating a Sensory Garden for Your Dog Step-by-Step Guide* [Online]. Available: https://www.rufflesnuffle.co.uk/creating-a-sensory-garden-for-your-dog-step-by-step-guide/ [Accessed 22 September 2022]

Samuel L, Arnesen C, Zedrosser A, Rosell F (2020) Fears from the past? The innate ability of dogs to detect predator scents. Anim Cogn 23:721–729
Santos NR, Beck A, Maenhoudt C, Fontbonne A (2020) Influence of ADAPTIL(R) during the Weaning Period: A Double-Blinded Randomised Clinical Trial. Animals, 10.
Sargisson RJ, Butler R, Elliffe D (2011) An evaluation of the aboistop citronella-spray collar as a treatment for barking of domestic dogs. ISRN Vet Sci 2011:759379
Sayorwan W, Siripornpanich V, Piriyapunyaporn T, Hongratanaworakit T, Kotchabhakdi N, Ruangrungsi N (2012) The effects of lavender oil inhalation on emotional states, autonomic nervous system, and brain electrical activity. J Med Assoc Thai 95(4): 598–606
Schilder MB, van der Borg JA (2004) Training dogs with help of the shock collar: short and long term behavioural effects. Appl Anim Behav Sci 85:319–334
Semin GR, Scandurra A, Baragli P, Lanata A, D'Aniello B (2019) Inter-and intra-species communication of emotion: Chemosignals as the neglected medium. Animals 9
Sheppard G, Mills DS (2003) Evaluation of dog-appeasing pheromone as a potential treatment for dogs fearful of fireworks. Vet Rec 152:432–436
Shiverdecker MD, Schiml PA, Hennessy MB (2013) Human interaction moderates plasma cortisol and behavioral responses of dogs to shelter housing. Physiol Behav 109:75–79
Silva K, Bessa J, de Sousa L (2012) Auditory contagious yawning in domestic dogs (Canis familiaris): first evidence for social modulation. Anim Cogn 15:721–724
Simon T, Guo K, Frasnelli E, Wilkinson A, Mills DS (2022) Testing of behavioural asymmetries as markers for brain lateralization of emotional states in pet dogs: A critical review. Neurosci Biobehav Rev 143:104950
Siniscalchi M, Sasso R, Pepe AM, Dimatteo S, Vallortigara G, Quaranta A (2011) Sniffing with the right nostril: lateralization of response to odour stimuli by dogs. Anim Behav 82:399–404
Siniscalchi M, Padalino B, Aubé L, Quaranta A (2015) Right-nostril use during sniffing at arousing stimuli produces higher cardiac activity in jumper horses. Laterality: Asymmetries of Body. Brain Cogn 20:483–500
Siniscalchi M, D'Ingeo S, Quaranta A (2016) The dog nose "KNOWS" fear: Asymmetric nostril use during sniffing at canine and human emotional stimuli Behav Brain Res 304:34
Siracusa C, Manteca X, Cuenca R, Alcala MD, Alba A, Lavin S, Pastor J (2010) Effect of a synthetic appeasing pheromone on behavioral, neuroendocrine, immune, and acute-phase perioperative stress responses in dogs. JAVMA 237:673–681
Site, T. P. 2014. *The Pig Site* [Online]. Available: https://www.thepigsite.com/ [Accessed 26 September 2022 2022].
Spiers JG, Chen H-JC, Lavidis NA (2015) Stress alleviating plant-derived 'green odors': behavioral, neurochemical and neuroendocrine perspectives in laboratory animals. Phytochem Rev 14:713–725
Stanghellini AL (2019) Effect of lavender (Lavandula angustifolia) essential oils on sheltered dog behavior: preliminary results. Dog Behavior 5:19–22
Steiss JE, Schaffer C, Ahmad HA, Voith VL (2007) Evaluation of plasma cortisol levels and behavior in dogs wearing bark control collars. Appl Anim Behav Sci 106:96–106
Sümegi Z, Oláh K, Topál J (2014) Emotional contagion in dogs as measured by change in cognitive task performance. Appl Anim Behav Sci 160:106–115
Sutoh M, Ito S, Kasuya E, Yayou K-I (2013) Effects of exposure to plant-derived odorants on behavior and the concentration of stress-related hormones in steers isolated under a novel environment. Anim Sci J 84:159–164
Taylor S, Madden J (2016) The Effect of Pet Remedy on the Behaviour of the Domestic Dog (Canis familiaris). Animals 6:64
Taylor K, Mills DS (2007a) A placebo-controlled study to investigate the effect of Dog Appeasing Pheromone and other environmental and management factors on the reports of disturbance and house soiling during the night in recently adopted puppies (Canis familiaris). Appl Anim Behav Sci 105:358–368
Taylor KD, Mills DS (2007b) The effect of the kennel environment on canine welfare: a critical review of experimental studies. Anim Welf 16:435–447

Taylor S, Webb L, Montrose VT, Williams J (2020) The behavioral and physiological effects of dog appeasing pheromone on canine behavior during separation from the owner. J Vet Behav-Clin Appl Res 40:36–42

Tilley LP, Smith FWK Jr (2015) *Blackwell's Five-Minute Veterinary Consult : Canine an Feline*, New York, USA. John Wiley & Sons, Incorporated

Titulaer M, Blackwell EJ, Mendl M, Casey RA (2013) Cross sectional study comparing behavioural, cognitive and physiological indicators of welfare between short and long term kennelled domestic dogs. Appl Anim Behav Sci 147:149–158

Tobajas J, Gómez-Ramírez P, María-Mojica P, Navas I, García-Fernández AJ, Ferreras P, Mateo R (2019) Selection of new chemicals to be used in conditioned aversion for non-lethal predation control. Behav Proc 166:103905

Tobajas J, Gómez-Ramírez P, Ferreras P, García-Fernández AJ, Mateo R (2020a) Conditioned food aversion in domestic dogs induced by thiram. Pest Manag Sci 76:568–574

Tobajas J, Ruiz-Aguilera MJ, López-Bao JV, Ferreras P, Mateo R (2020b) The effectiveness of conditioned aversion in wolves: Insights from experimental tests. Behav Proc 181:104259

Tobajas J, Descalzo E, Villafuerte R, Jimenez J, Mateo R, Ferreras P (2021) Conditioned odor aversion as a tool for reducing post-release predation during animal translocations. Anim Conserv 24:373–385

Tod E, Brander D, Waran N (2005) Efficacy of dog appeasing pheromone in reducing stress and fear related behaviour in shelter dogs. Appl Anim Behav Sci 93:295–308

Trauner G, Khom S, Baburin I, Benedek B, Hering S, Kopp B (2008) Modulation of GABA_A Receptors by Valerian Extracts is Related to the Content of Valerenic Acid. Planta Med 74:19–24

Uccheddu S, Mariti C, Sannen A, Vervaecke H, Arnout H, Gutierrez Rufo J, Gazzano A, Haverbeke A (2018) Behavioral and cortisol responses of shelter dogs to a cognitive bias test after olfactory enrichment with essential oils. Dog Behavior 4:1–14

Unwin SL, Saunders RA, Blackwell E-J, Rooney NJ (2020) A double-blind, placebo-controlled trial investigating the value of Pet Remedy in ameliorating fear of handling of companion rabbits. J Vet Behav-Clin Appl Res 36:54–64

Vallortigara G, Rogers LJ, Bisazza A (1999) Possible evolutionary origins of cognitive brain lateralization. Brain Res Rev 30:164–175

Vallortigara G, Chiandetti C, Sovrano VA (2011) Brain asymmetry (animal). Wires Cognit Sci 2:146–157

Van Bourg J, Patterson JE, Wynne CD (2020) Pet dogs (Canis lupus familiaris) release their trapped and distressed owners: Individual variation and evidence of emotional contagion. PLoS ONE 15:e0231742

Van den Berghe F, Paris MCJ, Sarnyai Z, Vlamings B, Millar RP, Ganswindt A, Cozzi A, Pageat P, Paris D (2019) Dog appeasing pheromone prevents the androgen surge and may reduce contact dominance and active submission after stressful interventions in African wild dogs (Lycaon pictus). Plos One, 14

van Haaften KA, Grigg EK, Kolus C, Hart L, Kogan LR (2020) A survey of dog owners' perceptions on the use of psychoactive medications and alternatives for the treatment of canine behavior problems. J Vet Behav-Clin Appl Res 35:27–33

Wasser SK, Davenport B, Ramage ER, Hunt KE, Parker M, Clarke C, Stenhouse G (2004) Scat detection dogs in wildlife research and management: application to grizzly and black bears in the Yellowhead Ecosystem, Alberta, Canada. Can J Zool 82:475–492

Wells DL (2001) The effectiveness of a citronella spray collar in reducing certain forms of barking in dogs. Appl Anim Behav Sci 73:299–309

Wells D (2003) Comparison of two treatments for preventing dogs eating their own faeces. Vet Rec 153:51–53

Wells DL (2006) Aromatherapy for travel-induced excitement in dogs. J Am Vet Med Assoc 229:964–967

Wilson C, Campbell K, Petzel Z, Reeve C (2022) Dogs can discriminate between human baseline and psychological stress condition odours. PLoS ONE 17:e0274143

Woszczylo M, Jezierski T, Szumny A, Nizanski W, Dzieciol, M (2020) The role of urine in semiochemical communication between females and males of domestic dog (Canis familiaris) during Estrus. Animals 10

Yuan C-S, Mehendale S, Xiao Y, Aung HH, Xie J-T, Ang-Lee MK (2004) The Gamma-Aminobutyric Acidergic Effects of Valerian and Valerenic Acid on Rat Brainstem Neuronal Activity. Anesth Analg 98:353–358

Applications of Dog Olfactory Capabilities

Forensic and Security Applications of Substance Detection Canines

Paola Prada-Tiedemann, Lauryn DeGreeff, and Craig Schultz

Abstract

This chapter presents the use of substance detection canines from a forensic and security viewpoint. It provides the reader with a foundational framework of forensic principles that can be extended and applied to the use of canine teams as a forensic tool for odor detection purposes. It draws a special focus on key aspects of legal and security considerations to better educate the reader on optimal and efficient uses of canine teams within operational settings. Furthermore, it provides a basic introduction to the chemical aspects of odor analysis as it relates to the detection of canine performance. Lastly, it provides a synopsis of detection disciplines within substance detection canines of relevance to the forensic practice.

Keywords

Forensic • Detection canines • Odor • Security • Evidence

P. Prada-Tiedemann (✉)
Forensic Analytical Chemistry & Odor Profiling Laboratory, Department of Environmental Toxicology, Texas Tech University, 1207 S. Gilbert Drive, Lubbock, TX 79416, USA
e-mail: Paola.tiedemann@ttu.edu

L. DeGreeff
Global Forensic and Justice Center, Florida International University, Miami, FL, USA

C. Schultz
Evidence Response Team Unit/Forensic Canine Consulting, Federal Bureau of Investigation, Laboratory Division, Quantico, VA, USA

L. Lazarowski (ed.), *Olfactory Research in Dogs*,
https://doi.org/10.1007/978-3-031-39370-9_11

1 Introduction

Forensic science is a unique discipline that provides an interdisciplinary connection across various areas of expertise with the common goal of providing a service to legal matters. The role of forensic science includes assisting either a criminal or civil investigation by: establishing necessary elements of the case (Corpus delicti), aiding the "trier of the fact" in reaching a just decision, and aiding the process of evaluating the credibility and reliability of testimony (Harris et al. 2019). There are many disciplines that can routinely play a role in forensic applications which include medical sciences, natural sciences, engineering, and technology (See Fig. 1). With the advancement of technology and digital capabilities, more types of evidence can be processed and collected from crime scenes and analyzed in the forensic laboratory. Novel sensor systems provide rugged and robust capabilities for the recovery of evidence items that can provide help with the identification, individualization, reconstruction, and association of evidentiary material.

Within technological advancements, both instrumental and biological sensor systems have changed the dynamics of evidence recovery. Instrumental technologies provide the ability for rapid on-site analysis of trace evidence items, while laboratory instrumentation provides cutting-edge results on the chemical composition and physical characteristics of recovered items. However, another tool in the forensic practitioner's toolbox is that of biological detection, specifically, canines as detection systems. The use of canines within law enforcement and forensic applications ranges from contraband detection (as seen with explosives, narcotics, and firearms), human remains, live human scent, human evidence collection (remains, blood, semen, etc.), arson (accelerant), and emerging applications such as electronic storage device detection. Substance detection canines play essential roles in routine criminal investigations as they provide a robust, rugged, and highly mobile detection mechanism as compared to conventional instrumentation. The highly developed olfactory capability of the canine nose allows for its wide applicability in substance detection purposes. The canine's nasal cavity has millions of sensory neurons within the olfactory epithelium. During the inhalation process, volatile molecules enter the nasal passage where they are directed in their path by complex tissue called turbinates located in the lining of nasal cavity wall.

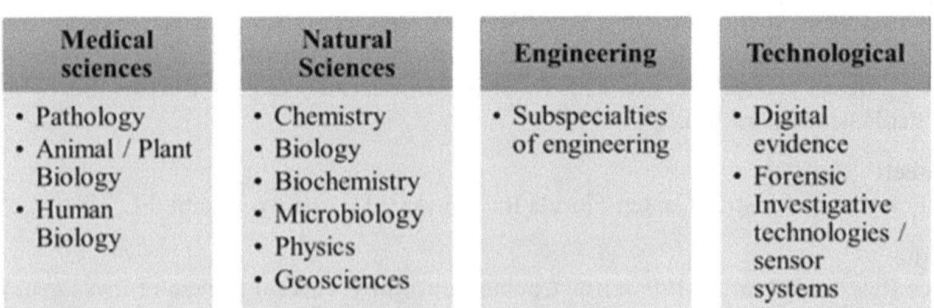

Fig. 1 Forensic science application discipline

Overall, the natural physical configuration of the olfactory sensory region of the canine nose allows for optimized nasal airflow patterns during sniffing that makes it an advantageous sensor system for odorant delivery and detection (Craven et al. 2010). Thus, canines offer a practical detection mechanism to locate and identify target odor sources in operational environments given their innate physiological and biological advantages with respect to olfaction capabilities.

Within a forensic and security perspective, there is a wide range of forensic specimens that emit volatile organic compounds (VOCs), thus depicting another source of evidentiary material—forensic odor evidence. Odor evidence has witnessed novel uses in forensic science applications to include toxicology, drug identification and profiling, explosive and arson investigations, and human odor (both live and deceased) (Furton et al. 2015; Titus et al. 2022). The use of canines as sensors has been explored for its utility in locating hidden or obscured physical evidence (and even live human scent when it is used to corroborate other evidence). Additionally, as a parallel effort analytical chemistry techniques to understand the chemical odor signatures emitting from these ranges of sources are gaining recognition in the courts as evidence (Anthony 2013; Torrez 2020). The scent or odor profile of a particular substance and/or subject can be characteristic of a particular specimen, thus making the odor trace an effective investigation tool for identification and differentiation purposes. As with any other type of forensic evidence, it is imperative to follow valid and reliable scientific approaches to safeguard the integrity of the crime scene and its associated evidence. Thus, in recent years, efforts have been geared to promote and enhance the understanding of detection canine performance by clearly delineating important variables for research test designs (Lazarowski et al. 2020) as well as promoting and creating standards for routine training and certification purposes (OSAC Dogs and Sensors Subcommittee 2022).

Trace evidence is fragile, transitory in nature, and most often not highly visible or recognizable. Odor evidence is not an exception to this rule and must be treated as any other type of trace evidence. The three "Rs" of trace evidence are known to be recognition, recording, and recovery (Houck 2016). As it relates to trace odor evidence, the recovery relies on understanding what odor is, how odor is transferred, how it persists in an environment, and how to optimally train the dog to locate such odor(s). Once located, the recording and recovery of physical evidence relies on the handler/operator's understanding of canine behavior and responses given to trained odor(s), and the joint work of other forensic professionals to recover evidence once it is found (Fig. 2). As such, the work of a detector canine team parallels that of any other scientific discipline in the forensics field. It is important to understand that, while the canine is crucial in locating these odor traces, the complementary use of technological instrumentation such as chromatographic methods can provide a final confirmatory analysis of the substances in question. Hence, the canine detector system can be viewed as a highly mobile and rugged screening tool for efficient operational contexts.

When used for legal purposes, canines may be utilized for (1) detection of substances to indicate the presence of an illegal material (explosives, narcotics,

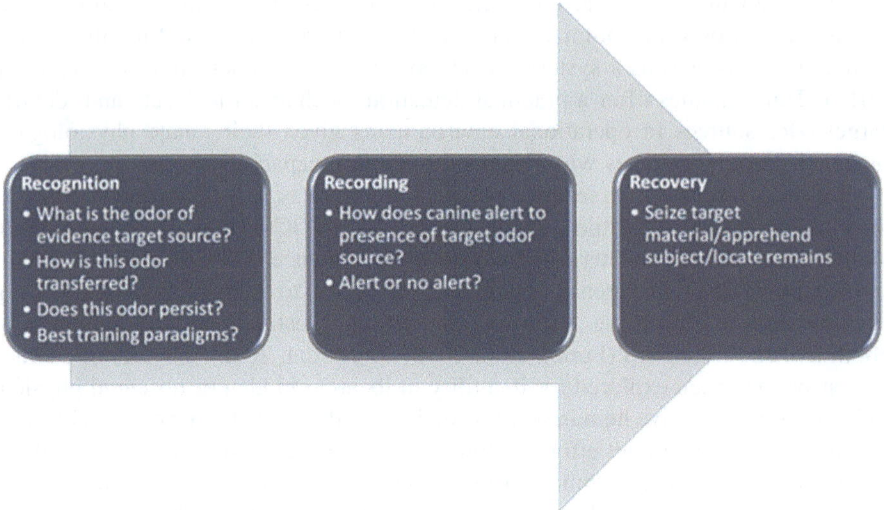

Fig. 2 The "3 R's of canine odor evidence

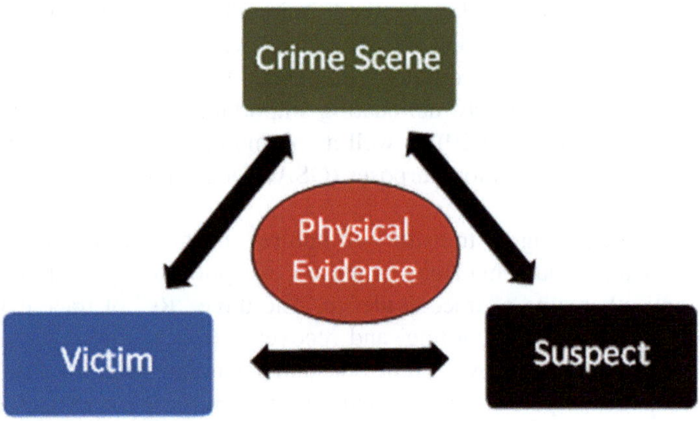

Fig. 3 Locard's exchange principle illustrated to show the linkages between crime scenes, suspects, victims, and physical evidence

contraband, weapons, digital storage devices, human remains, etc.), and (2) the formation of linkages that tie people to evidence and crime scenes, such as blood, trace odors associated with human decomposition, human scent in match-to-sample applications, seminal fluid, etc., which occurs when two items come into contact with one another, cross transfer of material occurs and/or is often left behind (See Fig. 3). The intensity, duration, and nature of the materials in contact with one another determine the extent of the transfer. The transfer of material occurs in two ways. Physical contact results in direct or "primary transfer" of material, whereas a secondary transfer occurs when evidence is transported from one area to another on something (such as clothing) and then the evidence is deposited elsewhere (Deedrick 2000). This concept was proposed by a Frenchman with a background in medicine and law, Edmond Locard, in the first police crime laboratory in Lyons, France in 1910 and has come to be known as Locard's Exchange Principle. His principle was instrumental in solving several crimes and, after World War I, Locard's successes served as an impetus for the formation of police laboratories in Vienna, Berlin, Sweden, Finland, and Holland (Criminalistics: An Introduction to Forensic Science). His principle remains the cornerstone of modern forensic science today. Canines used forensically, like any other sensor tool used in crime scenes, can help investigators create linkages. For example, when canines locate trace odors associated with blood, or seminal fluid from a suspect which is left within a crime scene, if blood or trace odors of human decomposition are found on a suspect's clothing, in their vehicle or on tools/weapons; then the linkages that occur as a result of cross transfer may prove to be probative in trial. Furthermore, the odor from contraband, such as narcotics or explosives, or human scent (living or remains) may also be transferred in a similar manner to the physical evidence and can aid in recovery of the contraband or establishing linkages relating to the location of a human or a body. Canine responses to any of the aforementioned may aid in the collection of evidence to support the linkage or strengthen a chain of circumstantial evidence. Important to note is that, Locard's Exchange Principle also has implications in contamination of target odor training aids for detection canines when the components are substituted with environmental variables, which will be discussed later in this chapter.

In the field of forensics, there are many types of tests that identify the *potential* of a material to be something that may have evidentiary value. "Class" and "Individual" characteristics are often discussed in terms of the types of evidence encountered at crime scenes. "**Class characteristics**" are considered as a piece of evidence as it belongs to a group of like items. Like Nike Pegasus running shoes, there are thousands of pairs of them in circulation within a population. "**Individual characteristics**" are those which make that item very specific and unlike the others, for instance, when a rock alters the tread pattern of the Nike Pegasus running shoe and makes that shoe unique. Blood, like other bodily fluids, can be viewed similarly. Mammalian blood collectively has similar components and as such would be a class characteristic. It can further be separated into classifications such as canine blood, feline blood, and human blood, for instance. Once it is sent to the laboratory and its DNA is examined, it can be designated as specifically belonging to someone with a high statistical likelihood, thus making it individualistic (it belongs to John Doe).

There are several ways in which evidence is located and evaluated in crime scenes. To determine if something is of evidentiary value, certain types of tests are conducted. These tests help investigators or crime scene professionals determine if something should be collected and sent to a laboratory for further analysis. To determine if a type of material or substance belongs to a certain class, presumptive tests can be used. A "**Presumptive Test**," is one that is used in criminal investigations which is not conclusive but is used to screen for the presence of a substance (US Legal 2022). These are tests that simply identify the presence of something, but offers no specificity as to who it belongs. For example, blood can be identified using chemiluminescence (Luminol, BlueStar, Fluorescein, etc.) but it does not offer information as to whether it is animal or human blood, or who specifically contributed to the sample. In order for that to happen, the sample requires further examination at a laboratory by a trained examiner. A "**Confirmatory Test**" is conducted at the laboratory by a serologist to determine if the sample is human by looking at certain proteins and then further testing can be conducted to determine individuality using DNA.

It is important to understand that **detection canines in law enforcement applications are to be viewed as presumptive tests** only in that they are trained to detect an odor profile associated with a particular class of substances, just like other types of presumptive tests. A trained final response from a canine does not imply that a substance belongs to a particular individual. They are responding to a "class" of odorants that comprise a target odor profile collectively. For this reason, canine evidence should be used as corroboration with other forms of evidence or testing. This issue has been highlighted in several court cases where it has been determined that canine evidence must be considered with caution, is of slight probative value, and, if found reliable, cannot support a conviction in the absence of other direct evidence of guilt (People v. Perryman 1979; McDuffie v. State. 1992).

This chapter explores the use of substance detection canines from a forensic and security application viewpoint. It introduces legal and security considerations, provides a brief description of the chemical analysis of odor, and details selected detection applications covering the characterization of odor targets and the training and proficiency of canines in that given area of work.

2 Substance Detection Canines and the Law

Canine handling has long been regarded as an art; however, court proceedings have forced science to quantify the performance of working dogs to corroborate evidence in criminal cases. In doing so, it has essentially married the art of canine handling and training to the sciences and has forever changed the landscape of how canines are viewed through a forensic lens. This process has woven the interplay between canines, law, and science so tightly together that you simply cannot have one component without the others. Just as the biological concept of emergent properties in which an organism is comprised of many components that work together, forensic canines must be viewed as a triad of the three (Schultz et al. 2022) (Fig. 4).

As far back as the early 1800s, scientific hypothesis and testing provided the foundation for the forensic sciences to build upon. In 1814, the first forensic science discipline (toxicology) was introduced by Mathieu Joseph Bonaventure Orfilia where he made chemical analysis a routine part of forensic medicine.

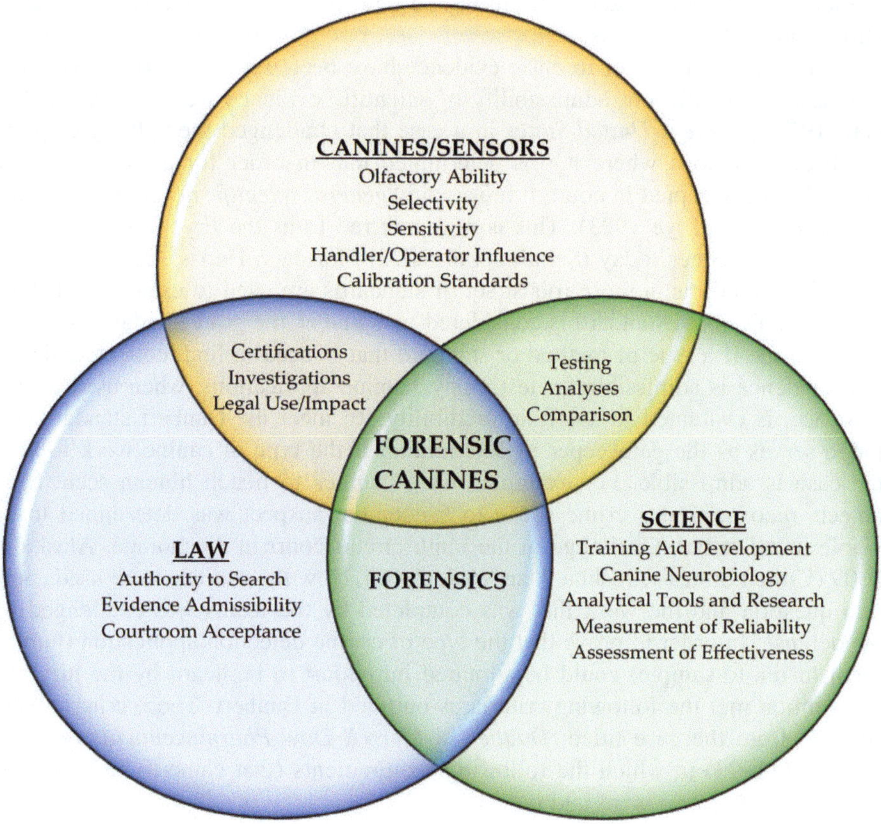

Fig. 4 The Forensic Canine Triad Venn Diagram illustrating the interplay between canines, science, and law (reproduced with permission from Jenny Stanford Publishing)

Throughout the 1800s and 1900s, contributors from various fields continued to advance the forensic sciences into what we know them to be today. Toxicology, fingerprints, accelerant and explosives analysis, DNA, firearms and toolmarks, anthropology, soil analysis, hair and fiber analysis, digital forensics, and more have come to gain acceptance. Since then, the methods, techniques, or procedures that are used to locate, collect, and examine evidence from crime scenes have become subject to rigorous evaluation in order to be used in court. In fact, in 1893 (129 years ago), Hans Gross wrote the first treatise describing the application of scientific disciplines to the field of criminal investigation (Criminalistics: An Introduction to Forensic Science).

When canines are used to aid in resolving criminal cases, their use may be called into question and brought before the court. In some cases, the utilization or technique in how the canines were used are disclosed and the defense and prosecution have the opportunity to prove or disprove the reliability of the application. This is done during evidence admissibility hearings and through Daubert or Frye challenges in which prosecutors are tasked with proving the reliability of canine utilization.

While various types of forensic evidence have been used in court proceedings, standards for courtroom admissibility of scientific evidence were not established until 1923 in *Frye v. United States* in a case that challenged the validity of polygraph examinations where it was determined that in order for a technique or a method to be accepted in court, it must gain "*general scientific acceptance*" within the community (Frye 1923). This is now referred to as the Frye Test and is still used by some states today (Schultz et al. 2022; Florida v. Harris 2013).

In federal courts, a more robust set of standards are used to establish admissibility. The Daubert Standard is considered by some as the gold standard by which to determine if a type of method or standard that is used to find, collect, and process evidence is admissible for testimony. Canine applications, when used to find evidence, is evaluated to ascertain credibility. To meet the Daubert standard, the judge serves as the gatekeeper in determining if the type of canine work used in the case is admissible. For example, using canines to match human scent from objects recovered at a crime scene to a potential suspect was determined to be viable in a Daubert Challenge in the ninth circuit court in Anchorage, Alaska in 2009 (United States v. Joshua Alan Wade 2009). How the canines were used came into question and the work that was completed by the teams was challenged by the defense. In order to prove that the type of canine detection application (human scent match-to-sample) could be admitted into court to be heard by the jury, the prosecution met the following criteria as outlined in Daubert. These criteria were adopted from the case titled, *Daubert v. Merrell Dow Pharmaceuticals Inc.* 509 U.S. 579 (1993) in which the following requirements (that came to be known as the Daubert Factors) were laid out:

1. The theory or technique in question has been tested.
2. The results of the testing have been subjected to peer review and publication.
3. The method or technique has established error rates.
4. There are standards for maintaining or controlling its operation.
5. The method or technique has gained acceptance within a relevant scientific community.

It is clear to see how research is expected and necessary for the Daubert Factors to be met in order to ensure court proceedings are fair and free of unjustifiable expense and delay. They exist today to promote further development of evidentiary law (Schultz et al. 2022). It is widely recognized that other disciplines within the field of forensic sciences have underpinnings deeply rooted in research to develop examination methods and provide outcomes in criminal cases. The use of canines in forensic settings has gained popularity, and thus research is needed in order to continue to support canine evidence admissibility in court. When they are utilized in forensic applications, the courts rely on science to help establish the reliability of the evidence. This is done during evidence admissibility hearings and through

Daubert or Frye challenges in which prosecutors are tasked with proving the reliability of canine utilization. To do so, research is used to support or refute legal inquiries to determine if canines are both capable and reliable detectors. Research may also be used to disprove the reliability of the methods, and care should be taken to ensure that research is conducted with an objective approach as much as possible. This should be done to mitigate the possibility of drawing erroneous conclusions which could create a false sense of capability or potentially establish poor case law. Researchers should be mindful of the purpose of the research and the potential impact it may have beyond the laboratory or field trials in various applications (Yeager 2014) as studies may be used to support or refute a method, technique, technology, or application to either convict someone and significantly change that person's life (potentially "end" their life if it is a capital crime with a death penalty sentence) or set a criminal free to potentially reoffend. Here, is it easy to see how research outcomes may impact legal proceedings.

Once or if canine evidence is deemed as admissible by the court, the method, technology, or application can be presented during the trial in which the prosecution presents its evidence before the judge and/or jury. The information is then released to the public. However, in national security cases, the judge can determine if an application would be harmful to national security if the information was publicly released and not allow the evidence to be presented or seal the testimony.

2.1 Implications of Research in Legal Settings

Another consideration to note is the quality of the research conducted. Not only may releasing the results of research have an impact on criminal cases or on national security, but poor-quality research can also be detrimental to an entire industry. If the experimental design is not robust or is executed without an understanding of how detection canines are trained, operated, or utilized, the effect can damage the credibility of an entire industry when erroneous conclusions are drawn.

An example of this occurred in a peer-reviewed paper published in Animal Cognition in 2011 titled "Handler beliefs affect scent detection dog outcomes" by Lisa Lit, Julie Schweitzer, and Anita Oberbauer (Lit et al. 2011). This research focused on how handler beliefs can impact detection outcomes. In this study, researchers carried target odor containers into the testing environment in order to lead handlers to believe that target odors were being placed, however, these containers were removed from the testing area shortly after. Handler–dog teams ran the testing exercise, and all responses were annotated. Canines responded in the testing environment and those responses were considered as false responses. The results of this study were published and subsequently used by defense councils to question the reliability of canine utilization in criminal cases in numerous cases throughout the United States. The broad conclusions drawn in this study did not consider other contributing factors aside from handler beliefs, such as the sensitivity of canine olfactory acuity, nor did they survey canine handlers to determine if limitations existed in their training methodology or certification standards.

Due to these concerns, the Scientific Working Group for Dogs and Orthogonal Detector Guidelines (SWGDOG) published a rebuttal on their website in 2011 stating that the authors could not draw the conclusions they did based on the work described in the publication, due to several flaws in the experimental design and execution of the study. In the rebuttal, SWGDOG's 55-member group of subject matter experts stated the following about the conclusions drawn from the study.

> The authors stated in their findings that "handler beliefs affect working dog outcomes and human indication of scent location affects distribution of alerts more than dog interest in a particular location." It is universally agreed that understanding how to minimize the manipulation of handler and canines is important to incorporate into best practices. However, the relevant scientific and canine community represented on SWGDOG believes there are a number of characteristics of the study presented that limit or invalidate the conclusions of this study and cannot be extended to working detector dog teams.
>
> This study was meant to focus on the behavior of the handlers and not the performance of the canines, but the baseline performance of the canines, critical in evaluating handler influence, was not addressed. The failure to evaluate the proficiency of the canine teams in a controlled blind setting at the time of testing prevents scientifically valid conclusions: a canine with a propensity to false respond during blind testing would render different results than a canine with no propensity to false respond. The authors speculated throughout the paper as to the possible explanations for the canine responses, but there is insufficient baseline data and insufficient controls to allow the readers to make meaningful conclusions. The authors did not indicate if the canine team's training records were reviewed to determine if the teams regularly engaged in documented maintenance training to include (multiple) blank areas, and distracters, such as food and experimenter's hand scent. Certification details are lacking and do not appear to conform to best practices. The use of two dual-trained drug/explosive canines described in this study as "certified" is problematic, this is universally recognized as an unsafe practice. The article failed to describe what certification standard was used for each detector dog team, when each canine team was initially certified and last certified, if distractors were included in the certification, and whether blank areas were included in the certification. There was also insufficient information regarding the handler's experience including number of historical deployments or searches conducted.
>
> The experimental design did not consider the complexity of using detection canines in a study and did not follow detection canine testing best practices. Drugs and explosives should never have been introduced into the blank search areas due to the potential for contamination of the test area and test materials. The authors stated that each day the experimenter carried a metal box and canvas bag containing drugs and explosive into the church. Though the containers were never opened in the test area, it does not alleviate the possibility of contamination. It is well known that drug and explosives odors can readily permeate even multiple layers of plastic and other materials. The authors should have mitigated the design problem by conducting a blind odor recognition assessment of participating canine teams prior to the start of the testing, as well as proofing the testing area with a non-participating certified canine team to ensure the absence of target odors such as drugs or explosives.
>
> The authors stated a "double-blind" test was conducted. In a conventional double-blind test, neither the tester nor the canine team would know any parameters of the test. The authors did not describe if the handlers were given instructions not to discuss their search results or whether they were sequestered until the conclusion of the study. In this study the handlers were instructed that "each condition might contain up to three target scents and that target scene markers consisting of a red piece of construction paper would be present in

> two conditions." It is SWGDOG's opinion that the authors would have provided some discussion of the extreme nature of the bias that was intentionally created relative to a typical detection dog scenario.
>
> The unusually high "false alert" rate indicates more than a cuing effect. There were 12 runs where handlers called more than the instructed maximum of odors present. SWGDOG members routinely involved in detection canine research indicated there are a number of factors that can contribute to higher than normal false response rates under testing scenarios. Failures that have occurred with dogs tested in an artificial environment have been attributed to factors including test site contamination, unfamiliarity and pressures associated with blind testing, extreme expectations, unfamiliar working procedures, and having canine teams search the same area multiple times. The authors did not describe if the handlers were instructed to search on or off-lead, or whether they were given a choice. If instructed to search in a particular method, the team must have demonstrated prior competency in that method (Letter in response to the Lisa Lit et al. research study).

To elaborate on the Lit et al. study (2011) and its limitations, the fact that containers housing target odors were brought into the blank testing area for a period of time and then removed makes it plausible that the canines could have responded to residual odors left behind after target source containers were removed from the environment. In this case, canines would have been correct in their responses. Interestingly, this is a point that two of the original authors acknowledge in a publication released in 2019 where they state "The olfactory abilities of dogs are far superior to those of humans. The problem is that detection dog alerts are not always corroborated by the actual presence of the detection target. For example, in some cases, a dog may detect the residual scent from an object that was previously (but is no longer) present" (Lit et al. 2019). An experimental design in which similar uncontaminated containers were brought into the blank testing environment would have been a better alternative.

It is well accepted that cueing does exist within all forms of animal training to include detection canine training, and the phenomenon dates as far back as the turn of the twentieth century with psychologist Oskar Pfungst and the horse, Clever Hans (Waterbury and Schultz 2022). While non-verbal handler cueing (whether intentional or unintentional) is recognized by the detection canine industry, this publication failed to acknowledge what, if any, other variables impacted the performance of the teams. A more heuristic design with an ethological approach to identifying specific handler behaviors, non verbal communication, or movement and how they impacted response rates would have been a good step in identifying why false positive responses occurred and would have created a more robust study. The information gleaned from such a study could help identify limitations associated with specific occurrences and correlations and lead to more rigorous canine training to reduce handler dependence and handler training protocols to eliminate such cueing. Research should *not only validate the efficacy of a technique, method, or a process* but it should *aim to identify the limitations of the process and offer insight into how to mitigate the limitations by providing solutions.* Making broad conclusions without identifying other contributing factors is incomplete. To determine that a limitation exists is the first step in understanding why it occurs, but robust experimentation identifies how the limitations developed with the ultimate aim of improving the process.

While the conclusions of the Lit et al. publication (2011) have not been widely accepted by practitioners in detection canine training, it did bring to light the need to adopt methodologies to train canines for additional stimulus control and independence in training to mitigate the fallout from its use in court proceedings. To prevent the challenge of ascertaining whether a canine is responding to non-verbal handler cues or a target odor, some agencies have adopted and implemented an "independence assessment" into their certification procedures and assessments. This is accomplished by placing a target odor source in an environment in which the canine is released into, but the handler remains out of sight. Here, the canine is unable to rely on visual cues from the handler as to the whereabouts of the target odor source. The canine is required to offer its trained final response without the line of sight or any prompting from the handler. Once the canine responds, the handler is notified and can then enter the space and reinforce the final response behavior. This mirrored a certification procedure utilized by the Federal Emergency Management Agency (FEMA) for disaster live-find and disaster cadaver canines in which the canines are sent into a limited access rubble pile and are required to respond to targets without visual contact with their handlers. Handlers are not allowed to access the rubble pile until the canine responds. This ensures that the canine is responding to the odor stimulus and not handler stimuli (The State of Florida Urban Search Rescue Response System Canine Search Specialist Training Student Manual 2004; SWGDOG 2010).

2.2 Security Considerations

It is important to note that not all research is, or should be, released for public consumption. When conducting research with canines used in forensic, law enforcement, or military applications, there are several things to consider, particularly when the information gleaned from the research will impact the safety of citizens. If the information impedes a law enforcement agency's ability to resolve cases or protect its citizens, then the information may be deemed as "Unclassified: Law Enforcement Sensitive," "Unclassified: For Official Use Only," or even "Classified." A summary of the United States classification system is given in Fig. 5. Furthermore, if a technique is discovered or a technology is developed based on research that may prove to be formidable in combating acts of terrorism, it may be assigned a higher classification such as secret or top secret. Then, the question that should be asked is "could that information be used to circumvent detection or reverse engineered and used against us?" Thus, putting human lives in jeopardy. Hence, it is important to understand where the research falls within the respective classification systems used in government and how the information obtained from the research will be or could be used.

For these reasons, there is likely additional applied research in the areas of forensic evidence recovery or security-related substance detection, particularly in

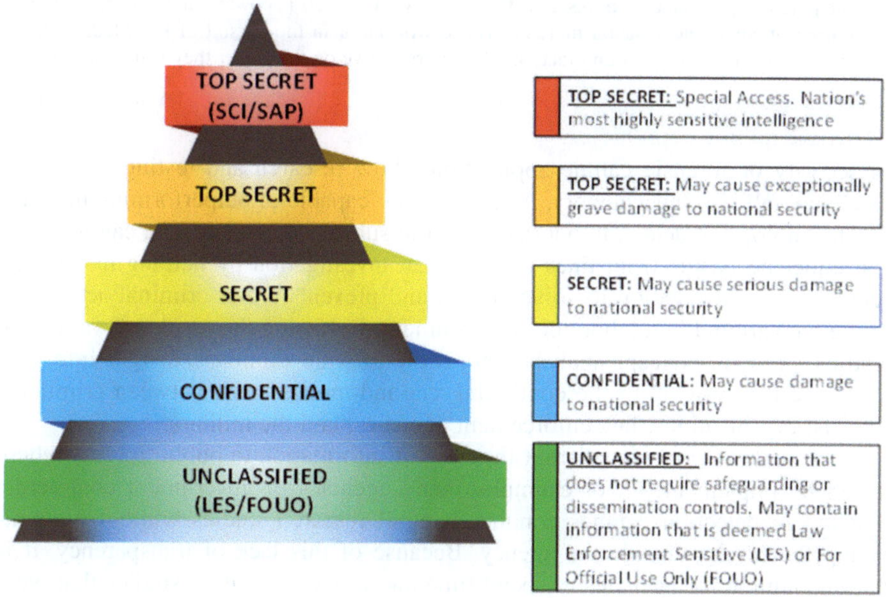

Fig. 5 Illustration depicting the types of classification and their definitions that are commonly used by the federal government

explosives detection, than is presently available in the public domain. Frequently, such research has security implications as research data have the potential to relay capability gaps that could be exploited by bad actors. Furthermore, there is potential for criminals to exploit publicly available research for the purpose of defeating forensic recovery techniques. Examples of this have played out in several terrorist bombings on U.S. soil whereby individuals obtained information from open sources to create explosive devices. Examples include Eric Rudolph's attack at the 1996 Summer Olympics in Atlanta and the Boston Bombing in 2013 (Yeager 2014). This is particularly alarming in today's digital environment where scientific articles are easily transitioned from printed publications to online venues, thus increasing their accessibility to those with nefarious intentions. This must be something taken into account while conducting such research in an unsecured capacity, for instance, academic research. Great care should be taken to not publicly release data that could become dangerous in the wrong hands.

Reasons for why information is pushed out may stem from scientists who rush to publish in efforts necessary to meet department requirements and to bolster bona fides in the laboratories of academia where the mantra "*publish or perish*" runs rampant. This is understandable because often, research scientists rarely have the ability to gain "boots on the ground" experience in live theater giving them first-hand knowledge of how the information can be used pervasively. In an editorial piece published in 2014, Dr. Kirk Yeager so eloquently writes,

> The phrase "**ipsa scientia potestas est**" or "knowledge itself is power" first appears in the writing of Sir Francis Bacon in 1597. Those who labor in the pursuit of knowledge can become jaded to this profound fact, and lose perspective on the power they truly possess.
>
> Yeager (2014)

In security or forensic canine applications, how research and testing results are used may fall into this category. As a biosensor capable of outperforming modern-day technology in detecting odors of specific substances or scents, a canine's role can either be on the front lines aiding in an investigation by helping narrow the scope of an investigation to solve crimes and prevent further criminal activity or in the background in a laboratory environment. Releasing the results of validation studies, proof-of-concept analysis, etc., may provide criminals insight into how to "game the system" in the continuing cat-and-mouse game between criminals/terrorist organizations, law enforcement agencies, and the military.

Such security concerns disrupt the flow of information to public forums where it can be implemented. For example, many agencies or government contractors that utilize detection canines conduct internal research studies without disseminating the results outside the agency. Because of this lack of transparency, it is not uncommon for agencies to spend time and money repeating studies that were already completed by other groups. A 2018 search of classified research database for canine detection research related to olfactory detection produced no output, though the authors had knowledge of such research activity occurring, indicating that classified or secured research is not being shared outside of the individual agency, even among different government entities with the appropriate security clearances (Peranich and DeGreeff 2018). Improved communication would benefit both the operational community and also stretch the small amount of funding available for canine detection research by preventing duplicative research. The authors herein strongly encourage those conducting research relating to canine detection, particularly of contraband such as explosives, to find mechanisms for safely disseminating research to the community. This may be by creating searchable reports in classified databases or by removing some pertinent information, such as specific target types, manufacturer names, or participating operational canine teams, for the purposes of publishing publicly available peer-reviewed manuscripts.

3 Chemical Analysis of Odor

When one considers detection canine research, one might first think of behavioral research relating to training, veterinary studies on canine fitness, or neurobiological studies with olfactory receptors. Perhaps, it is not necessarily intuitive that chemical analysis works hand-in-hand with the detection of canine performance. Testing of the animal alone only conveys half of the story; one must also understand the odor, more specifically, the *odorants*, that the canine is being asked to detect. Here, chemists focus on characterizing the odor that the canine is to detect instead of studying the animal itself. Chemical analyzes may be carried out to define the compounds that comprise an odor, ascertain where or when an odor is most concentrated, or develop and validate methods for controlling what or how much odor is presented to the canine for training or testing purposes. Chemical studies of odor have included the determination of the odorant availability from a given target confirming the particular compounds which have olfactory interest for detection (odor activity) (Cerreta and Furton 2015; Cannon 2020; Francis et al. 2019; Furton et al. 2002; Harper et al. 2005a; Kranz et al. 2014; Lorenzo et al. 2003; Macias and Furton 2011; Rice and Koziel 2015; Johnston et al. 1998), measurement of residual, contaminating, or background odor (Oesterhelweg et al. 2008b; Hallowell et al. 1997; Alexander et al. 2015), assessment of odor changes with time or environmental condition (Hudson et al. 2009; DeGreeff et al. 2017b), and measurement of the change in amount or characteristics of the odor due to manufacturer or form (DeGreeff et al. 2017b; Harper et al. 2005b; DeGreeff and Peranich 2021; Crespo Cajigas et al. 2019), to name a few. The chemical analysis of odor allows the canine handler or trainer to better understand the olfactory challenges the canine will face and the knowledge can be applied in the choice of training aid materials and training aid handling. This would be done with the goal of increasing canine proficiency and effectiveness in the field.

An odor is comprised of a collection of (mostly) volatile compounds. Volatile compounds are chemicals that are readily available in the vapor phase (as a gas) under ambient conditions. Chemical analysis of odor focuses on the characterization of these volatile compounds as, most often, odor enters the nose and interacts with the olfactory system as a gaseous vapor. A volatile compound that binds to an odorant receptor in the olfactory system and produced a recognized odor is termed an *odorant*, while the term *odor* is actually the perception of the odorant in the olfactory system] (Pickenhagen 2017). Many times, a perceived odor is made up of a collection of many odorants, and this is referred to as the odor profile (e.g., vapor profile, volatile profile, VOC [volatile organic compound] profile).

There are many types of instrumentation that can detect gases, but very few have the sensitivity and resolution to characterize and quantify complex vapor samples containing only trace levels of analyte. The most common is gas chromatography with mass spectrometry (GC/MS) owing to its superior selectivity and sensitivity for samples in the gas phase. As the hyphenated name implies, GC/MS is actually two instruments run in tandem. The gas chromatograph (GC) separates the individual components in a mixture, while the mass spectrometer (MS) is used for identification. Detailed information about the function and types of GC/MSs can be found elsewhere and a full discussion is beyond the scope of this chapter. In short, an odor sample is injected onto a GC column. The column has a stationary phase that coats the sides of the column. A mobile phase, most commonly helium or hydrogen, moves the molecules through the column, and as the individual analyte molecules interact with the stationary phase, they are separated into bands based on their affinity for the stationary phase (See Fig. 6). The GC alone does not identify these bands of odorant molecules. Instead, each discrete band enters the mass spectrometer (MS) where they are bombarded with high energy,

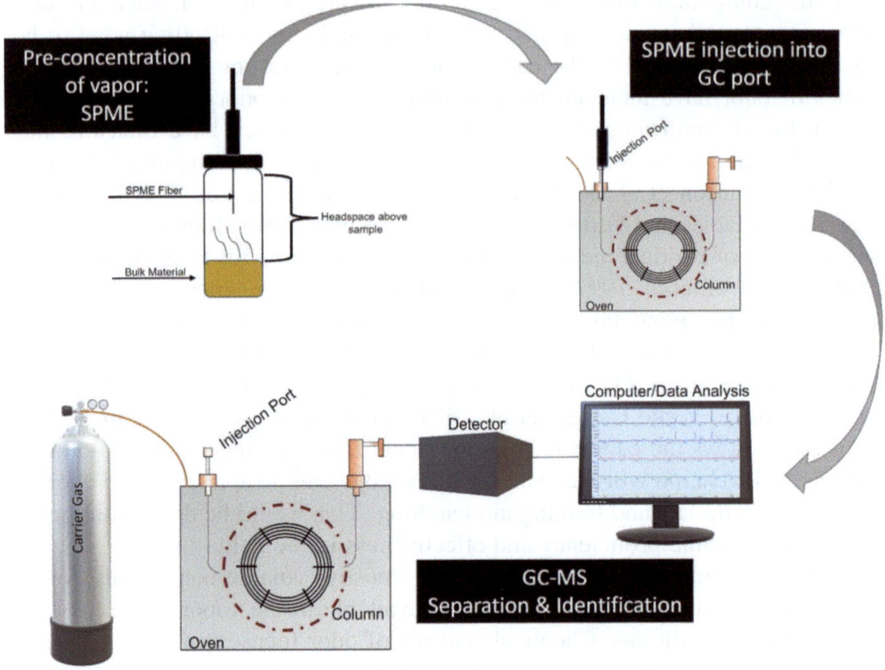

Fig. 6 Schematic of commonly used odor analysis laboratory instrumentation. Vapor is extracted via a pre-concentrated approach using solid phase microextraction (SPME). The SPME fiber is desorbed into the inlet of the gas chromatograph (GC). Separation and identification of volatile analytes by gas chromatography with mass spectrometry (GC–MS)

producing charged molecule fragments. The type and ratios of these charged fragments (e.g., fragment ions) are unique to each analyte. The resulting fingerprint of fragment ions is termed the mass spectra, and can be used for identification of a compound by comparison to standards or a mass spectral library. While many crime investigation shows would like you to believe otherwise, each compound may be identified by GC/MS, the origin or source of the odor is not. For example, we may identify cyclohexanone and 2-ethyl-hexanol from a material, but one would only know that this odor is likely from the explosive C4 through prior research and experience (Frank et al. 2022).

For injection of a vapor sample into the GC/MS, direct headspace injection systems are the simplest. With this type of system, a small plug of air containing the analytes of interest, such as a gas-tight syringe, is injected directly onto the head of the GC column. Sensitivity with direct headspace analysis is quite limited because the GC inlet can only accommodate a very small volume, and thus, only a small amount of analyte enters the instrument for detection. As such, this method is generally reserved for high-volatility analytes present at high parts per billion (ppb) levels or above (Wojnowski et al. 2017). For optimal sensitivity, the goal is to instead inject a bolus of analyte at the head of the GC column, normally by employing some type of pre-concentration step. To improve sensitivity, pre-concentration is often used, either by analyte extraction onto a sorptive medium or collection of the whole air sample onto a chilled or adsorbent trap. Sorbent extraction methods, such as solid-phase microextraction (SPME), are most common as they are simple to use, field portable, and require no additional instrumentation or modifications. SPME is a simple, needle-like apparatus, with a polymer-coated fiber that is extended from a protective sheath by depressing a plunger at the top (Fig. 6). The fiber is placed in the headspace, or gaseous space above a sample of interest. Any odorants in the headspace are adsorbed onto the fiber. The fiber can then be transferred into the heated injection port in the GC/MS where the odorants are desorbed and analyzed. Other methods of vapor pre-concentration include thermal desorption tubes where air is pulled using a pump from a sample or from the environment onto a sorbent material packed in a glass tube, or by direct collection of volatiles onto a chilled trap at the head of the GC column. Unlike SPME, both of these methods involve additional instrumentation and modification of the GC inlet.

Because sample introduction of vapors into the GC/MS instrument is not always straightforward, analysis can be time-consuming making near real-time analyzes impossible (Blake et al. 2009; Biasioli et al. 2011; Majchrzak 2018), and quantitation is often tedious. A move toward real-time analysis with high sensitivity, MS-based techniques that do not require prior sample pre-concentration or chromatographic separation may be considered. Direct Analysis in Real Time (DART), Selected-Ion Flow-Tube (SIFT), and Proton Transfer Reaction (PTR)—mass spectrometry, for example, allow for real-time analysis of vapor samples directly injected at or near the ion source. DART-MS is an atmospheric ionization technique that can handle a wide range of analytes. DART-MS experiments are largely carried out in the open laboratory, and though it is most commonly used for analysis of solids, liquids, or compounds that have been deposited or adsorbed to a substrate (Bylinski et al. 2017), analysis of vapor or headspace is possible, though it is less common and generally requires a custom interface to entrain vapors. This is essentially a brief (8–10 s) pre-concentration step, before analyte vapors are delivered to the instrument (Sisco and Robinson 2020). While ambient vapor detection may be possible, it is limited to high-volatility compounds and, because this would involve sampling an open environment, data can be greatly affected by contaminants or humidity in the room (Emmons and Gionfriddo 2021; Simon et al. 2021). Both PTR-MS and SIFT-MS yield accurate trace vapor analysis in real time with the capability for immediate quantitation. They also both have good sensitivities and wide analytical ranges (Sulzer 2014; Smith and Španěl 2011), though these pieces of equipment are extremely expensive, often costing more than $500,000, and are thus hard to come by for this application.

More novel instrumentation that is capable of real-time monitoring of trace vapors has been applied for the measurement of vapors from explosives and other threat materials in support of or working in tandem with canines. Nevertheless, these instruments are still in the development phase or have been recently commercialized and are not widely available, and are very costly prohibiting more widespread use. They are also not currently capable of full characterization of vapor profiles. Karsa Ltd. (http://karsa.fi) deployed a high-sensitivity atmospheric pressure time-of-flight mass spectrometer with a relatively small footprint for the vapor detection of explosives. With this system, the explosive vapor in air can either be collected onto and subsequently thermally desorbed from an adsorbent filter (Simon et al. 2018) or the vapor can analyzed directly in real-time. Another highly sensitive real-time vapor analysis mass spectrometer was developed by Ong et al. (2017) with the expressed goal of supporting canine detection training and deployments. It has the reported capability of detecting a range of explosives in the parts-per-trillion to parts-per-quadrillion range, rivaling the assumed detection limits of a canine. The author endeavors to deploy this instrument in locations where canines are working to verify the presence of explosive odor and detect contamination of training aids (Ong et al. 2017). This instrument is not currently commercially available and has not been widely deployed due to its high cost, requirement for the operator to have a high skill level, and logistical difficulties with field sampling in an open environment. More advancement to either of these mass spectrometers, or an alternative analytical method, has the potential to not compete with, but enhance canine work bringing not just detection but substance identification to threat mitigation.

4 Applications

4.1 Contraband Detection

Contraband detection is likely the most well-known type of detection canine discipline today. The Organization of Scientific Area Committees for Forensic Science (OSAC), Dogs and Sensors Subcommittee defines contraband as "an article or substance that is prohibited by law or regulation withing a specific area" (Academy and of Forensic Science Standards Board, Dogs and Sensors 2017). For detection canine teams, this most commonly refers to drugs or explosives, but may also include anything that may be illegally smuggled across borders or into controlled facilities. The list is long and includes, but is not limited to, firearms, digital storage devices and cellphones, tobacco, currency, agriculture, wildlife, in addition to drugs and explosives.

Canines were not regularly utilized for contraband detection until after World War II, when they were first deployed in Germany to locate smuggled coffee and, with success of that program, they eventually included the detection of illegal drugs. It was not for almost two more decades that the use of canines for contraband detection, namely drugs, became more widely utilized; and it was not until 1970 that the U.S. Customs established a drug detection canine training program (Drug detection dogs 1976). Today, the U.S. Customs and Border Protection, for example, deploys 1500 canine teams in disciplines including concealed human and narcotic detection, search and rescue, tracking/trailing, currency and firearms detection, and human remains detection (Customs and Protection 2022).

Of these disciplines, explosives detection has been the most widely represented in the peer-reviewed research literature. In 2018, a literature survey of canine detection topics turned out 102 unique citations on canine explosives detection, compared to 31 citations for narcotics detection, and literature covering canine detection of other types of contraband is even more limited (Peranich and DeGreeff 2018). While the general canine detection literature covers a wide range of topics and fields relating to odor, olfaction and the olfactory system/receptors, breeding, fitness, cognition, and learning; research explicitly related to contraband detection applications tends to be focused on a subset of these topics. These include detection proficiency, training aids—development and effectiveness, and characterization of the odor itself. More specifically, at the time of this publication, the contraband detection research community has a great interest in the topics of characterization and/or development of non-detonable explosives training aids (particularly for homemade explosives), generalization between like materials to improve detection proficiency in the field, as well as dog/puppy selection and breeding and training paradigms for the specific task of explosives detection. These targeted research initiatives are rooted in the growing trends in the manufacture of explosive devices, emergence of novel drugs, and increases in mass shootings and domestic terrorism events that call for enhanced tools in the detection of evolving threats.

4.1.1 Homemade Explosives—Training Aids and Generalization Across Detection Targets

A 2018 report by the Naval Research Laboratory identified the need to shift focus from the detection of traditional, military, and commercial explosives to homemade explosives (HMEs), and with that necessitates shifting canine training materials from solely the once more common explosives, such as TNT, C-4, and gun powders, to new aids representing the major classes of HMEs, namely fuel-oxidizer mixtures and peroxide-based explosives (Simon et al. 2018). Unlike the explosive materials to which the majority of explosives detection canines were previously trained, HMEs are not fabricated or mixed to set standards and thus potentially pose a greater safety risk. This is especially true with the peroxide explosives, triacetone triperoxide (TATP) and hexamethylene triperoxide diamine (HMTD), which are classified as primary explosives, meaning they are highly sensitive and can readily detonate with applied heat, shock, or friction. As such, training on bulk material is often impractical leading to a push for the development of non-detonable training aids as well as detection generalization across the many variations of each of these explosives.

A growing body of research is invested in the development of non-detonable HME training aids. The most common mechanism for fabricating a training aid that is both non-detonable, and thus safe to handle, and omits an odor realistic of the true material is through either an odor soak or the application of trace residue on a solid substrate. TATP, having a high vapor pressure for an explosive, is readily amenable to odor soaks, where the odor of the contraband is adsorbed onto a secondary material which will then off-gas the odorants related to the target (Simon

2020). Jeunieau et al. (2022) and Moore et al. (2011) utilized filter paper and cotton gauze or polydimethylsiloxane (PDMS, a rigid gel-like polymer), respectively, to adsorb and subsequently release vaporous TATP (Moore et al. 2011; Jeunieau et al. 2022). Others mix solid explosive with a substrate to dilute and render the explosive insensitive, or phlegmatize the explosive material, while allowing the release of the odorants of the solid explosive (Vu 2014; Wilhelm et al. 2022).

HMTD, unlike TATP, has a very low vapor pressure and, as such, molecular HMTD is not likely to be found as a vapor under normal environmental conditions (Oxley et al. 2009). However, also unlike TATP, HMTD degrades into highly volatile (and highly odorous) compounds that can be readily detected in the vapor phase. Inconveniently, the presence and quantity of these compounds change with time, storage condition, and method of synthesis (DeGreeff et al. 2017), making training difficult and the creation of training aids problematic. For example, if one looks to develop a non-detonable mimic of HMTD, should the manufacturer mimic laboratory or clandestine-made HMTD? Furthermore, should it be the odor profile from fresh material or aged (Simon and DeGreeff 2019)?

Like TATP, HMTD aids may be made from odor soaks, mixing into a substrate, or, alternatively, by imparted chemicals that represent the HMTD odorants onto a substrate in such a way that it imitates the odor of the actual material without the actual explosive material being present. Because of the complexity and instability of the HMTD odor profile, existing HMTD training aids have proven to be quite dissimilar from one another. Both Simon and DeGreeff (2019) and Buckley (2022) have shown that the odor profiles of HMTD from different manufacturers are distinct both immediately upon opening and even more so over time and regular usage (Simon and DeGreeff 2019; Combat Capabilities Development Chemical Biological Center Report 2022); however, extensive canine testing is required to determine the effect of this changing odor and which training aid variety best depicts the odor.

There are no widely adopted standards to which commercial training aid manufacturers must adhere (Combat Capabilities Development Chemical Biological Center Report 2022); however, for any novel training aids, it is necessary to validate the effectiveness of the aids by both chemical analysis and canine testing. Chemical analysis, as discussed in the previous section, is used to characterize the odor profile from the training material in comparison to true material and will be discussed in more detail below. Chemical analysis can confirm the presence of the odorants of interest and measure the quantity of confounding odorants that may exist from the manufacturing process or the substrate. Analysis of TATP training aids is relatively simple. The main odorant of TATP is TATP itself; as such, the analysis would simply need to confirm that this odorant is present at a high abundance and that any confounding odorants are present at significantly lower abundances. Several studies have assessed the odor profiles of peroxide explosive training aids (Simon et al. 2021; Simon and DeGreeff 2019; Combat Capabilities Development Chemical Biological Center Report 2022). Buckley (2022) assessed all current commercially available TATP and HMTD training aids. The study showed the presence of TATP vapor from all but one (out of seven) commercial training aid tested, and most had minimal interference from extraneous odorants. HMTD, however, was more complex with each training aid yielding a differing quantity and collection of odorants (Buckley 2022).

To further validate the utility of the training aids, it is imperative that they are tested with canines, with third-party testing being optimal. Data for canine testing relating to homemade explosive training aids available in the public domain is minimal. This is primarily due to the high cost of testing, as well as the availability of canines for testing, the level of labor necessary for training, and availability of facilities for both housing and testing, or the studies may not be publicly available due to security concerns. While it is useful to validate materials with canines previously trained for explosives detection, it is far more useful to carry out testing with "green" dogs, or dogs that have not previously been trained in the detection of explosives, though may have other operational detection experience. The report by Buckley (2022) is a good example of such a study. Here, a cross-over study was conducted with a control group of five canines imprinted and trained with true TATP explosives and an experimental group of another five canines trained with a commercial TATP training aid. Both groups were then tested on their ability to correctly choose the TATP training aid or explosive from other odors in a carousel. Results showed that canines trained solely on the commercial aid did spontaneously respond to real TATP, though at a lower rate than those in the control group, although recognition of the explosive was gained through repetition (Combat Capabilities Development Chemical Biological Center Report 2022). As an increasing number of commercial training aids are developed representing difficult-to-obtain contraband, such as homemade explosives or other emerging threats, it is imperative that research continue to validate these aids prior to or when they are made commercially available, both by characterization of the odor profiles as well as their effectiveness as a training tool measured by canine detection testing.

Due to the nature of the materials in HMEs being homemade or clandestine, there are commonly variations in the odor profiles of the materials making proficient operational detection of all variations challenging (DeGreeff et al. 2017b; DeGreeff and Peranich 2021). The first study examining the canine ability to generalize from the odor of a single explosive used for training to other variations of the same target material was reported by Cerna et al. (2011). In this study, explosives detection canines ($n = 5$) were trained solely with flake TNT supplied by a single manufacturer and were then tested in their ability to detect this flake TNT as well as a novel German TNT sample, and discriminate these from a number of distractor odors. All canines responded correctly to the flaked TNT, but none identified the novel TNT. This study indicated that canines trained solely to one variation of a given target, may not readily generalize to other variations of that target. More recently both DeGreeff and Peranich (2021) and Lazarowski et al. (2015) studied generalization across varying sources of ammonium nitrate ($n = 17$ and $n = 15$, respectively). Though the studies had differing experimental designs and testing compounds, both concluded that while there was some spontaneous generalization to variants of ammonium nitrate to which the canines were not originally trained, this generalization was weak and detection rates were lower than that of their trained target. Furthermore, both studies noted that canines showed individual differences in the particular variants to which they were more likely to detect, with some canines being more likely to generalize as a whole than others (DeGreeff and Peranich 2021; Lazarowski et al. 2015). On the other hand, Lazarowski et al. (2021b) studied generalization across variations of TATP formulations and found that the canines (n = 11) did readily generalize from a laboratory-prepared formulation to other clandestine-prepared formulations that were less pure (Lazarowski et al. 2021b). The discrepancy between the studies is quite possibly due to the dissimilarities between TATP and ammonium nitrate odor. TATP has a much higher abundance of odor with a limited number of background contaminants compared to ammonium nitrate with very low vapor availability and a number of compounds imparted by the manufacturing process (Frank et al. 2022), the higher odor abundance and more simplistic odor signature potentially making it easier for the canines to detect their trained odorant in the clandestine formulations of TATP than ammonium nitrate.

Aviles-Rosa et al. (2022) conducted research using canines (n = 5) trained to detect the odor of a double-base smokeless powder from the outlet of an olfactometer. Similar to the other studies described above, detection of a different, single-base smokeless powder was statistically lower than their trained odor of double-base smokeless powder (Aviles-Rosa et al. 2022). This result is supported by previous work showing distinct differences in the vapor profiles of different brands of smokeless powder (Harper et al. 2005b). Aviles-Rosa also examined generalization from the odor of pure smokeless powder to the odor imparted to various substrate materials. They found that canines were more likely to correctly respond to smokeless powder odor imparted to more adsorbent materials, such as fabrics and polymer tubes (Getxent tube 2022), than metallic substrates including a gun and bullets (Aviles-Rosa et al. 2022).

4.1.2 Emerging Threats in Contraband Detection

Other emerging threats, not relating to explosives, should also be considered. Examples include synthetic drugs, such as fentanyl and its analogs, 3-D printed guns, agriculture (food and safety/tax evasion), wildlife product (e.g., ivory, bushmeat, etc.) or chemical/biological weapons. Canines are a valuable tool in the detection of emerging threats due to their ability to be rapidly trained to new odors. Indeed, a canine previously trained in odor detection can be taught a new odor in less than a day. In comparison, instrumental field detectors often require, at minimum, arduous changes in software or hardware when new threats arise (Furton and Winialski 2022). Research on canine detection of emerging threats, outside of the realm of explosive detection, is often limited due to lack of funding and insufficient access to materials for testing.

To aid in the detection of new materials, researchers and canine trainers must first select appropriate training aids. This may be complicated by access to pure materials or the variety of the target materials necessary to encourage generalization across all versions of the target which may be encountered in live theater. As an example, Singletary et al. (2022) explored a novel method of training canines to detect emerging biological threats. Handling and storage of these materials present great health hazards and security concerns, limiting access to materials to be used for canine training. As such, the study proposed the use of a non-hazardous polymer-based training aid, which "captures" the odor from the target virus and then releases the same odor slowly over time for canine training. Following training, using this novel training aid, canines ($n = 6$) were able to detect and discriminate a culture of the threat material from other relevant distractor odors with a sensitivity of 97.22%. The canines were also tested on their ability to generalize from the training odor (virus culture) to the threat odor in other biological sample types, including blood, urine, fecal matter, nasal secretions, and saliva. Detection ranged from approximately 65–91%, indicating the novel polymer-based training aid was successful in imparting the threat odor for training but further research to improve general proficiency should be carried out (Singletary et al. 2022).

Prior to large-scale training of canine teams to detect a new threat material, it is valuable to have an understanding of the vapor profile of the threat, how the profile changes with factors such as synthetic method or manufacturing process, storage conditions, and degradation over time. The recent growth of the illegal synthetic drug market is a good example. Novel synthetic drugs are often designed to have similar effects to controlled substances, but are often not controlled due to the lag time between appearance on the black market and DEA scheduling. In an attempt to circumvent drug scheduling laws, novel analogs constantly appear on the illegal market often with only small modifications to their structure compared to the controlled substances. These drugs also account for a large number of drug overdoses. For instance, fentanyl, a synthetic opioid legally used as an analgesic, has surpassed heroin as the deadliest drug in the United States, with a potency 50–100 times greater than morphine. Fentalogs (fentanyl analogs) may be even more deadly, with carfentanil having an estimated potency one hundred times that of fentanyl (U.S. Drug Enforcement Agency 2017). Similarly, synthetic cathinones (bath salts) have increased in popularity though risks of intoxication and overdose are high. Training for the detection of solely fentanyl, or a single synthetic cathinone, by a canine (or instrumental detector) could mean many other dangerous analogs to which the canines were not trained are missed. Headspace analyzes of both confiscated fentanyl and fentalogs (Vaughan et al. 2021a) and illegal synthetic cathinones (Francis et al. 2019) indicated that, while the full odorant profiles differ, there are volatile compounds common to each group of drugs. The commonality implies that there is potential to train canines to not only detect a single analog, but also related analogs as well. In fact, a study where certified narcotic detection canines were trained to locate a single synthetic cathinone was able to generalize to another cathinone to which they were not previously trained, indeed implying that there were commonalities between the vapor signatures (Francis et al. 2019). To date, canines' tendency to generalize from fentanyl to other related analogs has not been studied; however, headspace analysis has shown that, while there are significant variations in the vapor profiles of various confiscated fentanyl-containing exhibits, there is at least one volatile compound in common to which canines could potentially be trained (Vaughan et al. 2021a).

4.1.3 Odor Characterization

As described, odor characterization in support of canine detection has been carried out on a wide variety of canine detection targets, from explosives and human remains to more novel applications such as bedbugs and COVID-19 to name a few. The aims of these studies vary and include measuring odorant availability, establishing feasibility of detection, elucidating the target odorants for mimic training aids, characterizating similarities and differences between variations in the target material, or testing for the presence of contamination (Lazarowski et al. 2020).

To the authors' best knowledge, the first research to recognize that working detection canines use volatile compounds, odorants, associated with the target of interest for detection, but not necessarily the parent molecule of that target was reported by Hallowell et al. (1994). The aim of the research was to assess an olfactometer vapor generator used to deliver explosive vapor to canines for detection threshold testing. The main explosive component of C-4, a plasticized explosive, is RDX. During the assessment, it was noted that no trace of RDX explosive could be detected instrumentally from the outlet of the olfactometer, though the canines, previously trained with actual C-4, were able to successfully detect an odor from the olfactometer as C-4. Vapor analysis found significant amounts of contamination from cyclohexanone, a solvent associated with C-4, in the system. Researchers drew the conclusion that the dogs may actually use this cyclohexanone and other volatile compounds present in the headspace of C-4 in the detection of C-4 instead of the explosive, RDX, itself (Hallowell et al. 1994).

Further work by researchers at Auburn University combining headspace analysis and canine testing odor profiles from C-4, as well as smokeless powder, dynamite, and TNT, confirmed that detection dogs learn a trained target based on its dominant volatile compounds and not necessarily the parent molecule of the explosive of interest. In this series of studies, the dominant volatile compounds from the explosives, commonly solvents, compounds from plasticizers, or other highly volatile compounds identified by headspace analysis were presented to the canines. The findings showed that, while the specific compound of interest to an individual dog may differ, only a few dominant compounds elicited positive responses from the dogs, contradicting the previously held belief that dogs respond to the energetic compound in the explosive just because that is of interest to the human, and confirming Hallowell's hypothesis. The authors further explain the importance of understanding odor signatures in selecting training aids, understanding false responses, and measuring contamination (Johnston 1999).

Since these early studies, headspace analysis and odor characterization in support of canine detection has been carried out for a number of additional explosive analytes. A full summary is beyond the scope of this chapter, but can be found in Frank et al. (2022). Initially, military and commercial high explosives (C4, TNT, dynamite, detonation cord, etc.), and low explosive powders (smokeless and black powders) were characterized. Today, research in this area is more focused on the characterization of homemade explosives, as mentioned above. TATP and HMTD, are homemade explosives synthesized from peroxide and acetone or hexamine with an acid catalyst. Though similar in the synthesis methods, their headspaces are quite distinct. Molecular TATP itself has a relatively high vapor pressure making it readily detectable at room temperature, while HMTD has a very low vapor pressure making it an unlikely olfactory detection target (Ewing et al. 2013), though HMTD does decompose under ambient conditions to yield a number of highly volatile and odorous compounds that are likely detection targets (DeGreeff et al. 2017b). Which of these compounds are the active odorants have yet to be determined.

The oxidizers in fuel-oxidizer mixtures are most often inorganic salts. Salts do not inherently have a vapor pressure; thus, they generally cannot be found in the vapor phase except at exceptionally high temperature. Odor from these oxidizer salts is derived either from its degradation or the presence of contaminants. Ammonium nitrate, the most common oxidizer salt, naturally degrades in ambient humidity forming ammonia and nitric acid vapors. The nitric acid is sequestered by ambient water leaving only the ammonia and any contaminant odors available for detection. As such, the amount of ammonia present is highly dependent on humidity with higher humidity yielding more ammonia (Steinkamp et al. 2016). Temperature and purity of the ammonium nitrate also changes the quantity of ammonia, which is thought to have implications on canine detection threshold for ammonium nitrate (DeGreeff and Peranich 2021). Potassium chlorate, on the other hand, does not readily degrade, and has very little associated odor. For some time, the canine detection target for potassium chlorate was unknown. Although, Crespo-Cajigas et al. (2019) detected low levels of chlorine in the headspace which is, currently, the only hypothesized active odorant (Crespo Cajigas et al. 2019).

At the same time the odor profiles of explosives were first being explored, that of narcotics, namely, cocaine, were also considered. Another early instance of odor characterization to understand canine detection was in 1997. Researchers used headspace analysis to determine the dominant odorants from cocaine, illicit and pharmaceutical grade. A number of volatile compounds were detected in the headspace of these samples, though methyl benzoate was found to be the most common across all varieties tested, and indeed, in a series of tests with trained drug detection dogs ($n = 15$) confirmed that methyl benzoate was the active odorant for cocaine detection (Furton et al. 1997, 2002).

The active odorants from drugs including heroin, methamphetamine, and MDMA (Ecstasy) have also been established through a combination of instrumental analysis and canine testing. These compounds include acetic and acetylsalicylic acids for heroin (Huertas-Rivera 2016), benzaldehyde for methamphetamine (Vu 2001), and piperonal for MDMA (Macias and Furton 2011). More recently, the headspace of fentanyl, a highly potent and dangerous synthetic opioid, has been characterized (Vaughan et al. 2021a; b). A number of volatile compounds were identified in the headspace of fentanyl, but in comparison between many confiscated samples, one compound, N-phenylpropanamide, was identified in the majority (Vaughan et al. 2021a). It has yet to be established if this compound will elicit a response from trained canines.

Odor profiles from natural products tend to be more complicated making interpretation of active odorants more complex. Several studies have been carried out to elucidate the odor profile of marijuana, but due to the complex profile, no active odorants have been identified. In her graduate dissertation research, Huertas-Rivera compared the headspace of ten samples of confiscated marijuana and identified six compounds that were common to all samples, and of those, caryophyllene and limonene contributed the largest abundance to the odor profile, followed by b-myrcene and a-caryophyllene. Eight trained drug detection canines were tested on three occasions on single marijuana odorants including limonene and caryophyllene. The alert rates for limonene and caryophyllene alone were 57.1 and 28.6%, respectively, compared to 100% for the positive control (real marijuana), 16.7% to linalool, and 42.9% to a-pinene. 1:1 ratios of combinations of these compounds were then presented to the dogs, though the highest alert rate was still only 43.7% (Huertas-Rivera 2016). This study exemplifies the difficulty in creating synthetic ("pseudo") training aids for targets with such complex odor profiles. Indeed, in another study the commercially available Sigma Pseudo Marijuana scent was not readily detected by any of the seven certified drug detection canines tested (Ewing et al. 2013); as such, pseudo-training aids that have not been validated using the scientific method should be regarded with caution.

The legalization of marijuana in much of the United States has made the need for pseudoscent training aids less necessary; however, canines at military bases and in parts of the country where it is illegal, as well as at the border and ports of entry, are still trained to detect marijuana. The introduction of the 2018 U.S. Farm Bill further complicated the landscape by creating a distinction between legal hemp, having 0.3% or less THC (tetrahyrdrocannabinol—the main psychoactive compound on marijuana), and [federally] illegal marijuana, having greater than 0.3% THC. This bill implied that canine handlers needed to train their canines to differentiate between the two, though they are visually and olfactorily similar. In his dissertation research, Frank compared the headspaces of eight hemp samples to five marijuana samples. While the majority of the dominant headspace compounds were in common, some small differences between each group were identified, indicating the potential for canines to also be able to distinguish between hemp and marijuana. Further experiments with detection dogs supported this hypothesis. Canines ($n = 24$) originally trained to detect marijuana often did generalize to hemp, with alert rates ranging from 25 to 75% for each of the eight hemp varietals. Handlers then trained the canines specifically to not alert to hemp by training them to discriminate between marijuana and hemp, after which the canines ($n = 11$) gave only three alerts to the hemp material out of a possible 132 total chances, a 2% alert rate with no deterioration of their marijuana detection proficiency (Frank).

While there are many other applications of canines in the detection of other contraband not relating to drugs or explosives, studies characterizing the odor profiles of such materials are minimal. For example, canines have been used to locate concealed firearms for some time; however, very few studies have examined the vapor signatures of firearms, and science has currently not yet formed a consensus on what odorants elicit recognition. In recent work, Ong et al. focused on the odor of double-base smokeless powder (the propellant used in firearms), unburned, and as gunshot residue. The work looked at the main smokeless powder-specific volatile compounds identified in earlier studies (Harper et al. 2005b; Joshi et al. 2009), including nitroglycerin, diphenylamine, dibutyl phthalate, and dinitrotoluene (DNT). Using the National Center for Forensic Science smokeless powders database, nitroglycerin and diphenylamine are the most common volatile compounds of smokeless powder with diphenylamine in all entries and nitroglycerin in all but two of the 899 entries. Dinitrotoluene occurs in the remaining two entries. The researchers suggest that training on these three compounds in specific ratios will enhance generalization across the broad class of smokeless powders (Ong et al. 2022). Nettles et al., however, collected headspace samples directly from firearms ($n = 15$) and ammunition magazines ($n = 30$). The major smokeless powder compounds were not detected from these samples, indicating the canines may use alternative compounds to locate firearms as opposed to smokeless powder used in explosives. Instead, the five compounds that were the most commonly identified across the samples were nonanal, decanal, octanal, tetradecane, and tridecane (Nettles et al. 2022). More research should be done to determine the active odorants for canine detection.

Another more recent use of canine contraband detection is for the detection of digital evidence to include items such as thumb drives, hard drives, SIM cards, and cellular phones. Digital evidence detection canines are used to detect smuggled electronic devices in prisons or other high-security facilities, or by law enforcement for evidence recovery. Identification of the odor profile associated with different devices can assist in the choice of training materials that will maximize recovery of the target by encouraging generalization, and minimize false alerts to other electronics not of interest to the canine handler. The goal of the first study that examined the odor signature of digital evidence items, specifically SIM and SD cards and USB drives, was simply to assess the feasibility of canines to discriminate those devices from other electronic devices, including batteries, circuit boards, cables, and whole cell phones. Several compounds were identified that were unique to these targeted digital evidence items and not detected in other electronics (DeGreeff et al. 2017a). While these data indeed support the canine capability to discriminate digital evidence from other electronics, more data, including canine testing, need be collected to further understand and optimize digital evidence detection, especially given the rapid and dynamic growth of digital media today.

This need is applicable to all analytical work in support of canine contraband detection. Laboratory testing alone can only be used to identify the compounds in the headspace of a sample, however, canine behavioral assessments are imperative in determining which of the identified compound(s) is the "active odorant" or the odorant (odorants). This issue is related to the compound's odor activity value (OAV). To briefly depart from the discussion of odor characterization, the OAV is a value determined by the ratio of the odorant concentration, calculated from the compound abundance in the headspace, to the natural olfactory detection threshold (ODT) of that odorant. Meaning, a compound will have a higher OAV and will be more likely to be perceived as the active odorant *if* that compound is found in high abundance in the headspace of that target *and* the animal has a natural ability to detect that compound at a very low level. Research has shown that the highest *abundance* compounds may not actually be the active odorant because the animal does not have a natural affinity to that compound in terms of detection (Rice and Koziel 2015; Grosch 1993). This has particular importance in the development of pseudo-scent formulation for use as surrogate training aids. Creating a training aid based solely on the most dominant compound(s) might mean the pseudo formulation excludes the compounds with the highest odor impact (Rice and Koziel 2015). The ODT for canines has been measured for very few compounds. Using ODT for humans in the calculation of OAV as an alternative when canine data is not available is possible, although the ODT for compounds that are operationally relevant to canine detectors.

4.1.4 Performance and Proficiency

The quality of the training aid and the odor is irrelevant if the dog is not properly trained or not capable of detection. Performance and proficiency testing of operational canines is not frequently published. This is not necessarily because the testing is not done, but because the studies are not shared, either because it was an informal assessment meant only for internal use or because of security guidelines that prevent the organization from publishing any data that could potentially be viewed as a capability gap. Jezierski et al. (2014) is one of few peer-reviewed publications reporting on the efficacy of drug detection canines. Researchers assessed 96 fully-trained, but not yet certified, Polish Police drug detection canines on their ability to detect street drugs including hashish, marijuana, amphetamine, cocaine, and heroin in a variety of double-blind search scenarios. They found that, on average, the canines made correct indications 87.7% of the time with a false alert rate of 5.3%. They further compared the difficulty level of the detection of different drugs, measured both by the time required to make a final indication and by a percentage of correct indications. They concluded that marijuana was the easiest to detect, followed by hashish, amphetamine, cocaine, and heroin (Jezierski et al. 2014). While this is useful data, the results may be oversimplified in that many factors affect the perceived "difficulty level" of detection including the availability of odor (greater odor availability presumably yielding greater ease in detection), which is related to the available surface area of the target, as well as drug purity, drug age, and wrapping or containment, among other factors (DeGreeff and Maughan 2022a).

Jezierski et al. also compared the performance of different dog breeds, again, based on mean detection time and percent of correct indications, misses, and false alerts. The terrier group ($n = 25$) had a significantly longer mean detection time ($P < 0.05$) and percent false alerts ($P < 0.01$), and lower correct indications than the other groups including German shepherds ($n = 61$), Englisher cocker spaniels (n = 10), and Labrador retrievers (n = 68). All other groups performed similarly (Jezierski et al. 2014).

Though it is of great interest to the community, few other studies have compared the performance of dog breeds, likely because obtaining enough subjects per breed group is challenging. A frequently referenced paper by Hall et al. (2015) compared the performance of pet dogs, pugs (n = 10), German shepherds (n = 10), and greyhounds (n = 10), not previously trained in odor detection on an odor-discrimination task of increasingly dilute odor samples. All but one of the greyhounds failed to meet the required motivation criteria prior to the start of testing, demonstrating that olfactory capability is not the only factor to consider when choosing a detection dog. Comparing the German shepherds and the pugs, the pugs, surprisingly, outperformed the shepherds at both the learning the discrimination task and their ability to perform the discrimination at increasingly low odor concentrations (Hall et al. 2015). Furthermore, a study investigating canine tendency to generalize from a trained odor to other similar odors compared groups of canines either classified as "working dog breeds," such as Labradors, German shepherds, spaniels (English springers and field spaniels), Belgian malinois, beagles, and "non-working dog breeds," which including breeds such as golden retriever, poodle, husky, pit bull, corgi, collie, and mixed breeds. That data showed that those in the working dog breed group were no more likely to generalize to similar odors than the non-working dog breed group (DeGreeff et al. 2018).

Contraband detection canines are often required to learn many target odors over the course of their careers. Several groups have tried to determine just how many unique odors can be learned and retained by canines. An early study sequentially trained canines to detect up to ten single-component odors, and found no decrement in performance as more odorants were learned. Canines were also shown to be able to retain this proficiency on all ten odors for up to four months (longest interval tested) with no additional training or other experience with the odors (Johnston 1999). In more recent, subsequent studies, canines (n = 9) were trained to up to 40 odors in sets of 10 over 16 months, and then asked to recall the subsets of trained odors. Duration from last exposure was as long as 12 months prior. The trained odors in this study were not simple, single-component odors, but items with complex odor profiles, such as explosives, food, and other household products. The data indicated that canines were capable of learning the 40 odors with no decline in proficiency as more odors were added, and the dogs were able to recall odors to which they were not exposed up to 12 months (Waggoner et al. 2022). Research studies have yet to challenge substance detection canines in their ability to learn and retain a large number of targets over an extended period of time. This ability? has important implications in the selection of training aids and training regimens. Due to safety and security concerns, some types of contraband that are important in canine detection have additional restrictions for use and require special precautions, personnel, etc., during training. These constraints can limit both the frequency and location of training on that target, making this research particularly valuable.

While researchers have shown that the impressive canine olfactory memory allows for long gaps in training with minimal decline in detection proficiency, this rule is not absolute. Research has also shown that some additional olfactory training is required to maintain a high level of proficiency. Two groups of canines ($n = 9$ per group) were trained on a set of ten odors, including mostly explosive-related odors. Over the following 12-month period, the first group of canines continued some odor detection training of only approximately 30 min a month on a single odor of the ten, while the other group continued only non-odor detection training. Following this period of time, all canines were reassessed on their ability to recall the ten odors. The first group increased their performance from 94% accuracy to 99%, while the second group decreased from 99% accuracy to 72%, indicating that canine olfactory memory can decline if not exercised regularly (Lazarowski et al. 2021a).

This type of performance decline has been shown to also occur due to repetitive, unproductive searches. With some disciplines of contraband detection, canines typically encounter low rates of operational finds, such as with explosives detection or more novel disciplines like biowarfare agents. This is not only due to the low prevalence of the threats on a daily basis, but also due to the inability of a trainer to place target items in an operational environment for safety and security reasons. As a result, it has been shown that canines will be more highly motivated to search in training scenarios where they have learned they are likely to find a target and be rewarded, than in their work environment (Porritt et al. 2015). However, inclusion of a non-hazardous target into the canine's olfactory detection repertoire and its use during daily operations in the work environment can mitigate this decline (Porritt et al. 2015; Thrailkill et al. 2018).

4.2 Human Scent

Human scent is a form of trace odor evidence that cannot be readily seen but is available on any surface and/or location where an individual has been in contact. The purpose of using canine detection teams to locate human scent is to establish an association between the human scent traces left at a given location or scented article to the human scent of the subject in question (whether a perpetrator/fugitive of criminal activity or a missing person). The ability of the canine to establish such an association relies on the theory that every individual has a characteristic odor signature that can be used as a physical trait for discrimination and identification purposes. Canines trained in this discipline help the forensic investigative team in several ways. Human scent canines can follow a subject's scent trail/track directly from the crime scene, assist in determining the direction of travel of the subject, identify a subject in a scent line-up procedure, or perform a location check by scent or aid in missing person recovery efforts (Hunt 1999). Within search and rescue operations, canines are trained to locate individuals trapped within debris following major, and sometimes dangerous, structural collapses. Forensic and law enforcement personnel gather human scent evidence by collecting odor traces from objects that the target subject may have handled during the commission of a crime or in the case of a missing victim, personal items that carry the donor's scent. This evidentiary material is then presented to the human scent-trained canine for pre-scenting procedures, comparing the odor traces from the presented item (or location) to that of the subject in question (Prada et al. 2015). Regardless of the application, a central gap in knowledge with respect to human scent detection is the chemical nature and identification of the human volatile markers perceived by dogs to make the identification. Thus, a necessary complementary tool to fully exploit canine scent detection is the need to fully understand human odor. Human scent detection research has encompassed several topics to include the definition and characterization of human odor origins, optimized odor collection techniques, persistence/survivability of human odor traces and canine–human scent matching and discrimination capabilities.

4.2.1 Human Scent Definition

The human "odor signature" is one of the most complex odorant mixtures as it relates to both its instrumental definition and related canine detection applications. The human odor print is comprised of many variables that include biological processes, genetic makeup, and microbial interactions (Prada-Tiedemann 2022). Characteristics of glandular activity and skin microbiota have been the subject of numerous studies targeting an understanding of the interplay between biochemical pathways and odor-releasing enzymes present in microorganisms on the human skin surface that yield a particular odor type (Baker and Wolfe 2020; Wilke et al. 2007; Natsch and Emter 2020; Smallegange et al. 2011). The dynamic chemical "topography" on the skin surface is crucial for microbial interactions. Human odor complexity is further heightened by the varying composition of the skin microbiota which is dependent on factors such as body location, skin pH, and moisture level, to name a few. Thus, occluded regions with limited air exposure such as the axillae and groin tend to favor microbial activity while regions of low water content (forearms) have reduced microbial density. Furthermore, lifestyle factors such as beauty and hygiene products can further affect microbial composition (Grice and Segre 2011; Turnbaugh et al. 2007; Costello et al. 2009; Holland and Bojar 2002). Genetic makeup has also been a widely studied area with respect to the definition of human scent. Foundational work in this area has targeted the Major Histocompatibility Complex (MHC) as a source of the individual odor signature (Kwak et al. 2010). The MHC is the most diverse part of the genome; in humans, this genetic coding region is referred to as the human leukocyte antigen (HLA) and represents a large locus containing closely linked polymorphic genes (Havlíček et al. 2020). Studies in this area are based on the premise that there is a specific odor compound along with a corresponding combination of MHC genes that can potentially function with respect to kin selection or recognition. Generally, it can be said that all studies propose mechanisms that result in a degradation product from the MHC molecules that interact with either peptides, specimens, volatiles, cells, or even microbes that generate a particular odor print (Prada-Tiedemann 2022). Recent studies, however, have made use of analytical advances to understand the human odor "volatilome." The human body is a constant source of volatile organic compounds (VOCs), which generate a complex and dynamic mixture of compounds that are often referred to as human odor. VOCs emitted by the human body are a useful tool in not only forensic science applications (Prada and Furton 2012, 2008) but also for disease and medical diagnostics (Shirasu and Touhara 2011; Broza et al. 2015; Buljubasic and Buchbauer 2015; Prugnolle et al. 2009; Olsson et al. 2014). A review article in 2014 collected a detailed review of VOCs from the human body of healthy individuals, reporting a total of 1,846 compounds from breath, saliva, blood, milk, skin secretions, urine, and feces (Lacy Costello et al. 2014). In 2021, an updated version of this compendium was published with the addition of over 900 new compounds that have been reported since 2014 (Drabińska et al. 2021). Recovered compounds depict a variety of chemical classes including aldehydes, short and long-chain hydrocarbons, carboxylic acids, alcohols, esters, ketones, and amines. It is important to highlight that while analytical methodology has advanced the understanding of the human volatilome, comparison of all this foundational work needs to be taken with caution as differences in headspace sampling approaches, participant characteristics (age, gender, diet, disease, ethnicity) and body regions produce distinctive odor signatures that can yield high variation and divergences in the results obtained. As can be seen, the definition of human scent is not a light one, as the variables that affect it range from the physiological to the environment in which the subject is in, hence, making this target odor source one of the most complex odor signatures for scent detection purposes via instrumental or canine means.

4.2.2 Odor Collection Techniques

A critical aspect of concern for valid protocols and performance is proper scent collection techniques. Currently, there is no uniform human scent evidence collection method established. However, regardless of the specific procedure employed, all variations make use of an absorbent substrate to gather human scent, yet the materials and detailed procedures (time, manner, personal hygiene for collection) have not been previously optimized or standardized among the canine scent detection community. Odor collection methods can be divided between direct and non-direct. The conventional manner of collecting human scent for canine pre-scenting is to allow the dog to smell the scent article/source directly. This procedure, however, introduces contamination risks and/or destruction of other forms of trace evidence such as fingerprints or DNA for further forensic analysis. Other direct methods of human scent sample collection include swiping the scented article with collection mediums. Generally, sterile gauze pads are used which are placed in direct contact with the object or person of interest. Given there is no standardization as to the type of collection medium (gauze may be cotton or cotton mixes), several studies have been performed to understand how textile chemistry affects the generated odor profile, which may indirectly affect canine performance (Prada et al. 2011, 2014). Non-direct methods include passive absorption (collection substrate in close contact but not touching) with object/person or the use of airflow dynamic devices during collection processes. The most used device specifically for human scent collection is the Scent Transfer Unit (STU-100). This device is a portable vacuum designed to draw air through 5 × 9-inch sterile gauze pads. The STU-100 allows for the ability to perform non-contact sampling and collection of human scent from objects without contaminating or altering the object of interest, as well as allows for large sample collection numbers on a vast array of target objects (steering wheels, doorknobs, etc.). Several studies have evaluated the STU-100 as a dynamic collection medium, either utilizing standard chemical mixtures and/or subjects (Eckenrode et al. 2006; DeGreeff et al. 2011) or from the perspective of optimal collection mediums in terms of generated chemical odor signature (Prada et al. 2011).

4.2.3 Aging and Persistence of Human Scent

An important concept in trace evidence analysis is that of persistence—how long does the evidence last on a particular location or object? In the case of canine human scent detection—how long does odor persist in a given trail or location? The answer is confounded by numerous variables that include environmental conditions (wind, terrain, humidity, temperature), scent deposition substrate (i.e., concrete vs. grass) as well as the innate biological definition of odor which is subject to bacterial degradation like any other biological evidence type. The passage of time can affect its availability as well as its integrity. While canines can match odor in human scent identification line-ups flawlessly with same-day collected samples, the performance drops over time periods from two weeks to six months. Possible causes include the volatile nature of scent and the distinctive vapor pressures of chemical compounds. Over time, scent samples sitting in glass jars may lose vapor concentration where no more volatile emission is available within the given containment volume, thus showing how storage protocols must be taken into consideration when handling scent samples (Schoon 2005). In another study, well-trained trailing canines were used to follow human trails up to 48 h old. This study was made with cross-trail contamination and varying environmental conditions (Harvey and Harvey 2003). To further understand the age of human scent, researchers have attempted to evaluate human scent odor signatures across a longitudinal study over a seven-week period at various environmental conditions to include room temperature, −80 °C conditions, dark storage, and UVA/UVB light exposure. Results suggested that human hand odor scent samples stored in glass containers and subjected to minimal UVA/UVB light exposure provided the most stable environment for containment purposes of human scent (Hudson et al. 2009). Persistence studies have also been evaluated in terms of the survivability capability of human scent. More specifically, as it relates to the ability of human scent to survive extreme thermal and mechanical effects such as those encountered in explosion and arson settings. In one study, 12 test subjects were selected and allowed to handle materials used for the construction of four pipe bombs using two low-explosive powders and two high-explosive materials. Following the explosions, scent pads were collected from the pipe-bomb debris using the STU-100. The study demonstrated the survivability of human scent which was still usable to the canine to make a positive scent match even after undergoing violent thermal conditions reaching temperatures that exceed 5000 K during the detonation of explosives (Stockham et al. 2004). Another study also evaluated the survivability of human scent using human scent samples from post-blast debris with an overall success rate of 73.5% (Curran et al. 2010). These types of studies aim to understand different environmental conditions on the preservation of a scent sample, although careful interpretation must be exercised given these experimental designs have taken into consideration double-blind setups and additional decoys, however, these studies do not depict canine search performance as it relates to using air scenting or ground disturbance as behavioral and olfactory cues when performing the scent detection task (Schoon 2022).

4.2.4 Canine Human Scent Matching and Discrimination Capabilities

Human scent detection canines are tasked with an array of different olfactory tasks that can range from location checks, scent identification line-ups, and man trailing (fresh and/or aged trails) activities. Variables to understand their performance are rooted not only in the definition and individualization of human scent but also lie in understanding how canines respond to the actual human odor cue when performing the identification and discrimination within that respective olfactory task.

A fundamental aspect to understand in canine scent detection capabilities includes the ability of the canine to determine the target scent source direction. Studies in this area suggest that for the dog to determine proper directionality along a selected trail, odor concentration plays a key role. The concentration of the target scent is a function of the source strength and its movement and dissipation in the air. The strength of the scent source is, in turn, directly related to factors such as compound vapor pressure, environmental conditions, and influenced by any affinity or binding key compounds may have for the substrate in which they are deposited (Cablk et al. 2008). Other findings in the area suggest that the canine's ability to determine the correct direction is related to the animal's age and sex whereas young male dogs are more efficient in determining the correct direction of a trail than older and/or female animals (Wells and Hepper 2003). Within practical operational situations, however, it is important to note that a canine's motivation to trail is the compounded effect of many training parameters such as reinforcement quality, enhanced canine assessment paradigms, and operational history. Moreover, practical implications of various studies have determined that canines trained to follow a track do not necessarily learn to determine the track's direction and that footsteps alone can provide enough olfactory information to determine directionality (a minimum of five sequential footsteps). Dogs do not determine the direction of a track by following each step from heel to toe thus being able to follow a scent source even if the individual walked backward (Wells and Hepper 2003; Steen and Wilsson 1990; Mackenzie and Schultz 1987). Canine behavior for scent detection purposes has been previously defined as consisting of three distinct phases: an initial searching phase, a deciding phase, and a final tracking phase. The deciding phase is usually characterized by a slow-moving speed and long sniffing period compared to the other two phases. Thus, the ability to determine the correct direction seems to rely on methods of sampling air and a specific sensitivity for certain substances (Thesen et al. 1993).

Other variables studied to understand human scent discrimination capabilities included the influence of genetics on canine discrimination ability. Studies have shown that subjects who are more genetically similar (kin related), fed the same diet, and live under identical environmental conditions do not produce highly discriminable odor cues for the canine (Harvey and Harvey 2003; Hepper 1988). With respect to gender differences, a study by Jerzierski et al. (2012) utilized scent line-up procedures with six dogs to evaluate human gender distinguishability to canine response. It was found that dogs in the study identified women's hand odors more accurately than those of men, thus suggesting a chemical difference in generated scent as a function of gender (Jezierski et al. 2012).

Generalization across body locations or regions has also been the subject of study with respect to scent-detection canines. In one study, more than 700 scent samples were utilized including both genders and a varying age range. Canines were presented with scented cloths which had been in contact with an unspecified body region and allowed to match on other body-scented cloths to identify the same target scent. The success rate was of 80% correct match recognitions for this part of the experiment. Consequently, a further experiment allowed canines to match odor from various parts of the body with the hand odor of the same subject on clean steel tubes. The percent correct matches were of 85% thereby depicting the canine's ability to generalize between individual human scents acquired from varying anatomical locations (Settle et al. 1994). Another study with respect to human scent generalization utilized Dutch police dogs to test if dogs were capable of matching hand scent to scent collected on the crook of the elbow as well as matching trouser pocket scent to collected hand scent, depicting how enhanced training and familiarization with target subject enhances canine performance to scent detection tasks (Schoon and Bruin 1994).

Other researchers have attempted to understand if human scent "fall-out" is enough for a scent discrimination task. In the context of this study, human "fall-out" is a term for a scent that may be collected or deposited without a subject touching the substrate. Hence, using two scent discrimination dogs, a control sample (cotton square touched by a human hand) versus test sample (hand 5 cm above the cotton square) was evaluated. Both canines satisfactorily matched the samples, suggesting the existence of a "fall out" component of human scent (Vyplelová et al. 2014).

An additional use of canines in human scent detection is that of identification line-up procedures. This task is focused on the canine correctly selecting the target odor in a line-up procedure, whereby the canine's behavior establishes an association between the crime scene's scented article given to the dog for pre-scenting, and the odor from the suspect in the row of odor samples being tested (Kaldenbach 2000). George Romanes performed the first experimental study depicting the canine's capability of scent discrimination in 1887. In the early 1900s, Inspector Bussenius from Germany was the first person to suggest the value of suspect discrimination in practical police work using individuals as the scent source, which led to the arrest of the suspect (Schoon and Haak 2002). The scent line-up procedure has undergone a lengthy developmental process (both operational and scientific) for the past 30 years. In the late 1990s, Schoon tested different experimental test designs using certified Dutch police canine teams. Important notions from this study depicted how incorporating a positive control check prior to the actual trial run reduced the number of mistakes, thereby introducing the concept of an initial "calibration" to uphold forensic scrutiny (Schoon 1996). In a recent study by Ferry et al. (2019), 11 countries were surveyed to compare methodological frameworks involved in routine line-up protocols, depicting a general lack of international standardization of the line-up procedure from the different countries that participated. Variables that were questioned included scent collection and handling, decoy characteristics, control and trial disqualification parameters, line-up experimental set-up, dog training, alerting and handling, handler qualifications, and degree of blindness.

Although scent evidence is a valuable tool in a range of human scent detection contexts, it is imperative that research moving forward focuses on valid experimental designs that aim to understand and bridge the scientific foundation of the definition of human scent with the canine capabilities in alerting to this complex target odor source. This includes evaluating the sensitivity and specificity of human scent detection canines in order to provide a better snapshot of their capability in both laboratory-designed experiments as well as in operational scenarios.

4.3 Human-Related Evidence

4.3.1 Human Remains Detection

Human remains detection (HRD) canines are an asset within forensic applications worldwide. Some interchangeable names for this specific application are also cadaver dogs or victim recovery dogs. Capabilities, where they prove useful can range from locating victims of homicide/suicide, accidents, mass disasters (natural or man-made), missing persons, or aiding in clandestine grave investigations (Titus et al. 2022). HRD canines are trained to identify the odor of human decomposition. Given the complexity and dynamic character of the human decomposition process, canines are trained on a wide spectrum of decomposition material (human blood, tissues, adipocere, body fluids, wet/dry bone, and burned human remains) to maintain optimal performance and proficiency (SWGDOG SC8 2009). The odor profile of human decomposition is one of the most complex as the decomposition process of the human body is affected by intrinsic and extrinsic factors that can include: environmental conditions, deposition ecosystems (land, water, urban, wilderness), victim characteristics (trauma, body size, containment), and ecological factors such as insect and scavenging activity (Titus et al. 2022; Irish and Dilkie 2022). In general, the total number of stages in the decomposition timeline can be grouped into five categories: fresh, bloat, active decay, advanced decay, and skeletonization or dry remain phase (Kaldenbach 2000). In recent years, there have been increased efforts in understanding the odor of death's volatilome, however, a significant challenge in this type of research is the ethical and legal restrictions of obtaining whole human cadavers. To circumvent this challenge, the development of the so-called "body farms" have pioneered taphonomic studies by using whole-body cadavers in different geographic locations and under different conditions to better understand the postmortem changes of the human body using more realistic and operational experimental designs. The first of these taphonomic facilities opened in 1980 at the University of Tennessee by Dr. William Bass, and since then a total of another seven have opened in the United States, one in Australia, and one in Holland (Wescott 2018). While there is no defined set of human decomposition-specific chemical odor signatures identified to date, numerous studies have utilized gravesite soil, human tissue samples, or human analogue models (whole animal carcass or parts) to understand the chemical odor profile at distinctive timepoints and/or under specific environmental conditions (Agapiou et al. 2015; Armstrong et al. 2016; Connor et al. 2018; DeGreeff and Furton 2011; Dekeirsschieter et al. 2012; Forbes et al. 2014b, 2016; Hoffman et al. 2009; Iqbal et al. 2017; Martin and Verheggen 2018; Perrault et al. 2014, 2015; Rosier et al. 2015; Stadler et al. 2015; Stefanuto et al. 2015).

4.3.2 Human Remains Canine Training Materials

Human remains are one of the most complex target odor materials for canine training purposes; as mentioned, the dynamic process of human putrefaction and degradation is highly variable and dependent on many elements within a given location. Globally, there is no set standard, which depicts the optimal training materials for this canine detection discipline, and while each country has made efforts in providing guidelines or best practices (Irish and Dilkie 2022), there is still a lack of consensus as to a finite set of materials for canine training purposes. Training materials for HRD canines can be divided into human remains, human analogue models (animal proxies), or synthetic training aids (pseudoscents). A detailed review by Dargan and Forbes (2021) provides a comprehensive list of VOCs previously detected in canine training materials to include blood, human remains, decomposition fluid, soil, burned remains, textile, and synthetic training aids (Dargan and Forbes 2021). It also provides a discussion on the reported capability of canines to respond to these target odor sources. In countries where it is either difficult to obtain human remains or laws prohibit their use as training aids, porcine material (pig) has been used as a proxy to facilitate the training of canines. Through stimulus generalization, the process was successful in the United Kingdom as evidenced by bodies being located. This more than likely occurred as a result of canines learning to respond to similar odor profiles and being rewarded for doing so. In essence, the canines were trained to initially respond to pig and rewarded for finding human remains, thus, both became part of their target odor repertoire. In the United States, this methodology sparked many a debate. The outcome of those discussions led to practitioners and scientists unifying to answer questions about VOC profiles to determine the best route to take when training human remains detection canines. In the 1990s the methods of how canines were trained started to work their way into the court system, furthering the need for knowledge and quantification of the VOC profile of human remains. This has shed light onto how canines are trained, with what odors human remains detection canines should be trained, and has even impacted law in areas where those items were once banned from possession for use to train canines with. Slowly, more state laws in the United States are being amended to accommodate the training of human remains detection canines with actual human remains and not that of other species as proxies.

Another option with respect to training materials are synthetic training aids, commonly known as pseudoscents. As discussed above in regards to contraband training aids, pseudoscent products are those that claim to synthetically replicate the decomposition odor profiles of target substances for which to train detection canines. Pseudoscent training aids are appealing to those with limited access to training material or where laws forbid their possession. While they have been effective in other disciplines of detection, pseudoscents have not proven to be useful in human remains detection canine training. Studies that tested canines trained using pseudoscents found they missed actual human remains, demonstrating the ineffectiveness of the training aids. Conversely, canines that have been trained only on human remains material would not respond to pseudoscent products. These findings are thought to be largely due to the fact that the odor profiles, when instrumentally analyzed, do not replicate that of actual human remains (Stadler et al. 2012; Tipple et al. 2014). Further complicating the matter is the fact that the odor of human remains is constantly changing throughout its various stages of decomposition and the environmental variables that aid in the stages. It is hypothesized that canines select a repertoire of similar compounds that remain consistent throughout the decomposition timeline. Until researchers can identify what the chemical odor profile is or what suite of compounds remain consistent throughout the process it is not plausible that an effective synthetic training aid could be developed that would yield accurate operational results (Irish and Dilkie 2022). Regardless of if those remains are whole bodies recovered on land, buried in different substrates, burned, submerged in water, scattered/dismembered, are thousands of years old (historical/archeological remains); or their odors are in trace amounts associated with minute amounts such as touch transfers of odors or drops of blood on tools and weapons, canine detection is an invaluable resource in this forensic capability. Countless accounts of canines detecting odors associated with human remains have been reported and have been recognized by the courts as being a reliable detection capability in criminal cases.

4.3.3 Residual Odor and Human Remains

Another area within human remains detection canine research that warrants more work is that of "Residual Odor," defined by the National Institute for Standards and Technology's Dogs and Sensors Subcommittee and later ratified by the American Academy of Forensic Sciences as: "Odor/scent that originated from any substance/ subject that may or may not be physically recoverable or detectable by other means. The odor/scent that originated from a target substance/subject and lingers after the target substance/subject has been removed or has totally decomposed (as in the case of human remains)" (OSAC Dogs and Sensors Subcommittee 2022). Another more brief definition from Daus defines it similarly as a detectable odor signature from an object that remains after the object has been removed (Daus 2013). The term itself sparks debate due to the belief that odor, if it is available to detect in an environment, is not a different type of odor and should not be classified as such. Thus, the belief that "odor is odor" is a topic of discussion in training circles because in essence the odor does not change types, it only changes in saliency or the amount that is available to detect. Beyond the consternation surrounding the term, the primary component of the debate that is in question is the amount of the odor that is available to detect and the canine's ability to detect it. This has been challenged in the courts by defense attorneys despite ongoing work over the past decade that has shown that odors emanating from sources can be captured and analyzed in laboratory and field settings after they have been removed. Canines have responded to those types of conditions, and responses have corroborated other circumstantial evidence in trials (Torrez 2020; Alexander et al. 2015; DeGreeff and Furton 2011; People v. Lane 2014; Braulio Marcelo Castillo v. Commonwealth of Virginia 2019). Other than a few studies, literature is absent that defines how long odor persists in certain environmental conditions and the limits of detection for human remains detection canines in these instances. Success with the detection of residual odor from burials where remains were removed and plots were aged 667 days have been reported. Here, canines responded with 75% accuracy or greater (Alexander et al. 2015). In another study, carpet squares were contaminated with remains odor through a non-contact transfer of the volatiles (no actual contact between the remains and the carpet squares was made). Following only two minutes of odor transfer, canines were able to detect the carpet squares above chance for up to 35 days, and with ten minutes of odor transfer this was increased to 65 days (Oesterhelweg et al. 2008a).

DeGreeff (2020b) reports on the retention of odor on various substrates once the odor source has been removed and the impact it could potentially have on detection rates. It is important to note that canines were not used in the study, but odor was measured with analytical instrumentation (namely, SPME-GC/MS). While this study was conducted with Nosework (non-operational, sport detection) canines, the implications could be applied to human remains detection teams and the methodology could be replicated. Here she discusses important considerations that arose out of three test groups in which odors were applied to various substrates (absorptive and non-absorptive) through either direct application, touch transfer, and non-contact (vapor) transfer. Results indicated that absorptive substrates of concrete, Cordura® (synthetic fiber-based fabric), and drywall interestingly, but not surprisingly, did not produce as much odor initially, though the odor took longer to dissipate and persisted longer due to absorption and trapping of odors. On non-absorptive materials (metal and tile), odor availability was high initially and was depleted more quickly. Also of interest was that on non-absorptive materials, odors from the non-contact group lasted as long as the touch transfer group (DeGreeff and Maughan 2022b; DeGreeff and Whittington 2020). This may provide insight into potential odor availability in certain crime scenes such as bathrooms with non-absorptive surfaces prone to high evaporation such as tile versus a concrete basement, a highly absorptive substrate if untreated. With this knowledge, handlers can adapt working strategies in different environments or scenes accordingly.

4.3.4 Canines Trained for Bodily Fluids

Blood

Blood detection is often considered part of human remains detection, but may also be a form of evidence detection in that canines can be trained to locate minute amounts or otherwise obscured blood which can then be collected and returned to the laboratory for DNA analysis, or may be used to presumptively locate crime scenes. Specialized blood detection canines are trained specifically to locate blood evidence and not human remains. Even though they have great utility in forensic investigations, unlike human remains detection dogs, they are not used widely around the globe and are predominantly deployed with law enforcement in the United Kingdom, Norway, South Africa, and Australia (Rust et al. 2022).

Limited research has been conducted in this field. The first piece of scientific research examining the capabilities of a blood-detection canine and the related VOC profiles was carried out by DeGreeff et al. (2012), where the VOC profile of blood was assessed over time as it aged in an open environment. The brief study demonstrated that the VOC profile of blood changes quite significantly over time as it ages, with fresh blood having an odor distinct from that of the decomposed blood. A single blood detection canine, the only trained in the United States at the time, was used to confirm the distinct shift in the odor signature from fresh to aged blood. The canine had only been trained to locate aged blood, and in a controlled, double-blind study the canine indeed demonstrated the trained response for aged blood, but did not give a trained final response for the fresh blood, indicating a notable difference in the blood odor. Others later confirmed the shift in VOC profile as blood ages which could be measured when the blood was dried on either porous or non-porous surfaces (Rust et al. 2016), as well as in blood stored at freezer, refrigerator, or room temperatures (Forbes et al. 2014a).

Since this initial study, further research probing blood-detection canines has been carried out (Rust et al. 2016, 2018; Forbes et al. 2014a; Schoon 2013; Rendine et al. 2019; Chilcote et al. 2018). Schoon compared the sensitivity of canine detectors to common tests used at crime scenes to presumptively identify blood including luminol, tetrabase, and Kastle-Meyer by depositing increasingly dilute blood samples on a substrate. Researchers determined that the two previously trained human remains detection canines outperformed the tests when blood was deposited on a porous surface (carpet), but the traditional presumptive tests were more sensitive for blood on the porous surface (vinyl) (Schoon 2013). Furthermore, in a study of the detection of blood from washed cotton fabric, canines could detect blood after as many as five washes, similar to the performance of luminol spray (Rust et al. 2018).

Blood detection is an underutilized capability of canine detectors, as studies have shown that they can be both very sensitive and specific, and thus aid in the recovery of small, obscured, or otherwise overlooked blood evidence. While the research in this specific field is currently limited, it is unlikely that it will see a large growth in literature in this area until the use of blood-detection canines become more common and widespread. Additional research should examine optimal ways of obtaining and storing blood for training materials, as well as expanded studies in canine sensitivity and specificity with greater numbers of participants allowing for statistical interpretation of the data.

Seminal Fluid

Seminal fluid detection by canines has become an application that is gaining attention largely due to work that has been initiated by the Netherlands National Police Agency and the University of Amsterdam. This partnership is a good example of practitioners and scientists working in concert to identify strengths and limitations, and have even helped identify practical workflows on crime scenes regarding when to utilize canines in the evidence collection process.

Sexual assault cases can be difficult to investigate due to the following complexities outlined by Schoon et al. (2022):

- Many are not reported as a result of shame, fear, lack of trust in the police investigative system
- No eyewitnesses because the nature of the crime is often conducted in seclusion
- Victims may be very young or mentally impaired and may not be able to express themselves

Traditionally, seminal fluid has been detected at crime scenes with the use of an Alternate Light Source (ALS), a high-intensity light focused on a scene through a fiber optic cable using various filters, or with a smaller ALS device with LED lights the size of a flashlight. These methods are used for detecting different types of organic materials which fluoresce and become visible through the use of filters and safety goggles. Latent fingerprints, fibers, and body fluids (saliva, urine, and semen) are easier to locate when ALS is used (Fish et al. 2015). Some of the limitations associated with ALS include some models being heavy and cumbersome to transport in a crime scene, only effective if technicians point the light in the right direction, have a narrow field of range, and if using the flashlight version they require multiple people to scan large areas. Additionally, it is not uncommon to miss areas if the stains are small and the light source is not pointed directly at the stain. The use of a specially trained canine can significantly reduce all of the issues stated above.

Detection of seminal fluid may be time sensitive but longitudinal studies are needed to quantify the temporal capabilities of detection in this relatively new application. In 2009, the Norwegian Police trained crime scene dogs on seminal fluid and blood and conducted research to determine how long canines could detect samples that varied in sample volume and length of time in outdoor environments. Seminal fluid was detected up to 2 days for lower volume samples and up to 4 days for larger volume samples. Spermatozoa was observed via microscopy between 32 h and 4 days which was also correlated with DNA results (Skalleberg and Bouzga 2016). Two case examples highlighted instances in which crime scene canines resulted in case resolution. In 2017, a canine responded to two spots that were sampled and tested positive with presumptive testing and ultimately produced a full DNA profile of the offender. In 2019, similar usage of crime scene dogs yielded positive results after one year and again produced DNA profiles that matched two suspects; even where ALS missed evidence previously (Schoon et al. 2022).

In a 2019 study, in which five canines were compared to three traditional methods of locating seminal fluid using ALS, a presumptive test called the Rapid Stain Identification (RSID) Kit, and the enzymatic Acid Phosphatase (AP) test for semen that was deposited on different types of fabrics highlighted the effectiveness of crime scene dogs. This study was done to determine the feasibility and efficacy of using canines in forensic settings. Much like blood, which typically requires visual detection in that if you cannot see it, you may miss it, seminal fluid stains can be difficult to detect at crime scenes due to several factors that include colors and patterns of fabrics and large areas that make searching with alternate light sources time-consuming and cumbersome (Dam et al. 2019). Based on the data that was presented, the authors came to the conclusion that canines were most effective with regards to sensitivity and selectivity in detection of semen samples, especially when it came to cotton and polyester fabric types with different colors and those fabrics that had patterns which make semen problematic in detecting visually. The only test that was comparable with canines was the AP test in the pattern testing experiment, in which visual acuity was not necessary. When Forensic light sources (FLS, also known as Alternate Light Sources ALS) were used, not all semen stains were detected due to the auto-fluorescent properties of some of the fabrics (six samples were missed). Canines detected 100% of seen stains on fabrics regardless of type, color, or pattern. However, they were not able to pinpoint the exact location of some samples, at which time other presumptive methods were used to identify exact locations (lending credence to the importance of using multiple presumptive methods for crime scene processing). Overall, the use of canines proved to be valuable when compared to other presumptive testing methods. This study combined with the case examples provided earlier has proven that crime scene canines can be an efficient search tool in locating primary or secondary crime scenes due to their mobility, sensitivity, and selectivity as a biological sensor. In some cases, canines were more effective than existing presumptive testing currently available.

There have been instances where the question has been posed as to whether human remains detection canines that have been trained on a full spectrum of odors associated with human decomposition could respond to other body fluids at a crime scene. The answer is difficult due to several factors. While a well-trained human remains detection canine that has been trained to generalize odor profiles among the different target odor types (decomposition odors associated with blood, tissue, bone, etc.), the odors associated with the different stages of decomposition, and the different thresholds or amounts of odor in various environments, whether they will respond to semen specifically has not been tested. Typically, in training, if canines show interest in saliva or locations where people have spat or urinated, they are often corrected or discouraged to do so because those types of odors are considered distractors. Once scientists have solved the odor profile of human decomposition, it will be interesting to see if similar VOCs are present in singular human bodily fluids. Anecdotally, it would not be outside the realm of possibility for human remains detection canines to respond to singular fluids if they have been trained with a high degree of generalization and to a low threshold, even if they have not been trained on semen samples specifically. Due to the material being organic, human in nature, and subjected to biological bacterial degradation and decomposition processes, the possibility should not be discounted. The question may arise as to whether human remains detection should serve like other detection canines that are trained to respond to multiple target odors, or, should canines only be trained and be used for specific substances? This application may be better served with single-purpose detection canines that are specialized, have documented exposures, certifications that are dedicated to the specialty, and handlers that are trained for the intricacies of such a search. Furthermore, from a research perspective, what would be useful are studies that shed light on the persistence of odor profiles in various environmental conditions, generalization capabilities of well-trained canines, and limits of detection to determine how low of a threshold canines can detect.

5 Conclusions and Future Direction

This chapter has covered much of the research related to the use of canine detectors from a forensic and security viewpoint. In recent decades the amount of research relating to canine detectors has grown exponentially as scientists and operational end-users alike begin to understand that, just like any other sensor technology that is used in forensics, the canine as a biosensor should be assessed both in the laboratory and the field using the scientific method. Such research should not only validate the efficacy of a technique, method, or a process but it should aim to identify the limitations of the process and offer insight into how to mitigate the limitations by providing solutions. Despite the growing field of research, there remain many gaps in knowledge that could improve the proficiency and deployment of these complex biological detectors.

As with any area of forensic science, canine odor detection represents an overarching convergence of disciplines that work in unison for optimal use. The forensic uses of a canine team must meet the same scrutiny as any other evidence type, with challenges stemming from the crime scene all the way to courts of law. Hence, it is important to remember that canine odor evidence must bear high-quality standards and best practices to protect the integrity with a solid scientific foundation. In order to continue in this path, it is important to understand basics of odor dynamics. These foundational blocks of odor detection include odor transport and movement, chemical characterization of target odor sources, persistence/survivability of odor across varying periods of time and under different environmental conditions, as well as evaluation of storage, containment, and handling practices to safeguard contamination issues and provide integrity of odor sample. With the development of synthetic training aids, it is also imperative to provide controlled delivery of target material, hence proper validation standards must be in place to maintain optimal canine performance. To further the discussion about the determining factors that should be used in decision-making paradigms surrounding their use, the following should be considered: Can the company that is producing the pseudoscent product provide peer-reviewed validation studies that illustrate the efficacy of the aids in double-blind testing scenarios with detection canines using cross-correlation? Here, canines trained on only actual target material should respond to pseudoscent without prior exposures and canines trained only with pseudoscent should respond to actual target material without prior exposures. Can the company provide peer-reviewed instrumental analysis of the VOC profiles that have elicited responses from trained detection canines over time and with consistency? If the answer is "No" to either of these questions, then buyer and trainer beware.

Another topic of great interest is the sensitivity of the canine olfactory system and olfactory threshold capabilities. Regardless of target substance detection discipline, it is important to understand the selectivity when trained target odors occur in mixtures or in the vicinity of other salient odors from which they have to discriminate. Additionally, the persistence of odors in different environments is also important to understand as it has faced scrutiny where canines have been used successfully in various cases in which unproductive responses have corroborated other circumstantial evidence (Torrez 2020; People v. Lane 2014; Braulio Marcelo Castillo v. Commonwealth of Virginia 2019). An additional area where research will prove to be beneficial in court proceedings is canine olfactory learning and behavior. For example, through canine behavioral assessments during detection, handlers will be better equipped to interpret subtle canine movements in response to target odors versus non-target odors. Research should be geared toward understanding performance from distinctive perspectives crucial to training paradigm developments such as reward motivations, distractions, reinforcement rates, and schedules, to name a few. The same field of study can also help identify relationships that exist between handler movement and canine performance. This will also allow handlers to augment training regimens to account for nuanced non-verbal body language communication that occurs during training. Here, the research would focus on identifying observable and measurable handler movements as environmental stimuli and the behavioral responses that occur as a result (Waterbury and Schultz 2022).

When considering research in the field of canine olfactory detection, one should bear in mind that canines are cognitive, learning beings, and that not all canines possess the same capabilities due to differences in training methodologies, exposures to odor mixtures, and operational experiences. Care should be taken not to characterize canines based on only a small sample size or over-generalize results from a group of canines trained in a particular way or with certain past experiences, to a different group of canines with other past experiences. For example, just because one group does not produce favorable results it should not mean that all canines are not capable of performing that particular task. It may simply mean that the wrong set of canines was selected for the study. Conversely, equal consideration should be given to research that does show favorable capability in that we should not assume that all canines can perform to the same standard based on the same reasons mentioned previously or if they are not trained accordingly to accomplish the task. Making broad conclusions from small sample sizes or a select group of canines can be dangerous. However, such studies do provide valuable information for proof-of-concept and serve as good launching points for future research. For this reason, the courts have determined that a canine's reliability should be evaluated on a case-by-case basis and have provided a framework by which they should be evaluated. Canine substance detection has proven to be an invaluable tool in the forensic practitioner's toolbox and although there has been a quest for research and dedicated teamwork between laboratory scientists and canine detection professionals, the heightened scrutiny of forensic evidence calls for continued validation and globally recognized standards to provide the full potential behind the wet nose.

This is publication number 23.02 of the Laboratory Division of the Federal Bureau of Investigation. Names of commercial manufacturers are provided for information only and inclusion does not imply endorsement by the FBI of the U.S. Government. The views expressed are those of the authors and do not necessarily reflect the official policy or position of the FBI or the U.S. Government.

References

Agapiou A, Zorba E, Mikedi K, McGregor L, Spiliopoulou C, Statheropoulos M (2015) Analysis of volatile organic compounds released from the decay of surrogate human models simulating victims of collapsed buildings by thermal desorption–comprehensive two-dimensional gas chromatography–time of flight mass spectrometry. Anal Chim Acta 883:99–108. https://doi.org/10.1016/j.aca.2015.04.024

Alexander MB, Hodges TK, Bytheway J, Aitkenhead-Peterson JA (2015) Application of soil in forensic science: residual odor and HRD dogs. Forensic Sci Int 249:304–313. https://doi.org/10.1016/j.forsciint.2015.01.025

American Academy of Forensic Science Standards Board, Dogs and Sensors (2022) Crime scene/death investigation—dogs and sensors—terms and definitions. ASB Technical Report 025. https://www.aafs.org/asb-standard/crime-scenedeath-investigation-dogs-and-sensors-terms-and-definitions. Accessed 09 Jul 2022

Armstrong P, Nizio KD, Perrault KA, Forbes SL (2016) Establishing the volatile profile of pig carcasses as analogues for human decomposition during the early postmortem period. Heliyon 2(2):e00070. https://doi.org/10.1016/j.heliyon.2016.e00070

Aviles-Rosa EO, Fernandez LS, Collins-Pisano C, Prada-Tiedemann PA, Hall NJ (2022) The use of an intermittent schedule of reinforcement to evaluate detection dogs' generalization from smokeless-powder. Anim Cogn. https://doi.org/10.1007/s10071-022-01648-y

Baker LB, Wolfe AS (2020) Physiological mechanisms determining eccrine sweat composition. Eur J Appl Physiol 120(4):719–752. https://doi.org/10.1007/s00421-020-04323-7

Biasioli F, Yeretzian C, Märk TD, Dewulf J, Van Langenhove H (2011) Direct-injection mass spectrometry adds the time dimension to (B)VOC analysis. TrAC Trends Anal Chem 30(7):1003–1017. https://doi.org/10.1016/j.trac.2011.04.005

Blake RS, Monks PS, Ellis AM (2009) Proton-transfer reaction mass spectrometry. Chem Rev 109(3):861–896. https://doi.org/10.1021/cr800364q

Broza YY, Zuri L, Haick H (2015) Combined volatolomics for monitoring of human body chemistry. Sci Rep 4(1):4611. https://doi.org/10.1038/srep04611

Braulio Marcelo Castillo v. Commonwealth of Virginia (2019)

Buckley P (2022) Training Aid Delivery Device (TADD) for use with Homemade Explosives (HMEs). Combat Capabilities Development Command Chemical Biological Center, Research and Technology Directorate

Buljubasic F, Buchbauer G (2015) The scent of human diseases: a review on specific volatile organic compounds as diagnostic biomarkers: the scent of human diseases. Flavour Fragr J 30(1):5–25. https://doi.org/10.1002/ffj.3219

Bylinski H, Gebicki J, Dymerski T, Namieśnik J (2017) Direct analysis of samples of various origin and compositioin using specific types of mass spectrometry. Crit Rev Anal Chem 47(4):340–358

Connor M, Baigent C, Hansen ES (2018) Testing the use of pigs as human proxies in decomposition studies. J Forensic Sci 63(5):1350–1355. https://doi.org/10.1111/1556-4029.13727

Chilcote B, Rust L, Nizio KD, Forbes SL (2018) Profiling the scent of weathered training aids for blood-detection dogs. Sci Justice 58(2):98–108. https://doi.org/10.1016/j.scijus.2017.11.006

Craven BA, Paterson EG, Settles GS (2010) The fluid dynamics of canine olfaction: unique nasal airflow patterns as an explanation of macrosmia. J R Soc Interface 7(47):933–943. https://doi.org/10.1098/rsif.2009.0490

Casey Marie Anthony v. State of Florida (2013)

Cannon C (2020) The volatile organic compound profile from Cimex lectularius in relation to bed bug detection canines. Forensic Chem 18:100214

Cablk M, Sagebiel J, Heaton J, Valentin C (2008) Olfaction-based detection distance: a quantitative analysis of how far away dogs recognize tortoise odor and follow it to source. Sensors 8(4):2208–2222. https://doi.org/10.3390/s8042208

Cerna K, Pinc L, Pachman J (2011) Ability of explosives detector dogs to generalize odor of TNT. New Trends Res Energ Mater, Czech Republic, pp 542–549

Cerreta MM, Furton KG (2015) An assessment of detection canine alerts using flowers that release methyl benzoate, the cocaine odorant, and an evaluation of their behavior in terms of the VOCs produced. Forensic Sci Int 251:107–114. https://doi.org/10.1016/j.forsciint.2015.03.021

Cid Torrez v. State of Florida (2020)

Costello EK, Lauber CL, Hamady M, Fierer N, Gordon JI, Knight R (2009) Bacterial community variation in human body habitats across space and time. Science 326(5960):1694–1697. https://doi.org/10.1126/science.1177486

Combat Capabilities Development Chemical Biological Center Report (2022) Training Aid Delivery Device (TADD) for use with Homemade Explosives (HMEs)—final report. Aberdeen Proving Ground, MD

Crespo Cajigas JM, Perez-Almodovar L, DeGreeff LE (2019) Headspace analysis of potassium chlorate using on-fiber SPME derivatization coupled with GC/MS. Talanta 205:120127. https://doi.org/10.1016/j.talanta.2019.120127

Curran AM, Prada PA, Furton KG (2010) Canine human scent identifications with post-blast debris collected from improvised explosive devices. Forensic Sci Int 199(1–3):103–108. https://doi.org/10.1016/j.forsciint.2010.03.021

Daus T (2022) When does drug odor become residual? Police K9 Magazine. https://www.police1.com/police-products/k9/k9-training/articles/when-does-drug-dog-odor-become-residual-5MI27HxiHFyk9afl/

Dargan R, Forbes SL (2021) Cadaver-detection dogs: a review of their capabilities and the volatile organic compound profile of their associated training aids. WIREs Forensic Sci 3(6). https://doi.org/10.1002/wfs2.1409

Deedrick D (2000) Hairs, fibers, crime and evidence Part 1. Forensic science communications. https://archives.fbi.gov/archives/about-us/lab/forensic-science-communications/fsc/july2000/deedric1.htm

de Lacy Costello B et al (2014) A review of the volatiles from the healthy human body. J Breath Res 8(1):014001. https://doi.org/10.1088/1752-7155/8/1/014001

DeGreeff LE, Furton KG (2011) Collection and identification of human remains volatiles by non-contact, dynamic airflow sampling and SPME-GC/MS using various sorbent materials. Anal Bioanal Chem 401(4):1295–1307. https://doi.org/10.1007/s00216-011-5167-0

DeGreeff LE, Whittington TC (2020) Elucidation and durability of odor profiles of K9 nose work materials. Internal Report No. 6180/0306 Naval Research Laboratory, Washington DC

DeGreeff LE, Peranich K (2021) Headspace analysis of ammonium nitrate variants and the effects of differing vapor profiles on canine detection. Forensic Chem 25:100342. https://doi.org/10.1016/j.forc.2021.100342

DeGreeff LE, Maughan M (2022a) Understanding the dynamics of odor to aid in odor detection. In: Canines: the original biosensors, 1st edn. Jenny Stanford Publishing, Singapore

DeGreeff LE, Maughan M (2022b) Understanding the dynamics of odor to aid in odor detection. In: Canines: the original biosensors, 1st ed. Jenny Stanford Publishing, pp 217–262

DeGreeff LE, Curran AM, Furton KG (2011) Evaluation of selected sorbent materials for the collection of volatile organic compounds related to human scent using non-contact sampling mode. Forensic Sci Int 209(1–3):133–142. https://doi.org/10.1016/j.forsciint.2011.01.010

DeGreeff LE et al (2012) Detection and identification of volatile organic compounds in dried human blood samples by instrumental analysis and canines. In: Presented at the American Academy of Forensic Science. DC, Washington, p D60

DeGreeff LE, Cerreta M, Rispoli M (2017a) Feasibility of canine detection of mass storage devices: a study of volatile organic compounds emanating from electronic devices using solid phase microextraction. J Forensic Sci 62(6):1613–1616. https://doi.org/10.1111/1556-4029.13472

DeGreeff LE, Cerreta MM, Katilie CJ (2017b) Variation in the headspace of bulk hexamethylene triperoxide diamine (HMTD) with time, environment, and formulation. Forensic Chem 4:41–50. https://doi.org/10.1016/j.forc.2017.03.001

DeGreeff LE, Peranich K, Holness HK, Furton KG (2018) Canine generalization to molecularly similar odors and odor mixtures. U.S. Naval Research Laboratory, Washington, DC, Memorandum Report NRL/MR/6181–18-9797

Dekeirsschieter J, Stefanuto P-H, Brasseur C, Haubruge E, Focant J-F (2012) Enhanced characterization of the smell of death by comprehensive two-dimensional gas chromatography-time-of-flight mass spectrometry (GCxGC-TOFMS). PLoS ONE 7(6):e39005. https://doi.org/10.1371/journal.pone.0039005

Drabińska N et al (2021) A literature survey of all volatiles from healthy human breath and bodily fluids: the human volatilome. J Breath Res 15(3):034001. https://doi.org/10.1088/1752-7163/abf1d0

Eckenrode BA, Ramsey SA, Stockham RA, Van Berkel GJ, Asano KG, Wolf DA (2006) Performance evaluation of the scent transfer unittm (STU-100) for organic compound collection and release. J Forensic Sci 51(4):780–789. https://doi.org/10.1111/j.1556-4029.2006.00178.x

Emmons RV, Gionfriddo E (2021) Minimizing transient microenvironment-associated variability for analysis of environmental anthropogenic contaminants via ambient ionization. Sci Total Environ 775:145789

Ewing RG, Waltman MJ, Atkinson DA, Grate JW, Hotchkiss PJ (2013) The vapor pressures of explosives. TrAC, Trends Anal Chem 42:35–48. https://doi.org/10.1016/j.trac.2012.09.010

Ferry B et al (2019) Scent lineups compared across eleven countries: looking for the future of a controversial forensic technique. Forensic Sci Int 302:109895. https://doi.org/10.1016/j.forsciint.2019.109895

Fish JT, Miller LS, Braswell MC, Wallace EW (2015) Crime scene investigation, 3rd edn. Routledge, New York, NY

Florida v. Harris (2013)

Forbes SL, Rust L, Trebilcock K, Perrault KA, McGrath LT (2014a) Effect of age and storage conditions on the volatile organic compound profile of blood. Forensic Sci Med Pathol 10(4):570–582. https://doi.org/10.1007/s12024-014-9610-3

Forbes SL, Perrault KA, Stefanuto P-H, Nizio KD, Focant J-F (2014b) Comparison of the decomposition VOC profile during winter and summer in a moist, mid-latitude (Cfb) climate. PLoS ONE 9(11):e113681. https://doi.org/10.1371/journal.pone.0113681

Forbes SL, Troobnikoff AN, Ueland M, Nizio KD, Perrault KA (2016) Profiling the decomposition odour at the grave surface before and after probing. Forensic Sci Int 259:193–199. https://doi.org/10.1016/j.forsciint.2015.12.038

Frye v. United States (1923)

Francis VS, Holness HK, Furton KG (2019) The ability of narcotic detection canines to detect illegal synthetic cathinones (Bath Salts). Front Vet Sci 6:98. https://doi.org/10.3389/fvets.2019.00098

Frank KJ, Holness HK, Furton KG, DeGreeff LE (2022) Explosives detection by dogs. In: Counterterrorist detection techniques of explosives, 2nd edn. Elsevier, pp 47–75. https://doi.org/10.1016/B978-0-444-64104-5.00004-7

Frank KJ Utilizng headspace solid-phase microextraction for the characterization of volatile organic copounds released from contraband and its implication for detector dog training. PhD dissertation, Florida International University, Miami, FL

Furton KG, Winialski D (2022) Comparing the olfactory capabilities of dogs with machines designed to detect odors. In: Canines: the original biosensors, 1st edn. Jenny Stanford Publishing, Singapore, pp 21–62

Furton KG, Hsu Y-L, Luo T-Y, Alvarez N, Lagos P (1997) Novel sample preparation methods and field testing procedures used to determine the chemical basis of cocaine detection by canines. In: Proceedings SPIE 2941. Forensic evidence analysis and crime scene investigation, Boston, MA

Furton KG, Hong YC, Hsu YL, Luo T, Rose S, Walton J (2002) Identification of odor signature chemicals in cocaine using solid-phase microextraction-gas chromatography and detector-dog response to isolated compounds spiked on U.S. paper currency. J Chromatogr Sci 40(3):147–155. https://doi.org/10.1093/chromsci/40.3.147

Furton KG, Caraballo NI, Cerreta MM, Holness HK (2015) Advances in the use of odour as forensic evidence through optimizing and standardizing instruments and canines. Phil Trans R Soc B 370(1674):20140262. https://doi.org/10.1098/rstb.2014.0262

Getxent tube (2022). Accessed 22 Sep 2022. www.getxent.com

Grosch W (1993) Detection of potent odorants in foods by aroma extract dilution analysis. Trends Food Sci Technol 4(3):68–73

Grice EA, Segre JA (2011) The skin microbiome. Nat Rev Microbiol 9(4):244–253. https://doi.org/10.1038/nrmicro2537

Harvey LM, Harvey JW (2003) Reliability of bloodhounds in criminal investigations. J Forensic Sci 48(4):811–816

Hallowell SF, Davies JP, Gresham GL (1994) Qualitative/semiquantitative chemical characterization of the Auburn Olfactometer. In: Presented at the SPIE's 1994 international symposium on optics, imaging, and instrumentation, San Diego, CA, pp 437–448. https://doi.org/10.1117/12.189192

Hallowell SF, Fischer DS, Brasher JD, Malone RL, Gresham GL, Rae C (1997) Effectiveness of quality-control aids in verifying K-9-team explosive detection performance. In: Presented at the enabling technologies for law enforcement and security, Boston, MA, pp 227–234. https://doi.org/10.1117/12.266782

Harper R, Almirall J, Furton K (2005a) Identification of dominant odor chemicals emanating from explosives for use in developing optimal training aid combinations and mimics for canine detection. Talanta 67(2):313–327. https://doi.org/10.1016/j.talanta.2005.05.019

Harper RJ, Almirall JR, Furton KG (2005b) Discrimination of smokeless powders by headspace SPME-GC-MS and SPME-GC-ECD, and the potential implications upon training canine detection of explosives. In: SPIE 5778, sensors, and command, control, communications, and intelligence technologies for homeland security and homeland defense IV

Hall NJ, Glenn K, Smith DW, Wynne CDL (2015) Performance of Pugs, German Shepherds, and Greyhounds (Canis lupus familiaris) on an odor-discrimination task. J Comp Psychol 129(3):237–246. https://doi.org/10.1037/a0039271

Harris HA, Lee HC, Gaensslen RE (2019) Introduction to forensic science and criminalistics, 2nd edn. CRC Press, Boca Raton, FL

Havlíček J, Winternitz J, Roberts SC (2020) Major histocompatibility complex-associated odour preferences and human mate choice: near and far horizons. Phil Trans R Soc B 375(1800):20190260. https://doi.org/10.1098/rstb.2019.0260

Hepper PG (1988) The discrimination of human odour by the dog Perception 17(4):549–554

Houck MM (2016) Materials analysis in forensic science. Academic Press/Elsevier, Amsterdam

Holland KT, Bojar RA (2002) Cosmetics: what is their influence on the skin microflora? Am J Clin Dermatol 3(7):445–449. https://doi.org/10.2165/00128071-200203070-00001

Hoffman EM, Curran AM, Dulgerian N, Stockham RA, Eckenrode BA (2009) Characterization of the volatile organic compounds present in the headspace of decomposing human remains. Forensic Sci Int 186(1–3):6–13. https://doi.org/10.1016/j.forsciint.2008.12.022

Hunt R (1999) The benefits of scent evidence. FBI Law Enforc Bull. https://indexarticles.com/reference/fbi-law-enforcement-bulletinthe/the-benefits-of-scent-evidence/

Huertas-Rivera AM (2016) Identification of active odors from illicit substances for hte development of optimal canine training aids. Dissertation, Florida International University, Miami, FL

Hudson DT, Curran AM, Furton KG (2009) The stability of collected human scent under various environmental conditions. J Forensic Sci 54(6):1270–1277. https://doi.org/10.1111/j.1556-4029.2009.01153.x

Irish L, Dilkie N (2022) The use of dogs for the detection of human remains. In: Canines the original biosensors, 1st edn. Jenny Stanford Publishing, pp 559–595

Iqbal MA, Nizio KD, Ueland M, Forbes SL (2017) Forensic decomposition odour profiling: a review of experimental designs and analytical techniques. TrAC, Trends Anal Chem 91:112–124. https://doi.org/10.1016/j.trac.2017.04.009

Jezierski T, Sobczyńska M, Walczak M, Gorecka-Bruzda A, Ensminger J (2012) Do Trained dogs discriminate individual body odors of women better than those of men? do trained dogs discriminate individual body odorS. J Forensic Sci 57(3):647–653. https://doi.org/10.1111/j.1556-4029.2011.02029.x

Jezierski T et al (2014) Efficacy of drug detection by fully-trained police dogs varies by breed, training level, type of drug and search environment. Forensic Sci Int 237.112–118. https://doi.org/10.1016/j.forsciint.2014.01.013

Jeunieau L, Simoens B, Lefebvre MH (2022) TATP: PREPARATION, characterisation and first tests of canine training aids. Forensic Chem 28:100409. https://doi.org/10.1016/j.forc.2022.100409

Johnston JM (1999) Canine detection capabilities: operational implications of recent R&D findings. Auburn University, Institute of Biological Detection Systems

Johnston JM, Williams M, Waggoner LP, Edge CC, Dugan RE, Hallowell SF (1998) Canine detection odor signatures for mine-related explosives. In: Presented at the aerospace/defense sensing and controls, Orlando, FL, p 490. https://doi.org/10.1117/12.324221

Joshi M, Delgado Y, Guerra P, Lai H, Almirall JR (2009) Detection of odor signatures of smokeless powders using solid phase microextraction coupled to an ion mobility spectrometer. Forensic Sci Int 188(1–3):112–118. https://doi.org/10.1016/j.forsciint.2009.03.032

Karsa Oy Karsa: ultrasensitive chemical threat detection. http://karsa.fi

Kaldenbach J (2000) K9 scent detection: my favourite judge lives in a kennel. Place of publication not identified: K9 & Law Enforcement

Kranz W, Kitts K, Strange N, Cummins J, Lotspeich E, Goodpaster J (2014) On the smell of composition C-4. Forensic Sci Int 236:157–163. https://doi.org/10.1016/j.forsciint.2013.12.012

Kwak J, Willse A, Preti G, Yamazaki K, Beauchamp GK (2010) In search of the chemical basis for MHC odourtypes. Proc R Soc B 277(1693):2417–2425. https://doi.org/10.1098/rspb.2010.0162

Lazarowski L et al (2015) Olfactory discrimination and generalization of ammonium nitrate and structurally related odorants in Labrador retrievers. Anim Cogn 18(6):1255–1265. https://doi.org/10.1007/s10071-015-0894-9

Lazarowski L et al (2020) Methodological considerations in canine olfactory detection research. Front Vet Sci 7:408. https://doi.org/10.3389/fvets.2020.00408

Lazarowski L, Waggoner P, Hutchings B, Angle C, Porritt F (2021a) Maintaining long-term odor memory and detection performance in dogs. Appl Anim Behav Sci 238:105301. https://doi.org/10.1016/j.applanim.2021.105301

Lazarowski L et al (2021b) Generalization across acetone peroxide homemade explosives by detection dogs. Front Anal Sci 1:797520. https://doi.org/10.3389/frans.2021.797520

Lit L, Oberbauer A, Sutton JE, Dror IE (2019) Perceived infallibility of detection dog evidence: implications for juror decision-making. Crim Justice Stud 32(3):189–206. https://doi.org/10.1080/1478601X.2018.1561450

Lit L, Schweitzer JB, Oberbauer AM (2011) Handler beliefs affect scent detection dog outcomes. Anim Cogn 14(3):387–394. https://doi.org/10.1007/s10071-010-0373-2

Lorenzo N et al (2003) Laboratory and field experiments used to identify Canis lupus var. Familiaris active odor signature chemicals from drugs, explosives, and humans. Anal Bioanal Chem 376(8):1212–1224. https://doi.org/10.1007/s00216-003-2018-7

Majchrzak T (2018) PTR-MS and GC-MS as complementary techniques for analysis of volatiles: a tutorial review. Analytica Chimica Acta 13

Mackenzie SA, Schultz JA (1987) Frequency of back-tracking in the tracking dog. Appl Anim Behav Sci 17(3–4):353–359. https://doi.org/10.1016/0168-1591(87)90158-4

Martin C, Verheggen F (2018) Odour profile of human corpses: a review. Forensic Chem 10:27–36. https://doi.org/10.1016/j.forc.2018.07.002

Macias MS, Furton KG (2011) Availability of target odor compounds from seized ecstasy tablets for canine detection. J Forensic Sci 56(6):1594–1600. https://doi.org/10.1111/j.1556-4029.2011.01854.x

Macias MS, Harper RJ, Furton KG (2008) A comparison of real versus simulated contraband VOCs for reliable detector dog training utilizing SPME-GC-MS. Am Lab 40(1):16–19

McDuffie v. State (1992)

Moore S, MacCrehan W, Schantz M (2011) Evaluation of vapor profiles of explosives over time using ATASS (Automated Training Aid Simulation using SPME). Forensic Sci Int 212(1–3):90–95. https://doi.org/10.1016/j.forsciint.2011.05.019

Natsch A, Emter R (2020) The specific biochemistry of human axilla odour formation viewed in an evolutionary context. Phil Trans R Soc B 375(1800):20190269. https://doi.org/10.1098/rstb.2019.0269

Nettles K, Ford C, Prada-Tiedemann PA (2022) Development of profiling methods for contraband firearm volatile odor signatures. Front Anal Sci 1:785271. https://doi.org/10.3389/frans.2021.785271

Oesterhelweg et al (2008a) Cadaver dogs—a study on detection of contaminated.pdf

Oesterhelweg L et al (2008b) Cadaver dogs—a study on detection of contaminated carpet squares. Forensic Sci Int 174(1):35–39. https://doi.org/10.1016/j.forsciint.2007.02.031

Olsson MJ et al (2014) The scent of disease: human body odor contains an early chemosensory cue of sickness. Psychol Sci 25(3):817–823. https://doi.org/10.1177/0956797613515681

Ong T-H, Mendum T, Geurtsen G, Kelley J, Ostrinskaya A, Kunz R (2017) Use of mass spectrometric vapor analysis to improve canine explosive detection efficiency. Anal Chem 89:6482–6490

Ong T-H, Ljunggren J, Mendum T, Geurtsen G, Kunz RR (2022) Vapor signatures of double-base smokeless powders and gunshot residues for supporting canine odor imprinting. ACS Omega 7(26):22567–22576. https://doi.org/10.1021/acsomega.2c01882

OSAC Dogs and Sensors Subcommittee (2022). https://www.nist.gov/osac/dogs-sensors-subcommittee

Oxley JC, Smith JL, Luo W, Brady J (2009) Determining the Vapor Pressures of Diacetone Diperoxide (DADP) and Hexamethylene Triperoxide Diamine (HMTD). Propellants Explos Pyrotech 5

People v. Perryman (1979)

People v. Lane (2014)

Perrault K, Stuart B, Forbes S (2014) A longitudinal study of decomposition odour in soil using sorbent tubes and solid phase microextraction. Chromatography 1(3):120–140. https://doi.org/10.3390/chromatography1030120

Peranich K, DeGreeff LE (2018) Canine research—past, present, future: analysis of gaps and user needs. Naval Surface Warfare Center Indian Head Explosive Ordnance Disposal Technology Division, Indian Head, MD

Perrault KA, Stefanuto P-H, Stuart BH, Rai T, Focant J-F, Forbes SL (2015) Detection of decomposition volatile organic compounds in soil following removal of remains from a surface deposition site. Forensic Sci Med Pathol 11(3):376–387. https://doi.org/10.1007/s12024-015-9693-5

Pickenhagen W (2017) History of odor and odorants. In: springer handbook of odor. Springer International Publishing, New York

Porritt F et al (2015) Performance decline by search dogs in repetitive tasks, and mitigation strategies. Appl Anim Behav Sci 166:112–122. https://doi.org/10.1016/j.applanim.2015.02.013

Prada-Tiedemann P (2022) Human odor: an overview of current knowledge and experimental designs. In: Odor in textiles: generation and control, 1st edn. CRC Press, pp 15–27

Prada PA, Furton KG (2008) Human scent detection: a review of its developments and forensic applications. Revista De Ciencias Forenses 1(2):81–87

Prada A, Furton KG (2012) Recent advances in solid-phase microextraction for forensic applications. In: Comprehensive sampling and sample preparation. Elsevier, pp 877–891. https://doi.org/10.1016/B978-0-12-381373-2.00117-4

Prada PA, Curran AM, Furton KG (2011) The evaluation of human hand odor volatiles on various textiles: a comparison between contact and noncontact sampling methods. J Forensic Sci 56(4):866–881. https://doi.org/10.1111/j.1556-4029.2011.01762.x

Prada P, Curran A, Furton K (2014) Characteristic human scent compounds trapped on natural and synthetic fabrics as analyzed by SPME-GC/MS. J Forensic Sci Criminol 1(1). https://doi.org/10.15744/2348-9804.1.S101

Prada PA, Curran AM, Furton KG (2015) Human scent evidence, 1st edn. CRC PRESS, Boca Raton, FL

Prugnolle F, Lefèvre T, Renaud F, Møller AP, Missé D, Thomas F (2009) Infection and body odours: evolutionary and medical perspectives. Infect Genet Evol 9(5):1006–1009. https://doi.org/10.1016/j.meegid.2009.04.018

Rendine M et al (2019) Decomposing human blood: canine detection odor signature and volatile organic compounds. J Forensic Sci 64(2):587–592. https://doi.org/10.1111/1556-4029.13901

Rice S, Koziel JA (2015) The relationship between chemical concentration and odor activity value explains the inconsistency in making a comprehensive surrogate scent training tool representative of illicit drugs. Forensic Sci Int 257:257–270. https://doi.org/10.1016/j.forsciint.2015.08.027

Rosier E, Loix S, Develter W, Van de Voorde W, Tytgat J, Cuypers E (2015) The search for a volatile human specific marker in the decomposition process. PLoS ONE 10(9):e0137341. https://doi.org/10.1371/journal.pone.0137341

Rust L, Nizio KD, Forbes SL (2016) The influence of ageing and surface type on the odour profile of blood-detection dog training aids. Anal Bioanal Chem 408(23):6349–6360. https://doi.org/10.1007/s00216-016-9748-9

Rust L, Nizio KD, Wand MP, Forbes SL (2018) Investigating the detection limits of scent-detection dogs to residual blood odour on clothing. Forensic Chem 9:62–75. https://doi.org/10.1016/j.forc.2018.05.002

Rust L, Nizio KD, Forbes SL (2022) Canine detection of blood evidence. In: Canines: the original biosensors, 1st edn. Jenny Stanford Publishing, pp 611–653

Saferstein R Criminalistics: an introduction to forensic science, 12th edn. Pearson Education

Schoon GAA, De Bruin JC (1994) The ability of dogs to recognize and cross-match human odours. Forensic Sci Int 69(2):111–118. https://doi.org/10.1016/0379-0738(94)90247-X

Schoon A (1996) Scent identification lineups by dogs (Canis familiaris): experimental design and forensic application. Appl Anim Behav Sci 49(3):257–267. https://doi.org/10.1016/0168-1591(95)00656-7

Schoon GAA (2005) The effect of the ageing of crime scene objects on the results of scent identification line-ups using trained dogs. Forensic Sci Int 147(1):43–47. https://doi.org/10.1016/j.forsciint.2004.04.080

Schoon A (2013) A comparison between canine detection of blood residue and some blood presumptive tests. JFI 63(3):255

Schoon A, Haak R (2002) K9 Suspect discrimination: training and practicing scent identification line-ups. Detselig Enterprises Ltd, Calgary Alberta

Schoon GAA (2022) The use of tracking/man trailing dog results as evidence in courts. J Vet Behav 52–53:14–20. https://doi.org/10.1016/j.jveb.2022.04.001

Schoon A, van Dam A, Staal D (2022) Using trained dogs in rape and assault cases. In: Canines: the original biosensors, 1st edn. Jenny Stanford Publishing, pp 655–681

Schultz C, Topoleski JJ, Eckenrode B, Tipple C, Warren W, Rispoli M (2022) How science supports and law influences forensic canine utilization in the United States: considerations for human scent evidence (Trailing) and human remains detection canines. In: Canines the original biosensor, 1st edn. Jenny Stanford Publishing, p 62

Scientific Working Group for Dogs and Orthogonal device Guidelines Letter in response to the Lisa Lit et al. research study. SWGDOG.org

Settle RH, Sommerville BA, McCormick J, Broom DM (1994) Human scent matching using specially trained dogs. Anim Behav 48:1443–1448

Shirasu M, Touhara K (2011) The scent of disease: volatile organic compounds of the human body related to disease and disorder. J Biochem 150(3):257–266. https://doi.org/10.1093/jb/mvr090

Sisco E, Robinson EL (2020) Determination of ethanol concentration in alcoholic beverages by direct analysis in real time mass spectrometry (DART-MS). Forensic Chem 18:100219

Simon A (2020) A review of the types of training aids used for canine detection training. Front Vet Sci 7:10

Simon AG, DeGreeff LE (2019) Variation in the headspace of bulk hexamethylene triperoxide diamine (HMTD): Part II. analysis of non-detonable canine training aids. Forensic Chem 13:100155. https://doi.org/10.1016/j.forc.2019.100155

Simon AG, DeGreeff LE, Maughan M, Gadberry J (2018) Canine detection of explosives: Shifting focus from traditional to homemade explosives. U.S. Naval Research Laboratory, Washington, DC, Memorandum Report NRL/MR/6181–18-9794

Simon AG, Van Arsdale K, Barrow J, Wagner J (2021) Real-time monitoring of TATP released from PDMS-based canine training aids versus bulk TATP using DART-MS. Forensic Chem 23:100315. https://doi.org/10.1016/j.forc.2021.100315

Singletary M et al (2022) A novel method for training the interdiction of restricted and hazardous biological materials by detection dogs. Front Med 9:847620. https://doi.org/10.3389/fmed.2022.847620

Skalleberg AG, Bouzga MM (2016) Detecting and collecting traces of semen and blood from outdoor crime scenes using crime scene dogs and presumptive tests. Forensic Sci Int 264:146–152. https://doi.org/10.1016/j.forsciint.2016.04.026

Smith D, Španěl P (2011) Direct, rapid quantitative analyses of BVOCs using SIFT-MS and PTR-MS obviating sample collection. TrAC Trends Anal Chem 30(7):945–959. https://doi.org/10.1016/j.trac.2011.05.001

Smallegange RC, Verhulst NO, Takken W (2011) Sweaty skin: an invitation to bite? Trends Parasitol 27(4):143–148. https://doi.org/10.1016/j.pt.2010.12.009

Stadler S, Stefanuto P-H, Byer JD, Brokl M, Forbes S, Focant J-F (2012) Analysis of synthetic canine training aids by comprehensive two-dimensional gas chromatography—time of flight mass spectrometry. J Chromatogr A 1255:202–206. https://doi.org/10.1016/j.chroma.2012.04.001

Stadler S, Desaulniers J-P, Forbes SL (2015) Inter-year repeatability study of volatile organic compounds from surface decomposition of human analogues. Int J Legal Med 129(3):641–650. https://doi.org/10.1007/s00414-014-1024-y

Steen JB, Wilsson E (1990) How do dogs determine the direction of tracks? Acta Physiol Scand 139(4):531–534. https://doi.org/10.1111/j.1748-1716.1990.tb08956.x

Steinkamp FL, DeGreeff LE, Collins GE, Rose-Pehrsson SL (2016) Factors affecting the intramolecular decomposition of hexamethylene triperoxide diamine and implications for detection. J Chromatogr A 1451:83–90. https://doi.org/10.1016/j.chroma.2016.05.013

Stefanuto P-H et al (2015) Exploring new dimensions in cadaveric decomposition odour analysis. Anal Methods 7(6):2287–2294. https://doi.org/10.1039/C5AY00371G

Stockham RA, Slavin D, Kift W (2004) Specialized use of human scent in criminal investigations. Forensic Sci Commun 6(3)

Sulzer P (2014) A Proton Transfer Reaction-Quadrupole interface Time-Of-Flight Mass Spectrometer (PTR-QiTOF): high speed due to extreme sensitivity. Int J Mass Spectrom 5

SWGDOG SC8—Substance Detector Dogs Human Remains Detection (HRD) Land and Water (2009). https://www.nist.gov/system/files/documents/2018/04/25/swgdog_substance_detector_dogs_-_human_remains_detection_hrd_land_and_water.pdf

SWGDOG SC 9—HUMAN SCENT DOGS Searching for live people in disaster environments. SWGDOG (2010) . https://www.nist.gov/system/files/documents/2018/04/25/swgdog_human_scent_dogs_-_searching_for_live_people_in_disaster_environments.pdf

The State of Florida Urban Search & Rescue Response System Canine Search Specialist Training Student Manual (2004) Department of homeland security DHS federal emergency management agency

Thesen A, Steen JB, Doving K (1993) Behaviour of dogs during olfactory tracking. J Exp Biol 180:247–251

Turnbaugh PJ, Ley RE, Hamady M, Fraser-Liggett CM, Knight R, Gordon JI (2007) The human microbiome project. Nature 449(7164):804–810. https://doi.org/10.1038/nature06244

Thrailkill EA, Porritt F, Kacelnik A, Bouton ME (2018) Maintaining performance in searching dogs: evidence from a rat model that training to detect a second (irrelevant) stimulus can maintain search and detection responding. Behav Proc 157:161–170. https://doi.org/10.1016/j.beproc.2018.09.012

Tipple CA et al (2014) Comprehensive characterization of commercially available canine training aids. Forensic Sci Int 242:242–254. https://doi.org/10.1016/j.forsciint.2014.06.033

Titus KC, Gallegos SF, Prada-Tiedemann PA (2022) Forensic odor analysis: current application in postmortem examinations. RRFMS 12.1–12. https://doi.org/10.2147/RRFMS.S272225

United Nations Office on Drugs and Crime (2020) Drug detection dogs. https://www.unodc.org/unodc/en/data-and-analysis/bulletin/bulletin_1976-01-01_3_page005.html. Accessed 19 Jun 2020

United States v. Joshua Alan Wade (2009)

U.S. Drug Enforcement Agency, Fentanyl: a briefing guide for first responders (2017). https://www.dea.gov/druginfo/Fentanyl_BriefingGuideforFirstResponders_June2017.pdf. Accessed 30 Jun 2017

U.S. Customs and Border Protection (2022) Canine Program. U.S. Department of Homeland Security. https://www.cbp.gov/border-security/canine-program. Accessed 09 Jul 2022

US Legal (2022). https://uslegal.com/

van Dam A, Schoon A, Wierda SF, Heeringa E, Aalders MCG (2019) The use of crime scene detection dogs to locate semen stains on different type of fabric. Forensic Sci Int 302:109907. https://doi.org/10.1016/j.forsciint.2019.109907

Vaughan SR, Fulton AC, DeGreeff LE (2021a) Comparative analysis of vapor profiles of fentalogs and illicit fentanyl. Anal Bioanal Chem 413(28):7055–7062. https://doi.org/10.1007/s00216-021-03670-4

Vaughan SR, DeGreeff LE, Forte L, Holness HK, Furton KG (2021b) Identification of volatile components in the headspace of pharmaceutical-grade fentanyl. Forensic Chem 24:100331. https://doi.org/10.1016/j.forc.2021.100331

Vu D (2001) SPME/GC-MS characterization of volatiles associated with methamphetamine: toward the development of a pseudomethamphetamine training material. J Forensic Sci 46(5):1014–1024

Vu T Process for producing non-detonable training aid materials for detecting explosives. US 2014/0097551

Vyplelová P et al (2014) Individual human odor fallout as detected by trained canines. Forensic Sci Int 234:13–15. https://doi.org/10.1016/j.forsciint.2013.10.018

Waterbury A, Schultz C (2022) When does handler perception become canine detection? how applied behavior analysis and ethology can improve handler interpretation of canine behavior. In: Canines: the original biosensors, 1st edn. Jenny Stanford Publishing, pp 439–474

Waggoner P, Lazarowski L, Hutchings B, Angle C, Porritt F (2022) Effects of learning an increasing number of odors on olfactory learning, memory and generalization in detection dogs. Appl Anim Behav Sci 247:105568. https://doi.org/10.1016/j.applanim.2022.105568

Wells DL, Hepper PG (2003) Directional tracking in the domestic dog, Canis familiaris. Appl Anim Behav Sci 84(4):297–305. https://doi.org/10.1016/j.applanim.2003.08.009

Wescott DJ (2018) Recent advances in forensic anthropology: decomposition research. Forensic Sci Res 3(4):278–293. https://doi.org/10.1080/20961790.2018.1488571

Wilke K, Martin A, Terstegen L, Biel SS (2007) A short history of sweat gland biology. Int J Cosmet Sci 29(3):169–179. https://doi.org/10.1111/j.1467-2494.2007.00387.x

Wilhelm I, Bikelytė G, Wittek M, Härtel MAC, Röseling D, Klapötke TM (2022) Phlegmatization of TATP and HMTD with activated charcoal as training aid for explosive detection dogs. Propellants Explo Pyrotec 47(2). https://doi.org/10.1002/prep.202100057

Wojnowski W, Majchrzak T, Dymerski T, Gębicki J, Namieśnik J (2017) Dynamic headspace sampling as an initial step for sample preparation in chromatographic analysis. J Aoac Int 100(6):1599–1606. https://doi.org/10.5740/jaoacint.17-0206

Yeager K (2014) The burden of knowledge. Propellants, Explos, Pyrotech 39(2):159–160. https://doi.org/10.1002/prep.201480150

Detection of Human Diseases for Medical Diagnostics

Astrid Concha

Abstract

For centuries, the medical community has recognized variations in human body odor as a symptom of diseases. The ability of dogs to detect human diseases using their sense of smell is linked to volatile organic compounds (VOCs). VOCs represent a wide range of chemicals, volatile at ambient temperature and pressure conditions, detectable in sweat, breath, urine, feces, and other fluids. In the last two decades, several peer-reviewed articles have been published reporting scientific evidence demonstrating the effectiveness of dogs in the detection of a variety of human diseases. Biomedical detection dogs have shown their ability to detect VOC signatures associated with human diseases as well as support the development of biomarker profiles for clinical diagnostic instruments.This chapter highlights the current state of research evaluating dogs' olfactory capabilities to detect human diseases. The majority of studies have evaluated dogs as detectors of a variety of forms of cancers including prostate, breast, lung, ovarian, bladder, and skin, among others. Similarly, a few infectious diseases have been assessed using detection dogs such as bacteriuria, clostridioides, and protozoal infections. Research on biomedical detection dogs has shown promising results, however, it is difficult to directly compare study results since they differ methodologically in many aspects. This chapter outlines key factors associated with training, testing, the type of odor sample materials, and the dogs' performance in terms of sensitivity and specificity. These key factors are used as points for comparison among the studies. Furthermore, methodological challenges and dogs' individual characteristics are discussed to

A. Concha (✉)
Canine Behavior Consultancy and Research, Individual K9, Santiago, Chile
e-mail: astridconcha@individualk9.com

L. Lazarowski (ed.), *Olfactory Research in Dogs*,
https://doi.org/10.1007/978-3-031-39370-9_12

be monitored for the success of biomedical dogs as a screening tool for the early detection of human diseases.

Keywords

Biomedical detection dogs • Olfaction • Human diseases • Volatile organic compounds • Cancer

1 Introduction

For a wide range of animals, especially social species, there is evidence showing their ability to recognize health status of their conspecifics through olfactory cues (Kavaliers et al. 2005; Arakawa et al. 2009). This olfactory communication not only allows the receiver individual to detect health status of potential rivals and mates, but also to detect those sick (Kiesecker et al. 1999; Arakawa et al. 2011; Ferkin 2018). Thus, the detection of olfactory cues indicating illness would enable the receiver to avoid potential infectious transmission. Nevertheless, the behavioral response of the healthy receiver toward a sick individual may vary from displaying social avoidance to affiliative behaviors (Renault et al. 2008, 2008; Boillat et al. 2015; Willette et al. 2007; Aubert and Richard 2008).

In the human medical community, variations in human body odor have been used as a symptom for the diagnoses of diseases for centuries. Physicians often came to recognize odors associated with specific diseases, for instance, a sweetish odor is described in cholera, sweet-fruity acetone-like breath suggesting unbalanced diabetes, a stale beer odor associated with tuberculosis, and a fishy reek with liver diseases (Phillips 1992; Shirasu and Touhara 2011). The ability of humans to diagnose disease by smelling has only rarely been the subject of quantitative studies. Observational studies have shown that healthy individuals perceive the body odor (e.g., axillar, saliva, and urine) of sick people as more intense and aversive compared to healthy ones (Moshkin et al. 2012; Olsson et al. 2014; Gordon et al. 2008). However, the ability to discriminate human body odor between healthy versus ill people depends on the degree of sickness symptoms, type of trigger of illness response, and multisensory integration combining sight and smell (Regenbogen et al. 2017).

A famous case reporting human capacity to smell disease in people is Joy Milne, a 72-year-old woman from Scotland, who noticed changes in her husband's body odor, primarily around his shoulders and the back of his neck. As the years passed, he was eventually diagnosed with Parkinson's disease. Joy Milne has hereditary hyperosmia (i.e., a heightened or increased sense of smell), a rare condition that gives her superior olfactory abilities compared to the normal human population. Humans are considered to have a poorly developed sense of smell that is clearly inferior to some nonhuman animals. This is probably, at least in part, why animals such as dogs have been employed to assist in the detection of human diseases (Edwards et al. 2017).

The first publication regarding the olfactory ability of dogs to detect human diseases was reported in The Lancet by Williams and Pembroke (1989). The authors described a case of a Border Collie and Doberman Pincher cross that spontaneously showed an inordinate interest in sniffing a skin lesion on the owner's leg, which thereafter was removed, and was found to be diagnosed as malignant melanoma. A decade later in the same journal, Church and Williams (2001) not only reported another anecdotic case of a pet dog constantly sniffing at a patch of eczema on his owner's thigh, which after excision was histologically diagnosed as a carcinoma, but also mentioned for the first time a dog trained for the detection of skin cancers suggesting that the dog's ability to detect unique odors of certain cancers can be taught and, therefore, it should be investigated through rigorously controlled experiments. In 2004, the first two experimental studies were published providing proof of the principle that dogs could be trained to detect cancer odors. Pickel et al. utilized two dogs already trained in explosive detection and hunting and trained them to detect samples from melanoma tissue. Following the training, dogs were tested sniffing at the skin of seven patient volunteers with some clinical suspicion of skin melanoma, showing high accuracy in confirming six positive patients out of seven (subsequently biopsy-proven). Similarly, Willis et al. published a controlled, blinded study where six trained dogs performed better than chance alone in detecting and discriminating bladder cancer urine samples from a range of controls obtained from age-matched diseased and healthy subjects.

These studies provided a benchmark and starting point for investigating the use of medical detection dogs. Over the past two decades, the scientific evidence has increased showing that trained dogs can successfully detect a variety of human diseases including different types of cancers such as skin (Picket et al. 2004), bladder (Willis et al. 2010), lung (Riedlova et al. 2022), prostate (Guest et al. 2021), breast (Kure et al. 2021), cervical (Yamamoto et al. 2020), ovarian (Horvath et al. 2010), and colorectal (Schoon et al. 2020), as well as diseases cause by virus, bacterial, and parasite diseases such as COVID-19 (Sakr et al. 2022), clostridium difficile (Bomers et al. 2012), urological infections (Maurer et al. 2016), and malaria (Guest et al. 2019).

The ability of dogs to detect diseases using their sense of smell is linked to volatile organic compounds (VOCs). VOCs represent a wide range of chemicals, volatile at ambient temperature and pressure conditions, detectable in sweat, breath, urine, feces, and other fluids. These VOCs are in general the end products of metabolism of carbohydrate and lipid, aerobic and anaerobic fermentation processes of microbial pathogens/commensals, produced by the body in response to microbial infections (e.g., inflammatory response), oxidative stress, glandular secretion, and external absorption, which result in the complex mixture composition of the human odor. In addition, VOCs emitted from different areas of the human body vary with age, sex, diet, genetics, and physiological/metabolic body status (Shirasu and Touhara 2011).

The investigation of VOCs using both instrumental methods such as SPME/GC–MS (solid-phase microextraction/gas chromatography-mass spectrometry) has

Table 1 Comparison of potential advantages and disadvantages between detection dogs and mass spectrometry (MS) methods for screening of human diseases (modified from Lippi and Haney 2019)

Characteristics	Dogs	MS
Initial cost	++	+++
Operating costs	+	++
Environment adaptation	+	+++
Portability	+++	−/+
Pre-analytical sample precautions	+	+++
Efforts for calibration (or dog training)	+++	+
Standardization	−	+/++
Possibility of quality control assessment	−/+	+++
Early diagnosis	++	+
Overall diagnostic performance	+++	+++
Inter-assay (or individual) imprecision	+++	+

shown correlation between the pattern and concentration of VOCs with some diseases, and therefore biomarkers from these VOCs can be developed (Phillips et al. 2004; Siegel et al. 2017; Probert et al. 2009; Hakim et al. 2012; Lima et al. 2019; Guest et al. 2021; Giró Benet et al. 2022). Biomarkers are defined characteristics that are measured as indicators of biological and pathological processes in the human body. These biomarkers can be obtained from blood, urine, breath, sweat, etc. (Farha and Salami 2022).

Biomarker VOCs not only provide a non-invasive and rapid method of diagnosis, but also may reflect disease progression and response to therapeutic intervention. However, despite progress in the identification of medical biomarkers using instrumental techniques, a wide practical use is still limited by costs and by the interfering effect of non-disease specific related VOCs present in the sample (e.g., breath and urine) such as inflammatory or metabolic products, but also physiological factors such as dietary and smoking habits (Fischer et al. 2015). Although the practical and wide application of dogs to assist in the diagnosis of human disease is still far from a routine screening method, scientific evidence has demonstrated the potential advantages of dogs over instruments (Table 1; Lippi and Heaney 2020). For instance, dogs' mobility enables detection in different sites outside of a laboratory and the response given by a dog to a positive sample is of a binary nature, meaning that it is a yes/no response facilitating results interpretation and data collection (Jezierski et al. 2015). Another advantage of using dogs for medical diagnosis of human diseases relies on the combination of the acuity of their sense of smell, such as threshold detection sensitivity (e.g., Concha et al. 2014), together with their ability to recognize subtle variations in odor background (Guest et al. 2020), and to learn by associative learning of odors leading to olfactory representation of a complex mixture of VOCs (Pirrone and Albertini 2017). The olfactory representations are the neural characterizations on which olfactory decisions are based and are processed by multisensorial cortical pathways (Invitto et al. 2019). Thus, olfactory representations are of features of molecules rather than of molecule identity per se (Weiss et al. 2012). This gives dogs a unique characteristic over instruments as an odor profile of a human disease might depend on a combinatorial mixture in perceptual representation rather than the identification of a mixture of molecules. Indeed, the study performed by Guest et al. (2021) using a multiparametric approach to create a non-invasive urine diagnostic method for prostate cancer showed that there may be information synthesized by trained dogs related to cancer odor signature that may not be identified by molecular currently available biomarkers analysis (e.g., GC–MS). The authors describe "the challenge remains on how to port canine intelligence into machine olfactors."

In general, biomedical or medical detection dogs are trained and tested under laboratory settings to search a determined number of odor samples collected from patients and healthy volunteers using exhaled breath, blood serum, urine, etc. The dog discriminates between the target sample (i.e., the conditioned odor of a human disease) and non-target samples (i.e., control odors) using a reward-based approach with food or toy rewards. Odor samples are placed next to each other on a non-restrictive searching system such as multichoice carousel and line-ups (Concha et al. 2019; Willis et al. 2004; Guest et al. 2021). The dog has to identify the target sample by showing a trained alert response (e.g., sit or down) and ignore the non-target samples. In a biomedical detection scenario, dogs detect disease biomarkers in human samples, which may relate to a particular cancer, bacterial or viral infections.

In the last two decades, several peer-reviewed articles have been published reporting scientific evidence demonstrating the effectiveness of dogs for the detection of a variety of human diseases. Biomedical detection dogs have shown their ability to detect VOC signatures associated with human diseases as well as support the development of biomarker profiles for clinical diagnostic instruments. However, it is difficult to directly compare study results since they differ methodologically in many aspects even when the same type of disease has been investigated; for instance, studies in lung cancer patients have been performed using breath, urine, and blood samples (e.g., Buszewski et al. 2012; Amundsen et al. 2014; Riedlova et al. 2022). These methodological differences are expected since the first decade of experiments in this field were focused on the dogs' ability to detect different types of human cancers, which showed some evidence of experimental reproducibility (see review Moser and McCulloch 2010). Since the last decade, studies not only began to investigate the use of dogs for the detection of other human diseases such as bacteria (Maurer et al. 2016; van Bomers et al. 2014; Davis et al. 2019a, b), viruses (Maughan et al. 2022; Sakr et al. 2022), and protozoa (Guest et al. 2019), but also how to develop a consistent training criterion for testing (Crawford et al. 2022).

This chapter discusses the current state of research in medical detection dogs for human diseases including different types of cancers and infectious diseases.

2 Cancer Detection

Cancer is a leading cause of death globally, with one in five people developing cancer during their lifetime (World Health Organization). There were an estimated 19.3 million new cases and 10 million cancer deaths in 2020 (Sung et al. 2021). Globally, the most frequently diagnosed cancer in women are breast and cervical cancers followed by colorectal and lung. Breast cancer and cervical also lead to the mortality rate. In men, prostate cancer is the most commonly diagnosed cancer, followed by lung, colorectal, and liver. Mortality is led by lung cancer followed by colorectal (Sung et al. 2021). According to the World Health Organization, at least 40% of all cancer cases could be avoided by prevention measures and further mortality could be reduced through early detection. Nevertheless, current early cancer detection is still limited, for instance, high-resolution scanning technologies and serum biomarker tests are highly accurate but can be costly and false positive results lead to unnecessary biopsy procedures.

The initial interest in the possibility that dogs might be able to detect cancer in humans began with case reports of people's pet dogs causing concern due to spontaneous excessive sniffing interest in skin lesions that were later determined to be cancerous (William and Pembroke 1989; Church and Williams 2001). Since then, vast robust scientific evidence has demonstrated the ability of dogs to screen samples from patients with different types of cancer from healthy and age-symptoms-matched controls including melanoma (Picket et al. 2004), bladder (Willis et al. 2004), lung (McCulloch et al. 2006), breast (McCulloch et al. 2006), colorectal (Sonoda et al. 2011), prostate (Gordon et al. 2008), ovarian (Horvath et al. 2008) and cervical (Yamamoto et al. 2020).

Here we discuss a variety of studies performed to assess the ability of dogs to detect cancer in humans. As parameters of the dog's detection accuracy, sensitivity, and specificity are described. Sensitivity is defined as the proportion of disease-positive samples (i.e., true positives) identified as positives by the dog, while specificity refers to the proportion of disease-negative samples or controls (i.e., true negatives) that are accurately identified as negatives by the dog. In other words, sensitivity measures how good the dog is at identifying the patients with cancer, whereas specificity assesses how good the dog is at ignoring the non-cancer patient samples; low specificity indicates a high rate of false positive indications (i.e., the dog indicates a patient's sample as cancerous/positive while it is negative).

2.1 Prostate Cancer

Prostate cancer is the second most frequent cancer and the fifth leading cause of cancer death among men, with an estimated 1.4 million new cases and 375,000 deaths worldwide in 2020 (Sung et al. 2021). The current standard screening method for prostate cancer is the prostate-specific antigen (PSA) blood test routinely used as an indicator, not diagnostic. Although PSA accuracy in symptomatic patients has been reported as high sensitivity and low specificity in the detection of prostate cancer, published scientific evidence shows that the sensitivity of PSA in symptomatic patients is unknown and likely to be lower than reported (Merriel et al. 2022). The low specificity of PSA tests results from false positives from non-cancer conditions affecting the prostate such as prostatitis and benign prostatic hypertrophy leading to unnecessary prostate biopsy. Given the limitations of PSA, additional diagnostic is needed for early accurate detection. Biomarkers of prostate cancer have been studied for the development of more specific and non-invasive methods of diagnosis and prognostic. Several molecular biomarkers are currently commercially available using different sample sources, such as serum, tissue, and urine. Further evaluation is needed for the selection of the appropriate biomarker depending on the clinical scenario (Farah and Salami 2022; Fujita and Nonomura 2018).

A potential non-invasive tool to assist in the improvement of prostate cancer diagnosis is trained detection dogs. To the best of the author's knowledge, six peer-reviewed publications on prostate cancer detection by trained dogs have been published so far (Table 2). The first study was published by Gordon et al. (2008), where four dogs were trained to detect urine samples from a patient with prostate cancer among six samples from healthy volunteers (age and sex-matched). The dogs were trained by their owners at different locations using clicker training and food as a reward for alerting the cancer samples. The results showed a success rate of 18%, none of the dogs performed significantly better than chance in identifying the positive cancer samples (sensitivity) and two dogs did better than chance in recognizing all the control samples as being negative (specificity). The authors mentioned that better management of the urine samples should have been done, but they do not discuss more about the reason for this statement and whether this might have been due to the storage conditions or sample handling. The urine samples were stored for a maximum of 5 months at −20C and then stored in freezers at the trainer's home for an undefined time. This management could have affected the quality of the urine samples as pH changes resulting from variations in $C0_2$ concentrations by freezing procedures and therefore, it is recommended that urine samples should be frozen at −20C and transferred to lower storage temperatures (−80C) within a week (Rist et al. 2013). Another reason for these unsatisfactory results discussed by the authors is a lack of a more rigorous training protocol for dogs. This could suggest that the training quality done by the owners could have been different, for instance in terms of timing and unintentional cues given to the dog. Furthermore, these results could also have been due to behavioral differences among the dog breeds trained for the study (Chihuahua mix, Pembroke Welsh Corgi, Goldendoodle, and Border Collie).

Cornu et al. (2011) used a Belgian Malinois belonging to the French Army Veterinary Department. The dog was professionally trained to detect prostate cancer in urine samples based on operant conditioning using a clicker and a ball as a reward for alerting to a positive sample. After the training phase of 16 months, a double-blinded test was performed; the dog discriminated one positive cancer sample among 5 controls (age and symptom matched). A total of 33 positive prostate

Table 2 Detection of prostate cancer by dogs

Author s	Year	Sample type	Number of dogs	Testing methods	Number of samples explored per run (+ Cancer, C Control)	Sensitivity	Specificity	Total positive samples & controls	Match symptoms control samples	Test design Double blind
Gordon et al.	2008	Urine	4	Lineup	1+ : 6 C	18% (None of the dogs better than chance	2 out of 4	11+ : 66C	Yes	Yes
Cornu et al.	2011	Urine	1	Lineup	1+ : 5C	91%	91%	33+ :33C	Yes	Yes
Elliker et al.	2014	Urine	2 (10 trained, 2 tested)	Lineup	1+ :3C	19%	73%	Test 1: 15+ : 45C Test 2 & 3: 16+ :48C	Yes	Yes
Taverna et al.	2015	Urine	2	Not indicated	0–6+ : 6–0 C	99%	97%	362+ : 540C	Y/N 23% of controls were samples from females	Yes
Urbanova et al.	2015	Urine	1	Lineup	1+ :2 C	94%	92%	45+ : 25C	Yes	Unclear
Guest et al.	2021	Urine	2	Lineup	1+ : 3 C	71.40%	76.20%	14+ : 41C	Yes	Yes

cancer samples against 33 controls were screened by the dog, positively identifying 30 out of the 33 cases (91% sensitivity). The dogs only made 3 false alerts (i.e., performed the trained alert response on non-cancer sample), however, one of these samples that initially was classified as healthy control was diagnosed with prostate cancer after biopsy was taken, confirming the dog's positive indication as correct. Despite only one dog being trained and tested, this study provided successful evidence in the detection of prostate cancer using urine samples. The authors indicated that some confounding factors in the study were the limited number of samples and the fact that they only selected positive patients and healthy controls over 50 years old, suggesting that results might not be generalizable to a wider population, therefore limiting the potential for clinical application.

A later study by Elliker et al. (2014) trained ten dogs using clicker training with a food reward to detect a wide range of urine samples from patients diagnosed with prostate cancer and controls all collected from Addembrooke's Hospital in the United Kingdom. The use of only one location for sample collection aimed to eliminate contamination or changes in the background odor associated with the ambient air of the clinical settings from where samples are collected. Dogs were trained to detect one positive cancer sample among 3 controls. A total of 50 prostate cancer samples with different degrees from small tumors to metastasized cancers, and 67 controls including healthy and matched symptom volunteers. Only two dogs (out of 10) succeeded in training, a 9-year-old Labrador and a 3-year-old Border Collie. The authors mentioned that the dogs failed to progress in training because they did not have optimal temperament for this type of detection task, though these dogs were recruited from a pool of dogs attending classes at a dog training center and were selected based on the professional opinion of dog trainers or behavioral scientists. The two remaining dogs were double-blind tested using 31 cancer samples and 93 controls. Dogs failed to indicate positive samples more frequently than expected by chance with a sensitivity of 13–25%, but specificity was relatively high (71–75%). Authors discussed that the unsatisfactory results could be the consequence of the memorization of samples used during the training phase and therefore, dogs were not able to generalize to a new sample set. A few studies have investigated dogs' capacity to remember odors; we know that dogs can learn 10, 20, and 40 different odors and retain them with an accuracy between 80 and 100% even after an interval of several months (William and Johnson 2002; Ka-ho et al. 2020; Waggoner et al. 2022). Thus, the maximum number of odors that a dog could learn to detect and memorize is still an open scientific question.

Taverna et al. (2015a, b) used two, three-year-old German Shepherds explosive detection dogs, and trained them to detect prostate cancer in urine samples using clicker training. Dogs were tested using 920 urine samples, 362 positive prostate cancer, and 540 controls. To identify confounding factors, prostate cancer samples were categorized into 5 groups as follows: (1) had been treated with surgery (n = 180), (2) had increased serum PSA (>2.5 ng/ml) and histological diagnosis of prostate cancer (n = 120), (3) had prostate cancer detected incidentally (n = 22), (4) had metastatic prostate cancer or received hormonal therapy (n = 29), and (5) had primary prostate cancer and another tumor (n = 11). Control samples included: (1) 120 samples from female volunteers including (i) healthy nonpregnant younger and older women ($n = 50$) and (ii) women with nonneoplastic conditions such as urolithiasis, neurological or metabolic disorders, urinary infections, and other cancers ($n = 72$); (2) 320 samples from males that were i) healthy and young ($n = 60$); (ii) older with no family history of prostate cancer, PSA < 2.5 ng/mL, and did not have urological or system disease ($n = 240$); (iii) treated with transurethral prostate resection for benign prostate hyperplasia ($n = 40$); (iv) non-prostatic cancer ($n = 78$). Controls from females against prostate cancer samples were used in the first phase of training to avoid dogs getting confused or learning to detect other non-prostatic cancer VOCs. The use of opposite sex has been avoided in other studies as dogs may be able to detect VOCs that differentiate males from females (Willis et al. 2004; Mazzatenta et al. 2021). This training protocol may have facilitated better discrimination and learning to detect specific VOCs associated with prostate cancer and therefore, successfully identifying the correct samples among controls. In the double-blind test phase, six samples were searched by the dogs, but the number of positive prostate cancer presented in each run varied from 0 to 6, meaning that blank runs or empty lineups which contained only controls (i.e., the positive target sample is not present) were presented to the dogs, and in other runs the whole lineup included only positive cancer samples without controls. This was the first study describing the use of blank runs or empty lineups in cancer detection dogs. A blank run trains the dog to avoid expecting to find a target every time, which may reduce the occurrence of taking a chance.

To reduce the likelihood of memorization of odors by the dogs, 55.25% ($n = 200$) of the positive prostate samples used during training were not reused during the testing phases. Similarly, 42.59% ($n = 230$) of the controls were not reused. The first dog achieved a sensitivity of 100 and 98.7%, specificity, and for the second dog sensitivity was 98.6% and specificity of 97.6%. Interestingly, when female controls were excluded the first dog's sensitivity dropped to 98%, and specificity did not show variation. Likewise, when controls of 45-year-olds were included the second dog slightly dropped in sensitivity and specificity to 96.9 and 96%, respectively. This study did not find an association between dogs' performance and the prostate cancer stage, nor different therapeutics that patients underwent.

Another study in prostate cancer detection also showed a high sensitivity (93.5%) and specificity (91.6%) using a 15-month-old German Shepherd, which was previously trained in scent work and obedience since the dog was 7 months old (Urbanova et al. 2015). This study showed that a dog can be trained as a biomedical detection dog from a young age to avoid undesirable behaviors arising from other trainings with different aims that could interfere with this type of detection task. However, as described by the authors, a motivated young dog can also bring some challenges such as inattentiveness and restlessness while working.

Finally, a more recent study by Guest et al. (2021) used a multiparametric approach to investigate a more effective and non-invasive early diagnostic method for the highly aggressive histology of prostate cancer (e.g., Gleason 9). This study combined medical detection dogs, VOCs, microbiota (i.e., group of microorganisms associated with a specific niche (Banerjee and Robertson 2019)), and profiling analysis as well as artificial neural network (ANN) trained to emulate the dog olfactory capability. Two dogs, a four-year-old Labrador and a seven-year-old Wire-Haired Hungarian Vizsla previously trained to detect prostate cancer in urine samples from a single source (positive and control samples collected from Milton Keynes University Hospital, UK) were trained using samples from another source. These samples included 7 positive prostate cancer (Gleason 9) and 21 controls. The dogs were able to detect Gleason 9 prostate cancer versus biopsy-negative controls with a sensitivity of 71.4% for both the dogs, and specificity of 76.2% in dog 1 and 70% in dog 2. This study identified differences in VOCs between cancer and biopsy-negative controls using gas chromatography–mass spectroscopy (GC–MS). Nevertheless, there were no differences in the microbiota profiling except by 16S rDNA sequence between the positive and negative samples. Finally, the trained ANN identified regions of interest in the GC–MS database on the dogs' detection performance. Results suggested that there may be information synthesized by the dog regarding the cancer VOCs that may be identified by conventional single molecular biomarker analysis. This is the first study providing evidence of the use of medical detection dogs together with a multiparametric approach to develop machine-based olfactory diagnostic tool for prostate cancer.

2.2 Breast Cancer

Female breast cancer is the fifth leading cause of cancer mortality worldwide. It was the leading cause of global cancer incidence in 2020, with an estimated 2.3 million new cases, equivalent to 11.7% of all cancer cases (Sung et al. 2021). Mammography (MG) is currently the most common screening test for breast cancer, with a reported sensitivity of 77% and specificity of 91% (Ohuchi et al. 2016) and it is known to be less sensitive in dense breast (Tohno et al. 2009). A noninvasive method for the diagnosis of breast cancer would be VOC biomarkers (Yuan et al. 2019; Huang et al. 2016; More et al. 2018). Phillip et al. (2010) analyzed breath VOCs in 54 women diagnosed with breast cancer and 204 cancer-free controls using GC–MS. They found a set of VOCs that differed between positive and control breath samples with a sensitivity of 78.5 and 84.4% specificity. Likewise, volatile organic metabolites (VOMs) were investigated in urine collected from 26 patients with breast cancer and 21 healthy controls. Six VOMs among 79 selected were statistically different between positive and control patients (Silva et al. 2012).

To date, four studies have evaluated the use of dogs for the detection of breast cancer in women using VOCs contained in exhaled breath and urine samples (Table 3). The first study published was done by McCulloch et al. (2006), where breath samples were collected using cylindrical, polypropylene, organic vapor sampling tubes (Defencetek, South Africa), which contained a silicon oil-coated polypropylene wool fiber that capture the VOCs in exhaled breath. A total of 31 positive breast cancer samples and 83 healthy volunteers (not age/symptom

Table 3 Detection of Breast cancer by dogs

Authors	Year	Sample type	Number of dogs	Testing methods	Number of samples explored per run (+ Cancer, C Control)	Sensitivity	Specificity	Total positive samples & controls	Match symptoms control samples	Test design double blind
McCulloch et al.	2006	Breath	5	Lineup	1+ : 4C	88%	98%	25+ : 66C	No	Yes
Gordon et al.	2008	Urine	6	Lineup	1+ : 6C	22% No dogs better than chance	2 out 6 dogs better than chance	9+ : 54C	Unclear	Yes
Walczak et al.	2012	Breath	3	Lineup	1+ : 4C	89–90%	84–88%	29+ : Unclear C	Unclear	Yes
Kure et al.	2021	Urine	1	Circle	1+ : 4C	100%	100%	40+ : 150C	Yes	Yes

matched) were available for use in training. Five dogs, three Labrador Retrievers and two Portuguese Water Dogs, were trained using clicker training with a food reward to detect both breast and lung cancer for three weeks. Following the training phase, a single-blind and then double-blind tests were performed; in both experiments, dogs alerted to one positive cancer sample among 4 controls. All samples used in the double-blinded test were a completely different set of subjects not previously encountered by the dogs. A total of 6 breast cancer-positive samples and 17 controls were used for the double-blind test. A difference from the previous training and the single-blind test was that the dogs were not rewarded when a positive sample was alerted to. After performing the trained alert response, the dogs were removed from the room and praised outside. This study found a sensitivity of 88% and specificity of 98%. The detection sensitivity for breast cancer was 10% lower than for lung cancer performed with the same dogs. Authors discussed that the specificity found in this study may be overestimated as they used only healthy controls. It is argued that dogs could have learned VOCs associated with the cancer process such as inflammation, necrosis, and infections. Therefore, further research should include symptom-matched samples. This study was the first one introducing the use of exhaled breath samples for cancer detection by dogs. Confounding factors that should be taken into consideration when using breath samples include smoking habits, dental infection, diabetes, and food as these can influence the VOCs present in the sample beyond the presence or absence of cancer.

Gordon et al. (2008) was the first study investigating breast cancer detected in urine samples. Six dogs were trained by their owners to detect one breast cancer-positive sample among six healthy controls (age and sex matched). In the testing phase, a total of 18 runs were carried out. However, none of the six dogs performed better than chance, the mean sensitivity was 22%. Similarly, two out of six dogs showed specificity better than chance, with a mean of 33.3%. One possible explanation for the unsatisfactory results of this study is the underlying variation in disease aggressiveness within the pool of patient samples presented to the dog. Very different results were obtained by Kure et al. (2021), overall showing both sensitivity and specificity of 100% in a double-blind setting. In this study, a 9-year-old Labrador Retriever trained to detect several types of human cancers was trained for 12 months to detect breast cancer in urine samples (Sonoda et al. 2011). Interestingly, the authors mentioned that under certain stressful conditions such as high temperature (26.3–26.8 °C) and high humidity (83% RH) the dog detection test could not be conducted because the dog could not focus on the task. A recent study in detection dogs showed that temperatures over 26.6 °C (80 °F) and humidity of 40% increase the time for a dog to engage in a search task and there is substantial reduction of search performance (DHS S&T 2022). Kure et al. (2021) bring up evidence that adjusting training and testing to the dog's needs, such as reducing training time or avoiding training during high temperature, results in performance with higher sensitivity and specificity. Authors also suggested that biomedical dogs should be trained in different conditions to habituate them to a variety of environments when the aim is operational scenarios.

2.3 Lung Cancer

Lung cancer is the second most commonly diagnosed cancer and the leading cause of cancer death in 2020, with an estimated 2.2 million new cancer cases and 1.8 million deaths. Lung cancer represents approximately 1 in 10 cancers diagnosed and 1 in 5 deaths. Lung cancer is the leading cause of mortality in men, while in women, it is the second cause of cancer deaths after breast cancer (Sung et al. 2021). Unfortunately, the lack of symptoms at early stages of the disease that would warn the patient to visit a doctor results in a poor prognosis (Christensen et al. 1997). An early diagnosis is fundamental to improve the five-year survival rate. If lung cancer is diagnosed earlier the rate of survival could increase to 70, or 54% when the cancer is still localized, but only 15% are identified at an early stage (American Cancer Society). The current diagnosis of lung cancer includes different types of imaging, such as low-dose computed tomography (LDCT), which has shown a sensitivity of 85% and specificity of 99% (Horeweg et al. 2014). Nevertheless, the overall false positive rate has been found to be between 81 and 96% (Gartman et al. 2018). This high false positive rate results in costly screening and invasive procedures (e.g., biopsy) in patients free of lung cancer. Another common method of diagnosing is the cytological examination of sputum which has been found to be a helpful screening method for early detection, however, peripheral tumors can be missed (e.g., adenocarcinoma arising from smaller airways) when this diagnostic method is used. The early diagnosis of lung cancer remains a challenge because most of the available clinical methods can detect cancer in advanced stages. Although biomarkers for lung cancer are an emergent field, more studies are needed to validate their use consistently (Nooreldeen and Bach 2021). Therefore, regular non-invasive and low-cost screening for the early diagnosis that facilitates a successful therapeutical intervention for reducing the mortality rate is still needed.

Studies investigating detection dogs for lung cancer screening started with McCulloch et al. (2006) (Table 4), where the same five dogs trained to detect breast cancer in exhaled breath samples were also trained in lung cancer following the same procedure (see Sect. 2.2). Dogs were tested in a double-blind trial to discriminate 55 non-small cell lung cancer (NSCLC) patients from 83 healthy controls. The overall sensitivity and specificity were both 99%, and no significant difference was found in detection accuracy among the dogs across the stages of lung cancer samples (i.e., stages 1–4 of adenocarcinoma and stages 2–4 of squamous cell carcinoma). The results of this study suggest that dogs may be able to detect early preclinical stages of lung cancer. These results contrast those reported by Walczak (2009) using the same breath sample polypropylene tubes (Defencetek, South Africa); 3 Labrador Retrievers were trained and tested, finding a sensitivity of between 53 and 58% in detecting lung cancer. This study evidenced that collecting exhaled breath samples from cancer patients outside of hospital rooms significantly decreased the odds ratio for positive indication by dogs and increased the odds ratio for false indication of healthy controls. Confounders can be a consequence of a characteristic "hospital odor" derived, for instance, from disinfectants

Table 4 Detection of lung cancer by dogs

Authors	Year	Sample type	Number of dogs	Testing methods	Number of samples explored per run (+ Cancer, C Control)	Sensitivity	Specificity	Total positive samples & controls	Match symptoms control samples	Test design double blind
McCulloch et al.	2006	Breath	6	Lineup	1+ : 4C	99%	99%	27+ : 66C	No	Yes
Buszewski et al.	2012	Breath	Unclear	Unclear	1+ : 4C	82%	82%	Unclear	No	Unclear
Ehmnan et al.	2012	Breath	4	Circle	1+ : 4C	71%	93%	35+ : 60C	Yes	Yes
Walczak et al.	2009	Breath	6 trained 3 tested	Circle	1+ : 4C	83–84% 53–58%	78–81%	118+ : 305C	No	Yes
Amundsen et al.	2014	Breath & urine	4	Circle	0–6+ ; 6–0 C	Breath: Test 1 65% Test 2 56% Urine: Test 1 74% Test 2 64%	Breath Test 1 8% Test 2 33% Urine Test 1 25% Test 2 29%	20+ : Unclear C	Yes	Unclear
Rudnicka et al.	2014	Breath	2	Unclear	1+ : 4C	86%	72%	118+ : 305C	No	Yes
Fisher-Tenhagen et al.	2018	Breath	2	Lineup	0–4+ : 4–0 C			9+ : 10C	No	Unclear
Junqueira et al.	2019	Blood serum	3	Unclear	1+ : 4C	96.70%	97.50%	10+ : 40C	No	Yes
Riedlova et al.	2022	Blood serum & breath	2	Lineup	1+ : 4C	Serum: 69 and 62% Breath: 62 and 75%	Serum: 67 and 97% Breath: 67 and 90%	47+ : 41C	No	Yes

or other odors in the room. These background odors can become part of the "odor signature" learned by the dog to discriminate positive samples from controls. To minimize background variability, controls should be collected from the same clinical environment as target odors and ideally at the same time. When this is not possible, maximum variability should be done to make sure that the dog is learning only the "disease" odor regardless of the background odor (Edward et al. 2017).

Another study in lung cancer detection by dogs by Ehmann et al. (2012) reported that trained dogs successfully discriminated lung cancers from chronic obstructive pulmonary disease (COPD) in breath samples. COPD is characterized by lung function deterioration, chronic airway inflammation, and structural changes in lung parenchyma (Bollmeier et al. 2020). Four family dogs were trained by a professional dog trainer for six months using reward-based training to indicate lung cancer in breath samples collected using cylindrical glass tubes closed at both ends by removable caps (Gassner Glastechnik GmbH, Germany). The tubes were filled with polypropylene fleece (one hydrophilic and one hydrophobic) that were impregnated with silicone oil to have either hydrophilic or hydrophobic absorbing properties. Three double-blind tests were performed to assess the dogs' ability to (1) detect positive samples of lung cancer patients ($n = 25$) among four healthy controls ($n = 50$); (2) discriminate lung cancer from COPD patients ($n = 50$); and (3) discriminate lung cancer from a mix of healthy controls and COPD patients. Overall sensitivity was 71% and specificity 93%. They found that dogs' performance was not affected by people's smoking and nutritional habits. However, the authors reported potential confounders resulting from medications used by patients such as metoclopramide, verapamil, metoprolol among another six drugs. Montes et al. (2017) followed the same methodology for sampling exhaled breath to train and test a three-year-old Labrador–Pitbull cross to detect lung cancer. The dog was exposed 785 times using 85 cancer-positive patients and 28 controls (11 patients without lung cancer and 17 healthy volunteers) showing an overall sensitivity of 95% and specificity of 98%.

Less optimistic results were obtained by Amundsen et al. (2014) investigating whether dogs can discriminate patients with malignant lung cancer from those with different benign lung conditions, including small cell lung cancer (SCLC), non-small-cell lung cancer (NSCLC), non-lung cancer (pulmonary, mesothelioma, carcinoid, and lung metastasis from other primary neoplasm), and non-cancer. Patient samples were collected from exhaled breath using sterile exhalation filters (Breath Filter Allegro Medical Inc, US) and urine samples. Four dogs were trained, a Belgian Shepherd, Border Collie, Dachshund, and a Rottweiler. Dogs were exposed to detect one positive cancer sample among 0–6 control samples placed in a circle. In an interim analysis of the first 46 patients, sensitivity was 70% for non-small-cell lung cancer (NSCLC) and 55.6% for small-cell lung cancer (SCLC), while the specificity for both lung cancer types was 8.3%. After intensive training, detection sensitivity for NSCLC decreased to 60% and specificity increased to 33.3%, whereas sensitivity for SCLC increased to 100% and specificity also increased to 33.3.%. Results of the double-blind tests showed a sensitivity between 56 and 76% and specificity between 8.3 and 33.3%, whereas test results from urine samples showed an overall sensitivity of 73.6–64.2%, and specificity of 25–29.2%. The dogs could not clearly discriminate between malignant and benign conditions during tests. The dogs were trained to discriminate lung cancer samples from healthy controls, but then tested with other non-malignant lung cancer symptom-matched controls; this suggests that dogs would need to have previous exposure to these samples to be able to have an olfactory perceptual learning to discriminate between VOCs of malignant lung cancer from benign ones.

Junqueira et al. (2019) investigated whether dogs can detect NSCLC from healthy controls using blood serum. Four two-year-old beagles were trained to detect one positive sample of lung cancer among four controls using clicker training and food reward for eight weeks. Authors reported that one dog was eliminated from the study because she was poorly motivated and did not respond well to any training method used. Following training, the three remaining dogs were tested in a double-blind procedure using 10 new NSCLC and 40 control samples. The dogs were able to detect lung cancer in blood serum with an overall sensitivity of 96.7% and specificity of 97.5%.

Similarly, Riedlova et al. (2022) compared the ability of dogs to detect lung cancer in blood serum and exhaled breath samples. A four-year-old Labrador Retriever and a five-year-old Australian Cattle Dog were trained to detect lung cancer-positive samples among three healthy controls in blood serum and urine. Following training, a double-blind test was performed using a set of 47 positive lung cancer samples and 41 healthy controls. Detection sensitivity using blood serum was found to be 69 and 62% in dogs 1 and 2, respectively, whereas specificity was 67 and 97%. Likewise, detection of lung cancer using breath samples showed a sensitivity of 62 and 75%, while specificity was 71 and 90% in dogs 1 and 2, respectively. Authors discussed that the better detection performance of dog 1 could have been because this dog was trained by an experienced dog trainer and for much longer than the second dog. According to the authors, the experience of the handlers, the selection of dogs, and the interaction between the handler and the dog are very important factors to be considered for obtaining good detection performance results.

Three studies have been published investigating the identification of volatile organic compounds present in human breath using GC–MS methods compared with trained detection dogs. Buszewski et al. (2012) and Rudnicka et al. (2014) examined exhaled breath samples that were collected from 108 patients with lung cancer, 121 control healthy volunteers, and 24 patients with other lung diseases. They found that the concentration of acetone, isoprene, ethanol, 1-propanol, 2-propanol, hexanal, and dimethyl sulfide were higher in patients with lung cancer than in healthy and non-cancer lung diseases. Buszewski et al. (2012) described that positive correlations were found between dog indications and the presence of ethyl acetate and 2 pentanone content in exhaled breath samples. The higher the concentration of these two compounds, the better they were indicated by the dogs. Contrary, the higher the concentration of 1-propanol and propanal in breath samples, the lower the percentage of correct indication by dogs. Rudnicka et al.

(2014) showed that dimethyl sulfide is the main compound enabling differentiation between patients with lung cancer and healthy volunteers. They prepared synthetic samples based on the exhaled air of cancer patients. The indication of the synthetic samples by the trained dogs was significantly worse (21%) than the indication of breath samples from cancer patients (86%). Fisher-Tenhagen et al. (2018) trained four dogs to detect one positive lung cancer among three healthy controls. In this study, breath samples were collected by an absorber technique using polypropylene fleeces, developed based on the work of Ehmann et al. (2012). After three months of training, two dogs were removed from training due to their lack of progress in discriminated breath samples ($n = 10$). Two tests were performed (1) to detect positive lung cancer ($n = 9$) among healthy controls ($n = 10$), and 2) to investigate the dog reaction to synthetic air samples spiked with 1-butanol, 2-butanone, 2-pentanone, and hexanal ($n = 4$) among healthy controls ($n = 10$). Results showed in test 1 a true positive rate of 9 out of 9 in dog 1, and 8 out of 10 in dog 2. Test 2 for both dogs indicated three out of four synthetic air samples as positives. Nevertheless, so far, no specific VOCs have reached clinical relevance for reliable recognition of lung cancer with sufficient sensitivity and specificity. The signature odor of human diseases learned by the dogs or even each individual dog maybe related to specific qualitative or quantitative olfactory perception produced by a mixture of VOCs (Buszewski et al. 2012; Guest et al. 2021).

2.4 Colorectal

More than 1.9 million new colorectal cancers were diagnosed, and 935,000 deaths were estimated to occur, in 2020. Colorectal cancer ranks third in terms of incidence, but second in terms of mortality (Sung et al. 2021). Globally, it is one of the cancers in which the incidence is increasing, comprising 11% of all cancer diagnoses (Wong et al. 2020). Unfortunately, 20% of patients with colorectal have metastatic disease at presentation for health care, and 25% who present with localized cancer will later develop metastases. The most common methods of diagnosis of colorectal cancer are invasive and uncomfortable for the patients including colonoscopy, guaiac fecal occult blood test (gFOBT), optical colonoscopy, flexible sigmoidoscopy, digital rectal exam, and fecal immunochemical tests for hemoglobin (FIT).

To the best of the author's knowledge only two peer-reviewed, published studies have used dogs for detecting colorectal cancer (Table 5). Sonoda et al. (2011) used an eight-year-old Labrador Retriever previously trained to detect several types of cancers including gastric, prostate, breast, bladder, and colorectal. For this study, the dog was tested to discriminate exhaled breath ($n = 33$) and watery stool ($n = 37$) cancer samples from healthy controls ($n = 132$ breath; $n = 148$ stools). The sensitivity and specificity in breath samples were 97 and 99%, whereas in watery stool samples were 97 and 99%, respectively. This study found a higher detection accuracy for early stages colorectal cancer, and there was no confounding by smoking habits, other colorectal diseases, nor the presence of hemoglobin nor

Table 5 Detection of colorectal cancer by dogs

Authors	Year	Sample type	Number of dogs	Testing methods	Number of samples explored per run (+ Cancer, C Control)	Sensitivity	Specificity	Total positive samples & controls	Match symptoms control samples	Test design double blind
Sonoda et al.	2006	Breath stool	1	Lineup	1+ : 4C	Breath: 91% Stool: 97%	Breath: 99% Stool: 99%	48+ : 258C	Unclear	Yes
Schoon et al.	2020	Stool	8	Carousel	0–2+ : 8–6C	Hit rate: 81.2–77.8%	False alert rate: <50%	18+ : 102C	Yes	Yes

transferrin in the stools. A recent study by Schoon et al. (2020) trained eight dogs to detect colorectal cancer using stool samples in an eight-choice carousel setup. This study investigated factors influencing the generalization of "colorectal cancer" odor signature to a new patient and control sample set, showing that monitoring the dogs' learning curves in training allows to follow-up every individual dog and adjust training to determine an appropriate timing for testing. For example, it is expected that some dogs increase their false alert rate when new controls are introduced in the search, which seems to be because of trial-and-error learning by the dogs. The dogs would use the "new" odor for a while in their learning process until they learned to ignore it, but also it could be odor similarity with the positive samples including background odors. Thus, the dog would use trial-and-error learning until able to learn the "cancer odor signature" and discriminate from the controls.

2.5 Cervical and Ovarian Cancers

Cervical cancer is the fourth most frequently diagnosed cancer, with an estimated 604,000 new cases. It is the fourth leading cause of cancer death in women with 342,000 deaths worldwide in 2020 (Sung et al. 2021). Cervical cancer is preventable primarily because of the Human Papillomavirus (HPV) vaccine and second by the screening prevention measures (annual Papanicolaou test or pap smear cytology). Nevertheless, an alternative, a non-invasive screening technique is still needed, and therefore, biomedical detection dogs can be a non-invasive and cost-effective screening tool for cervical cancers.

Canine detection of cervical cancer was first published in 2017 by Guerrero Flores et al. (2017) (Table 6a); a Beagle previously trained in drug detection was trained to detect cervical cancer using biopsies. The dog was first tested using fresh cytological smears from patients with cervical cancer and then tested using medical surgical bandage worn overnight in home by patients with invasive cervical cancer. This criterion was introduced to assess detection performance in samples with less cells available. Overall sensitivity of 92.78% and specificity of 99.1% was determined when a smear was tested (positive cancer smears = 97; healthy controls = 776), whereas a sensitivity of 96.36% and specificity of 99.5% was obtained when surgical bandages were tested (positive cancer = 55; healthy controls = 440). This study demonstrates that surgical bandages like sanitary pads could be used to collect several body metabolites and cancer-related molecules, which would be an inexpensive and non-invasive method of sample collection. However, the surgical bandage was worn for several hours (overnight) and only patients with invasive cervical cancer were sampled, therefore the concentration of VOCs may have been high compared to healthy controls, which might have overrated the dogs' performance. Yamamoto et al. (2020) used an experienced dog previously trained to detect other types of cancer in breath and urine samples (Sonoda et al. 2011). The dog was able to successfully discriminate urine samples from patients with cervical cancer (n = 83) from those with benign uterine

Table 6 Detection of cervical and ovarian cancers by dogs

Authors	Year	Sample type	Number of dogs	Testing methods	Number of samples explored per run (+ Cancer, C Control)	Sensitivity	Specificity	Total positive samples & controls	Match symptoms control samples	Test design double blind
a. Cervical										
Guerrero-Flores et al.	2017	Smear Surgical bandage	1	Circle	1+ : 9C	Smear: 92.78% Bandage: 96.36%	Smear: 99.1% Bandage: 99.5%	97+ : 776C 55+ : 400C	No	Yes
Yamamoto et al.	2020	Urine	1	Lineup	1+ : 5C	100%	100%	83+ : 112 C	Yes	Yes
b. Ovarian										
Horvath et al.	2008	Tissue	1	Lineup	1+ :5C 1+ : 8C	100%	97.50%	20+ : 80C	No	Yes
	2010	Tissue Blood plasma	2	Circle	1+: 5C	Tissue: 100% Blood: 100%	Tissue: 95% Blood: 98%	Unclear+ : 50C	No	Yes
	2013	Blood plasma	2	Circle	1–3+ : 9–7C	97%	99%	42+ : 210C	No	Yes
Kane et al	2022	Blood plasma	5	Carousel	Not indicated	85%	77%	Not indicated	Yes	Not indicated

diseases ($n = 49$) and healthy volunteers ($n = 63$) with an overall sensitivity and specificity of 100%.

Ovarian cancer is the 18th most common cancer overall in women, with an estimated 313,000 new cases in 2020. Horvath et al. (2008, 2010, 2013) (Table 6b) investigated the use of dogs for ovarian cancer detection. Their first study involved one dog trained to detect ovarian carcinoma of different stages using tissue and discriminate it from other tissues including abdominal fat, muscle, bowel, and two pieces of healthy postmenopausal ovary. To optimize the samples, the ovarian carcinoma tissue was placed next to a cotton wool pad into a closed glass for 24–168 h, then the cotton pads were used as the target odor during training. It is unclear whether the same methodology was followed in the test and was also done with controls tissues. In the double-blind test, the dog discriminated between ovarian carcinoma ($n = 20$) and healthy tissues ($n = 80$) with a sensitivity of 100% and specificity of 97.5%. This was the first study showing the ability of dogs to detect ovarian cancer, however, the use of histopathological samples as target odor and controls does not justify the use of dogs as screening method; histological samples are the confirmatory method of diagnosis. In their follow-up study (2010), blood and tissue samples were used as odor sources. Two dogs were trained (one dog used in the previous study), to detect ovarian carcinoma and borderline ovarian tumors. Tissue controls were similar to the described above, in addition to non-malignant cervical, vulvar, and endometrial carcinoma tissues. The dogs' performance results using tissue showed a sensitivity and specificity of 100 and 96% in dog 1, 100 and 94% in dog 2, respectively, while performance on blood samples resulted in a sensitivity and specificity of 100 and 96% in dog 1, and both 100% in dog 2. The authors concluded that the study strongly suggests that VOCs emitted by ovarian cancer samples are also present in the blood (plasma) taken from patients with the disease. Kane et al. (2022) used micro-preparative gas chromatography (MPGC) technology to separate VOCs from ovarian cancer and healthy plasma samples into three different fractions to identify potential biomarkers of the disease. These VOC fractions were presented to five dogs previously trained to discriminate between ovarian cancer and controls (healthy and benign ovarian tumors) in plasma. The duration of time spent by the dogs at the odor container port was measured to assess the presence of ovarian cancer VOCs in the fractionated samples. It was found that dogs spent significantly more time at the complete malignant VOC MPGC fraction, 0–15 min fraction, and 15–25 min fraction than the control sample, but there was no difference between the control sample and the 25–35 min MPGC fraction, suggesting that the VOCs that the dogs use to detect ovarian cancer in plasma may lie in these 0–25 range MPGC fractions.

The retrospective study of Horvath et al. (2013) aimed to investigate how primary surgery and chemotherapy treatment may affect the detection of cancer odor by dogs. The same two dogs from the previous study were used. One group of blood samples was collected from patients who had between five and six courses of chemotherapy, demonstrating an overall sensitivity and specificity of cancer detection of 97.6 and 99.5%, respectively. In the second group, the blood samples were collected from new patients at three and six months after the last (sixth) course of chemotherapy. The dogs indicated 3 positive samples out of the 10 patients. Unfortunately, all three patients had cancer recurrence. Thus, studies show that detection dogs can be used to monitor prognosis and relapse.

2.6 Bladder

Bladder cancer is the 10th most diagnosed cancer worldwide, with an estimated 573,278 new cases and 212,536 deaths in 2020. It is four times more common in men than in women. In men, it is the sixth most frequent cancer and the ninth leading cause of cancer death (Sung et al. 2021).

Bladder cancer tissue can change VOC composition in urine by excreting metabolites directly into it, therefore, cancer biomarkers can be used for their diagnosis. Such a vision conceived the idea of training dogs to discriminate people affected by cancer from those healthy by sniffing urine odor (Willis et al. 2004). This was a starting point for research groups to investigate chromatography and mass spectrometry VOCs in urine samples to identify biomarkers of different types of cancer (Ligor et al. 2022). Several studies have found urinary VOC markers showing a sensitivity between 70–77.5% and specificity from 89 to 93.25% (Pinto et al 2021; Zhu et al. 2019). Furthermore, 76 specific metabolites in urine have been identified, the presence of which allows to differentiate bladder cancer from healthy controls (Zhou et al. 2017).

The first study training dogs for the detection of bladder cancer was conducted by Willis et al. (2004) (Table 7). Six dogs were trained to discriminate between urine from patients with bladder transitional cell carcinoma from other urological diseases and healthy volunteers. The authors used two methods of sample presentation to the dogs: urine liquid, and "dry urine," which consisted of urine pipetted onto filter paper and placed into Petri dishes. After seven months of training, dogs were tested to detect one positive bladder cancer ($n = 9$) among controls ($n = 54$). Sensitivity of 50% was determined when urine samples were presented to the dog in liquid form, whereas it was 22% when "dry urine" was presented. This result was statistically better than the 14% that would have been expected by chance alone in a lineup of seven samples (1 positive cancer alongside 5 controls). Authors reported that the multivariate analysis did not show confounding factors related to measurable characteristics present in urine (e.g., blood, leucocytes, protein, ketones, etc.) and the selection of the urine samples by the dogs. In their follow-up research (2010), four dogs were used in a series of 30 double-blind test runs, like the previous study one positive sample was placed alongside 6 controls. The overall sensitivity was 64%, while the specificity of the dogs individually ranged from 56 to 92%. The lower specificity was obtained when controls used were urine samples from old patients with non-cancerous urological diseases, whereas the higher specificity was observed when they were urine from healthy young volunteers.

Table 7 Detection of bladder cancer by dogs

Authors	Year	Sample type	Number of dogs	Testing methods	Number of samples explored per run (+ Cancer, C Control)	Sensitivity	Specificity	Total positive samples & controls	Match symptoms control samples	Test design double blind
Willis et al.	2004	Urine	6	Lineup	1+ : 6C	Wet urine: 50% Dry urine: 22%	Not indicated	9+ : 54C	Yes	Unclear
	2010	Urine	4	Lineup	1+ : 6C	64%	89%	Unclear	Yes	Yes

2.7 Malignant Skin Melanoma

Melanoma skin cancer is less common than some other types of skin cancers, but it is more likely to metastasize. It was estimated to be 324,635 new cases and 57,043 cause of deaths in 2020 (Sung et al. 2021).

The first two reports of the ability of dogs to detect human cancer were in skin melanoma. Williams and Pembroke (1989) and Church and Williams (2001) reported anecdotal cases of untrained pet dogs spontaneously sniffing a skin lesion on their owners that after resection were histologically diagnosed as skin malignant melanoma and carcinoma, respectively. Campbell et al. (2013) also reported a pet dog that licked persistently at an asymptomatic lesion behind the owner's ear, which histology confirmed to be a malignant melanoma. These reports have inspired considerable research interest regarding dogs' olfactory capability as a potential non-invasive method for diagnosing malignant melanomas in humans (Table 8). To prove that dogs can be trained to detect melanoma under an experimental method, Picket et al. (2004) trained a Golden Retriever and a Standard Schnauzer, previously explosive detection dogs, to detect melanoma from tissue samples hidden in a lineup of boxes. As a following step in their training, the dogs reliably located melanoma tissue samples planted on the skin of healthy volunteers as preparation for the test phase. Dogs were tested sniffing at the skin of 7 patient volunteers with some clinical suspicion of skin melanoma in situ, showing an accuracy between 75 and 85.7%. The dogs detected melanoma in a patient that was considered negative in an initial pathological examination, being confirmed in a second examination. Walczak et al. (2009) trained three dogs to detect breath, lung, and melanoma cancers using breath samples. The detection sensitivity for melanoma ranged from 32.2 to 66.3%, being lower in comparison to the other cancers.

2.8 Other Human Cancers Detected by Dogs

A proof of concept for the potential use of trained dogs to detect osteosarcoma was reported by Ortal et al. (2022). Two dogs previously trained for search and rescue were trained for 16 months to detect a cultured cell of patients with osteosarcoma; diluted osteosarcoma culture cells and saliva were used as the positive samples. For testing, four stations were used in a row containing either one cancer sample (from cell lines or saliva) and three controls or four control samples. Blind tests using the same sarcoma culture cell employed during training among two different controls showed a sensitivity of 97.65% (Dog 1) and 98.57% (Dog 2), and a specificity of 90.90% in dog 1, whereas dog 2 was less efficient in discriminating with a specificity of 84.78%. Authors removed dog 2 from the following trials. The estimation of the limit of detection of osteosarcoma samples using dilutions of 1:5, 1:10, and 1:50 in the non-blinded tests showed a decrease in sensitivity from 95.29 to 50%; similarly, specificity declined from 98.76 to 92.86%. The decrease in accuracy was greater when a dilution of 1:50 was tested.

Table 8 Detection of skin melanoma by dogs

Authors	Year	Sample type	Number of dogs	Testing methods	Number of samples explored per run (+ Cancer, C Control)	Sensitivity	Specificity	Total positive samples & controls	Match symptoms control samples	Test design double blind
William and Pembroke	1989	Direct body contact-skin	1	Direct contact with the person	1	100%	NA	1+ :0C	No	Case report
Church and Williams	2001	Direct body contact-skin	1	Direct contact with the person	1	100%	NA	1+ :0C	No	Case report
Picket et al.	2004	Tissue Direct body contact-skin	2	Direct contact with the person	1	75–85.7%	NA	1+ :6C	No	Case report
Walczak	2009	Breath	3	Lineup	1 + : 4C	66–80%; 36–58%	77–87%	118+ : 305C	No	Yes
Campbell	2013	Direct body contact-skin	1	Direct contact with the person	1	100%	NA	1+ :0C	No	Case report

In a second test investigating six different osteosarcoma cell lines not previously exposed to the dog, sensitivity between 95 and 100% and specificity between 97.73 and 100% was found. These results suggested that osteosarcoma cell lines share a common odor signature. Then, the author investigated whether this odor signature can be detected in saliva samples from patients with osteosarcoma. The dog was able to discriminate osteosarcoma from healthy controls with a sensitivity between 98.46 and 100% and specificity from 99.22 to 100%. Altogether, this proof of concept showed that a dog can generalize osteosarcoma odor signature in cell lines, and this can be transferred to saliva samples, which may facilitate the development of rapid and cost-effective screening methods for early detection.

3 Detection of Human Infectious Diseases by Dogs

3.1 Bacteriuria

Urinary tract infections are the most common hospital-acquired infections for all patients and early diagnosis is difficult to achieve. Thus, a research group investigated the use of dogs to improve strategies for detecting early stages of bacteriuria (i.e., the presence of bacteria in urine) before the infection becomes a serious illness in hospitalized patients. It was expected that dogs identify VOCs which can be bacteria-specific fingerprints. Maurer et al. (2016) trained five dogs for eight weeks to detect patients with bacteriuria (bacterial-culture positive > 100,000 colony-forming units) among controls (bacterial-culture negative) in 1 ml of urine. Dogs were tested using new positive and control samples in double-blind experimental conditions. The dogs accurately identified samples containing bacteria *Escherichia coli* (sensitivity 99.6%, specificity 91.5%), *Klebsiella* (sensitivity 100%, specificity of 95.1%), *Enterococcus* (sensitivity 100%, specificity of 93.9%), and *Staphylococcus aureus* (sensitivity 100%, specificity of 96.3%). This study showed evidence of using dogs for the detection of early stages of bacterial infection that cause urinary tract infections. The authors anecdotally reported that one month after the study was done, one of the trained dogs spontaneously alerted to a person visiting the training center, the person visited the doctor, and a urine culture was performed on the next day and the presence of bacteriuria was confirmed.

3.2 Clostridium

Another nosocomial infection (hospital-acquired infection) is the *Clostridium difficile,* a Gram-positive bacterium, which causes symptoms such as formation of gas in the digestive system, diarrhea, and abdominal pain. *Clostridium difficile* is widely distributed in the intestinal tract of animals and humans and in the environment. Bomers et al. (2012) trained a Beagle to identify the presence of *Clostridium difficile* in stool samples absorbed in different materials and placed them in hidden places, and then tested in infected patients in a hospital setting. Sensitivity and Specificity were both 100% in identifying positive stool samples and 8% and 98% in hospital-infected patients. In the healthcare facility, the dog searched for positive patients by sniffing around them, there was no physical contact with patients. The dog correctly identified 25 out of 30 positive cases and 265 of 270 controls. The dog was able to detect *Clostridium difficile* in the air surrounding patients and the whole screening lasted less than 10 min. Thus, proof of concept showed the potential use of detection dogs in healthcare facilities to contribute to control and prevention of outbreaks. In a follow-up study (van Bomers et al. 2014) carried out during a *Clostridium difficile* outbreak in a hospital, the dog performed 651 screenings involving 371 patients during 9 hospital visits. The dog identified 12 out of 14 positive cases with 86% sensitivity and 97% specificity.

Environmental reservoirs of *Clostridium difficile* have been implicated in outbreaks, including the risk of a new patient acquiring this bacterium if a previous occupant of the room was infected with it (Bryce et al. 2017). Like the previous study, a dog was trained to detect odors released from pure culture and fecal samples of *Clostridium difficile.* Three container lineups (75 samples) and three clinical runs or room searches (53 samples) were conducted to test the dog. Sensitivity and specificity obtained from the container tests were 100 and 97%, respectively. When the dog was tested in clinical runs sensitivity was 67% and specificity was 91.5%. Dogs can be used for infection control and for assessing disinfectant and cleaning measurements in healthy facilities (Taylor et al. 2018).

3.3 Pseudomonas

Cystic fibrosis is a genetic disorder that causes abnormal thick secretion, which is not adequately cleared by the mucociliary system. This creates an environment for microbial colonization. The most prevalent bacterial infection is *Pseudomonas aeruginosa.* Davies et al. (2019a, b) investigated whether dogs can be trained to detect *Pseudomonas aeruginosa* in broth supernatant of bacterial culture. Three dogs were trained to discriminate *Pseudomonas aeruginosa*-positive samples among other bacterial and sterile broth. In this series of four tests, dogs were able to detect *Pseudomonas aeruginosa* with high levels of sensitivity, distinguishing them from other bacteria commonly encountered in the cystic fibrosis lung. The dogs' performance was generally maintained when the dogs encountered other organisms for the first time, or when *Pseudomonas aeruginosa* was present in mixed cultures. Sensitivity declined at dilutions of 1:10 000, equating to ~105 CFU $\cdot$ mL^{-1}. The results of the four tests were (1) *Pseudomonas aeruginosa* was tested alongside *Staphylococcus aureus*, *Moraxella catarrhalis*, and sterile broth; mean sensitivity was 94.2% and specificity was 98.5%; (2) In the second test, *Pseudomonas aeruginosa* presented alongside other bacteria previously unencountered by the dogs; Two of the three dogs maintained sensitivity above 90%, but the third dog was 62.5%; (3) In the third test, two dogs were tested on diluted broths; at dilutions of 1:1000 sensitivity was 93.8% and specificity 94.9%, whereas at dilution of 1:10,000, sensitivity decreased to 56.3% and specificity to 89.1%; (4) In the fourth test, a mixed multi-organism cultures were used. Dogs were still able to identify *Pseudomonas aeruginosa* with a sensitivity of 86% and specificity of 84.1%.

3.4 Malaria

For centuries, malaria has been a threat to global health with an estimated number of cases of 214 million, and several deaths of 627,000 in 2020 (World Health Organization, WHO). Malaria is an infection caused by *Plasmodium*, a protozoa parasite that is transmitted by *Anopheles* mosquito to humans. The diagnosis of Malaria commonly done by Rapid Diagnostic Testing (RDTs) is highly efficient for diagnosing febrile malaria patients, while the Loop-mediated isothermal amplification (LAMP) has a high negative predictive value (i.e., the likelihood that a person who has a negative test result indeed does not have the disease) (Picot et al. 2020). Nevertheless, efficacy in testing asymptomatic patients is still being investigated and the main challenge of diagnosing malaria is detecting asymptomatic carriers.

Recent studies on human malaria have identified changes in the VOCs emitted from the skin and breath of infected individuals (de Boer et al. 2017; Pulido et al. 2021). Malaria odor is detected by mosquitoes, consequently malaria mosquito feed on asymptomatic, malaria-infected individuals (Robinson et al. 2018). This change in the malaria-infected human odor could be used to train dogs as a non-invasive screening of infected asymptomatic individuals. Guest et al. (2019) performed the first study using trained dogs to distinguish between the odor of asymptomatic malaria-infected and uninfected individuals. Odor samples were obtained from foot odor from Gambian school children by wearing socks overnight. The study included 30 malaria-positive individuals and 145 uninfected children. Even in asymptomatic children, the dogs had a sensitivity of 70. 0 and 73.3% in dog 1 and 2, respectively, with a specificity of 91.0% in dog 1 and 90.3% in dog 2. The results of sensitivity and specificity using detection dogs were above the threshold required by the World Health Organization for malaria diagnostic based on the parasite density in the samples. This study was one of the pioneers in showing that dogs can detect infectious diseases even in asymptomatic people. Therefore, malaria detection dogs could be used to detect asymptomatic infected individuals at a port of entry in countries with *Anopheles* to control the spread of the disease (Kasstan et al. 2019).

3.5 Parkinson's

Parkinson's disease is a chronic, progressive neurodegenerative disease that affects mobility and muscle control causing difficulties to move, speak, swallow, and even breathe. There is currently no definitive test for Parkinson's diagnosis or cure.

However, developing a method for early detection would lead to treatments starting sooner. Mischley and Farahnik (2020) presented a pilot study where two Lagotto Romagnolo were trained to detect Parkinson's in samples collected from ear canal cerumen, sebum, and dermal microbiome. The dogs were exposed to a total of 46 samples, 28 individuals with Parkinson's and 18 healthy controls, identifying 27 out of the 28 positive samples (96.43% sensitivity), and 13 of the 18 controls identified as negative (72.22% specificity).

4 Methodological Challenges

Scientific studies investigating the use of dogs for the detection of human diseases have shown high accuracy in terms of sensitivity and specificity, and promising assistance for both non-invasive screening and the development of biomarkers for diagnostics. However, the lack of standardization of methodology in training and testing biomedical detection dogs makes it hard to compare these studies. Methods differ with reference to experimental setting, type of odor sample source (e.g., breath, urine, blood serum, tissue, etc.), sample collection and storage, source of dogs, and decision-making criterion to move from training to testing (see Moser and McCulloch 2010; Elliker et al. 2014; Jezierski et al. 2015; Edwards et al. 2017; Johnen et al. 2017; Lazarowski et al. 2020, Crawford et al. 2022 for review).

In general, biomedical detection dogs are trained to recognize a disease odor signature using positively reinforced operant conditioning with a clicker and food or toy as reward following training protocols similar to other types of detection tasks (Edwards et al. 2017). Horvath et al. (2008, 2010, 2013) reported the use of a different type of reward, the dog's natural hunting behavior, thus when the dog showed interest in the target odor, the trainer quickly snatched it away, strengthening the dog's motivation to select the target. The olfactory sensorial stimulation could be interpreted as a primary reinforcer, but also the attention/praise received by being pulled away could be the primary reinforcer depending on the dog's biological perception of it.

In a medical detection scenario, a dog is trained and tested in a controlled set-up with a fixed number of positions to be explored by the dog, for example, a line-up of four positions or a carousel with 8 positions (e.g. Walczak et al. 2012a, b; Schoon et al. 2020). The dog explores each position one at a time, in a "Go/no Go" task in which an independent decision of responding or withholding the response is required for each stimulus explored (Lazarowski et al. 2020). This presence/absence response is communicated to the trainer by performing a trained alert response, which is often a sitting or lying down behavior in front of the position with the target sample, or withholding the trained response. Traditionally, the behavior to be shaped into a trained alert response is often chosen by the trainer rather than spontaneously offered by the dog until becomes a default behavior. The study of Essler et al. (2020) suggested that the type of operant behavior response that a biomedical detection dog has been trained to perform to alert a positive sample, sit versus stand-stare, may influence the amount of time checking a sample and the duration of a dog's false positive indication. Thus, the effect associated with the decision to express an operant alert response can affect the reliability of the detection performance. In this way, sniffing behavior as an approach to determine olfactory alert performance beyond the trained alert response has been shown to be particularly valuable in differentiating true from false negative (i.e., where the target is present, but the dog fails to detect/indicate it) responses. Furthermore, dogs sniff twice the amount of time when the target odor is present or when it is indicated as present on a negative sample (i.e., false indication) (Concha et al. 2014). Similarly, Mancini et al. (2015) developed a pressure-sensitive odor sample holder that can be used to automatically detect dogs' responses to the samples placed in the holder. By recording the intensity and duration of pressure applied by the dogs, they found differences in pressure between true negatives and false negatives responses, but also found that dogs apply greater pressure on this sensor when they are certain the disease is present and therefore, pressure may indicate the level of certainty that the dog has. Although further research is needed to investigate sniffing behavior in complex odor samples, it can be used alongside the trained alert response to effectively assess detection performance that will support the development of biomarkers or even train artificial neural networks (ANN).

Several different dog breeds and crossbreds have been trained for the detection of human diseases including Labrador Retrievers, Schnauzers, Border Collies, Dachshunds, Beagles, Belgian, and German Shepherds, among other breeds (e.g., Jezierski et al. 2015). It is difficult to discuss if there is any particular breed more suitable for this type of detection task, since there are differences in other methodological aspects associated with the task, and therefore, it would be unrealistic to attribute the whole success or failure of detection performance reported in the literature to the dogs. As mentioned by Pirrone and Albertine (2017) the dogs' acuity sense of smell, combined with the ability to learn by operant conditioning, makes dogs excellent biodetectors for different kinds of purposes. However, individual differences play an important role in the dogs' trainability and detection performance. It is well known that differences in the genetic polymorphism of olfactory receptors influence detection acuity (Lesniak et al. 2008; Konno et al. 2018; Yang et el. 2022). Olfactory acuity measured in terms of threshold detection is a fundamental factor since biomedical dogs are trained to detect volatile organic compounds (VOCs), which are released in concentration ranges of parts per billion (ppb) to parts per trillion (ppt) in human blood and urine (Buszewski et al. 2012; Schmidt and Podmore 2015). These odor threshold concentrations are within the limit of olfactory detection described in dogs (Angle et al. 2016; Concha et al. 2019), but there is a significant inter-dog difference in detection thresholds. A study investigating dogs' olfactory detection threshold in ten dogs trained under the same conditions and worked without the assistance of the handler suggested that this inter-dog variability is likely influenced by other factors affecting the sense of smell such as age, diseases, medication, diet, and hydration (see Jenkins et al. 2018 for review), but also temperament and learning (Concha et al. 2019).

Dogs have to possess a temperament suitable for the high demands of detection training (Foyer et al. 2016; Maejima et al. 2007; Johnen et al. 2017), for instance in the study by Elliker et al. (2014) only three out of ten dogs that began training passed the initial phase, which involved indicating one single positive prostate cancer urine sample among empty containers. The author reported that dogs failed in progressing through training because of their temperament. Several reviews have been written about temperaments in detection dogs (e.g., La Toya et al. 2017; Brady et at. 2018; Bray et al. 2021), however, there are no published studies investigating temperament or behavior traits predicting performance specifically in biodetection dogs. The major difference with other types of detection tasks is the fact that biomedical dogs are tasked with comparing multiple odor sources, many with a similar odor profile, in close proximity to one another placed in a line-up or carousel. In addition, the search pattern is repeated several times within a session with unique slight variability in the odors presented to the dogs. Thus, the ability to focus on searching and motivation is another important factor to consider in dog selection (Porritt et al. 2015; Gazit et al. 2005). Kure et al. (2021) described that under certain stressful conditions, the dog trained to detect prostate cancer could not maintain concentration on the detection task and therefore, the session had to be postponed. Distractibility and lack of motivation to search and learn the detection task could negatively affect the performance results.

Olfactory perceptual learning may play a role in lower levels of detection thresholds observed in dogs after a period of training (Concha et al. 2019) and may be crucial for basic olfactory function because it sets the degree of discrimination between stimuli facilitated by separating the odorant from background odors (Wilson and Stevenson 2003). There is abundant evidence for the perceptual learning of meaningful odor representations, from their generalization properties (Waggoner et al. 2022; Moser et al. 2019; Cleland et al. 2009) to the mechanisms of odor learning and memory (Tong et al. 2014). Studies in animals have demonstrated that to develop complex discriminations and concepts, for example, "cancer odor," it is fundamental to train the animal with a wide range of example target samples (positive samples) and non-target stimuli (negative stimuli) (Wright et al. 2017; Bowman and Zeithamova 2020; Hall et al. 2021). Edwards et al. (2017) suggested an effective approach to training the targeted disease concept when the number of samples source is limited, as follows: (1) Train with a subset of the available sample sources until the animal reaches pre-established performance criteria or until performance is no longer improving; (2) Introduce samples from novel positive and control sources and evaluate performance with these samples prior to any reinforced indications of the positive samples; (3) Include these samples in the training set if performance with the novel samples is low and repeat these steps until performance with samples from novel sources is reliably high.

Biomedical detection dogs are able to make very subtle and complex discriminations of VOCs in human odor samples. If disease-positive samples and controls have systematic differences other than disease status, the dog learns to rely on additional olfactory cues, more than the disease itself. Therefore, it is fundamental to match controls to the characteristics of the target odor including age, gender, habits, symptoms, location of collection, etc. (Jezierski et al. 2015; Guest et al. 2020). This helps to minimize differences in background odors not associated with the disease state. Controls should be collected from the same clinical environment as the positive target samples (Edwards et al. 2017). If it is not possible to apply this standardization to avoid confounders, then variability needs to be included by using multiple collection methods, location, and processing procedures for both positive targets and controls, thus dogs learn to categorize based only on the disease state (Jezierski et al. 2015).

Besides discrimination of a specific disease from controls, biomedical dogs have to respond to positive targets that they have never been trained on before. Stimulus generalization is the tendency to respond not just to the stimulus trained but also to a new stimulus similar to the trained one (Waggoner et al. 2022; Moser et al. 2019). To enhance generalization, dogs should be trained with many exemplars of a target that vary by irrelevant dimensions (Hall and Wynne 2016). However, the size of the example set and how much variation in stimuli is required for the dog to optimize performance has not been investigated in biomedical dogs so far. Several studies have demonstrated that dogs could generalize and indicate a common disease odor when trained with a wide range of samples from different patient donors (Edwards et al. 2017), but also, evidence has shown that dogs can fail in generalizing the disease target odor from training to testing phase (Elliker et al. 2014). This could be a consequence of extensive training with the same sample set which tends to narrow generalization to other similar samples, despite increasing discrimination of the trained odors which can overstate sensitivity and specificity during training (Moser et al. 2019).

Unfortunately, there is usually a limited number of samples available to train and test biomedical detection dogs which may lead to memorization of the individual odors (Marchal et al. 2016) rather than promoting generalization (Jeziersky et al. 2015; Elliker et al. 2014). As mentioned by Guest et al. (2020) organizations using biomedical dogs are often forced to choose between using fewer samples and risking dog memorizing individuals or using differently sourced samples which can cause the dog to learn to discriminate target from control based on subtle cues given by the different source. It is important to consider every aspect of sample processing, collection, storage, and handling to avoid confounding factors associated with what might affect the dog's performance. Thus, when training a complex signature in a complex background, we need to ensure dogs learn to accurately discriminate and generalize disease odor signatures only.

5 Conclusion and Future Directions

The purpose of using trained dogs to detect human diseases is to assist as a non-invasive tool in the identification of early stages of diseases for diagnosis and control, as well as to identify the presence of biomarkers for the development of clinical diagnostic tests. Medical or biomedical detection dogs have been shown to have a high sensitivity and specificity in the detection of human diseases, mobility in both indoor and outdoor environments, and are time and cost-effective reporting screening results in real-time and rapid diagnostic. However, further research is necessary to standardize and validate training and testing methodology for the deployment of biomedical detection dogs in operational settings as an approved clinically valid health screening method. Moreover, studies investigating individual characteristics and selection of biomedical detection dogs are needed to predict suitability for this type of detection work and lifelong success.

References

Amundsen T, Sundstrøm S, Buvik T, Gederaas OA, Haarverstad R (2014) Can dogs smell lung cancer? First study using exhaled breath and urine screening in unselected patients with suspected lung cancer. Acta Oncol 53:307–315

Angle C, Waggoner LP, Ferrando A, Haney P, Passler T (2016) Canine detection of the volatilome: a review of implications for pathogen and disease detection. Front Veter Sci 3:47

Arakawa H, Arakawa K, Deak T (2009) Acute illness induces the release of aversive odor cues from adult, but not prepubertal, male rats and suppresses social investigation by conspecifics. Behav Neurosci 123:964–978. https://doi.org/10.1037/a0017114

Arakawa H, Cruz S, Deak T (2011) From models to mechanisms: odorant communication as a key determinant of social behavior in rodents during illness-associated states. Neurosci Biobehav Rev 35(9):1916–1928. ISSN 0149-7634. https://doi.org/10.1016/j.neubiorev.2011.03.007

Aubert A, Richard FJ (2008) Social management of LPS-induced inflammation in Formica polyctena ants. Brain Behav Immun 22:833–837

Banerjee S, Robertson ES (2019) Chapter 17. Future perspectives: microbiome, cancer and therapeutic promise. In: Robertson ES (ed) Microbiome and cancer. Humana Press, Switzerland, pp 363–389

Boillat M, Challet L, Rossier D, Kan C, Carleton A, Rodriguez I (2015) The vomeronasal system mediates sick conspecific avoidance. Curr Biol 25:251–255. https://doi.org/10.1016/j.cub.2014.11.061

Bollmeier S, Pharm D, Hartmann AP, Pharm D (2020) Management of chronic obstructive pulmonary disease: A review focusing on exacerbations. Am J Health-Syst Pharm 77(4):259–268. https://doi.org/10.1093/ajhp/zxz306

Bomers M, Van Agtmeal MA, Luik H, Van Veen M et al (2012) Using a dog's superior olfactory sensitivuty to identify clostridium difficile in stools and patients: proof of principle study. BMJ 345:e7396. https://doi.org/10.1136/bmj.e7396

Bowman CR, Zeithamova D (2020) Training set coherence and set size effects on concept generalization and recognition. J Exp Psychol Learn Mem Cogn 46:1442–1464. https://doi.org/10.1037/xlm0000824

Brady K, Cracknell N, Zulch H, Mills DS (2018) Factors associated with long-term success in working police dogs. App Anim Behav Sci 207:67–72. https://doi.org/10.1016/j.applanim.2018.07.003

Bray EE, Otto CM, Udell MAR, Hall NJ, Johnston AM, MacLean EL (2021) Enhancing the selection and performance of working dogs. Front Vet Sci 8:644431. https://doi.org/10.3389/fvets.2021.644431

Bryce E, Zurberg T, Zurberg M, Shajari S, Roscoe D (2017) Identifying environmental reservoirs of Clostridium difficile with a scent detection dog: preliminary evaluation. J Hosp Infect 97(2):140–145. https://doi.org/10.1016/j.jhin.2017.05.023

Buszewski B et al (2012) Identification of volatile lung cancer markers by gas chromatography-mass spectrometry: comparison with discrimination by canines. Anal Bioanal Chem 404(1):141–146

Campbell LF, Farmery L, George SMC et al (2013) Canine olfactory detection of malignant melanoma Case Rep 2013:bcr2013008566

Christensen ED, Harvald T, Jendresen M et al (1997) The impact of delayed diagnosis of lung cancer on the stage at the time of operation. Eur J Cardiothorac Surg 12(6):880–884. https://doi.org/10.1016/S1010-7940(97)00275-3

Church J, Williams H (2001) Another sniffer dog for the clinic? Lancet 358(9285):930. https://doi.org/10.1016/S0140-6736(01)06065-2. PMID: 11575380

Cleland TA, Narla VA, Boudadi K (2009) Multiple learning parameters differentially regulate olfactory generalization. Behav Neurosci 123:26–35

Concha A, Mills DS, Feugier A, Zulch H, Guest C, Harris R et al (2014) Using sniffing behavior to differentiate true negative from false negative responses in trained scent-detection dogs. Chem Senses 39:749–754. https://doi.org/10.1093/chemse/bju045

Concha AR, Guest CM, Harris R, Pike TW, Feugier A, Zulch H et al (2019) Canine olfactory thresholds to amyl acetate in a biomedical detection scenario. Front Vet Sci 5:345. https://doi.org/10.3389/fvets.2018.00345

Cornu JN, Cancel-Tassin G, Ondet V, Girardet C, Cussenot O (2011) Olfactory detection of prostate cancer by dogs sniffing urine: a step forward in early diagnosis. Eur Urol 59:197–201. https://doi.org/10.1016/j.eururo.2010.10.006

Crawford M, Perrone J, Browne C, Chang C et al (2022) Transitioning from training to testing with scent detection animals: application to lung cancer detection dogs. J Vet Behav 55–56:23–34. https://doi.org/10.1016/j.jveb.2022.07.004

Davies JC, Alton E, Simbo A, Murphy R, Seth I, Williams K et al (2019a) Training dogs to differentiate Pseudomonas aeruginosa from other cystic fibrosis bacterial pathogens: not to be sniffed at? Eur Respir J 54:1900970. https://doi.org/10.1183/13993003.00970-2019

Davies C, Alton E, Murphy R, Seth I, Williams K et al (2019b) Training dogs differentiate *Pseudomonas aeruginosa* from other cystic fibrosis bacterial pathogens: not to be sniffed at? Eur Respirat J 54(5):1900970. https://doi.org/10.1183/13993003.00970-2019

de Boer JG et al (2017) Odours of *Plasmodium falciparum*-infected participants influence mosquito-host interactions. Sci Rep 7:9283. https://doi.org/10.1038/s41598-017-08978-9

DHS (2022) Effect of heat and acclimatization on the capability of detection canines: a guide for detection canine handlers. https://www.dhs.gov/sites/default/files/2022-04/2022_0408_st_Battelle_HandlerEducation_508v2.pdf

Edwards TL, Browne CM, Schoon A, Cox C, Poling A (2017) Animal olfactory detection of human diseases: guidelines and systematic review. J Veter Behav (2017). https://doi.org/10.1016/j.jveb.2017.05.002

Ehmann R, Boedeker E, Friedrich U, Sagert J, Dippon J, Friedel G, Walles T (2012) Canine scent detection in the diagnosis of lung cancer: revisiting a puzzling phenomenon. Eur Respir J 39:669e676

Elliker KR, Sommerville BA, Broom DM, Neal DE, Armstrong S, Williams HC (2014) Key considerations for the experimental training and evaluation of cancer odour detection dogs: lessons learnt from a double-blind, controlled trial of prostate cancer detection. BMC Urol 14:9. https://doi.org/10.1186/1471-2490-14-22

Essler JL, Wilson C, Verta AC, Feuer R, Otto CM (2020) Differences in the search behavior of cancer detection dogs trained to have either a sit or stand-stare final response. Front Vet Sci 7:118. https://doi.org/10.3389/fvets.2020.00118

Farha M, Salami S (2022) Biomarkers for prostate cancer detection. Therapeut Adv Urol 14. https://doi.org/10.1177/17562872221103988

Ferkin MH (2018) Odor communication and mate choice in rodents. Biology (Basel) 7(1):13. https://doi.org/10.3390/biology7010013. PMID: 29370074; PMCID: PMC5872039

Fischer S, Trefz P, Bergmann A, Steffens M, Ziller M et al (2015) Physiological variability in volatile organic compounds (VOCs) in exhaled breath and released from faeces due to nutrition and somatic growth in a standardized caprine animal model. J Breath Res 9(2):027108. https://doi.org/10.1088/1752-7155/9/2/027108

Fischer-Tenhagen C, Johnen D, Nehls I, Becker R (2018) A proof of concept: are detection dogs a useful tool to verify potential biomarkers for lung cancer? Front Vet Sci 5:1–6

Foyer P, Svedberg AM, Nilsson E, Wilsson E, Faresjö Å, Jensen P (2016) Behavior and cortisol responses of dogs evaluated in a standardized temperament test for military working dogs. J Vet Behav. 11:7–12. https://doi.org/10.1016/j.jveb.2015.09.00659

Fujita K, Nonomura N (2018) Urinary biomarkers of prostate cancer. Int J Urol 25(9):770–779. https://doi.org/10.1111/iju.13734

Gartman EJ, Jankowich MD, Baptiste J, Nici L (2018) Providence VA lung cancer screening program: performance: comparison of local false positive and invasive procedure rates to published trial data. In: A98. Clinical strategies to improve lung cancer early detection: who is at risk here? American Thoracic Society International Conference Abstracts; American Thoracic Society, New York, NY, USA, p A2477

Gazit I, Goldblatt A, Terkel J (2005) The role of context specificity in learning: the effects of training context on explosives detection in dogs. Anim Cogn 8:143–150. https://doi.org/10.1007/s10071-004-0236-9

Giró Benet J, Seo M, Khine M et al (2022) Breast cancer detection by analyzing the volatile organic compound (VOC) signature in human urine. Sci Rep 12:14873. https://doi.org/10.1038/s41598-022-17795-8

Gordon RT, Schatz CB, Myers LJ, Kosty M, Gonczy C, Kroener J, Tran M et al (2008) The use of canines in the detection of human cancers. J Altern Complement Med 14:61–67. https://doi.org/10.1089/acm.2006.6408

Guerrero-Flores H, Apresa-García T, Garay-Villar Ó, Sánchez-Pérez A, Flores-Villegas D et al (2017) A non-invasive tool for detecting cervical cancer odor by trained scent dogs. BMC Cancer 17:1–8. https://doi.org/10.1186/s12885-016-2996-4

Guest C, Pinder M, Doggett M et al (2019) Trained dogs identify people with malaria parasites by their odour. Lancet Infect Dis 19:578–580

Guest C, Harris R, Sfanos KS, Shrestha E, Partin AW, Trock B et al (2021) Feasibility of integrating canine olfaction with chemical and microbial profiling of urine to detect lethal prostate cancer. PLoS One 16(2):e0245530. https://doi.org/10.1371/journal.pone.0245530

Guest CM, Rob H, Iqbal A, Concha Astrid R, Rooney NJ (2020) A lesson in standardization—subtle aspects of the processing of samples can greatly affect dogs' learning. Front Vet Sci 7. https://doi.org/10.3389/fvets.2020.00525

Hakim M, Broza YY, Barash O, Peled N, Phillips M, Amann A, Haick H (2012) Volatile organic compounds of lung cancer and possible biochemical pathways. Chem Rev 112:5949–5966

Hall NJ, Wynne CDL (2016) Canine olfactory learning and behavior. In: Jezierski T, Ensminger J, Papet LE (eds) Canine olfaction science and law. CRC Press; Taylor and Francis Group, pp 123–37. https://doi.org/10.1201/b20027

Hall NJ, Johnston AM, Bray EE, Otto CM, MacLean EL and Udell MAR (2021) Working dog training for the twenty-first century. Front Vet Sci 8:646022. https://doi.org/10.3389/fvets.2021.646022

Horeweg N, Scholten ET, de Jong PA, van der Aalst CM, Weenink C et al (2014) Detection of lung cancer through low-dose CT screening (NELSON): a prespecified analysis of screening test performance and interval cancers. Lancet Oncol. 2014(15):1342–1350

Horvath G, Andersson H, Paulsson G (2010) Characteristic odour in the blood reveals ovarian carcinoma. BMC Cancer 10:643

Horvath G, Andersson H, Nemes S (2013) Cancer odor in the blood of ovarian cancer patients: a retrospective study of detection by dogs during treatment, 3 and 6 months afterward. BMC Cancer 13:396. https://doi.org/10.1186/1471-2407-13-396

Horvath G, Järverud GAK, Järverud S et al (2008) Human ovarian carcinomas detected by specific odor. Integr Cancer Ther 7:76–80

Huang S, Chong N, Lewis NE, Jia WG, Xie G, Garmire LX (2016) Novel personalized breast cancer diagnosis. Genome Med 8:34. https://doi.org/10.1186/s13073-016-0289-9

Invitto S, Montinaro R, Ciccarese V, Venturella I, Fronda G, Balconi M (2019) Smell and 3D haptic representation: a common pathway to understand brain dynamics in a cross-modal task. A pilot OERP and fNIRS study. Front Behav Neurosci 13:226. https://doi.org/10.3389/fnbeh.2019.00226

La Toya JJ, Baxter GS, Murray PJ (2017) Identifying suitable detection dogs. Appl Animal Behav Sci 195:1–7. https://doi.org/10.1016/j.applanim.2017.06.010

Jenkins EK, DeChant MT, Perry EB (2018) When the nose doesn't know: canine olfactory function associated with health, management, and potential links to microbiota. Front Vet Sci 5:56. https://doi.org/10.3389/fvets.2018

Jezierski T, Walczak M, Ligor T, Rudnicka J, Buszewski B (2015) Study of the art: canine olfaction used for cancer detection on the basis of breath odour Perspectives and limitations. J Breath Res 9:027001. https://doi.org/10.1088/1752-7155/9/2/027001

Johnen D, Heuwieser W, Fischer-Tenhagen C (2017) An approach to identify bias in scent detection dog testing. Appl Anim Behav Sci 2017(189):1–12. https://doi.org/10.1016/j.applanim.2017.01.001

Junqueira,H, Quinn TA, Biringer R, Hussein M, Smeriglio C et al (2019) Accuracy of canine scent detection of non–small cell lung cancer in blood serum. J Am Osteopathic Assoc 119(7):413–418. https://doi.org/10.7556/jaoa.2019.077

Ka-Ho G, Macpherson K, MacDonald H, Roberts W (2020) A comparative study of memory for olfactory discriminations: Dogs (*Canis familiaris*), Rats (*Rattus norvegicus*), and Humans (*Homo sapiens*). J Comp Psych 134(2):170–179. https://doi.org/10.1037/com0000205

Kane SA, Lee YE, Essler JL, Mallikarjun A, Preti G, Plymouth VL, Verta A, DeAngelo A, Otto CM (2022) Canine discrimination of ovarian cancer through volatile organic compounds. Talanta 250:123729. https://doi.org/10.1016/j.talanta.2022.123729. Epub 2022 Jul 11. PMID: 35839605

Kasstan B, Hampshire K, Guest C, Logan JG, Pinder M, Williams K, D'Alessandro U, Lindsay SW (2019) Sniff and tell: the feasibility of using bio-detection dogs as a mobile diagnostic intervention for asymptomatic malaria in sub-Saharan Africa. J biosoc Sci 51(3):436–443

Kavaliers M, Choleris E, Pfaff DW (2005) Recognition and avoidance of the odors of parasitized conspecifics and predators: differential genomic correlates. Neurosci Biobehav Rev 29:1347–1359. https://doi.org/10.1016/j.neubiorev.2005.04.011

Kiesecker JM, Skelly DK, Beard KH, Preisser E (1999) Behavioral reduction of infection risk. Proc Natl Acad Sci USA 96:9165–9168. https://doi.org/10.1073/pnas.96.16.9165

Konno A et al (2018) Effect of canine oxytocin receptor gene polymorphism on the successful training of drug detection dogs. J Hered 109:566–572. https://doi.org/10.1093/jhered/esy012,Pubmed:29538681

Kure S, Iida S, Yamada M, Takei H et al (2020) Breast cancer detection from a urine sample by dog sniffing. Research Square. https://doi.org/10.21203/rs.3.rs-89484/v1

Kure S, Iida S, Yamada M, Takei H, Yamashita N et al (2021) Breast cancer detection from a urine sample by dog sniffing: a preliminary study for the development of a new screening device, and a literature review. Biology 10(6):517. https://doi.org/10.3390/biology10060517

Lazarowski L, Krichbaum S, DeGreeff LE, Simon A, Singletary M, Angle C, Waggoner LP (2020) Methodological considerations in canine olfactory detection research. Front Vet Sci 7:408. https://doi.org/10.3389/fvets.2020.00408

Lesniak A et al (2008) Canine olfactory receptor gene polymorphism and its relation to odor detection performance by sniffer dogs. J Hered 99:518–527. https://doi.org/10.1093/jhered/esn057

Ligor T, Adamczyk P, Kowalkowski T, Ratiu IA, Wenda-Piesik A, Buszewski B (2022) Analysis of VOCs in urine samples directed towards of bladder cancedetection. Molecules 2022(27):5023. https://doi.org/10.3390/molecules27155023

Lima AR, Pinto J, Azevedo AI, Barros-Silva D, Jerónimo C, Henrique R et al (2019) Identification of a biomarker panel for improvement of prostate cancer diagnosis by volatile metabolic profiling of urine. Br J Cancer. 121(10):857–868. pmid:31588123

Lippi G, Heaney LM (2020) The "olfactory fingerprint": can diagnostics be improved by combining canine and digital noses? Clin Chem Lab Med 58:958–967. https://doi.org/10.1515/cclm-2019-1269

Maejima M, Inoue-Murayama M, Tonosaki K, Matsuura N, Kato S, Saito Y et al (2007) Traits and genotypes may predict the successful training of drug detection dogs. Appl Anim Behav Sci 107:287–298. https://doi.org/10.1016/j.applanim.2006.10.005
Mancini C, Harris R, Aengenheister B, Guest C (2015) Re-centering multispecies practices: a canine interface for cancer detection dogs. In: Conference on human factors in computing systems—proceedings, pp 2673–2682. https://dl.acm.org/doi/proceedings/10.1145/2702123
Marchal S, Bregeras O, Puaux D, Gervais R, Ferry B (2016) Rigorous training of dogs leads to high accuracy in human scent matching-to-sample performance. PLoS One 11(2):e0146963
Maughan MN, Best EM, Gadberry JD, Sharpes CE, Evans KL,Chue CC, Nolan PL and Buckley PE (2022) The use and potential of biomedical detection dogs during a disease outbreak. Front Med 9:848090. https://doi.org/10.3389/fmed.2022.848090
Maurer M, McCulloch M, Willey AM et al (2016) Detection of bacteriuria by canine olfaction. Open Forum Infect Dis 3:ofw051
Mazzatenta A, Pokorski M, Giulio C (2021) Volatile organic compounds (VOCs) in exhaled breath as a marker of hypoxia in multiple chemical sensitivity. Physiol Rep 9(18):e15034. https://doi.org/10.14814/phy2.15034
McCulloch M, TurnerK Broffman M (2010) Lung cancer detection by canine scent: will there be a lab in lab? Eur Respir J 39:511–512
McCulloch M, Jezierski T, Broffman M et al (2006) Diagnostic accuracy of canine scent detection in early- and late-stage lung and breast cancers. Integr Cancer Ther 5:30–39
Merriel SWD, Pocock L, Gilbert E et al (2022) Systematic review and meta-analysis of the diagnostic accuracy of prostate-specific antigen (PSA) for the detection of prostate cancer in symptomatic patients. BMC Med 20:54. https://doi.org/10.1186/s12916-021-02230-y
Mischley L, Farahnik J (2020) Canine aromatic detection of parkinson's disease: can dogs identify prodromal parkinsonism? International Congress of Parkinson's Disease and Movement Disorders. Available online at: https://www.mdsabstracts.org/abstract/canine-aromatic-detection-of-parkinsons-disease-can-dogs-identify-prodromal-parkinsonism/ (accessed August 28, 2023)
Montes GA, Molins López-Rod6L, Ramón Rodríguez I et al (2017) Lung cancer diagnosis by trained dogs. Eur J Cardio-Thorac Surg: J Eur Assoc Cardio-Thorac Surg 52(6):1206–1210. https://doi.org/10.1093/ejcts/ezx152
More TH, Choudhury RS, Christie J, Taunk K, Mane A, Santra MK, Choudhury K, Rapole S (2018) Metabolomic alterations in invasive ductal carcinoma of breast: comprehensive metabolomic study using tissue and serum samples. Oncotarget 9:2678–2696. https://doi.org/10.18632/oncotarget.23626
Moser E, McCulloch M (2010) Canine scent detection of human cancers: a review of methods and accuracy. J Vet Behav: Clin Appl Res 5(3):145–152. https://doi.org/10.1016/j.jveb.2010.01.002
Moser A, Bizo L, Brown W (2019) Olfactory generalization in detection dogs. Animals 9:702. https://doi.org/10.3390/ani9090702
Moshkin M, Litvinova N, Litvinova EA, Bedareva A, Lutsyuk A, Gerlinskaya L (2012) Scent recognition of infected status in humans. J Sex Med 9:3211–3218. https://doi.org/10.1111/j.1743-6109.2011.02562.x
Nooreldeen R, Bach H (2021) Current and future development in lung cancer diagnosis. Int J Mol Sci 2021(22):8661. https://doi.org/10.3390/ijms22168661
Ohuchi N, Suzuki A, Sobue T, Kawai M, Yamamoto S, Zheng YF, Shiono YN, Saito H, Kuriyama S, Tohno E, Endo T, Fukao A, Tsuji I, Yamaguchi T, Ohashi Y, Fukuda M, Ishida, T (2016) Sensitivity and specificity of mammography and adjunctive ultrasonography to screen for breast cancer in the Japan Strategic Anti-cancer Randomized Trial (J-START): a randomised controlled trial the lancet 387(10016):341–348. https://doi.org/10.1016/S0140-6736(15)00774-6
Olsson MJ, Lundström JN, Kimball BA, Gordon AR, Karshikoff B et al (2014) The scent of disease: human body odor contains an early chemosensory cue of sickness. Psychol Sci 25(3):817–823. https://doi.org/10.1177/0956797613515681
Ortal A, Rodríguez A, Solis-Hernández MP et al (2022) Proof of concept for the use of trained sniffer dogs to detect osteosarcoma. Sci Rep 12:6911. https://doi.org/10.1038/s41598-022-11013-1

Phillips M (1992) Breath tests in medicine. Sci Am 267(1):74–79. https://doi.org/10.1038/scientificamerican0792-74
Phillips M, Cataneo RN, Cheema T, Greenberg J (2004) Increased breath biomarkers of oxidative stress in diabetes mellitus. Clin Chim Acta 344:189–194
Phillips M, Cataneo RN, Saunders C, Hope P, Schmitt P, Wai J (2010) Volatile biomarkers in the breath of women with breast cancer. J Breath Res 4(2):026003. https://doi.org/10.1088/1752-7155/4/2/026003
Picket DP, Manucy GP, Walker DB, Hall SB, Walker JC (2004) Evidence of canine olfactory detection of melanoma. Appl Animal Behav Sci 89(1–2):107–116
Picot S, Cucherat M, Bienvenu AL (2020) Systematic review and meta-analysis of diagnostic accuracy of loop-mediated isothermal amplification (LAMP) methods compared with microscopy, polymerase chain reaction and rapid diagnostic tests for malaria diagnosis. Int J Infect Dis 98:408–419. https://doi.org/10.1016/j.ijid.2020.07.009
Pinto J, Carapito Â, Amaro F, Lima AR, Carvalho-Maia C, Martins MC (2021) Discovery of volatile biomarkers for bladder cancer detection and staging through urine metabolomics. Metabolites 11:199
Pirrone F, Albertini M (2017) Olfactory detection of cancer by trained sniffer dogs: a systematic review of the literature. J Vet Behav 19:105–118
Porritt F, Shapiro M, Waggoner P, Mitchell E, Thomson T, Nicklin S et al (2015) Performance decline by search dogs in repetitive tasks, and mitigation strategies. Appl Anim Behav Sci 166:112–122. https://doi.org/10.1016/j.applanim.2015.02.01362
Probert CSJ, Ahmed I, Khalid T, Johnson E, Smith S, Ratcliffe N (2009) Volatile organic compounds as diagnostic biomarkers in gastrointestinal and liver diseases. J Gastrointestin Liver Dis 18:337–343
Pulido H, Stanczyk NM, De Moraes CM et al (2021) A unique volatile signature distinguishes malaria infection from other conditions that cause similar symptoms. Sci Rep 11:13928. https://doi.org/10.1038/s41598-021-92962-x
Regenbogen C, Axelsson J, Lasselin J et al (2017) Behavioral and neural correlates to multisensory detection of sick humans. PNAS Biol Sci 114(24):6400–6405. https://doi.org/10.1073/pnas.1617357114
Renault J, Gheusi G, Aubert A (2008) Changes in social exploration of lipopolysaccharides-treated conspecific in mice: role of environmental cues. Brain Behav Imm 22(8):1201–1207. https://doi.org/10.1016/j.bbi.2008.05.008
Riedlova P, Tavandzis S, Kana J, Tobiasova M (2022) Olfactometric diagnosis of lung cancer by canine scent—a double-blinded study. Complement Ther Med 64:102800. https://doi.org/10.1016/j.ctim.2022.102800
Rist MJ, Muhle-Goll C, Görling B, Bub A, Heissler S, Watzl B, Luy B (2013) Influence of freezing and storage procedure on human urine samples in NMR-based metabolomics. Metabolites 2013(3):243–258. https://doi.org/10.3390/metabo3020243
Robinson A, Busula AO, Voets MA et al (2018) Plasmodium-associated changes in human odor attract mosquitoes. Proc Natl Acad Sci USA 115:e4209–e4218
Rudnicka J, Walczak M, Kowalkowski T, Jezierski T, Buszewski B (2014) Determination of volatile organic compounds as potential markers of lung cancer by gas chromatography–mass spectrometry versus trained dogs. Sens Actuat B: Chem 202:615–621. https://doi.org/10.1016/j.snb.2014.06.006
Sakr R, Ghsoub C, Rbeiz C et al (2022) COVID-19 detection by dogs: from physiology to field application—a review article. Postgrad Med J 98:212–218
Schmidt K, Podmore I (2015) Current challenges in volatile organic compounds analysis as potential biomarkers of cancer. J Biomark 2015:981458. https://doi.org/10.1155/2015/981458
Schoon A, De Jonge D, Hilverink P (2020) How dogs learn to detect colon cancer—optimizing the use of training aids. J Vet Behav 35:38–44
Shirasu M, Touhara K (2011) The scent of disease: volatile organic compounds of the human body related to disease and disorder. J Biochem 150(3):257–266

Siegel A et al (2017) Analyzing breath samples of hypoglycemic events in type 1 diabetes patients: towards developing an alternative to diabetes alert dogs. J Breath Res 11. https://doi.org/10.1088/1752-7163/aa6ac6

Silva CL, Passos M, Câmara JS (2012) Solid phase microextraction, mass spectrometry and metabolomic approaches for detection of potential urinary cancer biomarkers—a powerful strategy for breast cancer diagnosis. Talanta 89:360–368. https://doi.org/10.1016/j.talanta.2011.12.041

Sonoda H et al (2011) Colorectal cancer screening with odour material by canine scent detection. Gut 60:814–819

Sung et al (2021) Globocan Global Cancer Statistics 2020: GLOBOCAN estimates of incidence and mortality worldwide for 36 cancers in 185 countries. A Cancer Journal for Clinicians, Wiley Online Library, CA

Taverna G, Tidu L, Grizzi F, Torri V, Mandressi A, Sardella P, La Torre G et al (2015a) Olfactory system of highly trained dogs detects prostate cancer in urine samples. J Urol 193:1382–1387. https://doi.org/10.1016/j.juro.2014.09.099

Taverna G, Tidu L, Grizzi F, Stork B, Mandressi A, Seveso M, Bozzini G et al (2015b) The ability of dogs to detect human prostate cancer before and after radical prostatectomy. EC Vet Sci 2:47–51

Taylor M, McCready J, Broukhanski G, Kirpalaney S, Lutz H, Powis J (2018) Using dog scent detection as a point-of-care tool to identify toxigenic *Clostridium difficile* in Stool. Open Forum Infect Dis 5(8), August. https://doi.org/10.1093/ofid/ofy179

Tohno E, Ueno E, Watanabe H (2009) Ultrasound screening of breast cancer. Breast Cancer 16:18–22

Tong MT, Peace ST, Cleland TA (2014) Properties and mechanisms of olfactory learning and memory. Front Behav Neurosci 8:238. https://doi.org/10.3389/fnbeh.2014.00238

Urbanova L, Vylmankova V, Krisova S, Pacik D, Necas A (2015) Intensive training technique utilizing the dog's olfactory abilities to diagnose prostate cancer in men. Acta Vet BRNO 84:77–82. https://doi.org/10.2754/avb201585010077

van Bomers MA, Agtmael HL, Vandenbroucke-Grauls CMJE, Smulders YM (2014) A detection dog to identify patients with Clostridium difficile infection during a hospital outbreak. J Infect 69(5):456–461. https://doi.org/10.1016/j.jinf.2014.05.017

Waggoner P, Lazarowski L, Hutchings B, Angle C, Porritt F (2022) Effects of learning an increasing number of odors on olfactory learning, memory and generalization in detection dogs. Appl Animal Behav Sci 247:105568. https://doi.org/10.1016/j.applanim.2022.105568

Walczak M (2009) Operant conditioning of dogs for detection of odour markers of cancer diseases (in Polish). PhD thesis. Institute of Genetics and Animal Breeding of Polish Academy of Sciences

Walczak M, Jezierski T, Górecka-Bruzda A, Sobczyńska M, Ensminger J (2012a) Impact of individual training parameters and manner of taking breath samples on the reliability of canines as cancer screeners. J Vet Behav Clin Appl Res 7:283–294

Walczak M, Jezierski T, Górecka-Bruzda A, Sobczyńska M, Ensminger J (2012b) Impact of individual training parameters and manner of taking breath odor samples on the reliability of canines as cancer screeners. J Vet Behav Clin Appl Res 7(5):283e294

Weiss T, Snitz K, Yablonka A, Khan R, Gafsou D, Schneidman E et al (2012) Perceptual convergence of multi-component mixtures in olfaction implies an olfactory white. In: Proceedings of the national academy of sciences of the United States of America, vol 109. pmid:23169632

Willette AA, Lubach GR, Coe CL (2007) Environmental context differentially affects behavioral, leukocyte, cortisol, and interleukin-6 responses to low dose of endotoxin in the rhesus monkey. Brain Behav Immun 21:807–815

William M, Johnston JM (2002) Training and maintaining the performance of dogs (*Canis familiaris*) on an increasing number of odor discrimination in a controlled setting. Appl Anim Behav Sci 78:55–65

Williams H, Pembroke A (1989) Sniffer dogs in the melanoma clinic? Lancet (London, England) 1(8640):734. Epub 1989/04/01. https://doi.org/10.1016/s0140-6736(89)92257-5

Willis CM, Britton LE, Harris R, Wallace J, Guest CM (2010) Volatile organic compounds as biomarkers of bladder cancer: sensitivity and specificity using trained sniffer dogs. Cancer Biomark 8:145–153. https://doi.org/10.3233/cbm-2011-0208

Willis CM, Church SM, Guest CM, Cook WA, McCarthy N, Bransbury AJ et al (2004) Olfactory detection of human bladder cancer by dogs: proof of principle study. BMJ (Clinical Research ed). 329(7468):712. Epub 2004/09/25. https://doi.org/10.1136/bmj.329.7468.712

Wilson DA, Stevenson RJ (2003) Olfactory perceptual learning: the critical role of memory in odor discrimination. Neurosci Biobehav Rev 2003(27):307–328. https://doi.org/10.1016/S0149-7634(03)00050-2

Wong MCS, Huang J, Lok V, Wang J, Fung F, Ding H, Zheng ZJ (2021) Differences in incidence and mortality trends of colorectal cancer worldwide based on sex, age, and anatomic location. Clin Gastroenterol Hepatol 19(5):955–966.e61. https://doi.org/10.1016/j.cgh.2020.02.026. Epub 2020 Feb 21. PMID: 32088300

Wright HF, Wilkinson A, Croxton RS, Graham DK, Harding RC, Hodkinson HL, Keep B, Cracknell NR, Zulch HE (2017) Animals can assign novel odours to a known category. Sci Rep 7:9019

Yamamoto A, Kamoi S, Kurose K, Ito M, Takeshita T, Kure S, Sakamoto K, Sato Y, Miyashita M (2020) The trained sniffer dog could accurately detect the urine samples from the patients with cervical cancer, and even cervical intraepithelial neoplasia grade 3: a pilot study. Cancers (Basel) 12(11):3291. https://doi.org/10.3390/cancers12113291

Yang M, Zhang HX, Geng GJ, Wang FJ, Liu CW, Liu JL (2022) Artificial selection drives SNPs of olfactory receptor genes into different working traits in labrador retrievers. Genet Res (Camb) 8319396. https://doi.org/10.1155/2022/8319396. PMID: 35185392

Yuan BW, Schafferer S, Tang QQ, Scheffler M, Nees J, Heil J et al (2019) A plasma metabolite panel as biomarkers for early primary breast cancer detection. Int J Cancer 144:2833–2842. https://doi.org/10.1002/ijc.31996

Zhou Y, Song R, Ma C, Zhou L, Liu X, Yin P, Zhang Z, Sun Y, Xu C, Lu X et al (2017) Discovery and validation of potential urinary biomarkers for bladder cancer diagnosis using a pseudotargeted GC-MS metabolomics method. Oncotarget 8:20719–20728. https://doi.org/10.18632/oncotarget.14988

Zhu S, Corsetti S, Wang Q, Li C, Huang Z, Nabi G (2019) Optical sensory arrays for the detection of urinary bladder cancer-related volatile organic compounds. J Biophotonics 2019(12):e201800165

Detecting Physiological Changes in Humans: Medical Alert and Assistance Dogs

Catherine Reeve and Clara Wilson

Abstract

In this chapter, Reeve and Wilson provide a thorough review and discussion of medical alert dogs: dogs that alert people to physiological changes. The authors first orient the reader with a discussion of the terminology surrounding, and the regulation of, medical alert dogs. Next, they discuss how odours emanating from the body signal physiological change, and how the canine olfactory system is well suited to perceive these odours. The chapter then includes a review of empirical studies examining dogs' ability to detect odours associated with physiological change and how medical alert dogs impact their owners' health and well-being. The authors explore the current training and selection of medical alert dogs and, finally, present a discussion of medical technology and the future of medical alert dogs.

Keywords

Medical alert dog • Assistance dog • Canine olfaction • Health • Well-being • Dogs and humans

1 Introduction

Dogs' evolutionary history alongside humans has resulted in a profound interspecies relationship. Research has elucidated dogs' sensitivity to human social and emotional cues and, more recently, their sensitivity to human physiological and psychological changes. Anecdotal reports of dogs' behaviour changing in response to their owners' experiencing a health event (e.g. Chen et al. 2000)

C. Reeve (✉) · C. Wilson
Animal Behaviour Centre, School of Psychology, David Keir Building, Queen's University Belfast, Belfast BT7 1NN, UK
e-mail: c.reeve@qub.ac.uk

L. Lazarowski (ed.), *Olfactory Research in Dogs*,
https://doi.org/10.1007/978-3-031-39370-9_13

have spurred a burgeoning field of research examining dogs' ability to detect and alert to human physiological change. In a recent survey, dog owners reported that their dogs alerted to a host of physiological changes (Reeve et al. 2021), including hypo- and hyperglycemia, epileptic seizures, postural orthostatic tachycardia syndrome (PoTS), allergic reactions, asthma, chronic obstructive pulmonary disorder (COPD), migraines, and complications associated with Addison's disease. Empirical, lab-based research has further supported aspects of this phenomenon (see Sect. 3). Alongside such reports, charities now train and place dogs that detect and alert people to a wide range of medical conditions. These dogs assist an individual with a specific health condition and have come to be known as Medical Alert Dogs (hereafter referred to as MADs). MADs are a specific type of assistance dog, as they primarily use their sense of smell to alert people to changes in physiological status and impending medical crises.

In the following chapter, the authors will present a discussion on the current state of research on MADs. To begin, we will briefly examine the terminology and regulation surrounding assistance dogs and MADs. This will be followed by a discussion of the role of volatile organic compounds (VOCs) and dog olfaction in MADs' detection of human physiological changes. Next is a discussion of the main conditions to which dogs are reported to alert and the empirical research on MADs for these conditions, as well as their impact on the well-being of their owners. The penultimate section will present a brief discussion of the characteristics of dogs that alert to medical conditions. Lastly, the authors present a view of the future of MADs and their place in humans' lives.

2 Terminology and Regulation

Dogs are used globally for an increasing number of assistance roles. The term "assistance dog" encompasses dogs trained for a wide range of practical tasks that allow people to live more independently and with more confidence than they might otherwise. For example, assistance dogs are trained to help people with sight loss, hearing loss, physical mobility problems and psychiatric conditions, among others. This chapter will focus on MADs, who are trained to respond to owners with medical conditions such as epilepsy, diabetes, and more.

The term, "Medical Alert Dog" is used to describe dogs that alert people to impending physiological changes, for example, an impending seizure. They are sometimes differentiated from "Medical Response Dogs", which are dogs trained to respond in certain ways after a medical crisis has occurred. For example, a MAD trained for seizure alert would signal to a person an impending seizure, whereas a Medical Response Dog would lie with or on a person during a seizure in such a way as to keep them physically safe. MADs can be further differentiated from "Medical Detection Dogs", or "Biological Detection Dogs", which are the terms commonly used to label dogs that are trained to detect disease in a lab setting (e.g. Edwards et al. 2017).

The training and certification of MADs are not widely regulated. In fact, there is no national-level regulation for the training of MADs in America, Canada, or the United Kingdom.

Within the United Kingdom, the Equality and Human Rights Commission (2017) that oversees human rights in England, Scotland, and Wales states that assistance dogs should be highly trained, but by whom, and with what accreditation(s) is not specified. MADs are included under the wider umbrella of Assistance Dogs, and are required to be under control in public. It is suggested that these dogs are instantly recognisable by the harness or identifying dog jacket (Assistance Dogs 2022). Likewise, the Americans with Disabilities Act states that assistance dogs (including MADs for seizure and hypoglycemia alert) are not required to be certified through a professional program and are not required to wear a vest or ID identifying them as a service dog (Americans with Disabilities Act 2022). This lack of regulation has, unfortunately, led to some reports of poor business practices and insufficiently trained dogs (AP News 2021).

Despite the lack of government regulation, Assistance Dogs International (ADI) is a charity that seeks to bring some level of standardisation to assistance dog programs. ADI is a global community of not-for-profit groups that train and place assistance dogs and is considered to be the foremost authority on assistance dogs. Their mission is to apply rigorous standards to the training and treatment of assistance dogs and the clients they aim to help. As such, ADI offers accreditation to charities that train MADs to their standards. To be accredited, an individual charity is evaluated in a number of areas including, but not limited to, the humane training of the dogs, and the ethical treatment of clients. Charities seeking accreditation must also provide ongoing documentation to demonstrate the dogs' continued performance. Currently, there are seven charities within the United Kingdom documented by ADI as accredited providers of Assistance Dogs, two of which train MADs, and 78 accredited providers within the United States, seven of which train MADs (Assistance Dogs International 2022). While accreditation is not required for public access under law, it does mean that the dog is recognised as meeting the criteria for the Equality and Human Rights Commission's definition of an assistance dog more readily.

The demand for assistance dogs has been shown to increase substantially in the last decade (Walther et al. 2017). It is therefore important to document the field at this stage, and specify what has, and has not, been established in the scientific literature. As the premise underlying MADs is dogs detecting changes within the human body, it is first necessary to understand this process.

3 Volatile Organic Compounds and Canine Olfaction

As touched on in the introduction, dogs are reported to detect and alert to a wide range of physiological changes in humans. Before we discuss in detail the conditions they are reported to alert to, we will first present and discuss the most accepted hypothesis regarding what it is that medical alert dogs are detecting.

The current theory is that dogs are detecting changes in volatile organic compounds (VOCs) associated with physiological change. VOCs are compounds with high vapour pressure that are gaseous at room temperature (Dewulf et al. 2002) and are emitted from the human body. Humans emit a wide variety of different VOCs. In fact, a recent review documented over 4,000 different VOCs emitted in the breath, and found in the headspace of blood, faeces, milk, saliva, semen, skin, and urine of healthy humans (Drabińska et al. 2021). This high number of VOCs is due to the fact that each individual has a VOC profile consisting of the presence or absence of different VOCs and VOCs in different concentrations depending on their age, diet, metabolism, medications, microbiota, and more (Shirasu & Touhara 2011). In addition to inter-person variability in VOC profiles, the VOC profile of a single individual will also change throughout the day (Phillips et al. 2014) due to metabolic processes (endogenous, or internally created VOCs), and external factors such as ambient VOCs in the air, and/or the food they have recently eaten (exogenous, or externally created VOCs) (Pleil et al. 2013). Another factor that will contribute to changes in endogenous VOCs is physiological changes in the form of metabolic change or disease (Shirasu & Touhara 2011).

When the cells of the body change, new VOCs may be generated, and/or the concentrations of VOCs already present may change. The changes in VOCs can therefore, theoretically, serve as biomarkers, or "signatures", for specific conditions (Shirasu & Touhara 2011). Despite the wide variety of VOCs emitted by people and their fluctuating nature, modern analytical techniques (see Jalal et al. 2018 for a review) have allowed researchers to document VOC profiles associated with specific conditions, and report differences in VOC profiles between people in a state of physiological change or pathology (e.g. hypoglycemia) versus healthy controls. For example, in people with Type 1 Diabetes, an increase in isoprene concentration has been documented during hypoglycemia (Los et al. 2017; Neupane et al. 2016) and during hyperglycemia, strong correlations between blood glucose levels and acetone concentration (Saasa et al. 2018) and blood glucose and methyl nitrate concentration (Novak et al. 2007) have been found. To account for inter- and intra-individual variability, other researchers have examined clusters of different VOCs rather than individual VOCs and found that a combination of specific breath VOCs may predict hypoglycemia before changes in blood glucose (Siegel et al. 2019) and other clusters that correlate with blood glucose levels (Dixit et al. 2021; Minh et al. 2011).

The analysis of VOCs has been shown to have diagnostic and predictive utility for epilepsy as well. Van Dartel et al. (2020) found that analysis of breath VOCs differentiated epileptic patients from controls and Catala et al. (2020a) found that analysis of VOCs in breath and sweat could be used to develop a model to accurately *predict* seizures. While Van Dartel et al. (2020) and Catala et al. (2020a) used overall VOC profiles, Davis (2017) found the compound menthone to be specifically related to seizure activity.

Evidence suggests that pulmonary diseases and exacerbations of these conditions can also be characterised by VOCs. In a review of the literature, Cavaleiro Rufo et al. (2016) revealed a long list of discriminant compounds for the diagnosis of asthma from controls, and van Bragt et al. (2021) found that a pattern of VOCs could indicate recent exacerbations in those with asthma. Bessa et al. (2011) found VOCs differed between people with COPD and healthy controls and Van Berkel et al. (2010) found a cluster of 13 different VOCs that classified participants as having COPD.

Further research suggests that physiological changes associated with psychological states are also identifiable by VOCs (see Zamkah et al. 2020 for a discussion of biomarkers for stress and emotion). Martin et al. (2016) and Tsukuda et al. (2019) analysed skin VOCs in people before and after an experimentally induced stressful experience (a paced auditory serial addition task and the Trier Social Stress Task, respectively) and found that the two psychological states were distinguishable. Martin et al. (2016) found that benzoic acid and *n*-decanoic acid were upregulated and a xylene isomer and 3-carene were down-regulated whereas Tsukuda et al. (2019) found that increases in a cluster of six different compounds (1,2-ethanediol, acetophenone, heptadecane, hexanedioic acid dimethyl ester, benzyl alcohol, and benzothiazole) served as a marker of psychological stress.

Considering MADs, evidence suggests that dogs can perceive changes in VOCs through their sense of smell. A canine's sense of smell is their primary mode of perception. Dogs gather critical information, essential for being aware of potential predators, locating food, identifying conspecifics (and their reproductive status), and enabling recognition of familial members (Horowitz 2020; McClanahan & Rosell 2020; Samuel et al. 2020) through odours. Their olfactory acuity is the result of their nose, olfactory cavity, and epithelium (Craven et al. 2010), and millions of olfactory receptors (Lawson et al. 2012; Lesniak et al. 2008), that are incredibly well adapted to take in different kinds of olfactory information. Research has shown that dogs can detect target odours in mixtures and in the presence of extraneous odours in concentrations as low as parts per trillion (Concha et al. 2019; Waggoner et al. 1998; Walker et al. 2006), making them comparable in sensitivity to the analytical techniques used in the studies above (Lippi & Heaney 2020). When applied to human physiological changes, this incredible acuity allows them to detect the odour signatures associated with changes in the human body.

As will be discussed below, anecdotal evidence and empirical data support the hypothesis that dogs detect odours associated with human physiological changes. A dog's superior sense of smell, coupled with an ability to train dogs to carry out specific tasks, opens opportunities to shape a process whereby a dog is able to detect VOCs (odour cues) associated with a physiological change and then "alert" their owner to this change.

4 Conditions for Which Dogs Are Trained to Assist

MADs have been utilised to provide non-invasive alert and assistance for various health conditions, including the detection of hypo- and hyperglycemic episodes for people living with Type 1 Diabetes (Gonder-Frederick et al. 2017; Wilson et al. 2019), oncoming epileptic seizures (Strong et al. 2002), migraines (Marcus & Bhowmick 2013), episodes associated with Postural Orthostatic Tachycardia Syndrome (POTS) and Addison's Disease (Medical Detection Dogs 2022a). Furthermore, service dog training organisations supply dogs that either respond to those having an asthma attack (US Service Animals 2022a), or claim to alert prior to the onset of an asthma attack (Maltese Assistance Dogs Center 2022), Chronic Obstructive Pulmonary Disease (COPD) (SIT Service Dogs 2022), and narcolepsy (US Service Animals 2022b). For most of these roles, it is assumed that dogs are detecting VOCs associated with the conditions that signal a change in the owner's physiological state. A distinct type of MAD is an allergen detection dog, which is trained to detect airborne allergens for owners with severe allergies (Medical Detection Dogs 2022b). In this case, the dog is trained to detect the odours of the food items themselves, rather than the owner's VOC profiles. For some of the health conditions listed, empirical studies support the claim that dogs can detect an odour associated with the physiological changes that accompany the conditions (i.e. diabetes, seizures), whereas for others, controlled studies testing the efficacy of these dogs have yet to be carried out (e.g. PoTS, Addison's disease, narcolepsy, asthma attacks). The subsections below provide more details on each condition.

4.1 Diabetic Alert Dogs

There are an estimated 415 million people worldwide living with diabetes (International Diabetes Federation 2022). Of those, approximately 9 million are currently living with Type 1 diabetes, the incidence of which is increasing by around 4% each year (Green et al. 2021). Without extraneous insulin intervention, blood glucose levels are susceptible to becoming too low (hypoglycemia) or too high (hyperglycemia). Hypoglycemia is a serious and prevalent complication of both Type 1 and Type 2 diabetes. Mild episodes can interfere with everyday functioning, while a severe episode requires intervention from another person and, if left untreated, can be fatal (Brands et al. 2005; Emerging Risk Factors Collaboration 2011). People with Type 1 diabetes can become unaware of the symptoms of hypoglycemia over time, which is associated with an increased risk of mortality (Seaquist et al. 2012). Fear of hypoglycemia causes some people to restrict their lifestyle in efforts to reduce the likelihood of an episode, which negatively impacts both their psychological well-being and quality of life (Tenzer-Iglesias & Shannon 2012). Individuals may intentionally maintain hyperglycemia because of fear of a severe hypoglycemic episode (Fidler et al. 2011). This practice confers various associated health risks over time, such as nerve and kidney damage, and cardiovascular disease (Amiel et al. 2019; Zoungas et al. 2010). While an array of developing technologies is available to people with Type 1 diabetes, many are invasive, requiring either finger-pricks or sensor insertion, and can carry considerable physical equipment or financial burdens (e.g. sensor replacement) (Gonder-Frederick et al. 2017). Diabetic Alert Dogs (DADs) potentially offer a non-invasive method of assisting in the recognition of an oncoming hypo- or hyperglycemic episode by alerting their owner when they can still take action (Pesterfield & Guest 2015). As such, DADs have become increasingly popular over the last decade as a method to facilitate tightened glycaemic control (Walther et al., 2017).

Chen et al. (2000), O'Connor et al. (2008), and Tauveron et al. (2006) were among the first to document cases of dogs alerting people to hypo- and hyper-glycemic episodes in the literature. These authors reported case studies of dogs whose behaviour repeatedly changed during their owners' hypoglycemic episodes, often before the owners themselves were aware of their low blood glucose levels. The dogs were reported to nudge owners awake, hide from their owners (Chen et al. 2000), stare, or bark at their owners during hypoglycemia (Tauveron et al. 2006). Although these were case studies of a handful of dogs, a subsequent survey of 225 dog owners with diabetes revealed that a substantial proportion of owners (65.1%) reported that their dogs' behaviour changed in response to at least one hypoglycemic episode, and 31.9% of owners reported that their dogs' behaviour changed in response to 11 or more episodes (Wells et al. 2008). Following these reports, researchers began empirically examining DADs. Given the wealth of evidence for dogs' olfactory acuity, combined with the literature on VOCs and hypo- and hyperglycemia, researchers began exploring the role of olfaction in dogs' ability to detect fluctuations in blood sugar levels. At the same time, dog training organisations began training DADs and placing them in the homes of people with Type 1 Diabetes.

In one of the first laboratory-based studies of DADs, Dehlinger et al. (2013) sought to examine whether odour as an isolated variable (e.g. independent of potential behavioural cues) allowed dogs to identify hypoglycemia. Dehlinger et al. (2013) evaluated three dogs that were already trained and placed with people with Type 1 Diabetes using samples taken from individuals unknown to the dogs. Researchers gathered sweat samples, on cotton swabs, from these three people during hypoglycemic and euglycemic (normal blood sugar level) episodes. The samples were presented to the dogs and researchers observed whether the dogs performed their trained alert behaviour (ringing a bell beside the sample). The dogs' combined sensitivity and specificity were 55.5% and 52.8%, respectively. While these initial results would suggest that dogs were not very able to discriminate between low and normal blood sugar levels, the study methodology and dog training needs to be examined further. Firstly, the training history of the dogs was not provided. These dogs were paired with people with Type 1 Diabetes; however, the samples were collected from unknown individuals. Considering the variability in VOCs across people, hypoglycemia may smell different from one person to another, and it is possible that the dogs had not suitably been trained to generalise hypoglycemic odour as they were not presented enough different people's samples to be able to generalise. Indeed, in Reeve et al.'s (2020) study, dogs were directly tested to see whether they could generalise across samples taken from the same individual's hypo-, hyper-, and euglycemic samples. It was found that dogs varied in their ability to do so, with one dog able to generalise across samples (sensitivity 62%, specificity 86%) while the other did not. In Hardin et al.'s (2015) laboratory study, six dogs were trained to discriminate between hypo- and euglycemic samples. Here, dogs were shown to be able to identify the hypoglycemic sample with a sensitivity of 77.6% and a specificity of 95.8%. It should be noted that the same samples were used during training and testing which may mean that dogs were memorising odours for which they had previously been rewarded. Their ability to do so, even if this were the case, does, however, suggest that the odours were able to be discriminated between.

Diabetic Alert Dogs have received the most empirical examination as compared to other Medical Assistance roles. The previous studies consist of owner-informed case reports, in vitro laboratory testing, and records of the dogs once placed with an owner. The majority of the studies assessing placed DAD performance include owner-reported information in at least one element of data collection, which may be influenced by bias (excluding Wilson et al. 2019; 2020 which used owner-objective measures). Small sample sizes and inconsistent sampling methods make drawing confident conclusions from these studies' problematic. Further, reports on DADs' accuracy once placed with an owner are mixed. Los et al. (2017) found that a cohort of eight DADs from multiple training backgrounds performed variably, with an average sensitivity of 36% to hypoglycemic events and a Positive Predictive Value of only 12%. However, seven of the eight dogs tested had been trained to alert to hyperglycemia (and were rewarded for this), yet this study only considered alerts to hypoglycemia correct. Therefore, of the reported 88% 'incorrect' alerts, it is unknown what proportion were actual events where the dog was alerting to hyperglycaemia. Gonder-Frederick et al. (2017) collected continuous blood glucose data, finger-prick blood test readings, and owner reports of DAD alerts from 14 participants over 6 weeks, and found substantial variation in performance between dogs, with only three out of 14 dogs performing statistically above chance level. Wilson et al. (2019) conducted the first entirely objective assessment of DAD performance once placed with an owner by comparing flash glucose monitor readings, finger-prick blood test results, and CCTV footage of the owners in their homes or places of work. Results similarly found variation between dogs, with sensitivity to hypoglycemic episodes ranging between 33.3% and 91.7%. The cause of this variability is as yet undefined but is likely influenced by the continuation of good training practices, and is discussed in more detail in Sect. 6 Characteristics of MADs.

As a whole, there is evidence that fluctuating blood glucose levels produce a change in a person's VOCs which dogs are able to detect. Reports show varying levels of detection/alerting accuracy, both in the laboratory and once placed with an owner. Nevertheless, DADs are the most widely distributed type of MADs, and owners often report substantial increases in their quality of life because of their DAD (Lippi & Plebani 2019). These potential discrepancies are discussed further in Sect. 5 Impact of MADs on Owner Well-Being.

4.2 Seizures

Epilepsy is one of the most common neurological diseases globally, with approximately 50 million people worldwide experiencing epileptic symptoms (World Health Organisation 2022). Characteristics of seizures will depend on the brain region that the disturbance first starts and what brain areas it spreads to affect. Temporary symptoms include disturbance in movement, sensation (including taste, hearing, and vision), mood, cognitive functions, and loss of awareness (World Health Organisation 2022). Seizures are caused by abnormal electrical activity in the brain, and can be either focal, generalised, combination focal and generalised, or unknown (Pack 2019). These terms refer to whether the seizure site is localised to one area (focal) or rapidly becomes distributed across multiple neural networks (generalised). The stages of a seizure can be defined as prodrome or pre-ictal (prior to seizure symptom onset), ictal (the time from the first symptom until the end of seizure activity) and post-ictal (the recovery stage where physical after-effects of the seizure are felt) (Epilepsy Foundation 2022). A major concern surrounding an epileptic seizure is the associated injuries that may occur as a result of experiencing a seizure, such as head injuries from falling (Willems et al. 2018). A system whereby epileptic seizures are pre-empted, so that the person can move to a safe environment, would be hugely valuable to minimise the risk of secondary injuries. First reports of dogs potentially detecting epilepsy emerged two decades ago, where it was documented that untrained dogs were responding to owners who had epilepsy prior to the onset of a seizure, in some cases with behaviours indicating fear or aggression (Brown & Strong 2001). Since these anecdotal reports, the premise that dogs can detect epileptic seizures has been investigated more thoroughly. A recent study by Powell et al. (2021) found that untrained dogs responded with more attention to their owner (i.e. staring) when in the presence of a seizure odour as compared to a non-seizure odour, suggesting, much like Brown and Strong (2001), that found untrained companion dogs may be able to detect a scent associated with seizures. Laboratory-based olfaction studies have also been carried out, the findings of which support that there is a "seizure-odour", and that dogs can detect it from the scent alone (Catala et al. 2019; Maa et al. 2021). In Catala et al.'s (2019) study, dogs were able to detect an odour, taken during a seizure, with 86.8% sensitivity and 98% specificity. Further, Maa et al. (2021) found that dogs could detect a seizure odour prior to the ictal phase, at a probability of 82.2% positive detection predicting a seizure. The average warning phase of this pre-ictal odour was 68 min. This finding is important as it confirms the presence of a prodrome or pre-ictal odour, which is the most relevant odour if the applied aim is that the dog can pre-alert their owner of an upcoming seizure so that they can get to a safe place. Currently, Epilepsy Alert Dogs are being trained and placed with owners who experience epileptic seizures, with reports from suppliers of "100% accuracy up to 50 min prior to an oncoming seizure" (Support Dogs 2022). Although dogs have been demonstrated to be able to detect oncoming seizure odours, what is currently missing from the scientific literature is a formal assessment of the claims, such as the above, specific to the performance of these dogs once placed with their owner. Future studies may wish to investigate the accuracy of seizure detection once placed in the home, as we know from studies pertaining to other MADs (e.g. Diabetic Alert Dogs) that continued training likely impacts dogs' alerting behaviours once in situ.

4.3 Migraines

Migraine is considered a leading cause of disability worldwide (Peters 2019; Steiner et al. 2018). Migraines affect an estimated 1 billion people globally, can be episodic or chronic, and are reported to be approximately 3 times more common in women than men (Walter 2022). Migraine Alert Dogs are said to be able to detect and respond to the earliest signs of a migraine, in the prodrome phase (National Headache Institute 2020). This period may begin as early as 48 h before the full migraine symptoms are experienced. Generally, individuals suffering from migraine become aware of their symptoms in the aura phase, by which time symptoms begin to become severe. It has been shown that medication taken in the earlier prodrome phase is more effective and can act as a preventative of a full migraine (Luciani et al. 2000). Therefore, a dog that could reliably detect the onset of the prodrome phase would be hugely valuable. Marcus and Bhowmick (2013) found that 53.7% of people who suffer from migraines reported that their untrained companion dog's behaviour changed prior to the onset of a migraine, usually around two hours before onset. Further, a recent survey of MAD owners found that migraine was one of the conditions that owners reported their dogs to alert to most frequently, even if this was not the condition that they had been primarily trained to respond to (Reeve et al. 2021). While Migraine Alert Dogs are being formally trained and distributed (National Headache Institute 2020; Neher 2020), the exact mechanisms behind this role have yet to be established fully. Future studies may wish to investigate the presence of a migraine odour using in vitro laboratory studies. However, as these dogs are reported to alert their owner prior to the onset of any symptoms, the independent collection of samples (e.g. without the aid of a dog) could be complex. As it stands, it appears that dogs may be able to detect early onset of migraine and could be of substantial operational value, however further research and training protocols are required in this area.

4.4 Addison's Disease and Postural Orthostatic Tachycardia Syndrome (PoTS)

Addison's disease is a rare condition characterised by insufficient production of the hormones cortisol and aldosterone by the 2 outer layers of cells of the adrenal glands (Nieman & Turner 2006). Symptoms can be non-specific, but include weakness, fatigue, dizziness, abdominal pain, vomiting, tachycardia (heart rate of over 100 bpm), and postural hypotension (low blood pressure after standing up). Individuals with Addison's disease can experience an acute adrenal crisis, where they may experience low blood pressure, low blood sodium level, and low blood glucose level (hypoglycemia) (Munir et al. 2022). Postural Orthostatic Tachycardia Syndrome (PoTS) is a syndrome that is estimated to affect approximately 170 out of every 100,000 people (Low et al. 2009). Symptoms of PoTS include light-headedness, blurred vision, tremor, palpitations, weakness, and exercise intolerance and also non-postural symptoms including abdominal pain, bloating, nausea, diarrhoea, sleep issues, fatigue, migraines and "brain fog" (Olshansky et al. 2020). MADs are being placed to assist and detect oncoming episodes of symptoms for owners who have Addison's disease and PoTS (e.g. Medical Detection Dogs 2022a; Service Dogs of Virginia 2022). While these conditions are distinct, they are considered together in this case because little to no research has been done on the potential mechanisms that dogs may be used to detect the onset of an episode. It is suggested that dogs are able to detect (and pre-empt) episodes of acute low cortisol in a person with Addison's disease due to a change in their odour associated with their cortisol levels, and episodes of increased heart rate, light-headedness and collapse in a person with PoTS due to changes in their heart rate and associated odour (Medical Detection Dogs 2022a). Further research is required to determine whether there is an odour profile associated with these physiological changes (e.g. Rooney 2019), whether these odour profiles are distinct (e.g. are associated with the health condition) or overlap (e.g. are associated with the symptoms), and to establish what impact these dogs have once placed with an owner. A recent study conducted by the authors (CW, CR, see Wilson et al. 2022) showed that dogs can discriminate between the odours of a healthy individual before and after a stress-inducing task, with each person's stress validated by an increase in their heart rate and blood pressure. This study supports the premise that changes in metabolic processes associated with heart rate and blood pressure can be discriminated between by dogs through odour alone, however, samples would need to be obtained from those individuals with these specific health conditions to confirm that there is a discriminable odour associated with Addison's disease and PoTS episodes. Once again, the issue of dogs pre-alerting, or sensing a condition *prior* to its onset, warrants further investigation to assess what metabolic processes may be occurring at these early stages.

4.5 Asthma and Chronic Obstructive Pulmonary Disease (COPD)

Both Asthma and COPD are health conditions associated with the respiratory system. To date, no peer-reviewed studies have assessed MADs' accuracy at detecting either asthma attacks or complications due to COPD, nor have there been studies documenting the impact that these dogs have on their owner's quality of life. Currently, the available information on these dogs is primarily supplied by the organisations that train them, which has the potential to be biased. It should be noted that some organisations describe MADs for those with Asthma and COPD as service animals tasked with *responding* to an owner already in medical distress, carrying out daily tasks such as reminding them to take medication, and providing comfort and confidence in public spaces (SIT Service Dogs 2022; US Service Animals 2022a). This definition does not presuppose that dogs can detect an *oncoming* health complication (e.g. asthma attack) and alert their owner prior to onset. However, other organisations do state that their dogs can alert to asthma attacks at, or immediately prior to, their onset (Maltese Assistance Dogs Center 2022). This is based on the assumption that changes in breathing rate affect the VOC profile of an individual, which dogs can detect and respond to (Maltese Assistance Dogs Center 2022—note that two studies are mentioned, however, no references are supplied, and they do not appear to be published or accessible). While there are, to our knowledge, no peer-reviewed studies directly assessing dogs' abilities to recognise and respond to odours such as asthma attacks and COPD in an applied setting, a laboratory in-vitro study carried out by Ehmann et al. (2012) did find that dogs trained to detect lung cancer could differentiate samples taken from individuals who either had been diagnosed with COPD, lung cancer, or were a healthy control. This finding shows that COPD has a distinct odour profile from lung cancer, however does not directly inform us as to whether COPD has a distinct odour profile from healthy controls. It is clear that further research is needed in this area, however, given dogs' proficiency at detecting VOC changes associated with other processes in the human body, it is feasible that changes associated with these conditions are detectable by dogs. The application of this proposed ability to a MAD setting is a distinct area of interest, and it is imperative that this is additionally evaluated under controlled conditions.

4.6 Narcolepsy

Narcolepsy is a condition marked by excessive sleep during the daytime, or one or more of the following symptoms: hallucinations, vivid dream imagery, cataplexy (sudden muscle weakness in response to intense emotions), and sleep paralysis (episodes without being able to move, often within a dream) (National Health Service 2022). Narcolepsy is often caused by a lack of a brain chemical that regulates wakefulness; however, this is not the cause of all narcolepsy and the exact cause of the condition can remain unclear. Narcolepsy Alert Dogs have not yet been studied sufficiently. At present, there is a single study which shows that dogs were able to pick out the odour taken from someone with narcolepsy in 11/12 cases (Dominguez-Ortega et al. 2013), however testing of the timeline of when this odour may become detectable has not yet been carried out. As with other types of MADs, the majority of information is currently provided by the charities and suppliers who train and place these dogs. For example, it is reported that Narcolepsy Alert Dogs can warn their handler of an upcoming sleep episode up to five minutes beforehand, in addition to other service dog tasks such as summoning help during a sleep episode, bringing medications and other items, and providing "pressure" therapy (leaning against the handler to provide them with physical support) (US Service Animals 2022b). In sum, narcolepsy is a currently understudied emerging area for MADs, which warrants further scientific investigation.

4.7 Psychiatric Conditions

Psychiatric service dogs are a type of assistance dog that are specially trained to perform tasks directly related to a person's psychiatric disability (Rodriguez et al. 2020). Psychiatric service dogs are currently predominantly trained to respond to visual cues (Kloep et al. 2017), and provide services such as interrupting physical signs of anxiety and providing positional support when in public (Rodriguez et al. 2020). In addition to responding to changes in psychological condition, dogs have been shown to be able to actually pre-empt aggressive outbursts, although it remains unclear exactly how they were able to do this (Bakeman et al. 2019). While these dogs are considered predominantly as service dogs rather than MADs (as they are reported to use primarily physical or auditory cues), reports of dogs "sensing" their owner's anxiety and tension (Taylor et al. 2013) suggest there could be another sense at play.

A recent study by Wilson et al. (2022) demonstrated that the onset of acute psychological stress induced an odour change in participants' breath and sweat, which dogs were then able to discriminate from baseline samples with 93% accuracy. Knowing that there is a detectable odour component to psychological stress, a discussion of olfactory-based training for the onset of anxiety or panic attacks becomes valuable. For example, trainers could follow current MAD training protocols by taking samples from a person when relaxed and when experiencing acute stress and then positively reinforce a dog for attending to, or performing attention-seeking behaviours in response to, this odour (similarly to how Diabetic Alert Dogs are trained). Although this study provides strong support for an odour, detectable by dogs, and associated with psychological stress, the question of real-life applicability remains. For example, the timeline that these odours become present and at what point dogs can detect the odour in relation to when the person starts experiencing stress is still unknown. Moreover, as discussed in Reeve et al. (2021), how a physiological stress response relates to other conditions (e.g. stress prior to a seizure, Kimiskidis et al. 2012, and see Strong et al. 2002 discussed below) merits further investigation. This line of research raises the possibility that other emotional states could have detectable odours to which dogs could be trained to respond. In line with the previously discussed health conditions, more scientific research is needed.

As has been the case with anecdotal reports later being confirmed with scientific studies, it is likely that there are many more human health conditions that confer VOC profiles that dogs can detect and be trained to assist humans within a working capacity.

5 Impact of Medical Alert Dogs on Owner Well-Being

Studies directly assessing the impact of MADs on medical outcomes are limited and focus mostly on Diabetic Alert Dog (DAD) partnerships. In a United States-based sample, Gonder-Frederick et al. (2013) found that the majority of DAD owners surveyed reported a decreased worry about hypoglycemia, improved ability to participate in physical activities, and improved quality of life since receiving their dog. Similarly, in a United Kingdom-based sample, Rooney et al. (2013) found that DAD owners reported positive effects including decreased unconscious episodes, reduced paramedic callouts, and improved independence. Both of these studies use retrospective designs assessing owners' viewpoints post-allocation, and are based on subjective owner responses rather than objective measures of accuracy.

Medical and psychosocial benefits have been documented for people with seizure-alert dogs as well. In a study of ten people–seizure alert dog teams, Strong et al. (2002) found that, compared to the baseline frequency of seizures, those with a seizure alert dog reported significantly fewer seizures during a 4-week training phase and this decrease in seizure frequency was maintained at 24 weeks follow up. Participants in the study reported feeling less fear associated with their seizures and that they were able to confidently engage in activities more than they had before. Strong et al. (2002) speculated that this increased confidence and engagement in activities as a result of having an alert dog could have resulted in decreased frequency of seizures.

It must be noted that, although many benefits of MADs have been reported, downsides have been documented as well. In a qualitative study, Nieforth et al. (2021) asked owners of, and family members of someone with a mobility and medical alert service dogs, what were the drawbacks of having a service dog (amongst other questions). In the most common responses, respondents described the extra work that goes into caring for and maintaining a service/alert dog and how much effort is required to leave the house and travel with a service/alert dog.

The studies discussed above highlight a key question in MAD research: what measures best capture whether these dogs are "successful"? While some may argue that high levels of accuracy and clinical outcomes are the most important measures of a successful MAD, the psychosocial benefits of owning an assistance dog outside of alerting accuracy should additionally be considered. For example, Bibbo et al. (2019) conducted a cross-sectional survey including 97 individuals partnered with a mobility or medical service dog and 57 individuals on the waitlist to receive one. These dogs are not classified as MADs as they are not trained to alert to a condition on the basis of odour. Therefore, the aspect of "accuracy" is less able to be objectively quantified as compared to a MAD, whose performance can be directly compared to technology (e.g. Continuous Glucose Monitoring Systems: Los et al. 2017; Wilson et al. 2019). It was found that, compared to those on the waitlist, individuals with a service dog exhibited significantly better psychosocial health including higher social, emotional, and work/school functioning. When considering MADs, impact on owner well-being likely extends to beyond their alerting accuracy alone. For example, in a study of 223 young people (aged 9–19) with Type 1 Diabetes, Maranda and Gupta (2016) found that children who were active in caring for a pet in their home were 2.5 times more likely to have control over their glycemic levels compared to those that did not. As such, approaching owner-perceived benefits not only in terms of alert-specific benefits, but the broader benefits of service dog ownership, and dog ownership in general, may be of interest to understand these relationships more clearly. Gonder-Frederick et al. (2017) documents a "significant gap" between the positive experiences that owners report, and research findings on DAD accuracy. Understanding this relationship with greater nuance than alerting accuracy alone may be necessary to truly define the benefits to owner well-being.

6 Selection and Training of Medical Alert Dogs

Much research time has gone into examining dogs' characteristics and evaluating testing strategies to assess potential dogs for military and detection roles (e.g. Lazarowski et al. 2020). Conversely, there is a dearth of research examining the selection and training of MADs.

While there is no universal training method across MAD training establishments, dogs are most commonly first exposed to their "target odour" in the form of a sample (e.g. a sweat and/or breath sample taken from an individual experiencing hypoglycemia, or during a seizure). Through classical conditioning, the dog learns that the sample odour is associated with a food reward. Then, using operant conditioning, the dog is rewarded for paying attention to and sniffing the sample, often with the aid of a clicker as a secondary reinforcer, and food as a primary reinforcer. Once an interest in these samples has been established, they may be hidden in someone's person, and attention-gaining behaviours such as nudging or pawing are shaped with operant conditioning and positive reinforcement. Training will then begin with individuals who have the health condition for which the target odour sample is an exemplar of the condition's associated VOCs. Here, the dog moves from alerting to samples to alerting to the person themselves. Positive reinforcement training is imperative to strengthen not only the dog's behavioural alert response, but also to condition the dog to associate the target odour with positive outcomes. Untrained dogs have been documented responding negatively to their owners' medical events (e.g. hypoglycemia), possibly because they perceive the owner behaving in such a way that is unsettling or even aversive (Chen et al. 2000; Dalziel et al. 2003). Ensuring that a suitable basis of positive reinforcement training is in place is necessary to ensure that the dog does not associate their target odours with an aversive event, and that the behavioural response is appropriate (Brown & Strong 2001).

Maintaining training protocols has additionally been found to impact performance once allocated with an owner, as it was found that owners who adhered most to training guidelines had dogs with higher alert sensitivity and positive predictive value than those who were not following training guidelines as closely (e.g.

rewarding the dog for an alert before checking their blood glucose level to confirm that it was either too low or too high) (Wilson et al. 2019). As with other scent-based working dog roles, the maintenance of quality training throughout the dog's lifespan is likely imperative to maintain performance outcomes.

But, as discussed above, the scientific study of MADs was spurred by anecdotal reports of untrained dogs spontaneously alerting their owners to impending physiological changes (e.g. Chen et al. 2000; discussed by Strong et al. 1999). And research appears to support the idea that dog-specific characteristics or targeted training programs are not necessarily relevant to whether or not a dog alerts their owner to a medical condition. For example, Wells et al. (2008) found that, out of 212 dog owners (with dogs untrained for medical alert) that participated in their study, 138 (65.1%) reported that their dog had shown a behavioural reaction to at least one of their hypoglycemic events. And in a study with 72 dog owners (untrained for medical alert) Catala et al. (2020b) found that 22 dogs (30.6%) were reported to demonstrate seizure alert behaviours. Analyses by Catala et al. (2020b) revealed that there were no significant differences in pet dogs' and alerting dogs' general demographics (e.g. breed, sex, reproductive state, age, origin), but that the alerting dogs were rated significantly higher by their owners on measures of motivation and training focus and significantly lower on neuroticism, consistent with other evaluations of working dogs (Lazarowski et al. 2019). Owners of alerting dogs also reported spending significantly more hours with their dog per day, compared to non-alerting dogs and, on a scale of the dog-owner relationship (MDORS, see Dwyer et al. 2006) reported significantly lower perceived costs associated with their dog and a trend towards a higher rating of emotional closeness to their dog.

As discussed by Reeve et al. (2021), there are a number of other characteristics that could influence whether a dog alerts to a medical condition that has yet to be explored scientifically. Research suggests that the biological "hardware" a dog is born with influences their potential as an olfactory detection dog. For example, Lesniak et al. (2008) found that polymorphisms in the alleles that code for olfactory receptors result in varying success of professional trained police sniffer dogs. It is also possible that affective and temperamental states, combined with training strategies, could influence a dog's decision-making (e.g. whether or not to alert, Cimarelli et al. 2021). As MADs become more commonplace, these variables may be worth exploring as the results could provide promising avenues for MAD selection and training.

7 Future of Alert and Assistance Dogs

In a world where healthcare technology is continuously improving the detection and maintenance of medical conditions, one may question the value of MADs. In fact, alongside the growing use of MADs, electronic noses, or "e-noses", are being developed to make use of VOCs for the detection of medical conditions (Das et al. 2016). Considering those conditions to which MADs also alert, the monitoring of blood sugar levels for people with diabetes has been shown considerable attention.

Considering current maintenance technologies for diabetes, modern technologies like continuous glucose monitors and subcutaneous continuous insulin infusion pumps have been found to better control blood sugar levels and improve quality of life (Benioudakis et al. 2022). But continuous glucose monitors are prone to false alarms (Anhalt 2016) and their accuracy relies on technological issues such as battery power and regular calibration that is not always optimised. Moreover, they are considered invasive by some, and can cause irritation and infection at insertion sites. These downsides may result in frustration with the technology which can then further increase social impairments and feelings of social stigma (Patton & Clements 2012; Schabert et al. 2013).

Maintenance technologies that analyse breath VOCs associated with blood sugar fluctuations have been tested by Faiola et al. (2019) and Shrestha et al. (2019). Faiola et al. (2019) reported on a prototype for VOC detection of hypoglycemia using a sensor system incorporated into a smart device. Surveys of usability suggest that people with diabetes were moderately satisfied with the system and felt that in the future, it could positively impact the management of their diabetes. Shrestha et al. (2019) found the sensor system to be highly accurate at classifying low and high blood sugar, indicating the real potential for non-invasive detection of blood sugar fluctuations for people with diabetes. Similar technologies have been suggested for the detection of volatiles associated with stress, but, as far as we know, have not to-date been developed or tested (Zamkah et al. 2020).

However, the authors of this chapter do not feel that technology is opposed to or in competition with MADs. As discussed above, the sensitivity of dogs' olfactory system is comparable to analytical technologies used to identify VOCs associated with human physiological change (Lippi & Heaney 2020). With this knowledge, Lippi and Heaney (2020) proposed a combination of and collaboration between canine detection studies and analytical chemistry. Such collaborations could provide a valuable synergy whereby each field informs the other with regards to the potential for VOCs to aid in the detection of human physiological change. Such a pairing was demonstrated by Catala et al. (2019) after first showing that dogs could detect an odour associated with epileptic seizures (Catala et al. 2019) that was then followed by an analysis of VOCs that provided potential predictive utility (Catala et al. 2020a). In a similar vein, others propose adding technology to the life of an MAD such that they are trained to utilise touch screens to help with emergency response (Byrne et al. 2018; Robinson et al. 2014).

Lastly, and most importantly, regardless of modern healthcare technologies, MADs are reported to have a positive impact on owners' medical outcomes as well as their quality of life and well-being. It is unlikely that dogs will ever be the primary source of detection for physiological changes and serious medical events, but, combined with healthcare technology, MADs have shown to be a valuable addition to the lives of people with a number of different health conditions.

References

Americans with Disabilities Act (2022) Service Animals. Retrieved August 12 from https://beta.ada.gov/topics/service-animals/

Amiel SA, Aschner P, Childs B, Cryer PE, de Galan BE, Frier BM, Gonder-Frederick L, Heller SR, Jones T, Khunti K (2019) Hypoglycaemia, cardiovascular disease, and mortality in diabetes: Epidemiology, pathogenesis, and management. Lancet Diabetes Endocrinol 7(5):385–396

Anhalt H (2016) Limitations of continuous glucose monitor usage. Diabetes Technol Ther 18(3):115–117

AP News (2021) Agreement: Va. service-dog company ordered to pay $3 million. Retrieved August 12 from https://apnews.com/article/business-dogs-a4153c9fb1dba02e3ae694d3c62f3c37

Assistance Dogs International (2022) Member Program Statistics. https://assistancedogsinternational.org/members/member-program-statistics/

Assistance Dogs UK (2022) Information for Service Providers. Retrieved August 12 from https://www.assistancedogs.org.uk/the-law/

Bakeman U, Eilam H, Schild CM, Grinstein D, Eshed Y, Laster M, Fride E, Anavi-Goffer S (2019) Detection of impending aggressive outbursts in patients with psychiatric disorders: Violence clues from dogs. Sci Rep 9(1):1–11

Benioudakis E, Karlafti E, Kalaitzaki A, Kaiafa G, Savopoulos C, Didangelos T (2022) Technological developments and quality of life in Type 1 Diabetes Mellitus Patients: A review of the modern insulin analogues, continuous glucose monitoring and insulin pump therapy. Curr Diabetes Rev 18(7): 40–48. https://doi.org/10.2174/1573399818666211103163208

Bessa V, Darwiche K, Teschler H, Sommerwerck U, Rabis T, Baumbach JI, Freitag L (2011) Detection of volatile organic compounds (VOCs) in exhaled breath of patients with chronic obstructive pulmonary disease (COPD) by ion mobility spectrometry. Int J Ion Mobil Spectrom 14(1): 7–13. https://doi.org/10.1007/s12127-011-0060-2

Bibbo J, Rodriguez KE, O'Haire ME (2019). Impact of service dogs on family members' psychosocial functioning. Am J Occup Ther 73(3): 7303205120p7303205121–7303205120p7303205111

Brands AM, Biessels GJ, De Haan EH, Kappelle LJ, Kessels RP (2005) The effects of type 1 diabetes on cognitive performance: A meta-analysis. Diabetes Care 28(3):726–735

Brown SW, Strong V (2001) The use of seizure-alert dogs. Seizure 10(1):39–41. https://doi.org/10.1053/seiz.2000.0481

Byrne C, Zeagler C, Freil L, Rapoport A, Jackson MM (2018) Dogs using touchscreens in the home: a case study for assistance dogs operating emergency notification systems. In: Proceedings of the fifth international conference on animal-computer interaction

Catala A, Grandgeorge M, Schaff JL, Cousillas H, Hausberger M, Cattet J (2019) Dogs demonstrate the existence of an epileptic seizure odour in humans. Sci Rep 9(1):4103. https://doi.org/10.1038/s41598-019-40721-4

Catala A, Latour P, Martinez-Caja AM, Cousillas H, Hausberger M, Grandgeorge M (2020a) Is there a profile of spontaneous Seizure-alert pet dogs? A survey of French people with epilepsy. Animals 10(2):254

Catala A, Levasseur-Garcia C, Schaff J-L, Till U, Vitola Pasetto L, Hausberger M, Cousillas H, Violleau F, Grandgeorge M (2020b) Prediction and detection of human epileptic seizures based on SIFT-MS chemometric data. Sci Rep 10(1):1–6

Cavaleiro Rufo J, Madureira J, Oliveira Fernandes E, Moreira A (2016) Volatile organic compounds in asthma diagnosis: A systematic review and meta-analysis. Allergy 71(2):175–188

Chen M, Daly M, Williams N, Williams S, Williams C, Williams G (2000) Non-invasive detection of hypoglycaemia using a novel, fully biocompatible and patient friendly alarm system. BMJ 321(7276):1565–1566

Cimarelli G, Schoesswender J, Vitiello R, Huber L, Viranyi Z (2021) Partial rewarding during clicker training does not improve naive dogs' learning speed and induces a pessimistic-like affective state. Anim Cogn 24(1):107–119. https://doi.org/10.1007/s10071-020-01425-9

Collaboration ERF (2011) Diabetes mellitus, fasting glucose, and risk of cause-specific death. N Engl J Med 364(9):829–841. https://doi.org/10.1056/NEJMoa1008862

Concha AR, Guest CM, Harris R, Pike TW, Feugier A, Zulch H, Mills DS (2019) Canine olfactory thresholds to amyl acetate in a biomedical detection scenario. Front VetY Sci 5: 345. https://doi.org/10.3389/fvets.2018.00345

Craven BA, Paterson EG, Settles GS (2010) The fluid dynamics of canine olfaction: unique nasal airflow patterns as an explanation of macrosmia. J R Soc Interface 7(47): 933–943. https://doi.org/10.1098/rsif.2009.0490

Dalziel DJ, Uthman BM, Mcgorray SP, Reep RL (2003) Seizure-alert dogs: A review and preliminary study. Seizure 12(2):115–120

Das S, Pal S, Mitra M (2016) Significance of exhaled breath test in clinical diagnosis: a special focus on the detection of diabetes mellitus. J Med Biol Eng 36(5):605–624

Davis P (2017) The investigation of human scent from epileptic patients for the identification of a biomarker for epileptic seizures

Dehlinger K, Tarnowski K, House JL, Los E, Hanavan K, Bustamante B, Ahmann AJ, Ward WK (2013) Can trained dogs detect a hypoglycemic scent in patients with type 1 diabetes? Diabetes Care 36(7): e98–e99. https://doi.org/10.2337/dc12-2342

Dewulf J, Van Langenhove H, Wittmann G (2002) Analysis of volatile organic compounds using gas chromatography. TrAC, Trends Anal Chem 21(9–10):637–646

Dixit K, Fardindoost S, Ravishankara A, Tasnim N, Hoorfar M (2021) Exhaled breath analysis for diabetes diagnosis and monitoring: Relevance, challenges and possibilities. Biosensors 11(12):476

Dominguez-Ortega L, Díaz-Gállego E, Pozo F, García-Armenter SC, Comino MS, Dominguez-Sanchez E (2013) Narcolepsy and odor: Preliminary report. SEMERGEN-Med Fam 39(7): e41–e46. https://doi.org/10.1016/j.semerg.2013.06.002

Drabińska N, Flynn C, Ratcliffe N, Belluomo I, Myridakis A, Gould O, Fois M, Smart A, Devine T, Costello BDL (2021) A literature survey of all volatiles from healthy human breath and bodily fluids: The human volatilome. J Breath Res 15(3):034001

Dwyer F, Bennett PC, Coleman GJ (2006) Development of the Monash dog owner relationship scale (MDORS). Anthrozoös 19(3):243–256

Edwards TL, Browne CM, Schoon A, Cox C, Poling A (2017) Animal olfactory detection of human diseases: Guidelines and systematic review. J VetY Behav 20:59–73

Ehmann R, Boedeker E, Friedrich U, Sagert J, Dippon J, Friedel G, Walles T (2012) Canine scent detection in the diagnosis of lung cancer: Revisiting a puzzling phenomenon. Eur Respir J 39(3):669–676. https://doi.org/10.1183/09031936.00051711

Faiola A, Vatani H, Agarwal M (2019) Hypoglycemic detection by human breath: A mobile health app that alerts diabetics of low blood glucose

Fidler C, Elmelund Christensen T, Gillard S (2011) Hypoglycemia: an overview of fear of hypoglycemia, quality-of-life, and impact on costs. J Med Econ 14(5): 646–655. https://doi.org/10.3111/13696998.2011.610852

Epilepsy Foundation (2022) Seizure Phases. Retrieved May 10 from https://epilepsyfoundation.org.au/understanding-epilepsy/seizures/seizure-phases/

Equality and Human Rights Comission (2017) Assistance dogs. A guide for all businesses. Retrieved from: https://www.equalityhumanrights.com/sites/default/files/assistance-dogs-a-guide-for-all-businesses.pdf

Gonder-Frederick L, Rice P, Warren D, Vajda K, Shepard J (2013) Diabetic alert dogs: A preliminary survey of current users. Diabetes Care 36(4):e47–e47

Gonder-Frederick LA, Grabman JH, Shepard JA (2017) Diabetes Alert Dogs (DADs): An assessment of accuracy and implications. Diabetes Res Clin Pract 134:121–130. https://doi.org/10.1016/j.diabres.2017.09.009

Gonder-Frederick LA, Shepard JA, Grabman JH, Ritterband LM (2016) Psychology, technology, and diabetes management. Am Psychol 71(7): 577. https://doi.org/10.1037/a0040383

Green A, Hede SM, Patterson CC, Wild SH, Imperatore G, Roglic G, Beran D (2021) Type 1 diabetes in 2017: Global estimates of incident and prevalent cases in children and adults. Diabetologia 64(12):2741–2750

Hardin DS, Anderson W, Cattet J (2015) Dogs can be successfully trained to alert to hypoglycemia samples from patients with type 1 diabetes. Diabetes Ther 6(4): 509–517. https://doi.org/10.1007/s13300-015-0135-x

Horowitz A (2020) Discrimination of person odor by owned domestic dogs. Int J Comp Psychol 33. https://doi.org/10.46867/ijcp.2020.33.01.02, https://doi.org/10.1046/j.1468-2982.2000.00030.x

International Diabetes Federation (2022) Diabetes is spiralling out of control. Retrieved August 12 from https://www.idf.org

Jalal AH, Alam F, Roychoudhury S, Umasankar Y, Pala N, Bhansali S (2018) Prospects and challenges of volatile organic compound sensors in human healthcare. Acs Sensors 3(7): 1246–1263. https://doi.org/10.1021/acssensors.8b00400

Kimiskidis VK, Valeta T (2012) Epilepsy and anxiety: Epidemiology, classification, aetiology, and treatment. Epileptic Disord 14(3):248–256

Kloep ML, Hunter RH, Kertz SJ (2017) Examining the effects of a novel training program and use of psychiatric service dogs for military-related PTSD and associated symptoms. Am J Orthopsychiatry 87(4): 425. https://doi.org/10.1037/ort0000254

Lawson M, Craven B, Paterson E, Settles G (2012) A computational study of odorant transport and deposition in the canine nasal cavity: Implications for olfaction. Chem Senses 37(6):553–566

Lazarowski L, Waggoner P, Katz JS (2019) The future of detector dog research. Comp Cogn & Behav Rev 14.https://doi.org/10.3819/CCBR.2019.140008

Lazarowski L, Waggoner LP, Krichbaum S, Singletary M, Haney PS, Rogers B, Angle C (2020) Selecting dogs for explosives detection: behavioral characteristics. Front VetY Sci 597https://doi.org/10.3389/fvets.2020.00597

Lesniak A, Walczak M, Jezierski T, Sacharczuk M, Gawkowski M, Jaszczak K (2008) Canine olfactory receptor gene polymorphism and its relation to odor detection performance by sniffer dogs. J Hered 99(5):518–527. https://doi.org/10.1093/jhered/esn057

Lippi G, Heaney LM (2020) The "olfactory fingerprint": can diagnostics be improved by combining canine and digital noses? Clin Chem Lab Med (CCLM) 58(6):958–967

Lippi G, Plebani M (2019) Diabetes alert dogs: A narrative critical overview. Clin Chem Lab Med 57(4):452–458. https://doi.org/10.1515/cclm-2018-0842

Los EA, Ramsey KL, Guttmann-Bauman I, Ahmann AJ (2017) Reliability of trained dogs to alert to hypoglycemia in patients with type 1 diabetes. J Diabetes Sci Technol 11(3):506–512

Low PA, Sandroni P, Joyner M, SHEN, W. K. (2009) Postural tachycardia syndrome (POTS). J Cardiovasc Electrophysiol 20(3):352–358

Luciani R, Carter D, Mannix LA, Hemphill M, Diamond M, Cady R (2000) Prevention of migraine during prodrome with naratriptan. Cephalalgia 20(2): 122–126. https://doi.org/

Maa E, Arnold J, Ninedorf K, Olsen H (2021) Canine detection of volatile organic compounds unique to human epileptic seizure. Epilepsy Behav 115:107690. https://doi.org/10.1016/j.yebeh.2020.107690

Maltese Assistance Dogs Center (2022) Asthma Alert Dogs. Retrieved May 10 from http://assistancedogscenter.com.mt/index.php/assistance-dogs/asthma-alert-dogs

Maranda L, Gupta OT (2016) Association between responsible pet ownership and glycemic control in youths with type 1 diabetes. PLoS ONE 11(4):e0152332

Marcus DA, Bhowmick A (2013) Survey of migraine sufferers with dogs to evaluate for canine migraine-alerting behaviors. J Altern Complement Med 19(6):501–508

Martin HJ, Turner M, Bandelow S, Edwards L, Riazanskaia S, Thomas C (2016) Volatile organic compound markers of psychological stress in skin: A pilot study. J Breath Res 10(4):046012

McClanahan K, Rosell F (2020) Conspecific recognition of pedal scent in domestic dogs. Sci Rep 10(1):1–9

Medical Detection Dogs (2022a) Allergen detection dogs. Retrieved September 20 from https://www.medicaldetectiondogs.org.uk/how-to-apply-for-a-medical-alert-assistance-dog/allergen-detection-dogs/

Medical Detection Dogs (2022b) Medical alert assistance dogs. Retrieved September 20 from https://www.medicaldetectiondogs.org.uk/maad-information/maad-advanced-information/

Minh TD, Oliver SR, Ngo J, Flores R, Midyett J, Meinardi S, Carlson MK, Rowland FS, Blake DR, Galassetti PR (2011) Noninvasive measurement of plasma glucose from exhaled breath in healthy and type 1 diabetic subjects. Am J Physiol-Endocrinol Metablsm 300(6):E1166–E1175

Munir S, Rodriguez BSQ, Waseem M, Haddad LM (2022) Addison disease (Nursing). In StatPearls [Internet]. StatPearls Publishing

National Headache Institute (2020) What is a migraine alert dog? Retrieved August 15 from https://nationalheadacheinstitute.com/blog/what-is-a-migraine-alert-dog/

National Health Service (2022) Narcolepsy. Retrieved August 16 from https://www.nhs.uk/conditions/narcolepsy/

Neher M (2020) A migraine alert dog in action. Retrieved August 10 from

Neupane S. Peverall R, Richmond G, Blaikie TP, Taylor D, Hancock G, Evans ML (2016) Exhaled breath isoprene rises during hypoglycemia in type 1 diabetes. Diabetes Care 39(7): e97–e98. https://doi.org/10.2337/dc16-0461

Nieforth LO, Rodriguez KE, O'Haire ME (2021) Benefits and challenges of mobility and medical alert service dogs for caregivers of service dog recipients. Disabil Rehabil: Assist Technol: 1–9

Nieman LK, Turner MLC (2006) Addison's disease. Clin Dermatol 24(4):276–280

Novak B, Blake D, Meinardi S, Rowland F, Pontello A, Cooper D, Galassetti P (2007) Exhaled methyl nitrate as a noninvasive marker of hyperglycemia in type 1 diabetes. Proc Natl Acad Sci 104(40):15613–15618

O'Connor M, O'connor C, Walsh C (2008) A dog's detection of low blood sugar: A case report. Ir J Med Sci 177(2):155–157

Olshansky B, Cannom D, Fedorowski A, Stewart J, Gibbons C, Sutton R, Shen W-K, Muldowney J, Chung TH, Feigofsky S (2020) Postural orthostatic tachycardia syndrome (POTS): A critical assessment. Prog Cardiovasc Dis 63(3):263–270

Pack, A. M. (2019). Epilepsy overview and revised classification of seizures and epilepsies. CONTINUUM: Lifelong Learn Neurol 25(2): 306–321

Patton SR, Clements MA (2012) Continuous glucose monitoring versus self-monitoring of blood glucose in children with type 1 diabetes-are there pros and cons for both? US Endocrinol 8(1):27

Pesterfield C, Guest C (2015) Does the presence of a trained diabetes alert dog reduce the diabetes-related anxiety in people with diabetes mellitus and hypoglycaemia unawareness?(2378-PO). Ned Tijdschr Voor Diabetol 13(3): 83–83

Peters GL (2019) Migraine overview and summary of current and emerging treatment options. Am J Manag Care 25(2 Suppl):S23–S34. https://doi.org/10.1111/j.1526-4610.2012.02185.x

Phillips C, Mac Parthaláin N, Syed Y, Deganello D, Claypole T, Lewis K (2014) Short-term intra-subject variation in exhaled volatile organic compounds (VOCs) in COPD patients and healthy controls and its effect on disease classification. Metabolites 4(2):300–318

Pleil JD, Stiegel MA, Risby TH (2013) Clinical breath analysis: Discriminating between human endogenous compounds and exogenous (environmental) chemical confounders. J Breath Res 7(1):017107

Powell NA, Ruffell A, Arnott G (2021) The untrained response of pet dogs to human epileptic seizures. Animals 11(8): 2267. https://doi.org/10.3390/ani11082267

Reeve C, Wilson C, Hanna D, Gadbois S (2021) Dog owners' survey reveals medical alert dogs can alert to multiple conditions and multiple people. PLoS ONE 16(4):e0249191. https://doi.org/10.1371/journal.pone.0249191

Reeve C, Cummings E, McLaughlin E, Smith S, Gadbois S (2020) An Idiographic investigation of diabetic alert dogs' ability to learn from a small sample of breath samples from people with type 1 diabetes. Can J Diabetes 44(1): 37–43 e31. https://doi.org/10.1016/j.jcjd.2019.04.020

Robinson CL, Mancini C, Van Der Linden J, Guest C, Harris R (2014) Canine-centered interface design: supporting the work of diabetes alert dogs. Proceedings of the SIGCHI conference on human factors in computing systems

Rodriguez KE, LaFollette MR, Hediger K, Ogata N, O'Haire ME (2020) Defining the PTSD service dog intervention: Perceived importance, usage, and symptom specificity of psychiatric service dogs for military veterans. Front Psychol: 1638. https://doi.org/10.3389/fpsyg.2020.01638

Rooney NJ, Morant S, Guest C (2013) Investigation into the value of trained glycaemia alert dogs to clients with type I diabetes. PLoS ONE 8(8):e69921

Rooney N (2019) Medical alert assistance dogs for Addison's disease, narcolepsy & PoTS. Retrieved August 11 from https://www.hra.nhs.uk/planning-and-improving-research/application-summaries/research-summaries/medical-alert-assistance-dogs-for-addisons-disease-narcolepsy-pots/

Saasa V, Malwela T, Beukes M, Mokgotho M, Liu C-P, Mwakikunga B (2018) Sensing technologies for detection of acetone in human breath for diabetes diagnosis and monitoring. Diagnostics 8(1):12

Samuel L, Arnesen C, Zedrosser A, Rosell F (2020) Fears from the past? The innate ability of dogs to detect predator scents. Anim Cogn 23(4):721–729

Schabert J, Browne JL, Mosely K, Speight J (2013) Social stigma in diabetes. Patient–Patient-Center Outcomes Res 6(1):1–10

Seaquist ER, Miller ME, Bonds DE, Feinglos M, Goff DC Jr, Peterson K, Senior P, Investigators A (2012) The impact of frequent and unrecognized hypoglycemia on mortality in the ACCORD study. Diabetes Care 35(2):409–414

Service Dogs of Virginia (2022) Medical alert service dogs. Retrieved May 17 from https://www.servicedogsva.org/medical-alert

Shirasu M, Touhara K (2011) The scent of disease: Volatile organic compounds of the human body related to disease and disorder. J Biochem 150(3):257–266

Shrestha S, Harold C, Boubin M, Lawrence L (2019) Smart wristband with integrated chemical sensors for detecting glucose levels using breath volatile organic compounds. Smart Biomed Physiol Sens Technol XVI

Siegel AP, Daneshkhah A, Mather KJ, Agarwal M (2019) 968-P: Volatile organic compounds in breath predict hypoglycemia well before plasma glucose levels fall. Diabetes 68(Supplement_1)

SIT Service Dogs (2022) Medical alert dogs. Retrieved May 12 from https://www.sitservicedogs.com/medical-alert

Steiner TJ, Stovner LJ, Vos T, Jensen R, Katsarava Z (2018) Migraine is first cause of disability in under 50s: Will health politicians now take notice? vol. 19. pp. 1–4, Springer

Strong V, Brown S, Walker R (1999) Seizure-alert dogs—fact or fiction? Seizure, 8(1): 62–65. https://doi.org/10.1053/seiz.1998.0250

Strong V, Brown S, Huyton M, Coyle H (2002) Effect of trained Seizure Alert Dogs® on frequency of tonic–clonic seizures. Seizure 11(6): 402–405. https://doi.org/10.1053/seiz.2001.0656

Support Dogs (2022) Epilepsy Seizure Alert. Retrieved May 16 from https://www.supportdogs.org.uk/epilepsy-seizure-alert

Tauveron I, Delcourt I, Desbiez F, Somda F, Thiéblot P (2006) Canine detection of hypoglycaemic episodes whilst driving. Diabet Med: J Br Diabet Assoc 23(3):335

Taylor MF, Edwards ME, Pooley JA (2013) "Nudging them back to reality": Toward a growing public acceptance of the role dogs fulfill in ameliorating contemporary veterans' PTSD symptoms. Anthrozoös 26(4):593–611

Tenzer-Iglesias P, Shannon MH (2012) Managing hypoglycemia in primary care. J Fam Pract 61(Suppl 10):S1–S8

Tsukuda M, Nishiyama Y, Kawai S, Okumura Y (2019) Identifying stress markers in skin gases by analysing gas collected from subjects undergoing the Trier social stress test and performing statistical analysis. J Breath Res 13(3):036003

US Service Animals (2022a) Service dogs for asthma. Retrieved May 10 from https://usserviceanimals.org/blog/service-dog-for-asthma/

US Service Animals (2022b). Service dogs for narcolepsy. Retrieved May 12 from https://usserviceanimals.org/blog/service-dogs-for-narcolepsy/

Van Berkel J, Dallinga J, Möller G, Godschalk R, Moonen E, Wouters E, Van Schooten F (2010) A profile of volatile organic compounds in breath discriminates COPD patients from controls. Respir Med 104(4): 557–563. https://doi.org/10.1016/j.rmed.2009.10.018

van Bragt JJ, Principe S, Hashimoto S, Versteeg DN, Brinkman P, Vijverberg SJ, Weersink EJ, Scichilone N, Maitland-van der Zee AH (2021) Exhaled metabolite patterns to identify recent asthma exacerbations. Metabolites 11(12):872

Van Dartel D, Schelhaas HJ, Colon AJ, Kho KH, De Vos CC (2020) Breath analysis in detecting epilepsy. J Breath Res 14(3):031001

Waggoner LP, Jones MH, Williams M, Johnston J, Edge CC, Petrousky JA (1998) Effects of extraneous odors on canine detection. Enforc Secur Technol

Walker DB, Walker JC, Cavnar PJ, Taylor JL, Pickel DH, Hall SB, Suarez JC (2006) Naturalistic quantification of canine olfactory sensitivity. Appl Anim Behav Sci 97(2–4):241–254

Walter K (2022) What is migraine? JAMA 327(1):93–93

Walther S, Yamamoto M, Thigpen AP, Garcia A, Willits NH, Hart LA (2017) Assistance dogs: Historic patterns and roles of dogs placed by ADI or IGDF accredited facilities and by non-accredited US facilities. Front VetY Sci 4: 1. https://doi.org/10.3389/fvets.2017.00001

Wells DL, Lawson SW, Siriwardena AN (2008) Canine responses to hypoglycemia in patients with type 1 diabetes. J Altern Complement Med 14(10):1235–1241. https://doi.org/10.1089/acm.2008.0288

Willems LM, Watermann N, Richter S, Kay L, Hermsen AM, Knake S, Rosenow F, Strzelczyk A (2018) Incidence, risk factors and consequences of epilepsy-related injuries and accidents: A retrospective, single center study. Front Neurol 9:414

Wilson CH, Kane SA, Morant SV, Guest CM, Rooney NJ (2020) Diabetes alert dogs: Objective behaviours shown during periods of owner glucose fluctuation and stability. Appl Anim Behav Sci 223:104915

Wilson C, Morant S, Kane S, Pesterfield C, Guest C, Rooney NJ (2019) An owner-independent investigation of diabetes alert dog performance. Front VetY Sci: 91

Wilson C, Campbell K, Petzel Z, Reeve C (2022) Dogs can discriminate between human baseline and psychological stress condition odours. PLoS ONE 17(9): e0274143. https://doi.org/10.3389/fpsyg.2020.01638

World Health Organisation (2022) Epilepsy. Retrieved May 10 from https://www.who.int/news-room/fact-sheets/detail/epilepsy

Zamkah A, Hui T, Andrews S, Dey N, Shi F, Sherratt RS (2020) Identification of suitable biomarkers for stress and emotion detection for future personal affective wearable sensors. Biosensors 10(4):40

Zoungas S, Patel A, Chalmers J, De Galan BE, Li Q, Billot L, Woodward M, Ninomiya T, Neal B, MacMahon S (2010) Severe hypoglycemia and risks of vascular events and death. N Engl J Med 363(15):1410–1418

Considerations for Detection Dog–Handler Teams in Infectious Agent Surveillance

Amritha Mallikarjun and Cynthia M. Otto

Abstract

Due to the SARS-CoV-2 pandemic, there is increased interest in the use of dogs for infectious disease surveillance. While dogs have often been used to detect explosives, narcotics, and other contraband, live-find and human remains, and non-infectious medical detection, dogs' use for infectious disease surveillance is a much newer field. While there is evidence that dogs can be trained to detect different types of infectious disease, most of these studies have been done in laboratory settings rather than in operational settings, except for deployed C. difficile detection dogs in hospital settings and deployed laurel wilt and citrus disease detection dogs in groves. This chapter discusses recommendations for training a successful, deployable handler-detection-dog team, and how to apply these recommendations to potential infectious disease detection teams.

Keywords

Detection dogs • Canine handler • Detection team • Canine training

A. Mallikarjun (✉) · C. M. Otto
School of Veterinary Medicine, Penn Vet Working Dog Center, University of Pennsylvania, Philadelphia, PA, USA
e-mail: amritham@vet.upenn.edu

C. M. Otto
e-mail: cmotto@vet.upenn.edu

C. M. Otto
Department of Clinical Sciences and Advanced Medicine, University of Pennsylvania, Philadelphia, PA, USA

L. Lazarowski (ed.), *Olfactory Research in Dogs*,
https://doi.org/10.1007/978-3-031-39370-9_14

1 Introduction

For many years, society has heavily used and relied on the detection capabilities of dogs. Working detection dogs have been used in a variety of different environments for complex tasks such as narcotics and contraband detection (Jezierski et al. 2014; Rice and Velasco 2021), explosives detection (Adamkiewicz et al. 2013; Gazit and Terkel 2003; Lazarowski and Dorman 2014), live-find or human-remains detection in building collapses or in wilderness areas (Bulanda 2012), medical detection as a part of service work (e.g., hypo- or hyperglycemia (Hardin et al. 2015)), and endangered species surveillance (Jamieson et al. 2021; Smith et al. 2005). Despite the increasing use of chemical analysis and technology for detection work, dogs currently remain one of the least expensive and most accurate odor detectors for these particular deployment scenarios.

More recently, researchers have been exploring other potential avenues to utilize dogs' incredibly sensitive sense of smell; one such avenue is disease detection. Most studies in disease detection have explored non-infectious human diseases, such as different types of cancer (Buszewski et al. 2012; Cornu et al. 2011; Crawford et al. 2022; Elliker et al. 2014; Essler et al. 2020; Kane et al. 2022; Mathieu et al. 2018; Moser and McCulloch 2010; Willis et al. 2004), diabetic alert (Gonder-Frederick et al. 2017; Hardin et al. 2015; Hart et al. 2019), or sleep apnea (Koskinen et al. 2019). For some of these diseases, like diabetes and sleep apnea, one of the potential goals is to train a service dog to alert on the medical state of a specific person. For cancer studies, the goal is first to demonstrate that the cancer odor is differentiable from the odor of a healthy person. There is ample evidence that within the research laboratory, dogs can differentiate samples from patients with cancer from those with benign tumors or healthy individuals with high accuracy (Moser and McCulloch 2010). Next, the data from the dogs' detection work can be examined alongside chemical analyses to help identify potential biomarkers for the disease (Kane et al. 2022). For most programs, these cancer-detection dogs are not intended to deploy to hospitals or other public places; rather, they contribute to our understanding of the disease and its diagnosis.

Due to the severe acute respiratory syndrome coronavirus 2 (SARS-CoV-2) pandemic in 2019, interest has grown in the use of dogs for infectious disease surveillance. The ability of detection dogs to identify a variety of infectious diseases in lab settings has spurred interest in the use of dogs in operational settings (e.g., screening people at airports and stadiums). However, an operational setting introduces many variables that are not present in a lab setting, which could impact detection dogs' performance. In this chapter, we first discuss current studies and uses of dogs for infectious agent surveillance. We then discuss recommendations for selecting, training, and deploying infectious disease detection dog-and-handler teams to increase their likelihood of success in the field.

Current Uses for Dogs for Infectious Agent Surveillance

The use of dogs to detect infectious diseases is a relatively new endeavor as compared to their use in search and rescue, police work, and even non-infectious medical detection. Given dogs' ability to detect volatile organic compound (VOC) changes in breath (Koskinen et al. 2019), sweat (Gonder-Frederick et al. 2017; Hardin et al. 2015), and blood (Horvath et al. 2013; Murarka et al. 2019; Rooney et al. 2019) samples associated with non-infectious disease, it was postulated that dogs could detect changes due to infectious disease as well. Thus far, dogs have successfully detected infectious viral, bacterial, fungal, and prion-related diseases in multiple different mediums, including sweat, blood samples, fecal samples, and breath samples (see Table 1 for detailed information about these infectious disease detection studies). Dogs have been shown to detect infectious diseases in humans, as well as in plants (Gottwald et al. 2020; Mendel et al. 2018), bovine species (Angle et al. 2016), and cervid species (Mallikarjun et al. n.d.).

Given dogs' success in laboratory settings, during the pandemic, many organizations began to train and deploy COVID-19 detection dogs in several different scenarios. While dogs have not been frequently used in real-world infectious disease detection cases, dogs' lab performance is often used to justify dogs' potential deployment and use in real-life settings. However, performance in the lab or in training settings can greatly differ from performance in the deployed setting (Rutter et al. 2021). In a sample of current infectious disease detection studies, 8 tested dogs in a lab setting, 2 tested dogs in a pseudo-operational setting, and 2 tested dogs in a true operational setting. A pseudo-operational setting is a controlled environment that is meant to mimic the actual operational setting but allows for a more precise examination of the dogs' detection behavior. Prior to the SARS-CoV-2 pandemic, published reports of the use of operational infectious disease detection dogs focused mostly on non-human infectious diseases (e.g., citrus disease and laurel wilt) (Gottwald et al. 2020; Mendel et al. 2018) with the exception of trained C. difficile detection dogs in Canada (Bomers et al. 2012; Taylor et al. 2018), who are not widely deployed at this point, despite their high accuracy rate.

To promote the success of future deployed detection dog–hander teams for infectious disease, we present recommendations for training a deployable detection dog–handler team and apply this information to potential infectious disease detection dog–handler teams.

Considerations and Recommendations for Detection Dog Training

Many factors can affect a detection dog's ability to find its target odor accurately and consistently in a deployment setting. We have broken these factors into seven sections: the choice to use detection dogs for detection rather than other methods, dog selection considerations, handler selection considerations, dog training considerations prior to deployment (e.g., training sample selection, training locations, and training frequency), handler training considerations, operational maintenance post-deployment (e.g., regular training in the operational setting), and other operational considerations (e.g., length of time dogs can work in a session, creating redundancy in the system to identify the target odor). Each of these factors can impact the future success of an infectious disease dog–handler team.

Table 1 Recent infectious disease detection studies across different target diseases and disease types

Study authors	Target disease	Disease type	Target substance	# of dogs	# of training samples	Testing scenario	Sensitivity	Specificity
Maurer et al. (2016)	E. coli	bacteria	Urine	5	234+, 453−	lab setting	99.6%	91.5%
Koivusalo et al. (2017)	MRSA	bacteria	Cell lines	1	3 incubation times for one cell line	lab setting	89.4% averaged over tests	91.4% averaged over tests
Bomers et al. (2012)	C. difficile	Bacteria	Fecal matter	1	30+, 270−	lab setting	100%	100%
			Human odor on people	1	30+, 270−	Operational setting (hospital)	83%	98%
Koskinen et al. (2017)	E. coli	Bacteria	Urine	1	7+, 78−, 6 ambiguous	Lab setting	95% accuracy	
Gottwald et al. (2020)	Liberibacter asiaticus	Bacteria	Trees	10	Phase 1: 1+, 3− Phase 2: 1+, 10 to 20−	Pseudo-operational setting (grid of trees)	85.8%	99.6%
Angle et al. (2016)	Bovine viral diarrhea	Virus	Cell cultures	2	2 BVDV + cell cultures, 2 negative cell cultures	Lab setting	90.9%, averaged over dogs	98.5%, averaged over dogs
Grandjean et al. (2020)	COVID-19	Virus	Sweat	18	101+, 198−	Lab setting	93.9% success	
Essler et al. (2021)	COVID-19	Virus	Urine, saliva	9	38+, 24−	Lab setting	68%	99%
Mendel et al. (2018)	Laurel wilt	Fungus	Training aids/ trees	2	Training aid mixtures	Operational setting (grove)	99.4% accuracy	
Mallikarjun et al., under review	Chronic wasting disease	Prion	Fecal matter	3	57+, 111−	Lab setting	40%	92.3%
						Pseudo-operational setting (mock field)	72%	False alert rate: 17.9%

2 The Choice of Detection Dog Teams

While detection dogs are widely used to locate many different target substances, it is necessary in novel use cases, such as infectious disease detection, to assess whether adding a detection dog team would significantly improve surveillance and control. If a detection dog team could be a useful potential tool, interested organizations must evaluate whether the detection work would be safe for the dog, handler, and search subjects (e.g., potentially infected humans, animals, or plants), and whether the practice would be cost-effective. If these benchmarks are not met, a solution other than a detection dog team would be required.

Safety of Dog and Handler

Prior to the use of a detection dog team, it is necessary to assess all aspects of the operational scenario to ensure the safety of the dog and handler. First, it must be determined if the pathogen can infect dogs or humans, and if it can, the degree to which the pathogen would impact the health of the detection team. Many animal and plant pathogens cannot infect dogs or humans, and as such would pose little to no risk to the team's health.

If the disease does pose a risk to humans, the transmission mechanism, contagiousness of the disease, and the potential for safety precautions must be evaluated. Infectious disease can be transmitted in several different ways, including airborne (e.g., COVID-19), fomites, direct contact, or ingestion. Depending on the transmission mechanism, personal protective equipment may be required for the handler. Additionally, if dogs can contract or transmit the disease, the transmission mechanism may affect the way dogs are trained to alert on the target odor (e.g., a sit or down alert away from the target odor if the transmission is via ingestion or direct contact) or the particular source of the target odor on which they work (e.g., detecting COVID-19 in sweat rather than breath or saliva, since sweat does not contain viral particles). If dogs can contract the target disease, that may also suggest that dogs are not an ideal choice to manage the disease.

Another important factor is the rapidity of the disease spread. This is measured via an R0 score, which indicates how contagious the disease is; for example, the original SARS-CoV-2 ancestral strain had an R0 of 2.4–2.6 (Mukherjee and Satardekar 2021) (meaning, someone who gets the disease will infect between 2.4 and 2.5 people), while the SARS-CoV-2 Omicron BA.2 strain had an R0 between 10 and 12 (Szanyi et al. 2022). Diseases with high R0 will be riskier for the dogs and handlers if they are at risk to contract the disease.

Lastly, if diseases do pose a risk to dogs or humans, vaccination could mitigate this risk; for example, the canine distemper virus and the rabies virus can be deadly for dogs, but dogs vaccinated against these diseases would be unlikely to contract them. Similarly, the COVID-19 vaccination provides a measure of protection for humans against the COVID infection. However, vaccines do not always provide full immunity or protection against diseases, so the degree to which the vaccine mitigates the disease symptoms and severity must be assessed prior to team deployment.

Cost

The deployment of a detection dog team can be quite costly in both time and capital. Comparison of costs has been more prevalent in the wildlife detection literature, which is closer to the operational scenario of animal or plant infectious disease detection; however, the considerations are important for all types of detection work. While a dog–handler team can be more efficient than other methods of surveillance because they can rapidly screen targets, they require a great deal of initial training, which can be expensive and requires more time prior to deployment than some other methods (Duggan et al. 2011). The calculated cost of dog–handler teams should include procurement of the dog, travel, initial and ongoing training, and cost per day for the salary of the dog handler and maintenance of the dog (including food and healthcare) (Bennett et al. 2020), but this cost can vary depending on the particular detection need. For example, a wildlife survey study, which is a relatively similar target environment as plant or animal pathogen detection, found that while dog–handler teams finished the survey much faster, they were no more accurate than standard human survey, and were more expensive (Duggan et al. 2011).

In certain cases the cost of a detection dog team can be less expensive than other testing methods; for example, if tests for the disease are in development, scarce, and/or time-consuming (e.g., early PCR tests for SARS-CoV-2 (Hag-Ali et al. 2021)), traditional testing can be more expensive in both time and money to administer to large groups of people than to screen the group with a detection dog team.

Canine and Handler Selection

Once it has been determined that detection dog teams would be useful for a surveillance need, it is necessary to develop and formalize the process of handler and canine selection for the detection teams.

2.1 Dog Selection

Not all dogs have the aptitude to become detection dogs (Maejima et al. 2007), and within that set of successful dogs, certain dogs are better suited to specific types of detection work than others. It is crucial to select dogs that would perform well in the particular deployment scenario in which they are needed. For example, search and rescue dogs must be comfortable working at a distance from their handler (Hare et al. 2018), but this characteristic would not be necessary for most explosives or drug detection dogs, who tend to work closer to their handler.

While it is beyond the scope of this article to discuss the individual characteristics that can identify the best detection dogs for certain deployment settings, general selection guidelines should be followed, as well as more specific requirements as dictated by the operational setting (*Standard for Detection Canine Selection, Kenneling, and Healthcare* 2021). Abundant research has been done on the characteristics of successful working dogs, including explosives detection (Lazarowski et al. 2020; MacLean and Hare 2018), narcotics detection (Jezierski et al. 2014; Maejima et al. 2007), search and rescue (Hare et al. 2018), wildlife detection (Beebe et al. 2016; DeMatteo et al. 2019) and service (Bray et al. 2017; Weiss 2002; Weiss and Greenberg 1997; Wilsson and Sundgren 1997) (see Bray et al. for a review (Bray et al. 2021)), all of which require very different skills. The more unsuccessful dogs are selected for a particular program, the more time and money the program wastes, and there would be a higher potential for error in infectious disease surveillance.

2.2 Handler Selection

While much research has focused on traits of the ideal dog for particular detection tasks, much less research has examined ideal handler traits. Some studies have examined the Big Five personality traits in relationship to dog handling with mostly pet dogs and therapy dogs, and have found some evidence that dog owners and dog handlers with higher Neuroticism (characterized by sadness, moodiness, and instability) are less effective at training (Hoummady et al. 2016; Stevens et al. 2021). However, owners with higher Neuroticism scores spend more time with their dogs and view their dogs more as a partner (Kotrschal et al. 2009), which studies have suggested could support the dogs' well-being and comfort in the dog–handler team (Payne et al. 2015b; Zubedat et al. 2014). In contrast, one study of professional handlers found that while, in general, the handlers surveyed scored high on Agreeableness and low on Neuroticism, their individual personality traits did not significantly affect their eventual handler success (Jamieson et al. 2018b). These studies all agree, however, on the idea that dedication of the handler to the dog–handler relationship leads to success (Jamieson et al. 2018b; Payne et al. 2015a, b; Zubedat et al. 2014). Relationship-building behaviors like taking the dog home after work as well as doing dog sports outside of work were found to improve canine performance with military dogs, for example (Lefebvre et al. 2007). Additionally, matching handlers to the specific task in addition to the dog is crucial. For example, police K9 handlers may not be successful conservation detection dog handlers, and vice versa, due to different situation-dependent handling strategies (DeMatteo et al. 2019). A carefully selected handler can improve the quality and performance of the dog–handler detection team.

2.3 Canine Training Considerations Prior to Deployment

Detection dog training prior to deployment is vital to produce a dog who alerts on the target odor and ignores negative and distractor odors in the operational environment. Sample selection and maintenance, development of a training plan that prepares the dog–handler team for the operational setting, training the handler to work well with their dog, and development of a standardized certification program help ensure the quality of the detection team.

3 Training Sample Selection and Maintenance

Careful sample selection is crucial to detection dogs' success in their eventual deployment scenario. The characteristics of training samples, like sample quantity, source, proportion of odorant compounds, and age, can have a significant effect on dogs' later ability to identify target substances. Ensuring that the characteristics of the training samples match the eventual target population is crucial for optimal detection accuracy. Additionally, the handling and storage of samples must be done with precision and care. Given dogs' superior olfactory capabilities, miniscule differences in the way the odors were collected, and handled can lead to detection errors from the dog–handler team (Elliker et al. 2014). Once these errors occur, it can be difficult to trace these errors back to sample characteristics; rather, it is often attributed to the detection dog.

Sample Selection

Selecting samples to create a large, diverse training set is important to ensure that dogs successfully learn a category of odor and do not merely memorize training samples. Dogs have been shown to learn large numbers of individual odors (Krichbaum et al. 2020). To ensure that dogs successfully learn a category of odor and do not merely memorize samples, as many training odors should be used as possible. However, sample acquisition can be difficult, and researchers and trainers do not often have access to endless amounts of new samples. It is inadvisable to mix samples together to create "novel" target odors, as dogs can detect the individual components of these mixtures and after overtraining on these samples will fail to generalize to novel samples (Essler et al. 2021). Previous studies have had varied success with small sample numbers; while dogs have successfully learned target odor and generalized it to novel samples with as few as 20 target odors and 20 negative odors (Wright et al. 2017), they have also failed to generalize to novel samples after training on larger numbers of samples (Elliker et al. 2014; Essler et al. 2021). As such, further research is needed to determine the appropriate formula to calculate the ideal sample number for particular odors.

The proportion of the total number of positive samples and the total number of negative samples presented to the dogs should, at the final stages of training, reflect the prevalence of the disease. There is currently no hard evidence for this recommendation in the canine training literature; however, at the onset of most training procedures, dogs are exposed to many more positive samples than they might expect to encounter in a real-world scenario so that the dog can successfully learn the target odor (Grandjean et al. 2020; Jezierski et al. 2010; Kure et al. 2021). This approach makes sense early on, as many target odors have such low incidence rates that initially training at that rate would not allow the dog to learn the odor. However, if the dogs continued training at this higher ratio, they may have a higher false alert rate when they experience the lower real-world proportion. For example, one study trained pet dogs to search for myrrh odor on cotton buds in outdoor settings to assess the possibility of using pet dogs for conservation detection work (Rutter et al. 2021). The initial training included no "blank" trials, meaning there were no trials without odor present. Once a blank trial was included at the test, the dogs had a high false positive rate, as they had not expected the absence of target odor. As such, when the target odor prevalence rate is low, the dog will occasionally or often perform searches where the odor is not present. The dogs must be exposed to more negative samples and trials in which there are no positive samples present, or they may falsely alert.

The sample weight and volume must as closely reflect the range of target sample weights and volumes as possible. Dogs are sensitive to odor quantity (Aviles-Rosa et al. 2021b; Horowitz et al. 2013), and training on items that differ greatly in odor quantity from the target item can lower their eventual accuracy (Aviles-Rosa et al. 2021b). If the access to the correct sample weights and volumes is restricted or limited in some way, there are alternative methods to generating increased or decreased odor from the available sample. For example, some labs use olfactometers (Aviles-Rosa et al. 2021a), which blow air over samples and can control concentration and odor strength via airflow. Additionally, alternative odor-capture devices like cotton or polymers may be incubated with samples to generate different concentrations of odor (Simon et al. 2020). It is more difficult to control the concentration or dissipation rate of the odor with these alternative aids, however, as very few studies have been done to explore the rate of absorption/desorption and how these aids react with different types of odorants.

Sample Collection

Samples from both groups of test subjects, confirmed to be positive or negative for the disease of interest, must be processed and handled in similar ways to ensure that dogs are detecting the target odor and not an extraneous co-occurring odor. This is a crucial step in the training process, as researchers cannot easily determine what molecules dogs are using to identify substances. Judicious selection of samples (and distractor items, as discussed) can reduce the possibility that dogs are detecting something other than the target odor. Dogs can detect differences in processing between the odors from negative and positive samples that are not always readily apparent. For example, one study that examined dogs' detection of prostate cancer found high rates of sensitivity (93.5%) and specificity (87.9%) during training, but the dogs failed to generalize to novel samples during a test phase. The researchers found that the positive training samples were tested by dipping a urinalysis stick in the sample, while the negative training samples were tested by pouring the sample on the stick. The testing samples, however, were processed in the same manner—by pouring the sample on the stick. As such, the researchers concluded that the dogs likely learned urinalysis-stick odor rather than the prostate cancer odor signature (Guest et al. 2020).

Additionally, to ensure that the dogs are detecting the actual odor signature of positive samples and not an unrelated but co-occurring odor, distractor, and control items must be selected carefully. Ideally, these distractors should include every non-target item that has come into contact with the samples. For example, in Kane et al.'s ovarian cancer detection study (Kane et al. 2022), distractor items included nitrile gloves and vinyl gloves (both worn while handling samples), 80% isopropyl alcohol and paper towels (both used to clean the ports that held samples), and Sharpies (used to label sample jars and ports). Generating a list of these odors prior to training onset will allow for these odors to be included at the beginning of the dogs' learning process and will help the dog hone in on the actual target odor as opposed to using one of the distractor odors for differentiation.

Training Environment
Training successful detection dogs requires careful consideration of the eventual training environment and building dogs up to eventual success in that environment. There are a wide variety of different operational scenarios in which infectious disease detection dogs could work; for example, there are deployed COVID-19 detection dogs that currently work in airports (Coronavirus-Sniffing Dogs Unleashed at Miami International Airport to Detect Virus in Employees 2021)and stadiums (*Miami Heat to Welcome Back Fans with Help of Covid-Sniffing Dogs*, 2021), *C. difficile* detection dogs work in a hospital (Bomers et al. 2012), while laurel wilt detection dogs work outdoors in groves (Mendel et al. 2018). These spaces are vastly different in size, population density, and noise level. Prior to the start of training, the operational location and context must be clearly identified; this includes the search area size as well as procedural-based aspects such as the amount of time dogs are expected to search. Dogs are very sensitive to context effects and accuracy can decrease if the operational setting differs substantially from the training scenario (e.g., searching a much larger area than trained previously (Rutter et al. 2021)).

Training Procedures
It is crucial for regular double-blind training and testing to be included in the dog–handler training procedures prior to deployment. To train or test *double-blind* means that the handler does not know if there is a target odor present or where the odor is located, and any additional people in the search area are also unaware of its presence or location. This type of training mimics the operational setting, where target odors are also unknown by the handler and other surrounding people. One well-known study, Lit, Schweitzer, and Oberbauer (Lit et al. 2011), demonstrated that handlers can unknowingly influence their detection dogs if they believe a target odor to be present at a certain location. As such, double-blind training where the handler practices their handling behavior while not knowing the location of the target item is crucial for accurate detection.

Handler Training Considerations
A strong dog–handler relationship is an important factor for peak team performance. When dogs are familiar and comfortable with their handler, canine detection accuracy is higher (Jamieson et al. 2018a). Handlers should spend time

with their dogs both in and out of the detection context to build a strong bond. Spending time with the dog also allows the handler to better understand the way the dog searches in different scenarios. If the handler understands the dog's behavior, the handler's stress will be lower in the actual operational scenario, as they are adequately prepared to handle the dog in that situation; as a result, the dog will be more accurate in their detection work (Gillespie 2022).

Handlers must also be well-educated in the biology of the disease being detected. They must be familiar with appropriate use of personal protective equipment and methods of decontamination associated with the disease. For example, the C. difficile detection team consulted with the hospital and the doctors prior to development of an operational detection plan; they implemented careful hand hygiene as well as no contact between the patients and the dog for both the dog and patient safety (Bomers et al. 2012). Handlers must also accurately document any responses of the dogs, report results according to local jurisdictions, and be familiar with privacy restrictions on health care information.

Certification

It is necessary to develop a certification process (e.g., Standard for Training and Certification of Canine Detection of Humans: Patrol Canine Team 2021) for infectious disease handler-dog teams based on the operational scenario and goals of the deploying organization. This is to ensure that the teams operate in a similar way and that all deployed teams meet high detection standards. Without a standardized certification process, it is unclear whether any given dog–handler team meets an appropriate benchmark of performance for their specific task. The certification process should be conducted in a double-blind fashion (as discussed above in the *Training procedures* section) and the administrator of the test should be someone who did not participate in the training of the dog, to eliminate any potential bias.

Other handler-dog detection teams go through rigorous certification processes prior to official deployment. Search and rescue teams in the United States must pass a national certification exam from the Federal Emergency Management Agency (FEMA) under the Department of Homeland Security. Moreover, individual search-and-rescue task forces across the United States often have additional required certifications per team to ensure peak performance for each task force. Government drug and explosives detection canines must pass certification developed by their umbrella organization, such as the Bureau of Alcohol, Tobacco, and Explosives, Border Patrol, or the Transportation Security Administration (TSA). State and county law enforcement K9s have internal certifications as well as organizations like the United States Police Canine Association (USPCA) and the National Narcotic Detector Dog Association (NNDDA) that provide guidelines for certification of detection K9s. The National Institutes of Standards and Technology (NIST) (https://www.nist.gov/osac/dogs-sensors-subcommittee) has helped develop the original Scientific Working Group for Dog and Orthogonal Detector Guidelines (SWGDOG) into published standards that have been through a formal development process with public comment (see https://www.aafs.org/academy-standards-board under Dogs and Sensors for documents open to public comment as well as published standards).

It is especially worthwhile to examine the TSA guidelines (*Explosives Detection Canine Recommended Standards* 2020) for any human infectious disease detection work, as the deployment scenario will likely include screening large numbers of people or large crowds (e.g., COVID detection dogs at stadiums), and may even occur in the same types of areas as TSA canines (e.g., airports, train stations).

To develop a certification for animal and plant infectious disease detection work, examining the certification guidelines for wilderness search and rescue as well as the current protocols for various conservation detection dogs can also help to develop an appropriate deployment plan and certification.

Operational Maintenance Post-Deployment

Once initial training is completed and the dog–handler team is certified for deployment, continued training and assessment is needed to ensure that high levels of detection accuracy are maintained (*Explosives Detection Canine Recommended Standards* 2020; *Standard for Training and Certification of Canine Detection of Humans: Patrol Canine Team* 2021).

Maintenance Training

Dogs generally maintain memory for trained odor for long periods of time with minimal training; however, their operational performance declines without consistent practice (Lazarowski et al. 2021). As such, regular training sessions must be conducted with the dog and handler team. As with the original training, the training sessions should be conducted in a manner as close to the operational scenario as possible. In this way, both the dog and the handler practice accurate search mechanics and any deficits in performance can be identified.

Assessment

In addition to continued training, it is important to implement a regular re-certification program such that the handlers and dogs have to search in a standardized, double-blind scenario (*Standard for Training and Certification of Canine Detection of Humans: Patrol Canine Team* 2021). The samples used in this scenario should be novel, as dogs have incredibly strong odor memory (Krichbaum et al. 2020) and can remember prior training samples (Essler et al. 2021). A regular assessment program can provide data on the teams' performances and feedback on areas of improvement. If regular assessments are not included in an infectious disease detection program, decrements in canine and handler performance cannot be properly assessed, and teams could potentially fail to detect cases of their target infectious disease or provide a false alert on a negative case.

Data Collection

While the hope and intention of training infectious disease detection dogs is for these dogs to detect their target odor at a high level of sensitivity and specificity for a long period of time, situations can occur where dogs' performance can worsen for reasons that are not immediately apparent to the handler or the team. As such, it is important to regularly collect data on the dogs' performance and health. It is also extremely important to store this data in such a manner that problems can be quickly and efficiently identified and steps can be taken to help the dog–handler team to return to best operational performance.

Other Operational Considerations
In addition to the previous considerations, there are some other additional operational facets that must be considered prior to deployment of a detection dog team.

3.1 Practical Detection Considerations

Building redundancy and backups into the operational infectious disease detection system is necessary, as dog–handler teams do not maintain the same level of accuracy on a day-to-day basis. This can be due to illness or stress on the part of the handler or the dog; however, even when both handler and dog are at peak health, errors in detection work can happen. As such, it is important to use redundant detection systems, like utilizing multiple dog–handler teams and/or using disease testing as a back-up to double-check potential targets identified by the team, to ensure that cases are not missed.

3.2 Perception Considerations

One under-examined aspect of detection dog deployment is the need to inform the public and survey public opinion about detection dogs prior to the use of detection dog teams. There are many countries in which many members of the population fear dogs; in these locations, the use of dogs for human infectious disease surveillance would likely be viewed negatively, and there may be low compliance with the surveillance program. It is important to determine the public opinion about dogs and detection dogs such that people will be receptive to their use and participate in the surveillance. Conversely, some individuals may be biased toward the dogs and be unwilling to believe the results of other testing, emphasizing the importance of accuracy if dogs are used operationally.

4 Conclusion

Prior to the deployment and use of infectious disease detection dog–handler teams, several steps must be taken to assess the need for the team, to select ideal dogs and handlers, to create an operational deployment plan and training plan that supports the operational deployment scenario, and to continue to maintain the performance of the dogs and handlers in the operational setting. While dogs can be an extremely effective tool for the surveillance of disease, optimal performance requires careful forethought, consideration, and a well-developed plan.

References

Adamkiewicz E, Jezierski T, Walczak M, Górecka-Bruzda A, Sobczyńska M, Prokopczyk M, Ensminger J (2013) Traits of drug and explosives detection in dogs of two breeds as evaluated by their handlers and trainers. Anim Sci Pap Rep 31(3):205–217

Angle TC, Passler T, Waggoner PL, Fischer TD, Rogers B, Galik PK, Maxwell HS (2016) Real-time detection of a virus using detection dogs. Front Vet Sci 2(January). https://doi.org/10.3389/fvets.2015.00079

Aviles-Rosa EO, Gallegos SF, Prada-Tiedemann PA, Hall NJ (2021a) An automated canine line-up for detection dog research. Front Vete Sci 8. https://doi.org/10.3389/fvets.2021.775381

Aviles-Rosa EO, Mcguinness G, Hall NJ, Cogger N, Cobb M (2021b) Case study: an evaluation of detection dog generalization to a large quantity of an unknown explosive in the field. https://doi.org/10.3390/ani11051341

Beebe SC, Howell TJ, Bennett PC (2016) Using scent detection dogs in conservation settings: a review of scientific literature regarding their selection. Front Vet Sci 3(October). Frontiers Media S.A. https://doi.org/10.3389/fvets.2016.00096

Bennett EM, Hauser CE, Moore JL (2020) Evaluating conservation dogs in the search for rare species. Conserv Biol 34(2):314–325. Blackwell Publishing Inc. https://doi.org/10.1111/cobi.13431

Bomers MK, van Agtmael MA, Luik H, van Veen MC, Vandenbroucke-Grauls CMJE, Smulders YM (2012) Using a dog's superior olfactory sensitivity to identify Clostridium difficile in stools and patients: proof of principle study. BMJ (Online) 345(7888). https://doi.org/10.1136/bmj.e7396

Bray EE, Otto CM, Udell MAR, Hall NJ, Johnston AM, MacLean EL (2021) Enhancing the selection and performance of working dogs. Front Vet Sci 8. Frontiers Media S.A. https://doi.org/10.3389/fvets.2021.644431

Bray EE, Sammel MD, Cheney DL, Serpell JA, Seyfarth RM, Marshall-Pescini S, Parker KJ (2017) Effects of maternal investment, temperament, and cognition on guide dog success. PNAS 114(34):9128–9133. https://doi.org/10.5061/dryad.50fj0

Bulanda S (2012) Ready! Training the search and rescue dog. Fox Chapel Publishing

Buszewski B, Ligor T, Jezierski T, Wenda-Piesik A, Walczak M, Rudnicka J (2012) Identification of volatile lung cancer markers by gas chromatography-mass spectrometry: Comparison with discrimination by canines. Anal Bioanal Chem 404(1):141–146. https://doi.org/10.1007/s00216-012-6102-8

Cornu JN, Cancel-Tassin G, Ondet V, Girardet C, Cussenot O (2011) Olfactory detection of prostate cancer by dogs sniffing urine: a step forward in early diagnosis. Eur Urol 59(2):197–201. https://doi.org/10.1016/j.eururo.2010.10.006

Coronavirus-sniffing dogs unleashed at Miami International Airport to detect virus in employees (2021) The Washington Post. https://www.washingtonpost.com/nation/2021/09/09/covid-sniffer-dogs/

Crawford MA, Perrone JA, Browne CM, Chang CL, Hopping S, Edwards TL (2022) Transitioning from training to testing with scent detection animals: application to lung cancer detection dogs. J Vet Behav. https://doi.org/10.1016/j.jveb.2022.07.004

DeMatteo KE, Davenport B, Wilson LE (2019a) Back to the basics with conservation detection dogs: fundamentals for success. Wildlife Biol 2019(1). https://doi.org/10.2981/wlb.00584

Duggan JM, Heske EJ, Schooley RL, Hurt A, Whitelaw A (2011) Comparing detection dog and livetrapping surveys for a cryptic rodent. J Wildl Manag 75(5):1209–1217. https://doi.org/10.1002/jwmg.150

Elliker KR, Sommerville BA, Broom DM, Neal DE, Armstrong S, Williams HC (2014) Key considerations for the experimental training and evaluation of cancer odour detection dogs: lessons learnt from a double-blind, controlled trial of prostate cancer detection. BMC Urol 14(1). https://doi.org/10.1186/1471-2490-14-22

Essler JL, Wilson C, Verta AC, Feuer R, Otto CM (2020) Differences in the search behavior of cancer detection dogs trained to have either a sit or stand-stare final response. Front Vet Sci 7:1–8. https://doi.org/10.3389/fvets.2020.00118

Essler JL, Kane SA, Nolan P, Akaho EH, Berna AZ, DeAngelo A, Berk RA, Kaynaroglu P, Plymouth VL, Frank ID, Weiss SR, Odom John AR, Otto CM (2021) Discrimination of SARS-CoV-2 infected patient samples by detection dogs: a proof of concept study. PLoS One 16(4 April):1–14. https://doi.org/10.1371/journal.pone.0250158

Explosives detection canine recommended standards (2020)

Gazit I, Terkel J (2003) Explosives detection by sniffer dogs following strenuous physical activity. Appl Anim Behav Sci 81(2):149–161. https://doi.org/10.1016/S0168-1591(02)00274-5

Gillespie B (2022) Training considerations for operational performance: train the handler, train the dog, train the team. In: DeGreeff LE, Schultz CA (eds) Canines: the original biosensors. Jenny Stanford Publishing Pte. Ltd., pp 413–438

Gonder-Frederick LA, Grabman JH, Shepard JA, Tripathi AV, Ducar DM, McElgunn ZR (2017a) Variability of diabetes alert dog accuracy in a real-world setting. J Diab Sci Technol 11(4):714–719. https://doi.org/10.1177/1932296816685580

Gottwald T, Poole G, Taylor E, Luo W, Posny D, Adkins S, Schneider W, McRoberts N (2020) Canine olfactory detection of a non-systemic phytobacterial citrus pathogen of international quarantine significance. Entropy 22(11):1–35. https://doi.org/10.3390/e22111269

Grandjean D, Sarkis R, Lecoq-Julien C, Benard A, Roger V, Levesque E, Bernes-Luciani E, Maestracci B, Morvan P, Gully E, Berceau-Falancourt D, Haufstater P, Herin G, Cabrera J, Muzzin Q, Gallet C, Bacqué H, Broc JM, Thomas L, ... Desquilbet L (2020) Can the detection dog alert on COVID-19 positive persons by sniffing axillary sweat samples? A proof-of-concept study. PLoS One 15(12 December):1–19. https://doi.org/10.1371/journal.pone.0243122

Guest CM, Harris R, Anjum I, Concha AR, Rooney NJ (2020) A lesson in standardization—subtle aspects of the processing of samples can greatly affect dogs' learning. Front Vet Sci 7. https://doi.org/10.3389/fvets.2020.00525

Hag-Ali M, AlShamsi AS, Boeijen L, Mahmmod Y, Manzoor R, Rutten H, Mweu MM, El-Tholoth M, AlShamsi AA (2021) The detection dogs test is more sensitive than real-time PCR in screening for SARS-CoV-2. Commun Biol 4(1). https://doi.org/10.1038/s42003-021-02232-9

Hardin DS, Anderson W, Cattet J (2015) Dogs can be successfully trained to alert to hypoglycemia samples from patients with type 1 diabetes. Diab Therapy 6(4):509–517. https://doi.org/10.1007/s13300-015-0135-x

Hare E, Kelsey KM, Serpell JA, Otto CM (2018) Behavior differences between search-and-rescue and pet dogs. Front Vet Sci 5(June). https://doi.org/10.3389/fvets.2018.00118

Hart LA, Weng H-Y, Rooney NJ, Wilson C, Morant S, Kane S, Pesterfield C, Guest C (2019) An owner-independent investigation of diabetes alert dog performance. Front Vet Sci 6:91. www.frontiersin.org, https://doi.org/10.3389/fvets.2019.00091

Horowitz A, Hecht J, Dedrick A (2013) Smelling more or less: investigating the olfactory experience of the domestic dog. Learn Motiv 44(4):207–217. https://doi.org/10.1016/j.lmot.2013.02.002

Horvath G, Andersson H, Nemes S (2013) Cancer odor in the blood of ovarian cancer patients: a retrospective study of detection by dogs during treatment, 3 and 6 months afterward. BMC Cancer 13. https://doi.org/10.1186/1471-2407-13-396

Hoummady S, Péron F, Grandjean D, Cléro D, Bernard B, Titeux E, Desquilbet L, Gilbert C (2016) Relationships between personality of human-dog dyads and performances in working tasks. Appl Anim Behav Sci 177:42–51. https://doi.org/10.1016/j.applanim.2016.01.015

Jamieson LTJ, Baxter GS, Murray PJ (2018a) You are not my handler! impact of changing handlers on dogs' behaviours and detection performance. Animals 8(10). https://doi.org/10.3390/ani8100176

Jamieson LTJ, Baxter GS, Murray PJ (2018b) Who's a good handler? Important skills and personality profiles of wildlife detection dog handlers. Animals 8(12). https://doi.org/10.3390/ani8120222

Jamieson LTJ, Hancock AL, Baxter GS, Murray PJ (2021) How quoll-ified are northern and spotted-tailed quoll detection dogs? Wildl Res 48(4):376–384. https://doi.org/10.1071/WR19243

Jezierski T, Górecka-Bruzda A, Walczak M, Świergiel AH, Chruszczewski MH, Pearson BL (2010) Operant conditioning of dogs (Canis familiaris) for identification of humans using scent lineup. Anim Sci Pap Rep 28(1)

Jezierski T, Adamkiewicz E, Walczak M, Sobczyńska M, Górecka-Bruzda A, Ensminger J, Papet E (2014) Efficacy of drug detection by fully-trained police dogs varies by breed, training level, type of drug and search environment. Forensic Sci Int 237:112–118. https://doi.org/10.1016/J.FORSCIINT.2014.01.013

Kane S, Lee YE, Mallikarjun A, Preti G, Plymouth VL, Verta A, DeAngelo A, Otto CM (2022) Canine discrimination of ovarian cancer through volatile organic compounds. Talanta

Koivusalo M, Vermeiren C, Yuen J, Reeve C, Gadbois S, Katz K (2017) Canine scent detection as a tool to distinguish meticillin-resistant Staphylococcus aureus. J Hosp Infect 96:93–95. https://doi.org/10.1016/j.jhin.2017.03.005

Koskinen A, Koskinen H, Bäck L, Saxen H, Klockars T (2017) A detection dog for paediatric urinary tract infection caused by Escherichia coli. Infect Dis. https://doi.org/10.1080/23744235.2017.1331466

Koskinen A, Bachour A, Vaarno J, Koskinen H, Rantanen S, Bäck L, Klockars T (2019) A detection dog for obstructive sleep apnea. Sleep Breath 23(1):281–285. https://doi.org/10.1007/s11325-018-1659-x

Kotrschal K, Schöberl I, Bauer B, Thibeaut AM, Wedl M (2009) Dyadic relationships and operational performance of male and female owners and their male dogs. Behav Proc 81(3):383–391. https://doi.org/10.1016/j.beproc.2009.04.001

Krichbaum S, Rogers B, Cox E, Waggoner LP, Katz JS (2020) Odor span task in dogs (Canis familiaris). Anim Cogn 23(3):571–580. https://doi.org/10.1007/s10071-020-01362-7

Kure S, Iida S, Yamada M, Takei H, Yamashita N, Sato Y, Miyashita M (2021) Breast cancer detection from a urine sample by dog sniffing: a preliminary study for the development of a new screening device, and a literature review. Biology 10(6). https://doi.org/10.3390/biology10060517

Lazarowski L, Dorman DC (2014) Explosives detection by military working dogs: olfactory generalization from components to mixtures. Appl Anim Behav Sci 151:84–93. https://doi.org/10.1016/j.applanim.2013.11.010

Lazarowski L, Waggoner P, Hutchings B, Angle C, Porritt F (2021) Maintaining long-term odor memory and detection performance in dogs. Appl Anim Behav Sci 238:105301. https://doi.org/10.1016/J.APPLANIM.2021.105301

Lazarowski L, Waggoner LP, Krichbaum S, Singletary M, Haney P, Rogers B, Angle C (2020) Selecting dogs for explosives detection: behavioral characteristics. Front Vet Sci 7. Frontiers Media S.A. https://doi.org/10.3389/fvets.2020.00597

Lefebvre D, Diederich C, Delcourt M, Giffroy JM (2007) The quality of the relation between handler and military dogs influences efficiency and welfare of dogs. Appl Anim Behav Sci 104(1–2):49–60. https://doi.org/10.1016/j.applanim.2006.05.004

Lit L, Schweitzer JB, Oberbauer AM (2011) Handler beliefs affect scent detection dog outcomes. Anim Cogn 14(3):387–394. https://doi.org/10.1007/s10071-010-0373-2

MacLean EL, Hare B (2018) Enhanced selection of assistance and explosive detection dogs using cognitive measures. Front Vet Sci 5(October). https://doi.org/10.3389/fvets.2018.00236

Maejima M, Inoue-Murayama M, Tonosaki K, Matsuura N, Kato S, Saito Y, Weiss A, Murayama Y, Ito S (2007) Traits and genotypes may predict the successful training of drug detection dogs. Appl Anim Behav Sci 107(3–4):287–298. https://doi.org/10.1016/j.applanim.2006.10.005

Mallikarjun A, Swartz B, Kane SA, Gibison M, Wilson I, Collins A, Moore MB, Charendoff I, Ellis J, Murphy LA, Nichols T, Otto CM (n.d.) Canine detection of chronic wasting disease (CWD) in laboratory and field settings. Prion, under review

Mathieu KB, Bedi DG, Thrower SL, Qayyum A, Bast RC (2018) Screening for ovarian cancer: imaging challenges and opportunities for improvement. Ultrasound Obstet Gynecol 51(3):293–303. Wiley. https://doi.org/10.1002/uog.17557

Maurer M, Mcculloch M, Willey AM, Hirsch W, Dewey D (2016) Open forum infectious diseases detection of bacteriuria by canine olfaction. https://doi.org/10.1093/ofid/ofw051

Mendel J, Burns C, Kallifatidis B, Evans E, Crane J, Furton KG, Mills DE (2018) Agri-dogs: using canines for earlier detection of laurel wilt disease affecting avocado trees in south florida. HortTechnol 28(2):109–116. https://doi.org/10.21273/HORTTECH03791-17

Miami Heat to welcome back fans with help of Covid-sniffing dogs (2021) https://www.nbcnews.com/news/us-news/miami-heat-welcome-back-fans-help-covid-sniffing-dogs-n1255955. Accessed 27 Aug 2022

Moser E, Culloch M (2010) Canine scent detection of human cancers: a review of methods and accuracy. J Vet Behav: Clin Appl Res 5(3):145–152. https://doi.org/10.1016/j.jveb.2010.01.002

Mukherjee R, Satardekar R (2021) Why are some coronavirus variants more infectious? J Biosci 46(4). Springer. https://doi.org/10.1007/s12038-021-00221-y

Murarka M, Vesley-Gross ZI, Essler JL, Smith PG, Hooda J, Drapkin R, Otto CM (2019) Testing ovarian cancer cell lines to train dogs to detect ovarian cancer from blood plasma: a pilot study. J Vet Behav 32:42–48. https://doi.org/10.1016/J.JVEB.2019.04.010

Payne E, Bennett PC, Mcgreevy PD (2015a) Current perspectives on attachment and bonding in the dog-human dyad. https://doi.org/10.2147/PRBM.S74972

Payne EM, Arnott ER, Early JB, Bennett PC, McGreevy PD (2015b) Dogmanship on the farm: analysis of personality dimensions and training styles of stock dog handlers in Australia. J Vet Behav 10(6):471–478. https://doi.org/10.1016/j.jveb.2015.08.005

Rice BL, Velasco JG (2021) Accuracy comparison of single-purpose and dual-purpose narcotic detection canines. Appl Anim Behav Sci 243. https://doi.org/10.1016/j.applanim.2021.105434

Rooney NJ, Guest CM, Swanson LCM, Morant SV (2019) How effective are trained dogs at alerting their owners to changes in blood glycaemic levels?: variations in performance of glycaemia alert dogs. PLoS One 14(1):1–16. https://doi.org/10.1371/journal.pone.0210092

Rutter NJ, Howell TJ, Stukas AA, Pascoe JH, Bennett PC (2021) Diving in nose first: the influence of unfamiliar search scale and environmental context on the search performance of volunteer conservation detection dog–handler teams. Animals 11(4). https://doi.org/10.3390/ani11041177

Simon A, Lazarowski L, Singletary M, Barrow J, van Arsdale K, Angle T, Waggoner P, Giles K (2020) A review of the types of training aids used for canine detection training. Front Vet Sci 7(June):1–10. https://doi.org/10.3389/fvets.2020.00313

Smith DA, Ralls K, Cypher BL, Maldonado JE (2005) Assessment of scat-detection dog surveys to determine kit fox distribution. Wildl Soc Bull 33(3):897–904

Standard for Detection Canine Selection, Kenneling, and Healthcare (2021). www.asbstandardsboard.org

Standard for training and certification of canine detection of humans: Patrol canine team (2021). www.asbstandardsboard.orgANSI/ASBStandard027, First Edition

Stevens JR, Wolff LM, Bosworth M, Morstad J (2021) Dog and owner characteristics predict training success. Anim Cogn 24(2):219–230. https://doi.org/10.1007/s10071-020-01458-0

Szanyi J, Wilson T, Howe S, Zeng J, Andrabi H (2022) An integrated epidemiologic and economic model to assess optimal COVID-19 pandemic policy. MedRxiv. https://doi.org/10.1101/2022.08.01.22278262

Taylor MT, McCready J, Broukhanski G, Kirpalaney S, Lutz H, Powis J (2018) Using dog scent detection as a point-of-care tool to identify toxigenic clostridium difficile in stool. Open Forum Infect Dis 5(8). https://doi.org/10.1093/ofid/ofy179

Weiss E (2002) Selecting shelter dogs for service dog training. J Appl Anim Welfare Sci 5(1):43–52. https://doi.org/10.1207/S15327604JAWS0501

Weiss E, Greenberg G (1997) Service dog selection tests: effectiveness for dogs from animal shelters. 53:297–308

Willis CM, Church SM, Guest CM, Cook WA, McCarthy N, Bransbury AJ, Church MRT, Church JCT (2004) Olfactory detection of human bladder cancer by dogs: proof of principle study. BMJ 329(7468):712–714. https://doi.org/10.1136/bmj.329.7468.712

Wilsson E, Sundgren E (1997) The use of a behaviour test for the selection of dogs for service and breeding. I: Method of testing and evaluating test results in the adult dog, demands on different kinds of service dogs, sex and breed differences. Appl Anim Behav Sci 53

Wright HF, Wilkinson A, Croxton RS, Graham DK, Harding RC, Hodkinson HL, Keep B, Cracknell NR, Zulch HE (2017) Animals can assign novel odours to a known category. Sci Rep 7(1). https://doi.org/10.1038/s41598-017-09454-0

Zubedat S, Aga-Mizrachi S, Cymerblit-Sabba A, Shwartz J, Leon JF, Rozen S, Varkovitzky I, Eshed Y, Grinstein D, Avital A (2014) Human-animal interface: the effects of handler's stress on the performance of canines in an explosive detection task. Appl Anim Behav Sci 158:69–75. https://doi.org/10.1016/j.applanim.2014.05.004

A Critical Review of the Use of Wildlife Conservation Canines in Species-at-Risk Research

Simon Gadbois, Laura Elliott, and Meredith Flannery

Abstract

Dogs as research assistants in wildlife conservation have been used since the twentieth century. The experimental and quasi-experimental research on their efficiency, accuracy, and reliability is more recent and does not exceed a few decades. We start by discussing the potential and overlooked conceptual and methodological contributions of psychophysics to the training and assessment of wildlife conservation canines. We then briefly discuss issues around transects in search and question the validity and relevance of this anthropocentric approach. The remainder of this chapter is a comprehensive and critical review of the literature. An important part of this review is a summary table (appendix) of the main research we collated identifying the specific goals, species, methods, and the overall results and conclusions of each peer-reviewed publication.

Keywords

Wildlife conservation • Conservation canines • Detection • Discrimination • Identification

Dedication: To all our canine research assistants that have diligently worked and played with us in the field over the years. In particular (in order of appearance): Kafka, Sable, Boomer, Caya, Zyla, Flynn, Nutella, and Ivy.

S. Gadbois (✉) · L. Elliott
Department of Psychology and Neuroscience, Dalhousie University, Halifax, NS, Canada
e-mail: sgadbois@dal.ca

L. Elliott
e-mail: laura.elliott@dal.ca

M. Flannery
Nova Scotia Department of Environment and Climate Change, Halifax, NS, Canada
e-mail: Meredith.Flannery@novascotia.ca

L. Lazarowski (ed.), *Olfactory Research in Dogs*,
https://doi.org/10.1007/978-3-031-39370-9_15

1 Introduction

This chapter is about dogs as research assistants in wildlife conservation and related areas (e.g. invasive species). We will first review some broad and rarely discussed methodological and conceptual issues, mostly from the perspectives of animal olfactory psychophysics, zoosemiotics, and ethology, and the opportunities they may offer. Then we present a comprehensive (although still incomplete) literature review of peer-reviewed studies in the field. We are aware that many more non-published reports exist (including many from the lead author's laboratory at Dalhousie University[1]) that emerged from contracts or diverse agreements with government agencies or private companies. They are not cited or covered here, but they are no less important in defining the field. Many of these reports should likely be published or systematically reviewed as they report negative outcomes and would prevent duplication and multiplication of efforts. Many of us in the field are aware of "some" of these disappointments.

Scent-processing dogs[2] have been employed to improve search and detection efforts in many fields of research and professions. For example, dogs have provided aid in explosives and narcotics detection, medical diagnostics such as cancer detection (Oh et al. 2014), search and rescue (Jones et al. 2004), weapons detection (Holland 2018), inspection of parasitic and fungal levels in beehives, and even more diverse fields such as bed bug detection and telephone pole rot assessments (Gadbois and Reeve 2014). Scent-processing dogs have also been widely used as tools for wildlife research, which is the focus of this review. Dogs are extremely well-tailored to these search efforts, based on their natural olfactory abilities. While humans possess a mere six million olfactory receptors, dogs have hundreds of millions, with greater differentiation in the types of olfactory cells enabling them to detect the presence of odours with extreme specificity (Grimm-Seyfarth et al. 2021). Other common dog characteristics further enhance their success in search and detection tasks, such as their physical conformation, energy level, sensitivity to reward, and strong "play drive", all of which increase their ability to work on a wide variety of target species, in a wide variety of environmental conditions (DeMatteo et al. 2019).

The use of dogs to aid in wildlife research is not new. Zwickel (1980) authored a chapter on the subject and noted that as early as 1930 scientists were using dogs to aid in wildlife research; employing dogs was considered an effective way to reduce human-induced bias and increase the sample size of the study. Early work involving dogs was generally aimed towards driving and trapping animals for

[1] For example, our (Wildlife Ethology and Conservation Canine Lab) many attempts with no or mediocre success at training dogs to find nests and eggs of wood turtles and Blanding's turtles.

[2] There are many ways to refer to conservation canines: scent-detection dogs, detection dogs, scent dogs, tracking dogs, scent-processing dogs, etc.... We choose to use the term "scent-processing dogs" as it makes no assumptions about the sensory and cognitive processes at work. In other words, the work of a tracking and trailing conservation canine goes beyond mere detection. We are explaining these nuances in the section on canine olfactory psychophysics.

tagging, or for specimen collections. Dogs were used to assist wildlife research in other ways, with most of their applications being for intrusive purposes: flushing dogs, pointers, and retrievers aided with bird censuses (e.g. adult waterfowl or brood counts), big game censuses, assessing relative abundances (e.g. of raccoons, squirrels, and chipmunks), locating nests, tracking "scentless" young cervids (e.g. fawns), locating dens (e.g. of seals and polar bears), and tracking wounded or drugged animals (e.g. deer and moose). Dogs were also used periodically to study other aspects of wildlife ecology such as "running" rabbits to quickly determine their home range.

While earlier use of dogs in wildlife research was quite invasive, a growing interest in the use of scent-processing dogs to aid conservation efforts demonstrated evolved methodologies with dogs in a non-invasive role, allowing for the investigation of species at risk. This evolution of their deployment made it clear that well-trained dogs with the appropriate temperament could minimize the impact on habitat, and the animals themselves, during search efforts. From there emerged a strong case documenting how numerous challenges associated with studying rare or endangered wildlife species could be overcome by using scent-processing dogs. As an unobtrusive survey tool, dogs may locate live target animals while maintaining an appropriate distance so as not to put the dog or target animal at risk. There are also other signs a scent-processing dog may rely upon to signify species presence such as nests or scats.

As discussed by Gadbois and Reeve (2014) a zoosemiotic and post-cognitivist approach to scent processing takes into consideration the interactional processes between subject and environment, in a perception–action perspective. It is not only "sensation and perception", but "sensing and perceiving". Zoosemiotics offers a theory of the signs (or the stimulus). Interestingly, along with some modern post-cognitivist schools, zoosemitotics avoids the "poverty of the stimulus" problem. In other words, not limiting the study of olfaction (in our case) to a representationalist approach, but rather a direct perception approach, considering the importance of the natural motor or olfactomotor patterns of the dogs, and their interaction with the space they work with. Here, ethology and zoosemiotics merge in offering a more naturalistic perspective to scent processing in dogs.

A zoosemiotic approach would consider how communication, representation, and signification are all processes in which animals may exchange information with one another. While communication involves a signal for the exchange of information between a sender and a receiver, representation, and signification rely on signs to provide or receive information when a specific receiver or sender is not present (Gadbois and Reeve 2014). For example, signs and tracks left by a target animal are cases of signification: There is no assumption of an intended recipient, but a tracker (animal or human) can learn the relevance of such signs and tracks to find the target. These signs include things such as scent-marking with urine, scats, sheddings of fur or skin, or glandular secretions. Any of these signs can provide information to conservationists trying to determine presence–absence or to better understand population dynamics. For example, through the application of diverse techniques with scats collected in the field, which can be efficiently located by scent-processing dogs, information on species, sex, individual identity (via DNA), health factors (e.g. parasite load), hormones, and diet can be gained (Browne et al. 2006).

Dogs were documented as demonstrating high efficiency and reliability for locating individual species or signs of species with low-density populations, high sensitivity, and elusive behaviours, as well as species living in habitats with dense vegetation, challenging landscapes, or large territorial ranges (Long et al. 2007b; Smith et al. 2003, 2006). With such an acute sense of smell, dogs made detections that their human counterparts would overlook. For example, dogs can find much smaller individual animals (Cablk and Heaton 2006) and locate signs from much farther afield (Rolland et al. 2006; Wasser et al. 2004) compared to human surveyors. Additionally, dogs locate significantly more scat, of varying ages, relative to humans (Smith et al. 2003). Overall, using dogs has rapidly proven to be a promising method for increasing the efficiency and accuracy of wildlife conservation research across a wide range of species, including those that were elusive or endangered.

In performing a deep dive, even into the earliest literature, the increased efficiency of using dogs as a non-invasive method for wildlife research is undebatable. For example, Long et al. (2007b) performed a study where dogs were successfully trained to simultaneously identify black bear, fisher, and bobcat while discriminating among other sympatric forest carnivores such as coyote, striped skunk, and badger. Reindl-Thompson et al. (2006) were able to report high accuracy and efficiency rates of scent-processing dogs used to locate endangered black-footed ferrets, suggesting dogs produced more promising results than the traditional choice of spotlight surveys.

With the interest in dogs as a tool for wildlife research growing, the application of methodology involving scent-processing dogs to wildlife conservation research became increasingly varied across species and habitat types. Most of these applications began as pilot studies, to examine the feasibility, efficiency, accuracy, and reliability of using dogs to study wildlife in different settings. These applications suggested great promise for a variety of deployments of dogs in wildlife research. Pilot studies demonstrated dogs being significantly more accurate and effective than humans at locating bat fatalities at wind energy facilities regardless of vegetation (Arnett 2006; Mathews et al. 2013), and as being reliably more effective than humans at locating threatened desert tortoises, regardless of tortoises being juvenile, neonate, or adult (Cablk and Heaton 2006). Pilot studies highlighted the ability of dogs to identify individual wild Amur (Siberian) tigers by matching two scat samples from the same individual (Kerley and Salkina 2007), as well as correctly matching 25 maned wolf scat samples to the respective individuals (Wasser et al. 2009). These are examples of matching-to-sample tasks, which demonstrate a more exact understanding of the stimulus, whereby the dog is identifying the sample as the same as the target (Gadbois and Reeve 2014). This specific process will be explained below.

2 Olfactory Psychophysics: A Methodological Comment and Emerging Opportunities for Training and Assessment

It is interesting that a well-established field of study of olfaction (in animals and humans) is rarely (if at all) evoked in the modern literature on scent processing in dogs (including in the literature on Wildlife Conservation Canines). We will cover some of the points made by Gadbois and Reeve (2014) about the relevance of such concepts in our understanding of olfaction in dogs, and especially in training for accuracy and reliability.

Psychophysics (a field of experimental psychology and now sensory neuroscience) has a long methodological and conceptual tradition in discussing olfactory processing. First, we will discuss how the classical view of sensory processing brings some important nuances to the details surrounding methodology and training. As Gadbois and Reeve (2014) point out, there are four main processes that psychophysics identifies (in principle, in all the sensory modalities): Detection, discrimination, identification, and scaling.

2.1 Detection: Yes/No, Go/No-Go

The term "detection" in the industry of scent processing is used for almost any task requiring a dog to use its nose. But in psychophysics, the term is much more specific. It refers to the simple task of judging the presence or absence of a stimulus. The typical tasks used here are simply presenting one stimulus at a time to which the animal responds to the presence of the target odour, i.e. "odour present" or "odour absent". We can compare this method to a "true/false" exam in humans. There are two main methods to do this with animals.

2.1.1 Yes/No Tasks

Yes/no tasks require two possible responses to the two possible options of presence or absence. If the target odour is present ("yes"), the dog can, for example, execute a nose-hold for 5 s over the scent (or press a lever, lay down, etc.). If the target odour is absent ("no"), the dog can sit. These tests, if not performed in olfactometers, can be done in training rooms with one stimulus station being present in the room.

2.1.2 Go/No-Go Tasks

Go/no-go tasks are very similar with one important difference in the absence situation. As with the yes/no tasks, the dog responds yes (or "go") with a nose hold or any other trained response. But if the odour is absent, the dog simply does not respond (for example, they can walk away). The problem with this approach is that it may add some ambiguity when the dog is uncertain about the response to produce.

2.1.3 Considerations on Using Detection Procedures

It comes down to what overall analytical tools are used. For example, if Signal Detection Theory is used, it is preferable to have a clear "yes" and a clear "no". The "no" response commits the dogs to the judgement of absence of the target odour. There is equal effort in saying "yes" as in saying "no". It also facilitates the identification of the errors: false alarms (or type I errors) and misses (or type II errors). Not to mention the other useful Signal Detection Theory parameters such as the d' (as a measure of sensitivity) and the criterion (c, or beta) providing information about how liberal or conservative a dog is in its decision-making.

2.2 Discrimination: mAFC

Discrimination tasks require the comparison of two or more instances of scent, either a target with blanks (usually in initial training stages) or target and distractors. A human analog would be a multiple-choice exam. The target is always referred to as the S^+ or S^D (discriminative stimulus), and the blanks or distractors (non-targets) are always referred to as the S^- or S^Δ (S delta). The literature calls this procedure "alternative forced choice" or mAFC (the m indicating any number of instances or stimuli used in the task). For example, a commonly used line-up could be labelled a 6AFC. Note that for the discussion below, we will not distinguish between line-ups and wheels or carousels, as all those methods use a sequential procedure and set-up. We will simply refer to them by the number of stations visited by the dog, i.e. mAFC for a generic case, or 2AFC, 3AFC, etc., for specific sizes.

2.3 Identification: Matching-To-Sample

An identification task in humans would be a naming task. In other words, as a stimulus is presented, the human has to "name" the stimulus, e.g. "lavender" or "floral", in describing odours. In animals, we must resort to matching-to-sample procedures. Typically, there is a pre-set number of odours presented one-by-one to a dog (see Gadbois and Reeve 2014 for description of a task with 8 odours). The procedure requires the dogs to investigate one of the smells (the sample) and then proceed to a sequential presentation (line-up, carousel, etc.) to pick the matching odour.

2.3.1 Considerations on Using Discrimination and Identification Procedures

The dog literature is filled with 6AFC discrimination tasks. We can trace this tradition to Schoon and Haak (2002), or at least its popularization through their thorough historical review of the methods used with canines in forensic science. It is a sound method to initiate a dog to search (see the section below on search) a target among a large number of distractors. We use this technique a lot to initiate our dogs to the presence of S^-s, including some that could be very similar (or suspected of being similar) to the target odour (see Gadbois and Reeve 2014, 2016). It is also a great way to start a naïve dog on simple sequential search, moving later to non-linear odour stations, or room searches. Otherwise, as a procedure to understand the performance of one's dog, mAFCs need to be used with care. For instance, Gadbois and Reeve (2014) argue that if the task requires a mnemonic component (and it would for larger or longer line-ups), then a smaller (shorter) mAFC such as a 2AFC or 3AFC would be preferable.

Put simply, there are at least three levels of interference, two of which are levels of mnemonic interference that could be problematic when facing a long line-up.

Pure Sensory Interference

It is quite likely that while sampling each station of a line-up or wheel (carousel), molecules of the preceding station are still lingering in the nasal cavity of the dog's nose. This "cocktail" of odours may make the identification of the target odour difficult, especially if mixed with the odour of an S- that is very similar to the target (S^+). Interestingly, the size of the nasal cavity of the dogs can be a factor. As Gadbois and Reeve (2014) point out, there are two dimensions to the phenotypic advantage of a dog for scent processing: The "hardware" (basic olfactory anatomy and neuroanatomy) and the "software" (neurochemistry). It is often believed that dogs with larger nasal cavities may have an advantage (e.g. bloodhounds). Although this is speculative, it is possible that dogs with significantly smaller nasal cavities may have an advantage. A smaller nasal cavity may, in some cases, facilitate a faster turn-around and flush-out of volatiles, especially if the sampling or sniffing time between samples or stations is short. This may explain the unexpected accuracy found by Hall et al. (2015) in pugs.

Sensory Memory

Not unlike iconic (visual) or echoic (auditory) memory, olfactory sensory memory can be of very short duration. With echoic (auditory) and iconic (visual) sensory memory, the stimulation of the sensory system keeps a brief "after-effect" of the stimulus after its removal. The same happens with olfactory sensory memory, and arguably for longer because of the chemical nature of the stimulus. Residual molecules linger in the nasal cavity until a natural or forced clearance (nasal exhale) occurs, then a delay in signal termination at the receptor level should be expected. Dalton (2000), Dalton and Sherer (1999) described the clearance mechanisms that may be at play: nasal submucosal blood flow, nasal mucociliary clearance, and expiratory desorption. Dalton and Sherer (1999) are clear that species differences should be significant, and it seems reasonable to believe that breed differences in dogs could be impactful (as mentioned in the section above) considering the nasal cavity's own fluid dynamics: A Bloodhound should not be expected to produce the same rate of odourant clearance as a pug. We should also note that Dalton's group, although working with human olfaction, focusses on adaptation. This interesting psychophysical point does not seem to be discussed in the canine literature, but may suggest that quick sequential sampling, as often done in line-ups, wheels, and carousels could be problematic.

Working Memory

The concept of working memory in animals was developed in the 70s by Werner Honig and David Olton (see Honig 1978; Dudchenko 2004 for historical accounts). Gadbois and Reeve (2014) have also pointed out that in some tasks, working memory could be affecting performance. This problem would apply only to identification tasks. The idea is that when the sample is processed, the dog must retain in memory the characteristics of that specific smell. As the dog progresses through the sequential set-up, memory interference can impair the performance of the dog. As the dog progresses towards the end of the sequence (e.g. line-up) the memory of the sample experiences some interference from the early stations of the line-up. This deleterious impact is more salient if the target odour is later in the sequence. The lead author's lab did a short experiment to assess this effect. Two expert dogs were trained on 8 odours until reaching a criterion of 80% correct responses. The position of the target odours was randomly determined by the throw of a die (the "n" at the bottom of the bars in Fig. 1). When we looked at the data by sequential position, we were surprised to see that the performance changed dramatically. Indeed, both dogs were at 97.7% performance when the sample was in first position. At positions 2 and 3, the performance hovered around 92%. In position 4, both dogs were close to 80%. The most consequential drop occurred at positions 5 and 6 with accuracy below 30% (and as low as 10%). It is important to note that, at the time, we used very small quantities of essential oils: Very salient, and highly discriminable odours. This example highlights the importance of a pure mnemonic effect on performance when using matching-to-sample procedures. This has led us to use mostly 2AFC or 3AFC set-ups even in our simple discrimination tasks.

2.4 Scaling: Threshold

Scaling methods are an important part of sensory processing tasks in laboratory settings and may apply to a small number of situations in the field. Scaling mostly requires an animal to detect a certain quantity of a stimulus. It is a discrimination task within a specific stimulus category (e.g. the question may be, as mentioned by Gadbois and Reeve 2014, not if *Nosema* or *Varroa* are present in bee hives—as most North American hives are infected—but rather if the hive is infected above a certain threshold). The dogs would then be trained to alert to a high infection rate, but not a low infection rate. These problems are more likely to be encountered in biomedical detection (see Reeve in this volume), e.g. identifying samples on a continuum such as hypoglycaemia, normoglycaemia, and hyperglycaemia, or "high stress" versus "low stress". It is clear that more familiarity (i.e. training) is necessary to attain high performance in these tasks.

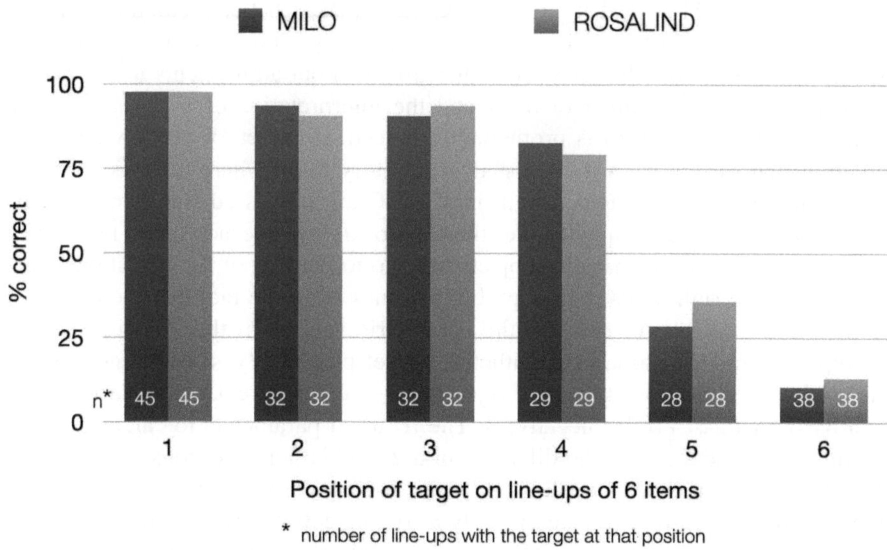

Fig. 1 Accuracy (in % correct) of two expert dogs in a matching-to-sample task in function of the position of the target in a 6AFC line-up

2.5 Classical Olfactory Psychophysics and Signal Detection Theory

The four classical psychophysical paradigms mentioned above (detection, discrimination, identification, and scaling) are not typically discussed in the literature on canine scent processing. Interestingly, the field often refers to "detection" as if that was the only process and type of task involved. We argue that knowing the literature in (olfactory) psychophysics is important. We have made the point elsewhere (Gadbois and Reeve 2016) that Signal Detection Theory (SDT) is a useful tool when using true detection tasks (yes/no preferably, as the dog is committed to a clear "no", but also go/no-go procedures). For instance, SDT provides both parametric and non-parametric quantitative analytical tools to define "sensitivity" (d' or A') as well as the very useful criterion or bias (c or Beta) informing us of the type of errors (false alarms or misses) produced by the dogs. Knowing the bias (liberal or conservative) of a dog can inform handlers and trainers on training remediation and fine-tuning, including adjusting the criterion of a dog for a specific task. For example, it is clear that a mine detection dog should be more liberal (saying "yes" when uncertain) than conservative (saying "no" when uncertain). Yet, in some other tasks, a conservative dog may be ideal: In a contract with the lumber industry and federal forestry stakeholders we were told that the methods used at the time to assess larval infestations were too "liberal", i.e. too many trees were destroyed and shown to be healthy when examined. In this case, a more conservative tool, with higher specificity was a welcomed detection tool.

One caveat of SDT is that it works better (and makes better predictions) if used in a strict decision task (as discussed above, either yes/no or go/no-go). It can in principle be used in 2AFC tasks (or other mAFCs) but adjustments to the parameters require a high number of trials and the interpretation of a "no" (or worse, a "no-go" or no response) is problematic. Indeed, in olfactory tasks, we are in a situation that differs from most visual or auditory tasks where there is a reasonable assumption that the two stimuli in a 2AFC are processed simultaneously (or in an extremely quick sequence, i.e. the time of an eye saccade). The problem in 2AFC with olfaction is that the subjects may go to the first of the two stations and decide it is the right answer. The problem comes from the fact that the data used to calculate a d' (just to mention the parametric version of the tool) is based on a simple matrix of responses to either a "target present" (yes) or "target absent" (no): Hits (true positives), correct rejections (true negatives), false alarms (false positives), or misses (false negatives). The required parameters to calculate d' are the hits and false alarms. The dilemma in a 2AFC is that a response to the target (S^+) that is not preceded or followed by an investigation of the S^- can hardly be interpreted as a (correct) rejection. Only a true detection task (where each trial is either the S^+ or the S^-) can identify unequivocal false alarms, correct rejections, or misses.

2.6 Why Consider Detection, Discrimination, Identification, and Scaling?

For the training of wildlife conservation canines, we can argue that most training and assessment tasks will benefit from both detection and discrimination training. Identification and scaling have more specialized applications. For identification, Sam Wasser's work on matching maned wolf (*Chrysocyon brachyurus*) or bear scats to specific individuals is a great example of an identification protocol (Wasser et al. 2004, 2009). Kerley and Salkina (2007) used a similar matching-to-sample procedure for the same purpose. Otherwise, we want to argue that both detection training and discrimination training (mAFC, and maybe preferably 2AFC or 3AFC) are useful in providing different information about the target scents. For one, "real life" in the field (and this applies to medical alerts as well) is more akin to a detection task. In other words, the dogs are searching for the presence of the target odour (S^+), but are not "comparing" (at least not in real time; maybe from memory) the S^+ and S^-. That being said, dogs in the field are not signalling "no" constantly either (unless required in a directed search spot check). For that reason, one must be aware of the pros and cons of each technique and select a methodology appropriate to the context and goals.

We are recommending a few general classic references in psychophysics, often focussing on SDT: Kingdom and Prins (2016), Macmillan and Creelman (2005), and McNicol (1972/2005) are great resources. For animal psychophysics specifically, the classics by Blough (1966), Blough and Blough (1977) are worth a read. Note that more formal methods exist, in particular, olfactometry and the pioneering work of Slotnick and Schellinck with rodents (e.g. Schellinck et al. 1991, 1997; Slotnick and Schellinck 2002); and Nathaniel Hall's group (e.g. Hall and Wynne 2018; Aviles-Rosa et al. 2021) with dogs.

In summary:

1. Detection as yes/no offers the best data to assess sensitivity (d' or A') and all the parameters that can be extracted from the matrix of hits, false alarms, correct rejections, and misses (such as sensitivity and specificity, precision, and general accuracy—the average of sensitivity and specificity; see Gadbois and Reeve 2016 for an example of the summary information obtained with SDT and associated parameters; Helton (2009), gives details on the extraction of other parameters). Yes/no offers a clear "no" to a S^- which is experimentally more desirable as it reduces the uncertainty and interpretation around a dog ignoring or simply walking away from a scent (as in go/no-go).
2. Detection as go/no-go is more realistic in a transition to fieldwork as typically a response to the target is all that needs to be shaped. For assessments of SDT parameters and response/error matrices and in cases of uncertainty we have seen dogs learning to ambiguously hover over a sample; So the option above brings more experimental rigour and balance in the response pattern, therefore, the response bias is easier to asses.
3. Discrimination in mAFC (we have made an argument for shorter ones above, i.e. 2 or 3AFC) is very useful when a dog needs to learn to suppress responses to just "any good smell" in the field, and more importantly, when having to discriminate between very phenotypically similar species (e.g. in Nova Scotia, the rare Ribbon Snake (*Thamnophis sauritus*) versus the very common Garter Snakes (*Thamnophis sirtalis*), see Gadbois and Reeve (2014). Also note that we have used errorless discrimination training (EDT) as described in Gadbois and Reeve (2014) in an adaptation of Terrace's fading-in procedure (Terrace 1963a, b, 1964, 1966). For some reasons that are not yet clear to us, the success of that technique is extremely dog-dependent. Some dogs train within half a day, others never seem to get the task until under an explicit 2AFC or 3AFC as described above.
4. Discrimination in large mAFCs: We are not suggesting avoiding large mAFCs. In fact, since the field is about air and ground scenting for the detection of an S^+ at any time, teaching the dog to investigate many different stations and odours in long sequences is part of the training. More formal search tasks indoors or outdoors can be carried out with explicitly defined stations (visually identifiable containers such as boxes) or relatively hidden scents (in holes, under objects, and in other random locations).

This brings us to the final and ultimate goal in the field: Search.

3 Search: Trailing and Tracking

In the dog training and scent sports world, there seems to be some confusion on the terms to differentiate between tracking, trailing, air scenting, etc. Interestingly, some languages do not distinguish semantically between tracking and trailing. For example, the first author never thought of a conceptual or methodological distinction between the two until he learned English. Indeed the French word "pistage" seems to cover both trailing and tracking. For the purpose of this text, we will keep the definition used in Gadbois and Reeve (2014): trailing defines the search for a source, i.e. before the source of the target is acquired (so often air scenting, "nose up" search), and tracking is specifically about following ("nose down" search) the source, until the target is found. Many excellent books cover the bases of tracking and trailing. We would like to recommend four that cover most of the information necessary to get started: (1) "Scent: Training to Track, Search and Rescue" by Pearsall and Verbruggen (1982), (2) "Secrets of the Snout" by Rosell (2014), (3) "Tracking Dogs: Scent and Skills" by Fält et al. (2015), and (4) "Detector Dogs and Scent Movement: How Weather, Terrain, and Vegetation Influence Search Strategies" by Osterkamp (2020).

The only point of discussion we want to bring forward is a contentious one: Transects, or the use of search grids or paths to survey plots of land. The review below will discuss some common practices with transects, and we have had multiple discussions, especially in the context of contracts, with stakeholders that insisted on using transects with our dogs. As mentioned below, working with transects can be challenging in some topographies. Other constraints, such as having to have dogs leashed (often the case in National Parks) or needing distance between the dog and the handler impair the line of sight. In a sense, much of the issues are around handling within the constraints of the transect. Here we want to take an ethological perspective on this problem. It seems to us that free searches are more natural for the dogs, and it is not clear that performance in find rates are affected. It is not easy to assess the difference between transect and free searches, especially not in the conditions where the implementation of the transects are the most problematic (dense vegetation, difficult topography, etc.). As we are not suggesting a solution or presenting data (other than the discussion below), we would like to present the problem from a different perspective. We argue that transects are an anthropocentric search strategy that would be most typical of very visual species. Much of the literature on olfactory behaviour and cognition in animals (e.g. Chap. 6 of Animal Communication by Bradbury and Verhencamp 2011) does not suggest scenting animals (reptiles or mammals) use systematic search patterns. In fact, a relatively recent literature suggests that search patterns in foraging animals follow Lévy walks or flights (in birds). Lévy walks or flights are a form of random walks based on a Markov process. Sims et al. (2019) go as far as framing this in a neuroethological perspective, suggesting that optimal searching behaviours may be generated intrinsically by Central Pattern Generators. Data suggests that such foraging patterns exist in jackals (Atkinson et al. 2002). To be clear, the alternative "natural" foraging style would be the use of "classical" random walks (see Benhamou 2007 for a discussion) or Brownian walks: In other words, "systematic" or deterministic or non-random search patterns, which are not discussed in the animal literature. This raises an interesting question: Why would we expect dogs to be comfortable and effective in a non-natural search pattern? What is the evidence suggesting transects yield a better outcome? We are merely proposing that more research needs to be done to understand optimal searching strategies for dogs in the field, and it is quite possible that systematic, non-random approaches are not it. One thing is sure, as many practitioners in the field could assert, free searches (leaving the dog to use a random search pattern) allow for much faster spatial coverage. If that is an important factor (maybe a trade-off), then free searches should be considered, at least until we have more data and continue to limit our dogs to the anthropocentric transects. With this frame of thought, we aim to further express the benefits and necessary considerations involved in using a scent-processing dog for wildlife conservation research, throughout the following review.

4 Recent Studies Using Wildlife Conservation Dogs

Before turning to the discussion of the recent literature on conservation canines, note that as an appendix, we include a summary table of some of the main studies mentioned in the text to give a quick overview of the scope of this paper. The table highlights the target species, type of study, study location, and applications to wildlife conservation. Note that, as for the main text, we are aware that this summary table is not exhaustive and that we most likely missed contributions.

In the past two or three decades, the employment of scent-processing dogs for wildlife research has increased exponentially. Dogs have become a frequently used tool to detect, identify, and monitor threatened and elusive species that traditional survey methods struggle to capture accurately and efficiently. As scent-processing dogs have been used more regularly for the identification of vertebrate species' presence (so far mostly mammals, birds, and reptiles), they have also been used for some extremely unique tasks with encouraging results.

With respect to wildlife conservation, the ability of dogs to search for a species with a wide habitat range efficiently and accurately is extremely valuable. Javan rhinoceros is one of the most highly threatened large mammal species, previously known to survive in two separate populations, however, it was through the work of dogs paired with DNA analysis that researchers were able to confirm the extinction of this species from Vietnam (Brook et al. 2012). Cougars are another species that pose problems for conservation research due to their elusive and solitary behaviour, yet with the aid of scent-processing dogs, Davidson et al. (2014) were able to produce a reliable population estimate, concluding that dogs were an efficient method for determining accurate cougar population density estimates. Furthermore, both DeMatteo et al. (2014) and Wultsch et al. (2014) provided evidence that dogs are an efficient and non-invasive sampling tool for the study of multiple wild felids, due to their ability to search for the scat of multiple species at once. Scent-processing dogs have provided detections of deer (de Oliveira et al. 2012), maned wolf (Vynne et al. 2012), fishers (Thompson et al. 2012), northern spotted and barred owls (Wasser et al. 2012), and black bears (Beckmann et al. 2015) all for population sampling in areas of high conservation interest.

Proving versatility, there are several publications that describe the use of scent-processing dogs to aid the conservation management of large aquatic mammals. Baleen whales communicate using low-frequency acoustic signals, which is overlapped by noise produced by large ships. In response to this acoustic disruption, baleen whales have been known to undergo habitat displacement which may be an indicator of increased physiological stress (Rolland et al. 2012). Scent-processing dogs were deployed on boats to locate scat samples floating on the surface of the water, allowing Rolland et al. (2012) to analyze chronic stress levels in this endangered population. This is not the only instance in which scent-processing dogs were applied for the collection of marine life scat samples: Lundin et al. (2015; 2016a, b) used dogs to locate killer whale scat for non-invasive collection of material that would offer insight into the trace levels of persistent organic pollutants, to provide useful information for population management. Furthermore, Wasser et al. (2017) took advantage of scent-processing dogs as a non-invasive sampling method for Southern Resident killer whale scat, to better understand the stage and health of pregnant individuals in the endangered population. In all instances of their use for detection of these large aquatic mammals, dogs performed impressively, guiding researchers to the scat, when it was upwards of one nautical mile away, by changing positioning on the bow of the boat, with the handler acting as an interpreter for the boat driver (Wasser et al. 2017).

Another domain of wildlife conservation that has recently experienced a surge of scent-processing dog applications is the study of reptiles. Dogs were trained to locate the Eastern Indigo snake by Stevenson et al. (2010), demonstrating high detection rates for both live snakes and snake skins. Savidge et al. (2011) also deployed scent-processing dogs to locate brown tree snakes on the Island of Guam, concluding that dogs are a promising method for the location of these snakes in the wild. Browne et al. (2015) examined the ability of dogs to identify tuatara and gecko scents, finding the dogs had high success rates for identifying fresh and old reptile scats, and could discriminate between the two reptile species. Tuataras are nocturnal and reside in underground burrows, and geckos are no less cryptic as they remain consistently well-camouflaged. Therefore, for these reptile species to be located by scent-processing dogs provides a significantly more accurate and efficient result, as compared to what is achieved from visual searches alone (Browne et al. 2015). Furthermore, dogs were applied in South Australia by Nielsen et al. (2016) for the location of pygmy bluetongue lizards, an endangered reptile that lives in burrows with very small entrance diameters making visual detection extremely challenging. The dogs were able to not only detect more than 30 occupied burrows in the field but discriminate the pygmy bluetongue lizard scent from other similar lizard species present in the area (Nielsen et al. 2016). This application of scent-processing dogs in the search for endangered reptiles continues to gain traction in different areas globally. Dogs were used to locate Blanding's turtles in Ontario (McIntosh Perry 2017), scat deposits from the blunt-nosed leopard lizard in California (Statham et al. 2019), Hermann's tortoises in Mediterranean scrublands (Ballouard et al. 2019), and, most recently, the giant bullfrog in South Africa (Matthew et al. 2021). In all instances, dogs proved to be an efficient and non-invasive survey tool to better understand the species' presence in the region of study.

While not the focus of this chapter, it is worth mentioning that beyond locating threatened and endangered species' presence with the intent of their conservation, scent-processing dogs have also played a role in locating invasive species' presence for the conservation of native resources. This is a role scent-processing dogs have maintained since their early use, such as detecting brown tree snakes in outgoing cargo from the Island of Guam (Engeman et al. 1998; Vice and Engeman 2000; Engeman et al. 2002), and now their target species and locations are broadening. The red palm weevil is the most detrimental pest to the date palms of Canary Island, with the adults being extremely elusive and challenging to detect. Scent-processing dogs successfully detected 78% of red palm weevil targets, providing promising support for dogs as a new method for identification of naturally infested trees (Suma et al. 2014). As well, the invasive brown marmorated stink bug, which causes significant losses in agricultural production in the United States, was investigated with scent-processing dog teams yielding greater than 84% accuracy in semi-field trials (Lee et al. 2014). The dogs also successfully located the overwintering sites of dormant populations, offering a great opportunity for early identification of future crop risk and management programs (Lee et al. 2014). Furthermore, dogs were able to detect spotted lanternfly egg masses, with high sensitivity and specificity, again providing evidence that scent-processing dogs could be a valuable tool for detecting egg masses while they lie dormant during the winter preventing their damaging spread before they become active (Essler et al. 2021). Scent-processing dogs have been used to locate the invasive Spotted Knapweed, and were compared to human surveyors, with researchers concluding dogs are more accurate and show promise for invasive plant detection (Goodwin et al. 2010). Additionally, laurel wilt disease, responsible for producing significant death in avocado trees in the United States, is yet another elusive disease that can be visually detected only when external symptoms become evident. However, at this point, the disease is too far advanced to stop its spread to adjacent trees. Scent-processing dogs were trained and evaluated for their ability to detect laurel-wilt-affected wood from avocado trees, and the results indicated scent-processing dogs could not only detect laurel-wilt, but did so with speed, and an accuracy of over 90% (Mendel et al. 2018). Dogs were also used to detect the presence of quagga mussels at the veliger larval stage. Invasive quagga mussels exert a massive strain on the ecosystem of Lake Mead, as well as an economic burden associated with their removal. When the quagga mussels are at the larval stage, they are microscopic and too small for a human surveyor to identify through visual searching alone. The scent-processing dogs trained on the veliger odour were able to correctly identify water samples containing the veliger larvae after the training period was completed, suggesting the use of dogs could be an advantageous management strategy for this invasive species (DeShon et al. 2016).

Scent-processing dogs have even more unique applications, with bodies of evidence growing to support their ability to detect endangered and threatened invertebrate species. A detection dog was trained on the nest material of a singular bumble bee species and was able to locate wild bumble bee nests from four different bumble bee species during field trials (O'Connor et al. 2012). This generalization is extremely beneficial, as it may be a challenge for researchers to gain access to training materials from rare bumble bee species. However, while dogs were able to generalize quite well, they did not perform any better than human searchers in this task. Mosconi et al. (2017) compared the ability of dogs to detect a threatened beetle species (*Osmoderma eremita*) to wood mould sampling. The scent-processing dogs were not only much more efficient, but produced accurate population estimates, and posed no risk to the target insect; unlike wood mould sampling, which is extremely invasive (Mosconi et al. 2017). Liczner et al. (2021) recently continued investigating the power of scent-processing dogs as a tool for bumble bee nest surveys, providing valuable insight about the limitations involved in using scent-processing dogs for this work. One concern is the difficulty in finding naturally occurring bumble bee nests for dog training. Furthering the use of scent-processing dogs in the search for endangered invertebrates, Rutter et al. (2021a) determined volunteer dog teams were able to identify 100% of the alpine stonefly targets when present for line-up searches following training. In the field, the dog teams alerted to the presence of the target species, even in instances the visual surveyors were entirely unaware of prior to dog detection (Rutter et al. 2021a), further emphasizing the benefit of employing scent-processing dogs for the analysis of endangered invertebrate species presence.

Recently, scent-processing dogs have recently stepped into a role for the detection of pathogens, toxicants, and endocrine disruptors in animal populations. This has a direct influence on wildlife conservation, as wildlife disease is a significant contributor to the loss of biodiversity, one such example being the threat that Chlamydia disease poses to koalas. Scent-processing dogs were trained to locate chlamydia-infected koala scats, to determine if dogs could become an appropriate non-invasive detection tool of disease. Dogs outperformed advanced laboratory sequencing techniques with 100% specificity and sensitivity (Cristescu et al. 2019). Another disease that has caused conservation concern is avian botulism, due to the speed at which it can spread through Hawaii's endangered water birds; once a water bird is killed by the disease its carcass provides a dangerous opportunity for the amplification of outbreaks and to kill greater numbers of water-birds (Reynolds et al. 2021). Early removal of carcasses is necessary to reduce the impact of the disease. Through the training and application of scent-processing dogs, Reynolds et al. (2021) found dogs to be a highly efficient surveillance technique for carcasses contaminated with avian botulism. Caron (2021) discusses the use of dogs by the Maryland Apiary inspection, through which dogs have been used to detect

American foulbrood (AFB) disease in honeybees since 1982. The dogs are reportedly incredibly efficient, and in a span of 3 weeks one dog was able to inspect over 1600 bee colonies that were meant to be shipped out of state, and located 100% of infected hives in the colonies (Caron 2021). The use of scent-processing dogs for AFB detection in honeybees is more efficient than when humans perform the detections, and allows the colony to be quarantined to reduce the spread of disease (Caron 2021). Scent-processing dogs have also located toxicants, which was demonstrated in Hungary to have great conservation importance. Pesticide poisoning has become one of the largest threats for raptors in Hungary and has been an issue since the beginning of the 2000s. Scent-processing dogs were trained on odours of common pesticides and decomposing raptor and mammal carcasses (Deák et al. 2021): once dogs were deployed in the field, they were able to locate greater numbers of poisoned raptors, and of poisoned baits, than human surveyors had found previously. Due to the findings of the dogs, the first search warrants were successfully executed to better protect and conserve the endangered raptor species in Central Europe (Deák et al. 2021).

As the use of scent-processing dogs continues to grow throughout a diverse range of conservation research, dogs continue to prove their ability to locate plants, reptiles, mammals, birds, and invertebrate species. Most recently, an important conclusion was drawn by Needs et al. (2021), where the researchers investigated whether scent-processing dogs could be effectively trained to detect live invasive weed targets when trained on dried and frozen samples. Dogs were successful in identifying both types of samples. As it is illegal to transport many invasive live weeds, due to the potential risk of their introduction, the successful training of dogs using dried and frozen targets provides a training methodology that significantly reduces ecological risk (Needs et al. 2021). Scent-processing dogs have also been used to scent-match individual Eurasian beavers with an accuracy of 88.9% and a specificity of 93.3% (Rosell et al. 2020), to locate cryptic brown hare leverets in dense vegetation (Karp et al. 2020), to maximize detections and population information during non-invasive surveying of koalas (Cristescu et al. 2020), to better understand the impact of wind farms on birds and bats (Dominguez Del Valle et al. 2020; Smallwood et al. 2020; Bernardino et al. 2022), and to establish more reliable group size estimates for populations of golden jackal (Hatlauf et al. 2021). In recent years, scent-processing dogs have been successfully trained to detect a number of live insects, such as the red imported fire ant (Chi et al. 2021), Asian longhorn beetle (Eyre and Barbrook 2021), and the citrus longhorn beetle (Arnesen and Rosell 2021) for better invasive species management and ecological monitoring. As well, scent-processing dogs have been employed to draw conclusions about species presence of the rare Morro Bay kangaroo rat (Villablanca et al. 2021), and for locating and identifying isolated populations of threatened little penguins (Cargill et al. 2022).

4.1 Comparison of Wildlife Conservation Dogs to Other Methods

When the use of scent-processing dogs for wildlife research began to gain traction as an efficient method, it was to be compared to more traditional methods of wildlife surveying. Several early studies compared the use of scent-processing dogs to the use of human surveyors, with dogs outperforming humans significantly on most accounts. Dogs detected twice the number of dead house sparrows compared to humans (Homan et al. 2001), located four times the number of scat samples as human surveyors (Smith et al. 2003), and were found to increase sperm whale scat collection fourfold (Rolland et al. 2006). In more recent studies, dogs continue to outperform human surveyors by consistently finding greater numbers of the target, or target signs (Nussear et al. 2008; Hurt and Whitelaw 2010; Paula et al. 2011; Mathews et al. 2013; Arandjelovic et al. 2015; Cristescu et al. 2015; Orkin et al. 2016). Dogs were also compared to commonly used technical survey methods, and still demonstrated better performance. Scent-processing dogs provided significantly higher detection rates than hair sampling stations and GPS radio collar tracking of black bear and grizzly bears (Wasser et al. 2004), were 3.5 times more effective in detecting the presence of multiple forest species than cameras (Long et al. 2007a), and were 10 times more effective in detecting bobcat than all other methods combined (Harrison 2006). This comparison of scent-processing dogs to more traditional methods has only increased as the use of dogs for wildlife research has become more common.

4.2 Human Search Teams

Scent-processing dogs have been repeatedly compared to human search teams, in numerous habitat types, for numerous target species. Dogs demonstrated twice the searching efficiency of human searchers for detecting passerine carcasses (Homan et al. 2001), bat carcasses (Arnett 2006; Mathews et al. 2013), bird carcasses (Paula et al. 2011; Bernardino et al. 2022), and bird and bat carcasses simultaneously (Dominguez del Valle et al. 2020; Smallwood et al. 2020) at wind farms. Dogs maintained increased efficiency and target detection under varying seasonal conditions and vegetation structures, and at low carcass densities, in all instances. Furthermore, when searching for bat carcasses at wind turbine sites in the United Kingdom, dogs located more than 4 times the number of carcasses compared to humans (Mathews et al. 2013). The high carcass detection rate of scent-processing dogs, regardless of environmental factors and target size, has important implications for better understanding the impacts of wind farms on regional species. As human searchers have proven to have significantly decreased carcass detection, the application of scent-processing dogs ensures a more accurate analysis of the repercussions of wind farms to surrounding wildlife.

Locating scat deposits of a target species, in a pre-selected environmental range, is a reliable non-invasive assessment for understanding species presence, range, and abundance. Furthermore, scat deposits can be collected for further analysis, such as determining the number of individual animals present in the selected range. However, through their comparison to scent-processing dogs, it has become evident that during human visual searches alone a large proportion of scats can be missed, resulting in unreliable population estimates. Scent-processing dogs have outperformed human surveyors in locating kit fox scats (Smith et al. 2003, 2006), scats from large mammals in Neotropical forests (de Oliveira et al. 2012), Cross River gorilla scat (Arandjelovic et al. 2015), koala scat (Cristescu et al. 2015), arboreal primate scat (Orkin et al. 2016), wide-ranging cheetah scat (Becker et al. 2017), otter scat (Grimm-Seyfarth et al. 2019), bilby scat (Thompson et al. 2020), and scat from northern and spotted-tailed quolls (Jamieson et al. 2021). Throughout these comparisons, scent-processing dogs recovered nearly four times the number of target scats compared to humans (Smith et al. 2006; Grimm-Seyfarth et al. 2019; Dominguez del Valle et al. 2020), found scats of varying ages while humans only identified those that were fresh (Arandjelovic et al. 2015; Orkin et al. 2016), located target species signs where humans failed to produce any detections (Becker et al. 2017), and in one study were recorded as being 153% more accurate than human surveyors (Cristescu et al. 2015). When tasked with locating a different sign of species presence, dogs were once again compared to human surveyors. Petroelje et al. (2021) used scent-processing dogs to locate prey remains to better understand carnivore movements, diet, and prey risk factors, concluding that dogs required 42% less survey time compared to human surveyors, and located prey remains at 2.3 times more study sites than humans. This provides valuable precedent to suggest that scent-processing dogs are not only useful for detecting scat signs, but their use can be broadened to locate other objects of conservation interest as well.

Dogs have outperformed human searchers in locating live targets as well. Dogs were able to find and identify a greater proportion of desert tortoises than humans, regardless of vegetation, and finished surveys faster by several hours (Nussear et al. 2008; Ballouard et al. 2019). Further research was performed to consider the impact of the dogs directly searching for the live target on the tortoises and confirmed that the use of dogs, as with human searchers, did not increase the risk of predation or alter movement patterns of the target species (Heaton et al. 2008). Scent-processing dogs have outperformed humans in searching for spotted knapweed (*Centaurea stoebe*), with more than twice the accuracy of human searchers when the targets were smaller or at greater distances (Goodwin et al. 2010), in locating snails (Hurt and Whitelaw 2010), and have matched the ability of human search teams to locate active bumble bee nests (O'Connor et al. 2012).

4.3 Traditional Survey Techniques

Scent-processing dogs have also been compared to more commonly used technical survey methods such as automatic or remote camera traps. Upon comparing the efficacy of scent-processing dogs to different forms of camera surveys, Harrison (2006) reported that dogs generated more than 10 times the bobcat detections compared to an automatic camera. Long et al. (2007a) compared the effectiveness of scent-processing dogs with remote cameras, concluding dogs were 3.5 times more effective in detecting the scat of bobcat, black bear, and fishers, and were more cost-effective. Scent-processing dogs have been compared to camera surveys for more species such as bush dogs, martens, and black-tailed antechinus, and have produced higher detection rates, increased efficiency, and indicated species presence where camera trapping had failed to detect presence for the past decade (DeMatteo et al. 2009; Moriarty et al. 2018; Thomas et al. 2020). Furthermore, when scent-processing dogs were compared to camera trapping for the detection of northern and spotted-tailed quolls, dogs were able to make detections of northern quoll scat where camera trapping failed to indicate species presence (Jamieson et al. 2021).

Other common technical survey methods are hair snares and scent stations. Dogs have been observed performing over 10 times better than hair snares and scent stations combined (Harrison 2006), and have outperformed hair snares for the detection of bobcat, black bear, and fishers (Long et al. 2007a). Along with the basic comparison of detection ability, scent-processing dogs are noted to have less bias than hair snares, as the snare and scent stations require attracting or baiting the target animals which can produce bias in the segment of the population attracted, which is avoided when using scent-processing dogs (Long et al. 2007a). Furthermore, hair snares require multiple study site visits to increase the likelihood of detection, whereas dogs only require one visit to the study site to properly survey the area: This is particularly beneficial when the study site is a remote location or challenging to get to. In only needing to visit a site once and being able to study a large survey area in a relatively short time, this can make scent-processing dogs more cost-effective than other methods, exclusively considering the time required for survey labor (DeMatteo et al. 2009).

Scent-processing dogs have also been compared to less common survey methods. When compared to live trapping, researchers recognized the dogs required only one hour per study site to yield the same detection rates achieved from two days of live trapping (Duggan et al. 2011). Vocalization surveys are used for the detection of owls, however, scent-processing dogs surveys were able to show a 28% higher detection probability for barred owls than that of vocalization surveys (Wasser et al. 2012). The presence of barred owl competitors or predators can significantly reduce the responsiveness of these owls to vocalization surveys, making scent-processing dogs a significantly improved survey protocol as they do not require a response from the owls to provide an accurate population estimate (Wasser et al. 2012). This demonstrates the use of dogs as a completely unobtrusive method in contrast to the invasiveness of call-back studies. The benefits of working dogs have been measured alongside radiotelemetry methods (Chambers et al. 2015), wood mould sampling with dogs taking a tenth of the time and providing nearly twice the detection probability (Mosconi et al. 2017), silt fences to which dogs were considered more cost-effective and reliable (MTO and McIntosh Perry 2017), and spoor-based surveys which produced zero detections in the presence of nearly 30 sample detections by dogs (Becker et al. 2017). Finally, in one of the most impressive comparisons documented, scent-processing dogs were found to outperform high-quality DNA analyses in the detection of Chlamydia in koala scats (Cristescu et al. 2019). DArTseq and qPCR techniques both rely on high-quality DNA presence for amplification, however there is generally only low amounts of Chlamydia DNA found in scats, particularly if the bacterial infection has been present for an extended amount of time, which makes the ability of a scent-processing dog to rely on scent from the volatile compound to assess the presence or absence of infection invaluable (Cristescu et al. 2019).

While scent-processing dogs have been gaining popularity for the investigation of threatened wildlife species, they have been compared to numerous survey techniques, in countless different environments. Across nearly every comparison performed, scent-processing dogs were either more efficient, more accurate, less biased, or all of the above. In instances where population estimates are being performed for the development of conservation protocols, accuracy of detection should be as high as possible. This is observed where bird and bat carcass surveys at wind farms were not producing accurate fatality estimates when performed by human visual surveyors, but dogs were able to locate more target carcasses and provide more accurate results (Homan et al. 2001; Arnett 2006; Paula et al. 2011; Dominguez del Valle et al. 2020; Smallwood et al. 2020; Bernardino et al. 2022). Furthermore, in instances where species have a wide habitat range or are in low density, a more efficient survey is preferable. Dogs have outperformed the efficiency of traditional survey techniques in numerous environments (DeMatteo et al. 2009; Moriarty et al. 2018; Thomas et al. 2020) and have further increased their value by locating numerous target species' scat at once (Long et al. 2007a). The accuracy and efficiency of scent-processing dogs for conservation-based surveys have challenged that of human surveyors and traditional survey techniques, supporting their importance as a tool for conservation research.

4.4 Existing Training Protocols

Significant training of scent-processing dogs is necessary in order to ensure accuracy and reliability of results. While their use has grown in popularity among conservation scientists, studies have not used consistent methodology for the training of scent-processing dogs. The majority of studies use detailed training protocols, but they differ in trial design as well as criteria for determining that dogs are ready for field testing.

Training for scent-processing dogs is well established in some fields (e.g. remote explosive scent training and search and rescue), but there are only a handful of organizations that offer dog training specifically for the purposes of wildlife conservation research. For example, PackLeader Dog Training (PDT) in Washington has been offering this type of training since 1997 and their training methods were the most commonly used in early research (e.g. Harrison 2006; Long et al. 2007a, b; Rolland et al. 2006; Smith et al. 2003; Wasser et al. 2004). PDT offers training for Comparable Species Detection—a method used following scat collection to validate species identification (see details in Harrison 2006; Smith et al. 2003). Working Dogs for Conservation is another organization that offers specialized training for scent-processing dogs and also gained recognition in the early years of scent-processing dogs in conservation work (Boydston 2005; Hurt et al. 2000; Reindl-Thompson et al. 2006). Both organizations use training programs that borrow from techniques used to train dogs in narcotics and forensic work, as well as search and rescue. Other programs such as the Washington State Department of Corrections Canine Narcotics Training Program at McNeil Island Correctional Center have been involved in wildlife research by providing training for dog handlers (Wasser et al. 2004).

When scent-processing dogs being used for conservation research are hired through a third-party company, that company is also responsible for training the dog on the specific targets relevant to the study. When this is the case, the authors generally do not comment on the training of the dogs in their methodology. For reference, Kretser et al. (2016), Becker et al. (2017), Wilbert et al. (2019), Statham et al. (2019), La Guardia et al. (2020), all hired Working Dogs for Conservation and do not mention the training protocol involved in preparing the dogs for the research. While this tends to be the norm, some researchers do include differing levels of detail; for example, Arandjelovic et al. (2015) stated the training protocol followed by Working Dogs for Conservation for their research is found in Wasser et al. (2004). This training method involves the use of a scent station which has a number of compartments that can each hold a different scent, and the dog is instructed to actively sniff each compartment and provide the trained response (usually sitting) at the compartment holding the target scent (Wasser et al. 2004). The dogs are rewarded with approximately 90 s of play and significant verbal praise; Wasser et al. (2004) explain this is to help the dog associate sample detection with the reinforcement of the reward. The training protocol detailed is an example of a "line-up" training method, where the dog learns to discriminate which of the present scents is the target. Liczner et al. (2021) also employed Working Dogs for Conservation and provides a brief table on the training methods involved, where it is evident a line-up method of training was also used in preparation for this research. This method is not only used by Working Dogs for Conservation, but by other companies as well. Lundin et al. (2016a, b) hired Washington's Center for Conservation Biology's Conservation Canines, and references Wasser et al. (2004) for the training methods that were used in preparing the dog for research. As well, Harrison (2006) employed PackLeader, and discusses that a line-up training method was also used.

As mentioned above, the line-up method is a discrimination task, as it requires the dog to identify the target stimulus as contrasted to other stimuli, making it a test of their ability to discriminate between samples. Gadbois and Reeve (2014) describe a detection task as being different from discrimination, as detection involves the identification of the stimulus among background noise and/or interference. While it may seem intuitive that a detection task be used for training, as this is more likely the type of searching that is performed by scent-processing dogs in the field, the majority of studies using scent-processing dogs for conservation research appear to rely on line-up training. The line-up training method, specifically that which is outlined in Wasser et al. (2004), is referenced frequently when discussing training for the use of conservation dogs, and the line-up method more generally has been applied even more frequently. One recent example is Holmstad Arnesen et al. (2020) using dogs to discriminate between rock ptarmigan (*Lagopus muta*) and the willow grouse (*Lagopus lagopus*) in a yes/no task.

There are many variations of the line-up training method that gained popularity through the work published by Wasser et al. (2004). For example, Lee et al. (2014), Oldenburg et al. (2016), DeShon et al. (2016), Mosconi et al. (2017), Rosell et al. (2020), Rutter et al. (2021a, b, c) provide detailed descriptions of the training that took place to prepare the dogs for their respective research applications, and while they differ in targets, as well as the types and numbers of trials, the researchers all describe the use of a line-up method for training. A common method (e.g. Essler et al. 2021, providing great detail about the training protocol) is a variation of the classical line-up training, where a stainless-steel scent wheel is used and the dog is trained to walk around the wheel checking each compartment for the target scent without a handler being present in the room. While this methodology uses a scent wheel instead of a scent box, it is still a sequential discrimination task requiring the dog to identify the target stimulus in contrast to other stimuli. Many different structures can be built to perform line-up training with scent-processing dogs. The structure may be a longboard with compartments that can hold individual scents, it may involve simply lining up scents in containers in a room, or as in Essler et al. (2021), a wheel may be built that is capable of holding the scents in compartments, allowing the dog to walk around the wheel and determine which arm holds the target odour.

Unfortunately, fundamental peer-reviewed articles that discuss training methodologies differing from line-up training are extremely limited. Suma et al. (2014) described an interesting training protocol where the target odour (Red Palm Weevil infested palm tissue) was inserted into a tennis ball, to allow the dog to associate the target odour with their retrieved toy. In the second part of training described by Suma et al. (2014), the ball was buried underground, and the dog was urged to go find the ball. This type of information, about protocols that are different from line-up training, is necessary to provide context for other training options. When line-up training is not applied, training protocols differ significantly from one study to the next. Yet, even more detrimental to the standardization of training methods, oftentimes the training protocol is not discussed at all. For example, Ballouard et al. (2019) performed preliminary training of dogs for finding tortoises by placing tortoises in a small enclosure and encouraging the dogs to search for the target, however, this is the only detail provided on this training plan. Similarly, Thomas et al. (2020) trained scent-processing dogs to locate Antechinus arktos by training the dog to associate the target scent with a reward, however, no specific protocol is detailed as to how this association was evaluated as being solidified, and very limited information about the remaining training in the field is provided. It is for this reason that training protocols need to not only be standardized, but also investigated further for their efficacy. An examination of training protocols should be performed, and standards for graduation from training should be set, to provide better uniformity of both training and evaluation of the dog prior to beginning research.

4.5 Cost of Employing Conservation Canines

The cost of employing scent-processing dogs has become one of their most discussed drawbacks. In considering prior literature discussing the use of dogs to aid in wildlife research, there is no doubt they can increase accuracy and efficiency of the search; however, the cost of acquiring, or hiring, and training the dogs has been discussed as outweighing their benefits in certain instances.

There are many factors that impact the total costs associated with a scent-processing dog. One of the first costs being encountered is the way the researcher chooses to acquire the dog. There are professional companies, such as Working Dogs for Conservation, PackLeader, and Washington Center for Conservation Biology, from which researchers can hire a scent-processing dog and their handler. By choosing the hired dog team method, members of the company will handle the training of the dog (based on specifications of the study provided by the researchers), the transportation of the dog, and the deployment of the dog in the field. Therefore, in utilizing this method, the researcher pays a predetermined rate and does not need to have in-depth knowledge of acquiring, training, and deploying a dog for wildlife research. While this can be a very worthwhile option for some studies, this does appear to be the most expensive way to include a scent-processing dog in a research study. For surveys that: are shorter; are concerned with target species that are more easily detected or are of higher abundance; or occur at locations that are smaller or without challenging terrain, the cost of hiring a detection dog team can be quite affordable. For example, Kretser et al. (2016) hired Working Dogs for Conservation to survey low-density moose in New York; due to the terrain and the size of the survey plots, the cost came to only $130.00 USD per sample. However, if the survey involves a more challenging environment that may reduce the efficiency of the search, or a less abundant target species reducing the number of samples collected per day, the cost of hiring a scent-processing dog team can increase rapidly. For example, considering the application of a scent-processing dog team in the search for Cross River gorillas, where dogs detected 1.37 samples per team per day, the cost per sample came to $1479.00 USD (Arandjelovic et al. 2015). Furthermore, the major professional companies that have dog teams for hire are located in the United States, meaning that hiring these companies for work in other countries quickly becomes more costly as travel costs for the dogs and accommodations for the handlers rise. Harrison (2006) hired dogs from PackLeader, with the cost of hiring the dog and handler team being $3000.00 USD for two 5-day work periods, one rest period, meals, and lodging. However, once costs required for training and travel were also included, the total cost for the dogs rose. The author notes in considering the situational factors, purchasing a dog would have been more cost-effective (Harrison 2006).

In purchasing a scent-processing dog there are costs associated with the acquisition, (maintenance) training, and care of the dog. However, if the researcher intends on completing multiple studies utilizing a scent-processing dog it may be worthwhile to make the investment. Dogs can be trained, and re-trained, on many different scents and therefore can be beneficial to many different projects. If a researcher performs considerable wildlife research that requires surveying for species presence, the ability to re-train a dog in-house for each project would be quite cost-effective, in comparison to other options. Long et al. (2007a) created a hypothetical scenario to compare the costs of hiring detection dogs, versus owning them, and suggested that if the dog were purchased at the beginning of the project the cost per search site would be a minimum of $59.00 (USD) less than if the dog were leased. The cost being even less for all following projects, as the purchasing of the dog is a one-time cost.

To further reduce the cost associated with acquiring a scent-processing dog, some researchers suggest adopting dogs from animal rescue shelters. Historically, specific purebred dog breeds were associated with specific search tasks, however, for the purpose of scent detection a purebred dog may not always be the better option. It is frequently noted that a key characteristic of a good scent-processing dog is to be highly focussed and have a strong play drive, because this makes the dog focussed and motivated to complete the task to receive the reward of play (Wasser et al. 2009). Dogs possess three types of drive: prey, hunt, and play (Zwickel 1980), and it is a combination of the latter two that make a scent-processing dog ideally suited for working with vulnerable wildlife populations (Cablk and Heaton 2006). Possessing a hunt drive means that a dog has a persistent desire to find something despite potential barriers such as difficult terrain or less-than-ideal environmental conditions, and play drive manifests itself with humans as the desire to tug or retrieve (Cablk and Heaton 2006). Cablk and Heaton (2006) argue that dogs that have the energy requirement for conducting scent work do not always make the best family pet, thus potential scent-processing dogs can be found at most shelters. Therefore, with the goal of identifying a dog with both a high hunt and play drive, but reduced prey drive, it is quite possible to find such a dog in an animal shelter. Adopting a dog from an animal shelter significantly reduces the upfront cost, when compared to purchasing a purebred dog, which would further decrease the average cost of the dog per project.

There are other, situation-specific, options for acquiring scent-processing dogs that have been discussed to reduce costs. Orkin et al. (2016) partnered with the Chinese Ministry of Public Security, where they trained the police scent-processing dogs for a specific project. The reported cost for the dog's training was $2,351.00 USD, with additional costs associated with the dog amounting to less than $500. The researchers suggest it is mutually advantageous for conservation research groups to partner with local police units, as it reduces the cost and there is no need for international transport of dogs or acclimation to the climate by the dogs, and it provides an organization that has already established protocols for travel, housing, and other logistical aspects of employing a dog (Orkin et al. 2016). A similar partnership was performed in Washington, where Wasser et al. (2004) partnered with the Washington State Department of Corrections Canine Narcotics Training Program, through which the dogs and associated costs of personnel and transport averaged to approximately $500.00 USD per found sample. Nussear et al. (2008) utilized a different method, where the researchers hired handlers that privately owned scent-processing dogs. This approach did not appear to have reduced the costs significantly compared to hiring dogs through a company. The reported cost to train the dogs and survey two passes through a 1×1 km plot cost $7,872.00 USD, with most costs being associated with the handlers (costing $120.00 USD per day).

Beyond the costs associated with acquisition of a scent-processing dog to be used for a research project, the planning of the study can significantly impact the costs directly associated with the dog as well. When DNA analysis of scats is included in the study plan, Wasser et al. (2004) suggest costs can be reduced for this expense by excluding the analysis of samples that will not amplify. Moreso related to the dogs, Reindl-Thompson et al. (2006) encountered factors in the habitat of the search area that reduced the efficiency of the dogs. The researchers encountered prickly pear cactus which caused difficulty for the dogs, however, if the dogs had been fitted with protective boots they likely could have moved through this terrain, as well, the dogs had not been acclimated to the environment prior to searching and therefore were distracted by the frequent barking of prairie dogs (Reindl-Thompson et al. 2006). Arandjelovic et al. (2015) also experienced challenges with the terrain of their study area, as the steep terrain and dense vine vegetation impeded the movement of both dogs and handlers to such an extent that a remapping of the grids and transects had to be performed before the study could continue. A good understanding of the barriers to a dog search team present in an environment prior to beginning the study is advised in order to avoid unexpected costs. Dogs can be trained to ignore certain environmental factors before working in the search environment and survey plots can be structured to optimize searching around challenging or dangerous terrain, both of which will optimize the efficiency of the dog and reduce costs associated with time spent searching. An additional barrier to optimizing the scent-processing dog in the field is climate. Nussear et al. (2008) report that while in desert environments dogs were able to survey the same distance as human searchers in 60% of the time, this did not allow the dogs to cover more ground per day because they experienced increased fatigue.

Performing surveys with dogs during specific times of day, or during specific seasons, can allow for optimization of dog performance, as they can survey more area per day when the climatic conditions allow. Duggan et al. (2011) suggested that in looking for Franklin's ground squirrels, performing the scent-processing dog survey earlier in the season may have allowed for more efficient surveys, as vegetation would be shorter allowing better airflow of scents and better visualization of burrow entrances. In another example of potential planning-related cost-savings, Mathews et al. (2013) used dogs to locate bat fatalities at wind energy facilities and made the point that if there is a particularly large survey area to be covered, or studies that are more long-term, the cost of a scent-processing dog tends to be much more rewarding, as these types of searches would take significant time if performed by the other commonly selected method; human surveyors. And finally, as scent-processing dogs can be trained to locate signs from more than one target species at a time, simultaneous verification of multiple species presence can be performed to optimize the costs associated with surveying a particular area with a dog (Smith et al. 2005; Arandjelovic et al. 2015).

When dogs were used to locate bat roosts, Chambers et al. (2015) hired dogs from PackLeader and found that 15% of the cost of using dogs was associated with their travel. This is a problem that was also faced by Arandjelovic et al. (2015), as they hired dog teams and handlers from the USA (Working Dogs for Conservation) and found the dogs and handlers required significant time to acclimatize to the environmental conditions. Finding dogs that are already within the region of study not only reduces their travel costs, but also reduces the amount of time required for the dogs to acclimate to the environmental conditions of the study region. Arandjelovic et al. (2015) discuss this issue in depth and emphasize the need for the establishment of regional detection dog programs where there is the demand for the use of scent-processing dogs for conservation research. Orkin et al. (2016) attempted, and succeeded, to surmount this challenge by partnering with local police canine teams to locate multiple primate species in China, while avoiding the usually high costs of performing research using scent-processing dogs internationally.

While using dogs is not expected to be the most economical approach, depending on research goals and the extent to which scent-processing dogs will be used, this approach may be equally if not more cost effective than mainstream methods. Both Long et al. (2007a) and Harrison (2006) provide detailed cost comparisons of scent-processing dogs and more traditional approaches to surveying wildlife species. The importance of standardization of their use is made evident when considering surveys where the lack of appropriate survey planning increased the cost of using scent-processing dogs, and decreased their efficiency as a survey tool. While there may be significant cost when hiring a dog and handler team, or costs associated with acquiring and training a dog, there are also many scenarios where a scent-processing dog may be the only appropriate choice. Examples of such being where camera trapping had failed to detect species presence for over a decade (Thomas et al. 2020), or when high-quality genetic sequencing failed to detect bacterial disease presence in koalas (Cristescu et al. 2019), but dogs were able to provide accurate population information in both instances.

4.6 Factors Influencing Dog Performance

There are several factors that may influence the performance of a dog-handler team and thus the accuracy and reliability of their work. These factors can be roughly grouped into three categories: external environment, internal environment, and training. The external environment refers to meteorological conditions or physical conditions of the environment; internal factors refer to the emotional and motivational state of the dog; and training, as discussed above, is performed by the handler prior to the beginning of field research.

4.6.1 External Environment

Few publications tackle environmental issues associated with the external environment at length except for Osterkamp (2020) and Pearsall and Verbruggen (1982). Bradbury and Verhencamp (2011) also discussed issues with a broader range of species depending on olfaction to find food and mates. The usual focus is on temperature, humidity, barometric pressure, wind, and soil. The complex interaction of these factors makes the assessment of "good field days" particularly challenging, including extreme conditions such as cross-overs (when the relative humidity of the air is lower than the ambient temperature in degrees Celsius). Atmospheric ionization (the amount of negative or positive ions in the air) is, as far as we know, never discussed in the context of field conditions and olfaction: We simply do not know if atmospheric ionization is a factor, yet, anecdotally, humans will often comment on scents being enhanced in some conditions associated with high levels of negative ions (for example, just before a rainfall). It is known for example that negative ions are more prevalent at sea level, near moving water (e.g. falls, rivers, the ocean), high winds, or mountain environments. Clearly, locations, time of year, time of day, and weather conditions are all factors that must be taken in consideration when working with our canine research assistants.

In considering what aspects of the external environment may influence the performance of a scent-processing dog, it is necessary to look at the aspects that will impact the movement or physical properties of the dog, as well as that which will impact the movement of the scent itself. A major environmental factor that will influence the performance of the dog is temperature. Heightened panting rate is the main behaviour utilized by a dog to reduce their body temperature. However, dogs are unable to pant and sniff simultaneously, and this increase in panting decreases a dog's ability to continuously follow a scent (Reed et al. 2011). In addition to increasing panting rate, heat can also result in nasal tissue dryness which may have implications on the olfactory abilities of the dog (Paula et al. 2011). Smith et al. (2003) found that a dog that panted excessively in hot conditions struggled to locate kit fox scat, however, the detection rate of the dog improved significantly in cooler temperatures. Beyond its ability to increase panting rate and reduce the sniffing ability of a dog, sustained heat can cause fatigue and poor performance (Harrison 2006; Homan et al. 2001), and some researchers have cautioned against using dogs in extreme climates (Long et al. 2007b). Nussear et al. (2008) discussed this point, when using scent-processing dogs in the Mojave Desert: they stated the dog could perform surveys in a shorter amount of time than human surveyors, but they often could not cover any more survey area per day due to fatigue of the dogs, brought on by working in sustained high temperatures.

While it is not possible to avoid environmental factors, specific and cautious study design may have the potential to minimize the impact it has on search efficiency. Dogs demonstrate shade-seeking behaviour, increased panting, and change in tongue colouration when they begin to overheat. Nussear et al. (2008) were able to avoid significant climatic effects on detection rate by closely monitoring for these observable changes and providing breaks for water, rest, and temperature regulation throughout the survey period. For certain locations, changing the season during which the survey is performed offers the most beneficial scenting conditions for dogs. Nussear et al. (2008) attempted to counteract the impacts of temperature by performing their study in the fall when desert air temperatures were slightly lower, rather than surveying in the spring when tortoises are most active. In the spring, the dogs would be under greater environmental stress, therefore limiting their performance and search time per day. If changing the season of search time is not an option, Nussear et al. (2008) suggest altering the time of day the search period begins, such as earlier in the morning or late in the evening when temperatures are slightly cooler.

There are clear challenges associated with high temperatures, however, cold temperatures do not provide much relief to this predicament. In colder temperatures, odour production is inhibited, and in turn can reduce the detection of targets (Gutzwiller 1990). As scent-processing dogs are relying on the propagation of a target scent to guide them to the target, any factors influencing odour production will also influence scent-processing dog performance. Even if certain weather, vegetation, or terrain conditions do not change the efficiency of the dog's movement, they may still impact the scent behaviour (Karp 2020). For example, Duggan et al. (2011) suggested their study was influenced by weak scent conditions, as they were locating ground squirrels with burrowing habits, preference for remaining in dense vegetation, and overall small body size. All of these factors reduce the strength of the scent, as well as the ability for the odour to propagate, making it more challenging for the dog to pick up on when searching. Decomposition of targets may also impact the odour properties (Gutzwiller 1990); however, this does not appear to affect successful target location by scent-processing dogs. Paula et al. (2011) performed an analysis on factors influencing dog performance and found target carcass decomposition did not produce a significant difference in the accuracy of the dog team. Furthermore, Orkin et al. (2016) used detection dogs to locate scat for arboreal primates, who defecate from elevations up to 30 m above ground causing much of the feces to be fragmented by leaves and branches, yet this challenge did not alter scat detection by dogs. As well, Dominguez del Valle et al. (2020) found carcass size and decomposition did not influence performance of detection dogs in locating bird and bat fatalities at a wind farm. While it does not appear to influence dog accuracy, the efficiency of dogs will decrease with increased carcass, or target, decomposition (Paula et al. 2011; Orkin et al. 2016).

Environmental factors other than temperature, such as precipitation, also play a part in scat detection. There is a level of humidity in the air that can optimize scenting conditions, by preventing nasal tissue dryness (Paula et al. 2011), but also by supporting the production of scent plumes from the target. However, strong precipitation events can degrade and remove scats from habitat ranges. Reed et al. (2011) noted detection rates for target scats made by the dogs increased with the number of days since the last precipitation event, suggesting the more time following a major precipitation event, the more time it allowed target scats to re-accumulate, increasing the accuracy of the data collected by a scent-processing dog survey.

Wind speed and direction are another environmental factor that have been speculated to play a role in dogs' detection functions. The direction of the wind may distort information about the location of the target, depending upon the location of the dog with respect to the wind. However, Reed et al. (2011) found that in all cases, wind speed and direction did not change the accuracy or efficiency of their scent-processing dog. The authors note it is likely this would not be the case had the dog been on-lead, as deploying the dog off-lead made it easier for the dog to search for scents in multiple directions and adjust its position in reference to wind directions (Reed et al. 2011). Leigh and Dominick (2015) also reported not encountering a relationship between wind speed and performance of the detection dog, when deploying the dog off-lead. Furthermore, in detecting the scat of Southern Resident Killer Whales, detection dogs locate scats up to one nautical mile away, regardless of high-speed wind currents (Wasser et al. 2017). Overall, previous evidence converges to suggest wind speed and direction may not have significant repercussions for a scent-processing dog, if the dog has adequate autonomy to manipulate the wind currents with their own movements.

Beyond meteorological conditions, the environment type is another external factor impacting survey ability. Scent-processing dogs can detect species' presence without relying on visual cues, which tends to be the most significant drawback to human-only search teams. Due to the remarkable ability of dogs to follow a scent, they have been reported to perform better than human searchers and thermal imaging in locating target species in dense vegetation (Nussear et al. 2008; Karp 2020). Habitat, which may significantly impact human accuracy, generally does not change the accuracy of a dog (Paula et al. 2011; Dominguez del Valle et al. 2020). This was supported when Leigh and Dominick (2015) tested the effects of habitat structure on scat detection dog performance, with dogs producing scat detection rates of 83% or higher in three habitats of differing complexity. However, in specific dense vegetation, such as that which contains thorny shrubbery, the efficiency of a dog can be decreased, not due to inability to locate the target producing a scent, but due to the impact shrubbery has on a dog's movement (Paula et al. 2011). Habitat structure produced a more significant impediment to the ability of Brook et al. (2012) to deploy scent-processing dogs, as a large area of the survey site contained impenetrable bamboo thickets, requiring the dog to remain on trail systems rather than moving throughout the entirety of the survey site. When the search environment consists of dense vegetation, it is necessary to consider if this dense vegetation only makes it challenging for visual searches but can still be appropriately searched by a dog, because this can, occasionally, be rectified by selecting an appropriate dog breed. If the vegetation is dense, but passable, a longer legged and more agile dog may be better at moving through the area efficiently (DeMatteo et al. 2019). Similarly, if there are many mid-height obstacles, a shorter dog may be more capable of searching efficiently, by being able to go under such obstacles (DeMatteo et al. 2019). However, if the vegetation actively impedes the ability of a dog to cover the entire survey area, due to impassable or dangerous terrain, this may drastically impact survey accuracy, as the dog must be able to move through the whole survey area to appropriately locate all targets present.

4.6.2 Internal Factors

While analyzing what factors may influence the performance of a scent-processing dog in the field, it is important to consider the emotional and motivational state of the dog. Gadbois and Reeve (2014) discuss motivational factors affecting working dogs: Interestingly, in all their work in biomedical alert and detection and wildlife conservation, they saw motivational issues only in lab-based work, and not surprisingly, biomedical work (which is exclusively lab-based). It is important to consider that fieldwork and searching for biologically active organisms seems to be self-reinforcing and intrinsically reinforcing. In much of our work, especially with ribbon snakes (*Thamnophis sauritus sauritus*) and coyotes, we rarely if ever noticed motivational issues. With wood turtles (*Glyptemys insculpta*), difficult terrain or weather conditions, maintaining the dog's interest may be difficult, especially with low occurrences of target species.

Dogs that are selected for scent work are chosen due to their high hunt and play drives (Cablk and Heaton 2006). This is noteworthy when considering motivation, because the scent-processing dog is highly motivated to find the target in order to get the reward of play. With the dogs being reward-driven, if the target species is in particularly low density in the target area, it may be important for the research team to plant target scats or biological material for the dog to find throughout the search area, to ensure the dog does not lose interest (Kerley 2010). If the dog spends a considerable amount of time searching, with no reward, it is understandable the dog could lose motivation to continue performing such a detailed, energetically demanding, search. DeMatteo et al. (2019) suggest when searching for a single rare species, it may be beneficial to also train the dog on another, more abundant, species in the environment. This allows the dog to stay focussed on searching for the entirety of the survey and reduces the possibility that the dog will become frustrated and lose motivation to search (DeMatteo et al. 2019).

Kerley and Salkina (2007) speculated the decline in performance of some dogs over their years worked was due to boredom, which may be due in part to the fact that dogs lived in kennels when they were not working. These authors emphasized the importance of rewarding dogs liberally with enthusiastic play and to avoid scolding dogs for incorrect choices. In a different study, dogs tended to show boredom or frustration in a lab setting after returning from the field (Smith et al. 2003). Harrison (2006) speculated that dogs made errors in training trials where there were no target samples out of frustration and desire for reward. It is important to note that while dogs can perform surveys significantly faster than humans, dogs also experience fatigue. As mentioned in the previous section, depending on the effort required to move through a certain external environment and the significance of temperature, a scent-processing dog search team may move at a faster pace than human teams, yet not be able to search more area each day than human surveyors (Nussear et al. 2008). Particularly, searches being performed in more complex habitats should account for the additional search effort required, and if the potential for fatigue of the dog is not considered this could lead to false negatives (Leigh and Dominick 2015).

Harrison (2006) felt that dogs were more likely to lose interest, or provide errors in judgement, when they were tired, hungry, or hot, and emphasized the importance of handlers being in tune with the dog's state. The way a handler interacts with, and around, the scent-processing dog will influence the dog greatly as the dog is looking to its handler for direction and reward in many instances. If a handler shows interest in a scat that may be similar to the target, it can teach the dog to search for this non-target scat (Vynne et al. 2011b). DeMatteo et al. (2014) suggest that improper handler training can hinder the dog's accuracy and negatively impact their field success. There is general agreement that a rigorous training regime and appropriate skill level is critical for both dogs and humans, although there appears to be an emphasis on the importance of handler competency (Wasser et al. 2004; Long et al. 2007b). Smith et al. (2003) provide a table summarizing points to consider minimizing training-related biases. It is important for dog handlers to engage in consistent communication with their dogs and training must ensure that dogs consistently indicate every time they identify a target scent (Cablk and Heaton 2006). Particularly when performing research on cryptic species, or locating signs that are challenging to visually detect, handlers must be extremely well trained. If handlers incorrectly reward a false alert performed by the scent-processing dog, they may confuse the dog, leading to increased false positive alerts and skew the results of the survey (Duggan et al. 2011; Vynne et al. 2011b).

4.6.3 Existing Search Protocols

When planning to use a scent-processing dog in research for wildlife conservation it is understood that a strict training protocol for both the dog and the handler should be followed. Beyond the training protocols, preparing search protocols tailored specifically to the search environment is also important. Many factors regarding search protocol require consideration when planning a study, such as whether both handler and dog will follow a transect search approach or walk free range, whether the dog will be on- or off-lead, the size of the survey area, the time spent searching, and whether additional targets will be planted to increase motivation of the dog throughout the search. Glen and Veltman (2018) discuss the need for greater standardization of scent-processing dog search protocols. In the literature, the decision to include or forgo a transect and whether the search pattern will involve the dog being on-lead or being free range appear to be the most varied, and researchers have contrasting opinions about what combination of these choices is ideal for scent-processing dog work. While there is a need for standardization for comparison of search results between projects, it is also necessary to ensure the search protocol for a particular project is maximizing accuracy and efficiency of the dog (Glen and Veltman 2018).

Potentially the largest decision to be made when planning search protocols, and before other plans can be made, is deciding whether transects will be performed. Transects can be used to introduce greater standardization within a search, and are most often applied with the intention of ensuring full survey area coverage. When a transect search is chosen, the dog and handler must walk up and down the evenly spaced linear sections throughout the study site. Smith et al. (2003) applied 17 transects spaced at 400 m intervals to locate San Joaquin kit fox scats, finding dogs to be 100% accurate in all of their searches. Smith et al. (2003) concluded that through the use of systematic searching dogs may be useful for determining species presence or absence when concerned with rare or cryptic species, a result that was similarly observed by multiple researchers (Boydston 2005; Smith et al. 2005, 2006; Beckmann et al. 2015). Wasser et al. (2004) also provided results in favor of the transect approach, suggesting transects can help to reduce bias in scent-processing dog surveys, by requiring the dog and handler to systematically cover the whole survey site. Homan et al. (2001) also used a transect search pattern when using dogs to locate passerine carcasses in dense vegetation and concluded it may not be feasible to set transect lengths that are considered an adequate proportion of the total area to be searched without increasing the speed the dog and handler team worked, which could reduce accuracy during a transect search. In accordance with this, in writing a final report for the U.S. Fish and Wildlife Service, Vesely (2008) stated that keeping the dog following a transect in close proximity to the handler will significantly limit their search speed, and dogs may be more efficient when a transect search pattern is not imposed. The amount of time it takes to complete transect searches, and the proportion of survey area necessary for empirical significance, can increase the cost of a field survey. This was observed by Kretser et al. (2016) when they used transects to search for moose, concluding the costs were significantly increased by the time required to execute transects. This was not the only instance where transects may have impeded the ability to properly search a full area for the most reliable population estimates. Ralls and Smith (2004) measured the distance kit fox scats were located from the transect lines, finding the furthest latrine to be approximately 40 m. Latrines are an accumulation of scats and/or urine (common especially in felids), and therefore may have a stronger odour. They however point out it may have been challenging for the dog to pick up on the scent of a single scat from that same distance while moving along the transect (Ralls and Smith 2004). Furthermore, Harrison (2006) suggested the transect method was beneficial as it produced 10 times the number of detections as the other methods. Harrison (2006) also commented that dogs can become frustrated if they search for extended periods of time without a target identification, and consequently a reward. While this is not a direct result of the transect method, continuing to move in a direction where a dog does not sense a likely target may increase frustration, as was observed in Harrison (2006), and lead to false positives performed by the dog in an attempt to receive a reward.

4.6.4 Transects on Versus Off Lead

When a transect is the search protocol selected, the dog can be on or off lead while still performing a transect based search. In many instances, dogs are kept on leads of varying lengths while the handler walks a transect (Homan et al. 2001; Smith et al. 2003; Ralls and Smith 2004; Smith et al. 2005; Smith et al. 2006; Long et al. 2007a; Long et al. 2007b; Long et al. 2010; Duggan et al. 2011; Waters et al. 2011; Lee et al. 2014; Beckmann et al. 2015; Smallwood et al. 2020; Reynolds et al. 2021). When a dog is on-leash, the amount of freedom they have can vary drastically. For example, Smallwood et al. (2020) had dogs on a 5 m lead while walking transects to locate bird and bat carcasses, however, even if a dog indicated detection of a scent as being beyond the transect, the dog was constrained from moving towards the scent until the transect physically intersected with the carcass location. In a very different variation of a transect search with a dog on a leash, Matthew and Relton (2021) worked the dog on a 5 to 10 m lead (depending on vegetation type) while searching a transect, and once the dog indicated picking up a scent, the handler would allow the dog to deviate from the predetermined path.

Dogs may also be deployed off-leash, while still following the transect pattern. There are many instances where dogs are sent into the field off-leash, the handler walks the transect line, and the dog is redirected towards the transect through voice commands as necessary. In Wasser et al. (2004), all transects were performed with the dog off-leash, and while the dog always remained in sight of the handler, the researchers suggest this allowed the dog to maximize the area covered. Along this thread, in a final report written for Presidio Trust by Boydston (2005), the handler walked four loop-shaped transects throughout the area, to ensure full coverage of the survey site, while the dog walked ahead of the handler and was allowed to move freely and typically chose to move side to side across the transect, frequently covering upwards of 25 m in either direction. As well, Harrison (2006) used transects to compare dogs to automatic cameras, hair snares, and scent stations, and the dog was allowed to move approximately 15 m away from the handler. The purpose of this distance constraint was to ensure the handler could see the dog at all times, while walking the transect (Harrison 2006). In a slightly more restrictive variation of this method, Arnett (2006) walked a transect line while the dog was only allowed to search 5 m to either side of the transect. The purpose of keeping the dog within a closer distance to the transect was to provide as close a comparison between the detection rates of dogs to human surveyors, who performed visual searches walking the transect lines (Arnett 2006). Even more prohibitive, Chambers et al. (2015) worked dogs off-leash while walking transects, and specifically states the dogs were redirected towards the transect and not allowed to roam freely during the survey.

To use transects during a field survey, while trying to avoid the drawbacks of target distance from transect lines and frustration of the dog, some researchers have chosen to further reduce the impact of the transect on the way the dog chooses to search. Generally, this involves the handler following the transect without putting in place limitations on the movements of the dog. One such example of this is where Reed et al. (2011) established a singular straight-line transect through the center of the survey site on which the handler walked, allowing the dog to move around the survey plot freely. Kretser et al. (2016) also offered significant leniency to the dogs' movements when searching for moose, having the handler walk along transects, while allowing the dog a significant amount of freedom in movement keeping the dog within 100 m of the transect. Becker et al. (2017) also relied on transects to spatially sample their survey area, and had dogs working off leash with the freedom to move across the 500 m width of the transects as needed to survey for target scent. Similarly, de Oliveira et al. (2019) deployed a scent-processing dog to locate Brazilian dwarf brocket deer, with the handler walking along transect lines and the dog roaming off-lead. The dog was allowed to move away from the transect to a maximum radius of 20 m, providing the dog with significant freedom to locate the target scent, while the handler maintains a level of order of the search (de Oliveira et al. 2019).

There are instances where dogs are not taken off-lead during the survey period. In an evaluation of the potential for scent-processing dogs to identify the brown marmorated stink bug in its overwintering sites (dry crevices in dead trees), the dogs were kept on-lead for the entirety of the search period (Lee et al. 2014). In this instance, it is quite reasonable the dogs were not released, as the researchers intended to search each dead tree in the survey area and were leading the dog to each tree, allowing the dog to search, and then moving on to the next (Lee et al. 2014). Another instance where scent-processing dogs were kept on lead was for the research performed by Reynolds et al. (2021) when dogs were used to search for carcasses of waterbirds that had succumbed to avian botulism. Dogs were kept on a 5 m lead for the entirety of the search and wore wire basket muzzles for the protection of birds in the environment, and to reduce the risk of the dogs making contact with the carcass, and consequently disease (Reynolds et al. 2021). While there are clear situational benefits to having a dog on-lead, there are also situations where it is not the most practical survey protocol. Cristescu et al. (2015) performed a study with both an experimental trial where the dog was leashed and a field trial during which the dog was not leashed. The researchers note that all failures to find scats occurred when the scent-processing dog was on lead, whereas once the dog was off-lead their accuracy reached 100% (Cristescu et al. 2015). Many potential implications of the leash are discussed, such as the physical constraint to the dog's movement produced by being on-lead, the potential for the connection to the handler to influence the ability of the dog to initially detect a scent cone to follow, the disruption to a dog's natural search patterns it imposes, and the decrease in the dog's search drive observed when leashed (Cristescu et al. 2015). In contrast, Homan et al. (2001) used scent-processing dogs to search for passerine carcasses, the dogs were kept on a 5 m lead, and the researchers do not state any issues with search efficiency when the dog was on-lead.

In considering the different methods for deploying scent-processing dogs in the field, it is evident that whether the dog be on- or off-lead is extremely contextually dependent. In some instances, having the dog off-lead is optimal, such as for cryptic species of which the researchers do not have strong assumptions about species range (Thomas et al. 2020). Whereas, in the need to systematically search every square foot of an area, it may be more beneficial to the researchers to keep the dog within the quadrant through voice commands (Thompson et al. 2012). Furthermore, there are benefits to keeping a dog on-lead, such as for the safety of the dog or the target (Deák et al. 2021; Reynolds et al. 2021), or directing the dog to specific biological objects of interest (Lee et al. 2014). However, there is a need for better understanding of when each protocol is most beneficial to provide standards as to when each practice should be performed, in order to minimize any bias introduced onto the dog by the search protocol.

4.6.5 Reducing Spatial Structure in Surveys

Whether a dog is released into a search area on- or off-lead appears to be extremely variable from one wildlife conservation project to the next. Another issue that must be tackled at the same time as making this decision when planning search protocols, is the level of freedom the dog will have. Thompson et al. (2012) suggest that scent surveys are by nature not spatially structured. Thompson et al. (2012) comment dogs must be given the freedom to identify and track a scent source in relation to wind currents. Similarly, DeMatteo et al. (2019) wrote a review concerning protocols involving the use of scent-processing dogs, and comment it is entirely unrealistic to assume a dog should stay within a set distance of a transect line. The researchers support this comment by discussing that dogs are constantly cataloging the scents around them while working, while they have no awareness of the transect line, therefore by restricting the dog to an invisible linear path it may confuse the dog and interfere with their searching abilities (DeMatteo et al. 2019). Along this thread, Thompson et al. (2012) developed small search grids, and used the dogs' free range off-lead within each small grid, to attempt to introduce some spatial structure to the search without inhibiting the scent-tracking ability of the dog. This was also observed in Wasser et al. (2012), where very precise survey polygons were planned to increase the likelihood for successful detection within the overall study site, and the dogs were allowed to search freely within these polygons off-leash. Similarly, Davidson et al. (2014) sectioned a 220 km squared survey site into 8 smaller survey quadrants, which were further sub-sectioned into quarters. The dogs were off-lead and able to travel in and out of the densely vegetated terrain as they searched for the target scent, while the handler walked the ridgelines of the subsection (Davidson et al. 2014). This method provided a collection of 272 scat samples. The reported capture probabilities at the time became the largest observed for any research conducted with wild felids, further highlighting the benefits of this systematic, yet less restrictive, search protocol (Davidson et al. 2014). This same "quartering" of search areas was observed in Mathews et al. (2013), where the handlers only redirected the dog to ensure the whole area was covered, and, when compared to humans, the success of the dogs was evident in that they located more than three times the number of bat carcasses.

In a unique use of handler cues to the scent-processing dog, Dominguez del Valle et al. (2020) used a dog trained to follow a specific search pattern. The handler walked along the transect lines and the dog was allowed to move off-lead 10 m to either side of the handler, however, the dog was trained to move in a zig-zag pattern perpendicular to the transect line (Dominguez del Valle et al. 2020). This is the only study design found in this literature review in which there was a specific search method imposed on the dog, however, the researchers conclude, like in many other instances, the dogs were more efficient and accurate than human searchers and this method made it possible for the researchers to homogenize coverage of the search area (Dominguez del Valle et al. 2020). The imposed search method did not appear to change the relative performance of the dog, compared to when there was no applied search method, therefore it may be a feasible method to ensure no survey area is missed; however, it did reduce the efficiency of the search. In a less strict version of this type of search, Nussear et al. (2008) deployed dogs off leash in a 1 by 1 km survey plot, within the plot the dogs were directed by voice commands to utilize a zig-zag or a contour search pattern, depending upon the terrain in the plot. Redirecting the dog towards the handler when they are working off-leash in a free-range method can be performed to ensure the safety of the dog (Deák et al. 2021), or to be sure they cover the entirety of the search area in question (Thompson et al. 2012; Mathews et al. 2013; Karp 2020; Petroelje et al. 2021). However, if the handler is focussed on following a specific transect pattern and redirects the dog to return to the planned search pattern, this may decrease the ability of the dog to find, or follow, a scent.

Duarte et al. (2016) did not impose any strict movement or spatial limitations on the working scent-processing dog with which they worked, by having handlers walk along natural trails and borders of rivers or forests they allowed the dog to work freely and indicate when it was necessary to stray from the border of the area. This methodology showed great success for Duarte et al. (2016), as they were able to support the extension of the geographical range of the target species (small red brocket deer) which is an essential piece of information for species conservation. Using a similar survey protocol, in a slightly different study design, DeMatteo et al. (2014) worked their dog off lead during surveys, while the handler walked along pre-selected trails. The dog was not redirected by the handler at any point but appeared to stay within 15 m of the trail in any direction throughout the survey, with great results; the dog located 588 target scats across two survey periods. In the dense Atlantic forest, this methodology was also applied; the handler and orienteer walked along trails while the dog was offered complete freedom in search patterns throughout the environment, and achieved great success locating target scats as well as effectively covering a large area (de Oliveira et al. 2012). In another similar approach, Thomas et al. (2020) used a dog that worked off lead, with no handler influence on the movement of the dog, and experienced significant success. While the researchers had estimated species presence in certain areas where it had already been identified, the detection dog was able to guide researchers to successful identifications in areas where species habitat range had not yet been confirmed (Thomas et al. 2020). Equally as free range, Deák et al. (2021) deployed dogs off-lead for the location of poisoned raptor carcasses, and only gave signals to redirect the dog back to the handler when they had been out of sight for 30 s or longer, to ensure the safety of the dog. With only the welfare of the dog being a limitation, Cargill et al. (2022) allowed the working dog to perform large zig-zag sweeps of the coastline of interest in the search for little penguins, only leashing dogs in close proximity to roads.

Performing opportunistic searches with scent-processing dogs has also been used when searching for multiple target species at once, as was seen in Wultsch et al. (2014) when the dogs performed free-range searches for the scat of jaguars, pumas, and ocelots, as well as when Orkin et al. (2016) deployed scent-processing dogs to locate three different primate species. Opportunistic searching also proved beneficial for improving detection of bird strike mortality, for multiple species, surrounding 10 wind turbines of interest (Paula et al. 2011).

4.6.6 GPS Locations of Dogs

Another factor involved in planning a search protocol for conservation research involves determining what, if any, tracking devices will be used. It is quite common for the orienteer or handler to carry a GPS with them during the survey period, which may be used for a variety of reasons. Handlers looking for fox scat used a GPS to log the coordinates of each scat collected (Ramsey et al. 2017). Whereas, when looking for northern spotted and barred owls, orienteers carried a GPS to ensure the search team stayed within the designated survey area (Wasser et al. 2012). As well, handlers held GPS devices to track their search tracks while conducting a large-scale carnivore study in the forests of Western Oregon (Barry et al. 2021). While all of this information is useful to the researcher, for a better understanding of sign distribution within their study area, or for a better understanding of the space covered during surveys, there is also the opportunity to gain even more insight by fitting the scent-processing dog with a GPS collar.

Thomas et al. (2020) took full advantage of this technology by having the handler carry a GPS device for recording information during the search, but also fitted the dog with a GPS collar in order to review the search patterns of the dog after the field trial. Along this line, Reynolds et al. (2021) fitted dogs with GPS collars and had the handler mark changes of the dog's behaviour in their handheld GPS device. This allowed the authors to make a detailed graphic demonstrating the first and second changes in behaviour observed as the dog began to switch from searching, to trailing, to tracking, until the dog alerted the carcass (Reynolds et al. 2021). Often, dogs are fitted with GPS collars with the sole intention of tracking the search path, to ensure the entire survey area was covered (Cablk and Heaton 2006; Nussear et al. 2008; Vynne et al. 2011a, b; Becker et al. 2017; Deák et al. 2021; Petroelje et al. 2021) or to track the distance and speed of the dog (Mutoro et al. 2021; Reynolds et al. 2021). However, there is a wealth of information about the search patterns used by dogs to most effectively locate a target, as well as for a better understanding of target locations, that could be taken advantage of if the dog and handler are both fitted with GPS devices.

5 Conclusion

There is a multitude of opportunities for the use of scent-processing dogs in future wildlife conservation research to aid with conservation management planning. Dogs have been used to aid in research concerning wind farm fatalities (Homan et al. 2001; Arnett 2006; Paula et al. 2011; Mathews et al. 2013; Reyes et al. 2016; Dominguez del Valle et al. 2020; Smallwood et al. 2020; Bernardino et al. 2022), carnivores with varying range sizes (Wasser et al. 2004; Ralls and Smith 2004; Smith et al. 2005, 2006; Harrison 2006; Kerley and Salkina 2007; Long et al. 2007a, b; DeMatteo et al. 2009; Wasser et al. 2009; Ralls et al. 2010; Kerley 2010; Long et al. 2011; Vynne et al. 2011a, b; Reed et al. 2011; Vynne et al. 2012; Thompson et al. 2012; DeMatteo et al. 2014; Davidson et al. 2014; Wultsch et al. 2014; Beckmann et al. 2015; Wilbert et al. 2015; Lundin et al. 2015; Oldenburg et al. 2016; Becker et al. 2017; Ramsey et al. 2017; Richards et al. 2018; Moriarty et al. 2018; Wilbert et al. 2019; Grimm-Seyfarth et al. 2019; La Guardia et al. 2020; Barry et al. 2021; Deák et al. 2021; Mutoro et al. 2021), rare or cryptic mammal species (Reindl-Thompson et al. 2006; de Oliveira et al. 2012; Brook et al. 2012; Cristescu et al. 2015; Chambers et al. 2015; Duarte et al. 2016; Kretser et al. 2016; Orkin et al. 2016; de Oliveira et al. 2019; Cristescu et al. 2019; Rosell et al. 2019, 2020; Karp 2020; Thompson et al. 2020; Cristescu et al. 2020; Thomas et al. 2020; Hatlauf et al. 2021; Villablanca et al. 2021; Matthew and Relton 2021), a diverse range of live birds (Gutzwiller 1990; Mosa 2004; Wasser et al. 2012; Leigh and Dominick 2015; Jamieson et al. 2021; Cargill et al. 2022), aquatic mammals (Rolland et al. 2012; Lundin et al. 2015, 2016a, b; Wasser et al. 2017), invasive animal and plant species (Goodwin et al. 2010; Lee et al. 2014; Suma et al. 2014; DeShon et al. 2016; Mendel et al. 2018; Rosell et al. 2019; Essler et al. 2021; Chi et al. 2021; Eyre and Barbrook 2021; Needs et al. 2021; Arnesen and Rosell 2021), invertebrates of conservation interest (Waters et al. 2011; O'Connor et al. 2012; Mosconi et al. 2017; Rutter et al. 2021a; Liczner et al. 2021), and reptiles (Engeman et al. 1998; Vice and Engeman 2000; Engeman et al. 2002; Cablk and Heaton 2006; Heaton et al. 2008; Stevenson et al. 2010; Savidge et al. 2011; Browne et al. 2015; Nielsen et al. 2016; Statham et al. 2019; Ballouard et al. 2019; Matthew et al. 2021). There is a wealth of evidence suggesting scent-processing dogs can be employed to better understand target species home range, diet, population density, health status, and even to aid in identifying individuals. With their diverse abilities being consistently supported, and continuously tested, the next important stage for this field of research is quickly approaching. More emphasis needs to be placed on not only publishing training protocols, but standardizing them. Standards should be put in place to evaluate the effect of training protocols and deem a dog ready to enter the field for conservation research. Furthermore, research should be performed to better consider different search protocols, and their impact on the efficiency and accuracy of the dog. To optimize their performance, more focus needs to be placed on these details, and on developing protocols for future research. With clear protocols, developed based on empirical evidence, the opportunities for applying scent-processing dogs will grow globally. By providing standardization to the practice, it can become more accessible internationally, and dogs could have a limitless and lasting impact on conservation research.

Appendix

See Tables 1 and 2.

Table 1 Summary of studies with notes on target species, type of study, study location and applications to wildlife conservation

Publication title	Author	Year	Target species	Type of study	Study location	Application to wildlife conservation
Application of removal sampling to estimate the density of Ringed Seals (Phoca hispida) in Barrow Strait, Nortwest Territories	Hammill and Smith	1990	Ringed Seal (Phoca hispida)	Application	Barrow Strait, Northwest Territories	Dogs were able to increase accuracy for searches of seal hole estimates
Minimizing dog—induced biases in Game Bird research	Gutzwiller	1990	Game Birds	Discussion	NA	Improving detection dog usage methods for conservation, and minimizing dog induced bias in research
Effectiveness of the detector dogs used for deterring the dispersal of Brown Tree Snakes	Engeman et al.	1998	Brown Tree Snake (Boiga irregularis)	Application	Island of Guam	Emphasized the importance of utilizing detection dogs to prevent the invasion of non—native snake species
Standardizing the evaluation of Brown Tree Snake trap designs	Vice and Engeman	2000	Brown Tree Snake (Boiga irregularis)	Application	Island of Guam	Detection dogs were able to detect snakes in outgoing cargo, with more being detected following a typhoon, emphasizing the importance of searches following certain environmental events
Dogs increase recovery of passerine carcasses in dense vegetation	Homan et al.	2001	House Sparrow (Passer domesticus)	Comparison	South Dakota and North Dakota	Improved detection rate and efficiency for locating passerine carcasses
Canine assistants for conservationatists	Smith et al.	2001	San Joaquin Kit Fox (Vuples macrotis mutica)	Pilot Study (Unpublished)	NA	Increased efficiency for performing scat searches, as compared to humans
Sustained evaluation of the effectiveness of detector dogs for locating Brown Tree Snakes in cargo outbound from Guam	Engeman et al.	2002	Brown Tree Snake (Boiga irregularis)	Application	Island of Guam	Ongoing controlled monitoring of invasive species is important to ensure quality of patrols, and this can be performed using detection dogs
Detection and accuracy rates of dogs trained to find scats of San Joaquin Kit Foxes (Vulpes macrotis mutica)	Smith et al.	2003	San Joaquin Kit Fox (Vuples macrotis mutica) and Red Fox (V. macrotis mutica)	Pilot Study	California	Increased accuracy for detecting presence of target species through scat

(continued)

Table 1 (continued)

Publication title	Author	Year	Target species	Type of study	Study location	Application to wildlife conservation
Scat detection dogs in wildlife research and management: application to Grizzly and Black Bears in the Yellowhead Ecosystem, Alberta, Canada	Wasser et al.	2004	Black Bear (Ursus americanus) and Grizzly Bear (Ursus arctos)	Comparison	Yellowhead Ecosystem, Alberta	Improved detection rate for the presence of endangered species, through scat detection
Impact of agriculture and grazing on Pale—Spotted (Northura darwinii) and Andean (Northoprocta pentlandii) Tinamous in the Lerma Valley, Salta Province, Argentina	Mosa	2004	Darwin's Pale Spotted Nothura (Nothura darwinii) and Andean Tinamou (Nothoprocta pentlandii)	Application	Lerma Valley, Argentina	Estimation of relative abundance through flushing, relevant to land use planning for conservation of specific bird species
Latrine use by San Joaquin Kit Foxes (Vulpes macrotis mutica) and Coyotes (Canis latrans)	Ralls and Smith	2004	San Joaquin Kit Fox (Vulpes macrotis mutica) and Coyotes (Canis latrans)	Application	NA	Detection dogs trained to locate more than one scent per search can be effective for identifying the presence of two or more species in an area, to determine the level of interaction between the two
Assessment of scat—detection dog surveys to determine Kit Fox distribution	Smith et al.	2005	San Joaquin Kit Fox (Vulpes macrotis mutica)	Pilot Study	San Joaquin Valley, California	Increased detection rate of an endangered species in areas of low population density, to determine species distribution, habitat use, and relative abundance
Assessing reliability of microsatellite Kit Fox faecal samples using genetic and GIS analyses	Smith et al.	2006	Kit Fox (Vulpes macrotis)	Application	Carrizon Plain National Monument, California	Conclusions were not made related to dogs
Relative abundance of endangered San Joaquin Kit Foxes (Vulpes macrotis mutica) based on scat—detection dog surveys	Smith et al.	2006b	San Joaquin Kit Fox (Vulpe macrotis), Red Fox (V. macrotis mutica), and Gray Fox (Urocyon cinereoargentueus)	Application	San Joaquin Valley, California	Improved methods for determining endangered species status and detecting different species' abundance in various survey areas

(continued)

Table 1 (continued)

Publication title	Author	Year	Target species	Type of study	Study location	Application to wildlife conservation
A comparison of survey methods for detecting Bobcats	Harrison	2006	Bobcat (Lynx rufus)	Comparison	National Forest land in Sandia, and Manzano Mountains, Central New Mexico	Improved detection rate, compared to previously employed population monitoring techniques
A preliminary evaluation on the use of dogs to recover Bat fatalities at wind energy facilities	Arnett	2006	Bats (Chiroptera)	Pilot Study	West Virginia and Pennsylvania	Bat carcass detection to estimate bat fatalities at a wind farm
Accuracy and reliability of dogs in surveying for Desert Tortoises: Results from the DT—K9 trials	Cablk and Heaton	2006	Desert Tortoise (Gopherus agassizii)	Pilot Study	Mojave Desert	Improved population assessment methods of tortoises (a threatened terrestrial species)
Efficacy of scent dogs in detecting Black—Footed Ferrets at reintroduction site in South Dakota	Reindl—Thompson et al.	2006	Black Footed Ferret (Mustela nigripes)	Pilot Study	South Dakota	Improved speed and efficiency of methods for detecting the presence of an endangered species
The use of scent—detection dogs	Browne, Stafford, and Fordham	2006	NA	Review	New Zealand	Discusses the great potential for the use of detection dogs in conservation research
Using scent matching dogs to identify individual Amur Tigers from scats	Kerley and Salkina	2007	Amur (Siberian) Tiger (Panthera tigris altaica)	Pilot Study	Russian Far East (Lazovsky State Nature Zapovendik)	Improved accuracy of individual identification for assessment of an endangered carnivore population
Comparing scat detection dogs, cameras, and hair snares for surveying carnivores	Long et al.	2007a	Black Bears (Ursus americanus), Fishers (Martes pennanti), and Bobcats (Lynx)	Pilot Study	Vermont	Compared efficiency and cost of three different survey techniques, identifying scent detection dogs as being 3.5 times more effective at locating forest carnivores than other methods
Effectiveness of scat detection dogs for detecting forest Carnivores	Long et al.	2007b	Black Bears (Ursus americanus), Fishers (Martes pennanti), and Bobcats (Lynx)	Pilot Study	Vermont	Effective detection of multiple species of forest carnivores

(continued)

Table 1 (continued)

Publication title	Author	Year	Target species	Type of study	Study location	Application to wildlife conservation
Comparison of effects of humans versus wildlife detector dogs	Heaton et al.	2008	Mojave Desert Tortoise (Gopheris agassizii)	Comparison	California	Detection dogs were not found to have any additional negative effects on desert tortoises, in comparison to human searchers alone
Are wildlife detector dogs or people better at finding Desert Tortoises (Gopherus agassizii)?	Nussear et al.	2008	Desert Tortoise (Gopherus agasizii)	Comparison	Mojave Desert, California	Dogs were more efficient at locating smaller individuals of the target species, and locating the target species in denser habitat, when compared to human searchers
Detection dogs: An effective technique for Bush Dog surveys	Dematteo et al.	2009	Bush Dog (Speothos venaticus)	Pilot Study	Upper Parana Atlantic Forest of Misiones, Argentina	Dogs showed improved detection / location ability compared to previous methods for Bush Dog conservation research
Scent—matching dogs determine number of unique individuals from scat	Wasser et al.	2009	Maned Wolf (Chrysocyon brachyurus)	Pilot Study	Centre for Conservation Biology, Seattle, Washington	Dogs have the potential to be used in wildlife conservation for individual identification through scats
Using a wildlife detector dog for locating Eastern Indigo Snakes (Drymarchon couperi)	Stevenson et al.	2010	Eastern Indigo Snake (Drymarchon couperi)	Pilot Study	Southern Georgia	Dogs showed high detection rates for both live snakes and snake skins, suggesting dogs are an effective tool for studying snake species presence
Changes in Kit Fox defecation patterns during the reproductive season: Implications for non—invasive surveys	Ralls et al.	2010	Kit Fox (Vulpes macrotis)	Experimental Study	Lokern Natural Area, San Joaquin Valley, California	Recognized that relative abundance of scat deposit changes throughout the year, which should be considered when employing detection dogs for scat surveys
Using dogs for Tiger conservation	Kerley	2010	Wild Tigers (Panthera tigris)	Review	NA	Reviews the strengths, weaknesses, and possible conservation application of wildlife detection dogs, concluding they are a promising tool for tiger research
Trained dogs outperform human surveyors in the detection of Rare Spotted Knapweed (Centaurea stoebe)	Goodwin, Engel, and Weaver	2010	Spotted Knapweed (Centaurea stoebe)	Comparison and Pilot Study	Gallatin Country, Montana	Dogs showed increased accuracy, compared to humans, for detection of small plant specimens, showing promise for use in invasive plant detection

(continued)

Table 1 (continued)

Publication title	Author	Year	Target species	Type of study	Study location	Application to wildlife conservation
Canine detection of free—ranging Brown Tree Snakes on Guam	Savidge et al.	2011	Brown Tree Snake (Boiga irregularis)	Pilot Study	Island of Guam	Concluded that detection dogs provide a promising method for finding snakes in the wild
Dogs as a tool to improve bird—strike mortality estimates at wind farms	Paula et al.	2011	Birds	Comparison and Pilot Study	Serra de Candeeiros Wind Farm, Portugal	Dogs showed increased accuracy, independent of vegetation, when compared to humans, supporting future use of detection dogs for bird mortality studies
Comparing detection dog and live trapping surveys for a cryptic Rodent	Duggan et al.	2011	Franklin's Ground Squirrels (Poliocitellus franklinii)	Comparison	Illinois, Indiana, Ipwa, Missouri and Wisconsin	Dogs showed increased efficiency compared to previous live trapping methods, due to cost of employing detection dogs it is suggested they be paired with other methodologies during conservation research
Testing a detection dog to locate Bumble Bee colonies and estimate nest density	Waters et al.	2011	Bumblebee (Bombus pensylvanicus)	Pilot Study	Island of Tiree, West Coast of Scotland	Dogs showed strong potential for the detection of bumble bee nests
Predicting carnivore occurrence with non—invasive surveys and occupancy modeling	Long et al.	2011	American Black Bears (Ursus americanus), Fishers (Martes pannanti), Bobcats (Lynx rufus)	Application	Rugged Vermont landscape	A good example of how the use of detection dogs as a survey methodology can compliment the use of existing non—invasive survey methods
Resource selection and its implications for Wide—Ranging Mammals of the Brazilian Cerrado	Vynne et al.	2011a	Giant Armadillo (Priodontes maximus), Giant Anteater (Myrmecophaga tridactyla), Maned Wolf (Chysocyon brachyurus), Jaguar (Manthera onca), and Puma (Puma conocolor)	Application	Emas National Park, Cerrado, Brazil	Dogs demonstrated increased detection rate of all elusive species in a variety of vegetation, as a non—invasive, multi—species conservation methodology
Detection distance and environmental factors in conservation detection dog surveys	Reed et al.	2011	Mountain Lion (Puma concolor), Bobcat (Lynx rufus), Red Fox (Vulpes vulpes), Gray Fox (Urocyon cinereoargenteus), and Kit Fox (Vulpes macrotis)	Controlled Experiment	California Oak Woodlands, Northern California	Concluded that considering environmental conditions (most importantly rain) is recommended for effective use of detection dogs

(continued)

Table 1 (continued)

Publication title	Author	Year	Target species	Type of study	Study location	Application to wildlife conservation
Effectiveness of scat—detection dogs in determining species presence in a Tropical Savanna landscape	Vynne et al.	2011b	Maned Wolf (Chrysocyon brachyurus)	Application	Tropical Savanna in the Brazilian Cerrado	Utilized detection dogs for the collection of scats having undergone different levels of degradation, suggesting their ability to detect scats present regardless of freshness, and discussing factors that impact the sample quality
Dogs can detect scat samples more efficiently than humans: an experiment in a continuous Atlantic Forest remnant	De Oliveira et al.	2012	Deer (Mazama spp.)	Comparison	Atlantic Forest, Paranapi Acaba and Sau Paula, Brazil	Provided evidence detection dogs performed better at a detection task, compared to humans, in Neotropical habitats
Factors influencing degradation of DNA and hormones in Maned Wolf scat	Vynne et al.	2012	Maned Wolf (Chrysicyon brachyurus)	Application	Cerrado, Brazil	Provided evidence dogs could identify target species scat in all seasons, in all stages of degradation
Integrated and novel survey methods for Rhinoceros populations confirm the extinction of Rhinoceros sondaicus annamiticus from Vietnam	Brook et al.	2012	Javan Rhinoceros (Rhinoceros sondaicus annamiticus)	Pilot Study	Vietnam	The use of detection dogs for the non—invasive survey of rhinoceros' is promising
Humans versus dogs; a comparison of methods for the detection of Bumble Bee nests	O'Connor, Park, and Goulson	2012	Bumble Bee (Bombus pensylvanicus) nests	Comparison	Stirling, Scotland	Dogs showed high detection rate, although they were no more accurate than humans
A framework for inference about Carnivore density from unstructured spatial sampling of scat using detector dogs	Thompson, Royle, and Garner	2012	Fisher (Martes pernianti)	Case Study	Sierra National Forest, California	The approach detailed in the article provides a flexible analytical framework for reconciling inconsistencies between detection dog survey data and density estimation procedures

(continued)

Table 1 (continued)

Publication title	Author	Year	Target species	Type of study	Study location	Application to wildlife conservation
Evidence that ship noise increases stress in Right Whales	Rolland et al.	2012	Baleen Whales (Mysticeti)	Application	Bay of Fundy, Canada	Detection dogs are able to detect scat from large aquatic mammals, which can then be analyzed for better understanding of stress levels due to changing environmental conditions
Using detection dogs to conduct simultaneous surveys of Northern Spotted (Strix occidentalis caurina) and Barred Owls (Strix varia)	Wasser et al.	2012	Northern Spotted Owl (Strix occidentalis caurina) and Barred Owls (Strix varia)	Comparison	South For Management Unit of Shasta—Trinity National Forest, Northern California	Results suggest that detection dogs can provide a reliable method for establishing occupancy of both owl species, which is important for meeting the objectives of the recovery plan for the Northern Spotted Owl
Using detection dogs and genetic analysis of scat to expand knowledge and assist felid conservation in Misiones, Argentina	DeMatteo et al	2014	Jaguar (Pathera onca), Puma (Puma concolor), Ocelots (Leopardus pardalis), Oncillas (Leopardus tigrinus)	Case Study	Misiones, Argentina	The effectiveness of detection dogs as a non—invasive sampling technique combined with genetic analyses produces opportunities for research in areas that were previously impossible to consider with standard survey techniques
Characterization of overwintering sites of the invasive Brown Marmorated Stink Bug in natural landscapes using human surveyors and detector canines	Lee et al.	2014	Halyomorpha halys	Application	Forested landscapes in the Mid—Atlantic region	Detection dogs show high detection rates for the invasive species, providing evidence for their ability to detect overwintering populations of invasive insects
Estimating Cougar densities in Northeast Oregon using conservation detection dogs	Davidson et al.	2014	Cougar (Puma concolor)	Application	Northeast Oregon	Estimating cougar densities using scat detection dogs could be feasible at a broader scale and allow for less effort than other methods currently in use
The use of sniffing dogs for the detection of Rhymchophorus ferrugineus	Suma et al.	2014	Red Palm Weevil (Rhynchophorus ferrugineus)	Experimental	Catania, Italy	Detection dogs could be an effective tool for detecting the elusive and invasive Red Palm Weevil, for successful management

(continued)

Table 1 (continued)

Publication title	Author	Year	Target species	Type of study	Study location	Application to wildlife conservation
Non—invasive individual and species identification of Jaguars (Panthera onca), Pumas (Puma concolor) and Ocelots (Leopardus pardalis) in Belize, Central America using cross species microsatellites and faecal DNA	Wultsch, Waits, and Kelly	2014	Jaguar (Panthera onca), Puma (Puma concolor), Ocelots (Leopardus pardalis)	Experimental	Belize, Central America	Provided evidence detection dogs could be used as an efficient, non—invasive sampling tool to study multiple, wild Neotropical felids
Using detection dogs and RSPF models to assess habitat suitability for Bears in Greater Yellowstone	Beckmann et al.	2015	Black Bears (Ursus americanus)	Application	Centennial Mountains, Idaho—Montana border	Detection dogs are a useful method for sampling large carnivores and when they are coupled with genetics and RSPF models, detection dogs are a very effective method for addressing habitat suitability in areas of high conservation importance
An assessment of the effects of habitat structure on the scat finding performance of a wildlife detection dog	Leigh and Dominick	2015	Spotted Tailed Quoll (Dasyurus maculatus)	Experimental	New South Wales, Australia	Detection dogs can work effectively across a diversity of habitats and can search in complex vegetation
Detection dog efficacy for collecting faecal samples from the critically endangered Cross River Gorilla (Gorilla gorilla diehli)	Arandjelovic et al.	2015	Cross River Gorilla (Gorilla gorilla diehli)	Comparison	Border of Nigeria and Cameroon	Detection dog teams located more fresh scats than humans, and showed less bias, producing a more reliable population estimate of the critically endangered target species
The detection and identification of Tuatara and Gecko scents by dogs	Browne, Stafford, and Fordham	2015	Tuatara (Sphenodon punctatus) and Forest Gecko (Hoplodactylus granulatus)	Experimental	New Zealand	Demonstrated that detection dogs can provide an alternative method for locating cryptic reptiles, to that of visual location currently being used
Non—invasive baseline genetic monitoring of the endangered San Joaquin Kit Fox on a photovoltaic solar facility	Wilbert et al.	2015	San Joaquin Kit Fox (Vulpes macrotis mutica)	Application	Topaz Solar Farms, San Luis Obispo County, California	Detection dogs are effective and accurate for determining presence of a target species which provides data that can be coupled with future surveys, to assess effects of man made facilities on endangered populations

(continued)

Table 1 (continued)

Publication title	Author	Year	Target species	Type of study	Study location	Application to wildlife conservation
Accuracy and efficiency of detection dogs: A powerful new tool for Koala conservation and management	Cristescu et al.	2015	Koala (Phascolarctos cinereus)	Experimental	Minjerribah (North Stradbroke Island)	Demonstrated that the use of detection dogs decreases false negatives and survey time, allowing for improvement in the quality and quantity of data collection, improving koala conservation
Persistent organic pollutant determination in Killer Whale Scat samples: Optimization of a gas chromatography / mass spectrometry methods and application to field samples	Lundin et al.	2015	Southern Resident Killer Whale (Orcinus orca)	Application	San Juan Island and Puget Sound, Washington	Suggests detection dogs could be a beneficial detection tool for large aquatic mammals that are of conservation interest
Polycyclic aromatic hydrocarbons in Caribou, Moose, and Wolf scat samples for three areas of Alberta oil sands	Lundin, Riffell, and Wasser	2015	Woodland Caribou (Rangifer tarandus caribou), Moose (Alces americanus), Grey Wolf (Canus lupus)	Application	Fort McMurray, Alberta	Detection dogs are efficient for locating scats of multiple species, which can then be analyzed to determine the presence of species threatening toxic substances in the environment
Efficacy of scent—detection dogs for locating Bat roosts in trees and snags	Chambers et al.	2015	Bat snag and tree roosts	Experimental	Northern Arizona	Detection dogs could be important for locating roosts of bat species that are imperiled or of conservation concern
Scat detection dogs, DNA and species distribution modelling reveal a diminutive geographical range for the vulnerable Small Red Brocket Deer Mazama Bororo	Duarte et al.	2016	Small Red Brocket Deer (Mazama boro)	Application	Ombrophilous Dense Forest, Brazilian Atlantic Forest	Indicates the importance of incorporating non—invasive sampling techniques with molecular tools, to determine geographical range
Scat detection dogs survey low density Moose in New York	Kretser et al.	2016	Moose (Alces alces)	Pilot Study	Adirondack Park, New York State	Scat detection dogs proved to be a reasonable, non—invasive, method to collect useful data from the low density moose population
Wildlife detection dog training: A case study on achieving generalization between target odour variations while retaining specificity	Oldenburg, Schoon, and Heitkonig	2016	Wild Otter (Lutra lutra)	Case Study	NA	Detection dogs are able to rapidly generalize to, and detect, variations of the target odor they were not trained on, while still retaining specificity

(continued)

Table 1 (continued)

Publication title	Author	Year	Target species	Type of study	Study location	Application to wildlife conservation
The ability of scent detection canines to detect the presence of Quagga Mussel (Dreissena rostrigormis bugensis) veligers	DeShon et al.	2016	Quagga Mussel Veliger (Dreissema rostriformis bugensis)	Pilot Study	NA	Detection dogs were able to detect the invasive mussel species at the larval stage, suggesting their utility as an early detection tool for the presence of this invasive mussel species
Cost effective scat detection dogs: Unleashing a powerful new tool for international mammalian conservation biology	Orkin et al.	2016	Western Black Crested Gibbons (Nomascus concolor), Indochinese Gray Langurs (Trachypithecus crepusculus), and Stump Tailed Macaques (Macaca arctoides)	Pilot Study	Yunnan, China	Demonstrated the detection dogs can locate fecal samples from primates with variable diets, locomotion, and grouping patterns, even in challenging field conditions, allowing for conservation and genetic monitoring of the target species
A nose for lizards; can a detection dog locate the endangered Pygmy Bluetongue Lizard (Tiliqua adelaidensis)?	Nielsen, Jackson, and Bull	2016	Pygmy Bluetongue Lizard (Tiliqua adelaidensis)	Experimental	Mid—North region of Southern Australia	Detection dogs demonstrated the ability to distinguish between the target reptile species, and other closely related reptile species, providing a new method for detection of burrows - other than labour intensive visual searches - which is more efficient, accurate, cost efficient, and lowers workload
Modulation in persistent organic pollutant concentration and profile by prey availability and reproductive states in Southern Resident Killer Whale scat samples	Lundin et al.	2016	Southern Resident Killer Whale (Orcinus orca)	Application	San Juan Island and Puget Sound, Washington	Suggests detection dogs could be a beneficial detection tool for large aquatic mammals that are of conservation interest
Using dogs to find cats: Detection dogs as a survey method for wide ranging Cheetah	Becker et al.	2017	Cheetah (Acinonyx jubatus)	Comparison	Liuwa—Mussuma Transfrontier Conservation Area, Western Zambia	Indicated that detection dogs were far superior to spoor surveys for detecting cheetah presence, suggesting detection dogs are an efficient at providing information on cheetah populations across large landscapes

(continued)

Table 1 (continued)

Publication title	Author	Year	Target species	Type of study	Study location	Application to wildlife conservation
Training of a dog for the monitoring of Osmoderma eremita	Mosconi et al.	2017	Osmoderma eremita larvae	Application	Latium, Italy	Employing detection dogs was much less time consuming than the standard procedure (Wood Mould Sampling), poses less risk to the target species, and was deemed an accurate method
Detecting rare carnivores using scats: Implications for monitoring a Fox incursion into Tasmania	Ramsey et al.	2017	European Red Fox (Vulpes vulpes)	Experimental	Tasmania	Detection dog teams had higher probabilities of detecting scat from the invasive target species, when compared to human searchers
Population growth is limited by nutritional impacts on pregnancy success in endangered Southern Resident Killer Whales (Orcinus orca)	Wasser et al.	2017	Southern Resident Killer Whale (Orcinus orca)	Application	Salish Sea	The use of detection dog greatly increases the sample size of large aquatic mammal scats collected, due to their ability to smell the scats up to one nautical mile away even in fast moving currents
Using scat detection dogs to monitor environmental contaminants in sentinel species and freshwater ecosystems	Richards et al.	2018	Mink (Mustela vision) and Otter (Lontra canadensis)	Application	Five Rivers, Montana	Detection dogs are highly effective at locating target species scat, making them an ideal non—invasive tool for analysis of the impact of heavy metals in ecosystems
Search strategies for conservation dogs	Glen and Veltman	2018	NA	Review	NA	Discuss how search theory may be used to standardise and optimise the use of detection dogs for conservation, which would make results more comparable across different locations and times
How behaviour non—target species affects perceived accuracy of scat detection dog surveys	DeMatteo et al.	2018	NA	Experimental	NA	Indicate that detection dogs are trained to locate odor independent of masking, which can lead to the collection of non—target samples, and provide ways to minimize the collection of non—target samples during conservation research

(continued)

Table 1 (continued)

Publication title	Author	Year	Target species	Type of study	Study location	Application to wildlife conservation
Seeking efficiency with carnivore survey methods: A case study with elusive Martens	Moriarty et al.	2018	Pacific Marten (Martes caurina humboldtensis)	Comparison	Coastal Oregon	For target species detection, the use of detection dog teams was more cost efficient, and using multiple techniques provided opportunity to avoid limitations of survey methods for elusive species detection and identification
An evaluation of scent—discriminating canines for rapid response to Agricultural Diseases	Mendel, Furton, and Mills	2018	Laurel Wilt Disease (Raffaelea lauricola)	Experimental	Homestead, Florida	Detection dogs can detect Laurel Wilt affected wood and could be useful in the early detection of this disease to reduce the spread to surrounding trees
Back to the basics with conservation detection dogs: Fundamentals for success	DeMatteo, Davenport, and Wilson	2019	NA	Review	NA	The use of conservation detection dogs is expanding quickly, and this article reviews fundamentals behind dog selection and training, handler training, testing trials, and study design to increase conservation benefits
Evaluating conservation dogs in search for rare species	Bennett, Hauser, and Moore	2019	NA	Review	NA	Inconsistent methods and incomplete reporting prevent meaningful evaluation of performance and comparisons
Non—invasive identification of Herpetofauna: Pairing conservation dogs and genetic analysis	Statham et al.	2019	Blunt—Nosed Leopard Lizard (Gambelia sila)	Application	San Joaquin Desert, California	Highlights the broad potential for the use of detection dogs for non—invasive sampling of reptiles
Distribution, fine—scale subdivision, and population size of San Joaquin Kit Foxes in the Ciervo—Panoche Natural Area, California	Wilbert et al.	2019	San Joaquin Kit Fox (Vulpes macrotis mutica)	Application	San Joaquin Desert, California	While non—invasive sampling can be challenging due to low quality and quantity DNA, aged scat detected by detection dog teams was still useful in detecting target species presence
Excellent performances of dogs to detect cryptic Tortoises in Mediterranean scrublands	Ballouard et al.	2019	Hermann's Tortoises (T estudo hermanni)	Pilot Study	Mediterranean Scrublands	Detection dogs were proven as very efficient in finding the target species, there is immense potential for dogs to be used to detect and protect Mediterranean reptiles

(continued)

Table 1 (continued)

Publication title	Author	Year	Target species	Type of study	Study location	Application to wildlife conservation
Distribution of the elusive and threatened Brazilian Dwarf Brocket Deer refined by non—invasive genetic sampling and distribution modelling	Leite de Oliveira	2019	Brazilian Dwarf Brocket Deer (Mazama nana)	Application	Southern Brazil	The Brazilian dwarf brocket deer is one of the least studied Neotropical deer as it is highly elusive, however, detection dogs were able to track and collect scat samples to better inform researchers about the geographical distribution of this threatened species
Performance of detection dogs and visual searches for scat detection and discrimination amongst related species with identical diets	Grimm—Seyfarth et al.	2019	Eurasian Otter (Lutra lutra) and American Mink (Neovision visión)	Experimental	Central Europe	Detection dogs were able to discriminate otter and mink scats, suggesting utilizing detection dogs for species identification will allow laboratory analyses to be species—specific and reduce time and money spent on laboratory work for non—target species
Developing non—invasive methodologies to assess Koala population health through detecting Chlamydia from scats	Cristescu et al.	2019	Koala (Phascolarctos cinereus)	Comparison	Queensland, Australia	Detection dogs could be a powerful tool to enhance disease surveillance in populations of interest
Scent—sniffing dogs can discriminate between native Eurasian and invasive North American Beavers	Rosell et al.	2019	Eurasian Beaver (Castor fiber) and North American Beaver (C. canadensis)	Experimental	NA	Suggests that detection dogs can be used to discriminate olfactory beaver signals, which makes detection dogs a promising tool to promote conservation of native beaver species and eradication of those that are invasive
A non—invasive environmental monitoring tool for brominated flame—retardants (BFRs) assisted by conservation detection dogs	La Guardia, Richards, and Hale	2020	Mink (Mustela vision) and Otter (Lontra canadensis)	Application	Tri River System, Missoula, Montana	The application of detection dog teams for collection of fecal matter for brominated flame—retardant (BFR) analysis is a valuable conservation tool, that is currently underutilized
Dogs can scent—match individual Eurasian Beavers from their Anal Gland secretion	Rossel, Kniha, and Haviar	2020	Eurasian Beavers (Castor fiber)	Experimental	Southeastern Norway	Scent detection dogs can be used as a reliable additional method to DNA analysis of biological samples, to improve identification of individuals within a target species

(continued)

Table 1 (continued)

Publication title	Author	Year	Target species	Type of study	Study location	Application to wildlife conservation
Detecting small and cryptic animals by combining thermography and a wildlife detection dog	Karp	2020	Brown Hare Leverets (Lepus Europaueus)	Application	Zurich, Switzerland	Detection dogs can be very effective in detection of small, cryptic, endangered species when there is dense vegetation cover not suitable to be searched using thermography
Conservation detection dog is better than human searcher in finding Bilby (Macrotis lagotis) scats	Thompson et al.	2020	Bilby (Macrotis lagotis)	Comparison	NA	Researchers strongly recommend conservation detection dogs are incorporated into the search protocol for Bilbies, as they are more accurate and faster than human searchers
Sniffing out solutions to enhance conservation: How detection dogs can maximise research and management outcomes, through the example of Koalas	Cristescu, Miller, and Frere	2020	Koala (Phascolarctos cinereus)	Application	New South Wales, Australia	Detection dogs can help researchers and land managers collect more data, to better inform conservation decisions
Determining the efficacy of camera traps, live capture traps, and detection dogs for locating cryptic small mammal species	Thomas et al.	2020	Antechinus arktos	Comparison	Scenic Rim of Eastern Australia and Border Ranges National Park	Detection dogs can be a highly effective method for locating threatened, cryptic species, when traditional methods have failed to detect low density populations
Factors affecting carcass detection at wind farms using dogs and human searchers	Del Valle et al.	2020	Birds and Bats	Experimental	Wind Farm in Spain	Carcass size, vegetation characteristics and weather conditions did not influence the ability for detection dogs to locate target species carcasses, making them ideal for monitoring fatalities of small or inconspicuous species in closed vegetation
Detection dogs in nature conservation: A database on their world—wide deployment with a review on breeds used and their performance compared to other methods	Grimm—Seyfarth, Harms, and Berger	2021	NA	Review	NA	Provides information about specific detection dog breeds being employed in different countries for various targets, and provides information about choosing the most suitable dog for conservation research

(continued)

Table 1 (continued)

Publication title	Author	Year	Target species	Type of study	Study location	Application to wildlife conservation
Buzzing with possibilities: Training and olfactory generalization in conservation detection dogs for an endangered stonefly species	Rutter et al.	2021a	Alpine Stonefly (Thaumatoperla alpina)	Pilot Study	Victoria, Australia	Detection dog teams can be trained to locate endangered invertebrate species
Diving in nose first: The influence of unfamiliar search scale and environmental context on the search performance of volunteer conservation detection dog handler teams	Rutter et al.	2021b	NA	Case Study	Australia	Volunteer detection dog teams can be trained to perform conservation related searches, if properly trained in environmental contexts similar to intended working conditions
Efficacy of detection canines for Avian Botulism surveillance and mitigation	Reynolds et al.	2021	Hawaii's Waterbirds	Pilot Study	Kaua'i Island	Detector dogs were able to detect and locate waterbird carcasses infected with Clostridium botulinum (avian botulism) for early carcass removal, to reduce the risk the carcasses pose to other endangered waterbirds
Conservation detection dogs increase efficacy for prey detection at carnivore GPS cluster sites during summer	Petroelje et al.	2021	Prey animal remains, left by Gray Wolves (Canis lupis), Coyotes (C. latrans), and Bobcats (Lynx rufus)	Comparison	Upper Peninsula of Michigan	Detection dog teams outperform human searchers when targets are objects of conservation interest that are visually cryptic but are traceable by scent
Training and usage of detection dogs to better understand Bumble Bee nesting habitat: Challenges and opportunities	Liczner et al.	2021	Bumble Bee (Bombus pensylvanicus) nests	Experimental	Belgium	Detection dogs are able to generalize and find natural nests after training on artificial nests, however there are limitations in the present study that need to be considered for detection dogs to effectively locate natural bumble bee nests

(continued)

Table 1 (continued)

Publication title	Author	Year	Target species	Type of study	Study location	Application to wildlife conservation
Egg masses as training aids for Spotted Lanternfly (Lycorma delicatula) detection dogs	Essler et al.	2021	Spotted Lanternfly (Lycorma delicatula) egg masses	Experimental	NA	Detection dogs were able to detect spotted lanternfly egg masses with high sensitivity and specific, suggesting dogs could be a valuable tool for the prevention of the spread of this species by detecting the dormant eggs during winter months

Table 2 Other, non-peer reviewed studies mentioned in the text

Publication itle	Author	Year	Target Species	Type of Study	Study Location	Article Type	Application to Wildlife Conservation
Use of dogs in wildlife biology	Zwickel	1980	Multiple Species	Review	NA	Book Chapter	Provides a detailed review of the use of detection dogs and their implications within wildlife conservation
Training dogs to distinguish between Black Bear (Ursus americanus) and Grizzly Bear (Ursus arctos) feces	Hurt, Davenport, and Greene	2000	Black Bear (Ursus americanus) and Grizzly Bear (Ursus arctos)	Experimental	Montana	Undergraduate publication	Detection dogs were successful at discriminating between the visually similar scat, emphasizing the promising application of detection dogs for individual species identification
Efficacy of scent dogs in detecting Black Footed Ferrets (Mustela nigripes) at a reintroduction site in South Dakota	Reindl	2004	Black Footed Ferret (Mustela nigripes)	Experimental	South Dakota	Masters Thesis	Detection dogs showed increased efficiency when compared to traditional surveys, showing potential to be an improved method for detecting presence of an endangered species
Behavior, ecology, and detection surveys of mammalian carnivores in the Presidio	Boydston	2005	Red Fox (V. macrotis mutica) and Gray Fox (Urocyon cinereoargentueus)	Experimental	California	Final Report	Detection dogs could have applications in population surveys, as well as determining presence of endangered species
Training of conservation detection dogs to locate Kincaid's Lupine (Lupinus sulphureus ssp. Kincaidii)	Vesely (Oregon Wildlife Institute)	2008	Kincaid's Lupine Plant (Lupinus sulphureus ssp. Kincaidii)	Pilot Study	Western Oregon and Southwest Washington	Final Report	Indicated detection dogs could offer valuable assistance conduction reconnaissance surveys for rare and endangered plants
Euglandina Rosea detection by dogs, February—March 2010	Hurt and Whitelaw	2010	Euglandina rosea	Pilot Training	Wai'anae Range on Oahu, Hawaii	Training and Fielding Summary Report	Results indicated that dogs can be trained to detect and locate wild rare snails in natural environments and communicate the location of the target to the handler for non—invasive sampling
Ecological monitoring using wildlife detection dogs: Bat carcass searches at the Wanlip Wind Turbine	Stanhope (Atkins Ltd.)	2015	Common Pipistrelle (Pipistrellus pipistrellus)	Pilot Study	Wanlip Wind Turbine, Leicestershire	Article	Detection dogs have been shown to be more effective than human searchers at finding bat carcasses and are an important tool for monitoring risk posed to bird and bat species by wind turbines

(continued)

Table 2 (continued)

Publication itle	Author	Year	Target Species	Type of Study	Study Location	Article Type	Application to Wildlife Conservation
McIntosh Perry Conservation Dogs	McIntosh Perry	2017	Blanding's Turtles (Emydoidea blandingii)	Pilot Study	Ontario, Canada	Executive Summary	Detection dogs provided an efficient and accurate method for determining presence or absence of the at risk turtles along the highway, allowing proper protocols to be followed to protect the species prior to road work commencing
Take it easy: Scat detection dog and conservation of Italian Hare	Fortebraccio et al.	2018	Italian Hare (Lepus corsicanus)	Application	Hunting Districts of L'Aquila Province	Conference Poster (Xi Congresso Italiano Di Teriologia)	Use of detection dogs greatly improved the scat collection by increasing efficiency and accuracy and provided adequate information to ban hare hunting in the area
Giant Garter Snake scent—detection dog team validation guidelines	H.T. Harvey and Associates	2018	Giant Garter Snake (Thammophis gigas)	Review	NA	Validation Guidelines	Provides guidelines for the effective us of detection dogs for the conservation of the elusive giant garter snake

References

Arandjelovic M, Bergl RA, Ikfuingei R et al (2015) Detection dog efficacy for collecting faecal samples from the critically endangered Cross River gorilla (Gorilla gorilla diehli) for genetic censusing. R Soc Open Sci 2:1–14. https://doi.org/10.1098/rsos.140423

Arnesen CH, Johnsen CB, Costanzi J-M et al (2020) Canines (Canis lupus familiaris) as biodetectors for conservation work: Can they discriminate the rock ptarmigan (Lagopus muta) from the willow grouse (L. lagopus) in a yes/no task? PLoS One 15(1):e0228143. https://doi.org/10.1371/journal.pone.0228143

Arnesen CH, Rosell F (2021) Pest detection dogs for wood boring longhorn beetles. Sci Rep 11:1–13. https://doi.org/10.1371/journal.pone.0228143

Arnett EB (2006) A preliminary evaluation on the use of dogs to recover bat fatalities at wind energy facilities. Wildl Soc Bull 34:1440–1445

Atkinson RP, Rhodes CJ, Macdonald DW et al (2002) Scale-free dynamics in the movement patterns of jackals. Oikos 98:134–140. https://doi.org/10.1034/j.1600-0706.2002.980114.x

Aviles-Rosa E, Gallegos SF, Prada-Tiedemann P et al (2021) An automated canine line-up for detection dog research. Front Vet Sci 8:1–17. https://doi.org/10.3389/fvets.2021.775381

Ballouard JM, Raphael G, Fabien R et al (2019) Excellent performances of dogs to detect cryptic tortoises in Mediterranean scrublands. Biodivers Conserv 28(14):4027–4045. https://doi.org/10.1007/s10531-019-01863-z

Barry BR, Moriarty K, Green D et al (2021) Integrating multi-method surveys and recovery trajectories into occupancy models. Ecosphere 12(12):1–26. https://doi.org/10.1002/ecs2.3886

Becker M, Durant SM, Watson FG et al (2017) Using dogs to find cats: detection dogs as a survey method for wide-ranging cheetah. J Zool 302(3):184–192. https://doi.org/10.1111/jzo.12445

Beckmann JP, Waits LP, Hurt A et al (2015) Using detection dogs and RSPF models to assess habitat suitability for bears in Greater Yellowstone. West N Am Nat 75(4):396–405. https://doi.org/10.3398/064.075.0410

Benhamou S (2007) How many animals really do the Lévy walk? Ecology 88(8):1962–1969. https://doi.org/10.1890/06-1769.1

Bennett EM, Hauser CE, Moore JL (2020) Evaluating conservation dogs in the search for rare species. Conserv Biol 34(2):314–325

Bernardino J, Martins RC, Bispo R et al (2022) Ecological and methodological drivers of persistence and detection of bird fatalities at power lines: Insights from multi-project monitoring data. Environ Impact Assess Rev 93:106707

Blough DS (1966) The study of animal sensory processes by operant methods. In: Honig WK (ed) Operant behavior: Areas of research and application Meredith Publishing Company New York, NY, pp 345–379

Blough D, Blough P (1977) Animal psychophysics. In: Honig WK, Staddon JER (eds) Handbook of operant behaviour, Prentice-Hall Englewood Cliffs, NJ, pp 514–539

Boydston EE (2005) Behavior, ecology, and detection surveys of mammalian carnivores in the Presidio. Final Report. U.S. Geological Survey, Sacramento: CA

Bradbury JW, Vehrencamp SL (2011) Principles of Animal Communication, 2nd Edition. Sinauer, NYC

Brook SM, van Coeverden P de Groot, Scott C et al (2012) Integrated and novel survey methods for rhinoceros populations confirm the extinction of Rhinoceros sondaicus annamiticus from Vietnam. Biol Conserv 155:59–67. https://doi.org/10.1016/j.biocon.2012.06.008

Browne C, Stafford K, Fordham R (2006) The use of scent-detection dogs. Irish Vet Jour 59(2):97–104

Browne CM, Stafford KJ, Fordham RA (2015) The detection and identification of tuatara gecko scents by dogs. J Vet Behav 10(6):496–503. https://doi.org/10.1016/j.jveb.2015.08.002

Cablk ME, Heaton JS (2005) Efficacy and reliability of dogs for surveying desert tortoises: Results from the DT-K9 trials. Desert Conservation Centre, Las Vegas: NV (p 57)

Cablk ME, Heaton JS (2006) Accuracy and reliability of dogs in surveying for desert tortoise (Gopherus agassizii). Ecol Appl 16:1926–1935

Cargill CP, Judkins AG, Weir JS (2022) Distribution of little penguins (Eudyptula minor) along the greater Kaikoura coastline, New Zealand. N Z J Mar Freshwater Res 56:43–58. https://doi.org/10.1080/00288330.2020.1844766

Caron D (2021) Dogs can smell honeybee disease. Bee Culture, The Magazine of American Beekeeping. https://www.beeculture.com/dogs-can-smell-honey-bee-disease/ Accessed 20 Feb 2022

Chambers CL, Vojta CD, Mering ED et al (2015) Efficacy of scent-detection dogs for locating bat roosts in tree snags. Wildl Soc Bull 39(4):780–787. https://doi.org/10.1002/wsb.598

Chi WL, Chen CH, Lin HM, et al (2021) Utilizing odor-absorbed filter papers for detection canine training and off-site fire ant indications. Animals 11(8):1–12. https://doi.org/10.3390/ani11082204

Cristescu RH, Foley E, Markula A et al (2015) Accuracy and efficiency of detection dogs: A powerful new tool for koala conservation and management. Sci Rep, 5:1–6. https://doi.org/10.1038/srep08349

Cristescu RH, Miller RL, Frere CH (2020) Sniffing out solutions to enhance conservation: How detection dogs can maximise research and management outcomes, through the example of koalas. Aust Zool 40(3):416–432. https://doi.org/10.7882/AZ.2019.030

Cristescu RH, Russell LM, Schultz AJ, et al (2019) Developing non-invasive methodologies to assess koala population health through detecting Chlamydia from scats. Mol Ecol Resour, 19(4):957–969. http://https://doi.org/10.1111/1755-0998.12999

Curry E, Skogen M, Roth T (2021) Evaluation of an odour detection dog for non-invasive pregnancy diagnosis in polar bears (Ursus maritimus): considerations for training sniffer dogs for biomedical investigations in wildlife species. J Zoo Aquar Res 9:1–7. https://doi.org/10.19227/jzar.v9i1.568

Dalton P (2000) Psychophysical and behavioral characteristics of olfactory adaptation. Chem Senses 25(4):87–492. https://doi.org/10.1093/chemse/25.4.487

Dalton P, Scherer PW (1999) A mass transport model of human olfactory adaptation. Chem Senses 24:561

Davidson GA, Clark DA, Johnson BK, et al (2014) Estimating cougar densities in Northeast Oregon using conservation detection dogs. J Wildl Manage 78(6):1104–1114

De Oliveira ML, do Couto HT, Duarte JM (2019) Distribution of the elusive and threatened Brazilian dwarf brocket deer refined by non-invasive genetic sampling and distribution modelling. Eur J Wildl Res 65(21):1-9. https://doi.org/10.1007/s10344-019-1258-6

de Oliveira ML, Norris D, Ramirez JFM et al (2012) Dogs can detect scat samples more efficiently than humans: an experiment in a continuous Atlantic Forest remnant. Zoologia (Curitiba) 29(2). https://doi.org/10.1590/S1984-46702012000200012

Deak G, Arvay M, Horvath M. (2021) Using detection dogs to reveal illegal pesticide poisoning of raptors in Hungary. J Vertebr Biol 69(3):20110–20111. https://doi.org/10.25225/jvb.20110

DeMatteo KE, Rinas MA, Sede MM et al (2009) Detection dogs: an effective technique for bush dog surveys. J Wildl Manag 73(8):1436–1440. https://doi.org/10.2193/2008-545

DeMatteo KE, Blake LW, Young JK et al (2018) How behavior of nontarget species affects perceived accuracy of scat detection dog surveys. Sci Rep 8:1–11. https://doi.org/10.1038/s41598-018-32244-1

DeMatteo KE, Davenport B, Wilson LE (2019) Back to the basics with conservation detection dogs: Fundamentals for success. Wildlife Biol 1:1–9. https://doi.org/10.2981/wlb.00584

DeMatteo KE, Rinas MA, Arguelles C et al (2014) Using detection dogs and genetic analyses of scat to expand knowledge and assist felid conservation in Misiones, Argentina. Integr Zool 9(5):623–639

DeShon DL, Wong WH, Farmer D et al (2016) The ability of scent detection canines to detect the presence of quagga mussel (Dreissena rostriformis bugensis) veligers. Manag of Biol Invasion 7(4):419–428. https://doi.org/10.1111/1749-4877.12113

Dominguez del Valle J, Peralta FC, Arjona MJ (2020) Factors affecting carcass detection at wind farms using dogs and human searchers. J Appl Ecol 57(10):1929–1935. https://doi.org/10.1111/1365-2664.13714

Duarte JMB, Talarico AC, Vogliotti A et al (2016) Scat detection dogs, DNA and species distribution modelling reveal a diminutive geographical range for the Vulnerable small red brocket deer Mazama bororo. Oryx 51(4):656–664. https://doi.org/10.1017/S0030605316000405

Dudchenko PA (2004) An overview of the tasks used to test working memory in rodents. Neurosci Biobehav Rev 28(7):699–709. https://doi.org/10.1016/j.neubiorev.2004.09.002

Duggan JM, Heske EJ, Schooley RL et al (2011) Comparing detection dog and livetrapping surveys for a cryptic rodent. J Wildl Manag 75(5):1209–1217. https://doi.org/10.1002/jwmg.150

Engeman RM, Vice DS, Rodriguez DV et al (1998) Effectiveness of the detector dogs used for deterring the dispersal of Brown Tree Snakes. Pac Conserv Biol 4:256–260

Engeman RM, Vice DS, York D et al (2002) Sustained evaluation of the effectiveness of detector dogs for locating brown tree snakes in cargo outbound from Guam. Int Biodeter Biodegrad 49(2–3):101–106. https://doi.org/10.1016/S0964-8305(01)00109-3

Essler JL, Kane SA, Collins A et al (2021) Egg masses as training aids for spotted lanternfly Lycorma delicatula detection dogs. PloS One 16(5):1–15. https://doi.org/10.1371/journal.pone.0250945

Eyre D, Barbrook J (2021) The eradication of Asian longhorned beetle at Paddock Wood, UK. CABI Agricul Biosci 2:1–17

Fält L, Gustavsson T, Frank J, Aberg J, Norin A (2015) Tracking dogs: scent and skills. SWDI Publishing

Gadbois S, Reeve C (2014) Canine Olfaction: scent, sign, and situation. In: Horowitz A (ed) Domestic dog cognition and behaviour. Springer-Verlag, Berlin Heidelberg, pp 3–29

Glen AS, Veltman CJ (2018) Search strategies for conservation detection dogs. Wildlife Biol 2018(1):1–9. https://doi.org/10.2981/wlb.00393

Goodwin KM, Engel RE, Weaver DK (2010) Trained dogs outperform human surveyors in the detection of rare spotted knapweed (Centaurea stoebe). Invasive Plant Sci Manag 3(2):113–121. https://doi.org/10.1614/IPSM-D-09-00025.1

Grimm-Seyfarth A, Harms W, Berger A (2021) Detection dogs in nature conservation: a database on their world-wide deployment with a review on breeds used and their performance compared to other methods. Methods Ecol Evol 12(4):568–579. https://doi.org/10.1111/2041-210X.13560

Grimm-Seyfarth A, Zarzycka A, Nitz T, et al (2019) Performance of detection dogs and visual searches for scat detection and discrimination amongst related species with identical diets. Nat Conserv 37:81–98. https://doi.org/10.3897/natureconservation.37.48208

Gutzwiller KJ (1990) Minimizing dog-induced biases in game bird research. Wildl Soc Bull 18:351–356

Hall NJ, Glenn K, Smith DW et al (2015) Performance of Pugs, German Shepherds, and Greyhounds (Canis lupus familiaris) on an odor-discrimination task. J Comp Psychol 129(3):237–246. https://doi.org/10.1037/a0039271

Hall NJ, Wynne C (2018) Odor mixture training enhances dogs' olfactory detection of home-made explosive precursors. Helyon 4(12):1–16. https://doi.org/10.1016/j.heliyon.2018.e00947

Harrison RL (2006) From the field: a comparison of survey methods for detecting bobcats. Wildl Soc Bull 34:548–552. https://doi.org/10.2193/0091-7648(2006)34%5B548:ACOSMF%5D2.0.CO;2

Hatlauf J, Böcker F, Wirk L et al (2021) Jackal in hide: detection dogs show first success in the quest for golden jackal (Canis aureus) scats. Mamm Res 66:227–236. https://doi.org/10.1007/s13364-020-00537-4

Heaton JS, Cable ME, Nussear KE et al (2008) Comparison of effects of humans versus wildlife-detector dogs. Southwest Nat 53(4):472–479. https://doi.org/10.1894/PAS-03.1

Helton WS (2009) Canine ergonomics: the science of working dogs. CRC Press

Holland K (2018) Sniffer dogs. Internat J of Secur Stud 1(1)

Homan HJ, Linz G, Peer BD (2001) Dogs increase recovery of passerine carcasses in dense vegetation. Wildl Soc Bull 29:292–296

Honig WK (1978) Studies of working memory in the pigeon. In: Hulse SH, Fowler H, Honig WK (eds) Cognitive processes in animal behavior. Erlbaum, Hillsdale, NJ, pp 211–248

Hurt A, Whitelaw A (2010) Euglandina rosea detection by dogs. Working Dogs for Conservation. https://manoa.hawaii.edu/hpicesu/DPW/2010_YER/025.pdf Accessed online on October 28, 2022

Hurt A, Davenport B, Greene E (2000) Training dogs to distinguish between black bear (Ursus americanus) and grizzly bear (Ursus arctos) feces. University of Montana Under-Graduate Biology Journal. http://ibscore.dbs.umt.edu/journal/Articles_all/2000/Hurt.htm. Accessed online in November, 2007

Jamieson LTJJ, Hancock AL, Baxter GS et al (2021) How quoll-ified are northern and spotted-tailed quoll detection dogs? Wildl Res 48(4):376–384. https://doi.org/10.1071/WR19243

Jones K, Dashfield K, Downend A, Otto C (2004) Search and rescue dogs: an overview for veterinarians. Veterin Med Today Disaster Med 225(6):854–860. https://doi.org/10.2460/javma.2004.225.854

Karp D (2020) Detecting small and cryptic animals by combining thermography and a wildlife detection dog. Sci Rep 10:1–11. https://doi.org/10.1111/csp2.12809

Kerley LL (2010) Using dogs for tiger conservation and research. Integr Zool 5(4):390–396. https://doi.org/10.1111/j.1749-4877.2010.00217.x

Kerley LL, Salkina GP (2007) Using scent-matching dogs to identify individual Amur tigers from scats. J Wildl Manage 71:1349–1357. https://doi.org/10.2193/2006-361

Kingdom FAA, Prins N (2016) Psychophysics. Elsevier, A practical introduction

Kretser H, Glennon M, Whitelaw A et al (2016) Scat-detection dogs survey low density moose in New York. Alces 52:55–66

La Guardia MJ, Richards NL, Hale RC (2020) A non-invasive environmental monitoring tool for brominated flame-retardants (BFRs) assisted by conservation detection dogs. Chemosphere 260:127401. https://doi.org/10.1016/j.chemosphere.2020.127401

Lee DH, Cullum JP, Anderson JL, Daugherty JL et al (2014) Characterization of overwintering sites of the invasive brown marmorated stink bug in natural landscapes using human surveyors and detector canines. PLoS ONE 9(4):1–9. https://doi.org/10.1371/journal.pone.0091575

Leigh KA, Dominick M (2015) An assessment of the effects of habitat structure on the scat finding performance of a wildlife detection dog. Methods Ecol Evol 6(7):745–752. https://doi.org/10.1111/2041-210X.12374

Liczner AR, MacPhail VJ, Woollett DA et al (2021). Training and usage of detection dogs to better understand bumble bee nesting habitat: Challenges and opportunities. PloS One 16(5):1–18. https://doi.org/10.1371/journal.pone.0249248

Long RA, Donovan TM, MacKay P et al (2007a) Comparing scat detection dogs, cameras, and hair snares for surveying carnivores. J Wildl Manage 71:2018–2025. https://doi.org/10.2193/2006-292

Long RA, Donovan TM, MacKay P et al (2007b) Effectiveness of scat detection dogs for detecting forest carnivores. J Wildl Manage 71:2007b-2017. https://doi.org/10.2193/2006-230

Lundin, JI, Dills RL, Ylitalo GM et al (2016a) Persistent organic pollutant determination in killer whale scat samples: Optimization of gas chromatography/mass spectrometry method and application to field samples. Arch Environ Contam Toxicol 70(1):9–19. https://doi.org/10.1007/s00244-015-0218-8

Lundin JI, Riffell JA, Wasser SK (2015) Polycyclic aromatic hydrocarbons in caribou, moose, and wolf scat samples from three areas of the Alberta oil sands. Environ Poll 206:527–534. https://doi.org/10.1016/j.envpol.2015.07.035

Lundin JI, Ylitalo GM, Booth RK et al (2016b) Modulation in persistent organic pollutant concentration and profile by prey availability and reproductive status in southern resident killer whale scat samples. Environ Sci Technol 50(12):6506–6516. https://doi.org/10.1021/acs.est.6b00825

Macmillan NA, Creelman CD (2005). Detection theory. A user's guide (2nd ed.). Lawrence Erlbaum Associates, Mahwah, NJ

Mathews F, Swindells M, Goodhead R et al (2013) Effectiveness of search dogs compared with human observers in locating bat carcasses at wind-turbine sites: a blinded randomized trial. Wildl Soc Bull 37(1):34–40. https://doi.org/10.1002/wsb.256

Matthew EE, Relton CE (2021) Training methodology for canine scent detection of a critically endangered lagomorph: a conservation case study. J Vertebr Biol 69(3):1–15. https://doi.org/10.25225/jvb.20092

Matthew EE, Verster R, Weldon C (2021) A case study in canine detection of giant bullfrog scent. J Vertebr Biol 69(3):1–12. https://doi.org/10.25225/jvb.20043

McNicol D (2005) A primer of signal detection theory. Lawrence Erlbaum Associates, Mahwah, NJ

Mendel J, Furton KG, Mills D (2018) An evaluation of scent-discriminating canines for rapid response to agricultural diseases. HortTechnol 28(2):102–108. https://doi.org/10.21273/HORTTECH03794-17

Moriarty KM, Linnell MA, Thornton JE et al (2018) Seeking efficiency with carnivore survey methods: a case study with elusive martens. Wildl Soc Bull 42(3):403–413. https://doi.org/10.1002/wsb.896

Mosa SG (2004) Impact of agriculture and grazing on pale-spotted (Nothura darwinii) and Andean (Nothoprocta pentlandii) tinamous in the Lerma valley, Salta Province, Argentina. Ornitologia Neotropical, 15(suppl.):309–315

Mosconi F, Campanaro A, Carpaneto GM et al (2017) Training of a dog for the monitoring of Osmoderma eremita. Nature Conserv 20:237–264. https://doi.org/10.3897/natureconservation.20.12688

MTO and McIntosh Perry (Conservation Dogs) (2017) Environmental achievement awards application: executive summary. https://www.tac-atc.ca/sites/default/files/conf_papers/mcintosh_perry_conservation_dogs-complete.pdf

Mutoro NM, Eberle J, Petermann JS et al (2021) Impact of weather conditions on cheetah monitoring with scat detection dogs. J Trop Ecol 37(5):222–227. https://doi.org/10.1017/S0266467421000316

Needs S, Bennett E, Mao B et al (2021) Do detection dogs respond differently to dried, frozen, and live plant targets? Appl Anim Behav Sci 236:105276. https://doi.org/10.1016/j.applanim.2021.105276

Nielsen TP, Jackson G, Bull CM (2016) A nose for lizards; Can a detection dog locate the endangered pygmy bluetongue lizard (Tiligua adelaidensis)? Trans R Soc S Aust 140(2):234–243. https://doi.org/10.1080/03721426.2016.1218698

Nussear KE, Esque TC, Heaton JS et al (2008) Are wildlife detector dogs or people better at finding desert tortoises (Gopherus Agassizii)? Herpetol Conserv Biol 3(1):103–115

O'Connor S, Park KJ, Goulson D (2012) Humans versus dogs: a comparison of methods for the detection of bumble bee nests. J Apicult Res 51(2):204–211. https://doi.org/10.3896/IBRA.1.51.2.09

Oh Y, Lee Y, Heath J et al (2014) Applications of animals as biosensors: a review. IEEE Sens J 15(2):637–645. https://doi.org/10.1109/JSEN.2014.2358261

Oldenburg C, Schoon A, Heitkonig IM (2016) Wildlife detection dog training: a case study on achieving generalization between target odor variations while retaining specificity. J Vet Behav 13:34–38. https://doi.org/10.1016/j.jveb.2016.03.008

Orkin JD, Yang Y, Yang C et al (2016) Cost-effective scat-detection dogs: Unleashing a powerful new tool for international mammalian conservation biology. Sci Rep 6(1):1–10. https://doi.org/10.1038/srep34758

Osterkamp T (2020) Detector dogs and scent movement: how weather, terrain, and vegetation influence search strategies. CRC Press

Paula J, Costa Leal M, Silva MJ et al (2011) Dogs as a tool to improve bird-strike mortality estimates at wind farms. J Nat Conserv 19(4):202–208. https://doi.org/10.1016/j.jnc.2011.01.002

Pearsall MD, Verbruggen MD (1982) Scent: training to track. Alpine Publications, Search and Rescue

Petroelje TR, Fowler NL, Kautz TM et al (2021) Conservation detection dogs increase efficacy for prey detection at carnivore GPS cluster sites during summer. Wildl Soc Bull 45(3):402–409. https://doi.org/10.1002/wsb.1203

Ralls K, Smith DA (2004) Latrine use by san joaquin kit foxes (Vulpes macrotis mutica) and coyotes (Canis latrans). West North Am Natura 64(4):544–547. https://www.jstor.org/stable/41717412

Ramsey DS, Barclay C, Campbell CD et al (2017) Detecting rare carnivores using scats: implications for monitoring a fox incursion into Tasmania. Ecol Evol 8:732–743. https://doi.org/10.1002/ece3.3694

Reindl-Thompson SA, Shivik JA, Whitelaw A et al (2006) Efficacy of scent dogs in detecting Black-footed ferrets at reintroduction site in South Dakota. Wildl Soc Bull 34:1435–1439. https://doi.org/10.2193/0091-7648

Reindl SA (2004) Efficacy of scent dogs in detecting black-footed ferrets (Mustela nigripes) at a reintroduction site in South Dakota. Thesis, South Dakota State University, Brookings: USA (p 73)

Reed SE, Bidlack AL, Hurt A et al (2011) Detection distance and environmental factors in conservation detection dog surveys. J Wildl Manag 75(1):243–251. https://doi.org/10.1002/jwmg.8

Reynolds MH, Johnson KN, Schvaneveldt ER et al (2021) Efficacy of detection canines for avian botulism surveillance and mitigation. Conserv Sci Pract 3(6):1–18. https://doi.org/10.1111/csp2.397

Richards NL, Tomy G, Kinney CA et al (2018) Using detection dogs to monitor aquatic ecosystem health and protect aquatic resources. Palgrave Macmillan

Rolland RM, Hamilton PK, Kraus SD et al (2006) Faecal sampling using detection dogs to study reproduction and health in North Atlantic right whales (Eubalaena glacialis). J Cetacean Res Manage 8:121–125

Rolland RM, Hunt KE, Doucette GJ et al (2007) The inner whale: hormones, biotoxins, and parasites. In Kraus SD, Rolland RM (eds), The Urban Whale: North Atlantic Right Whales at the Crossroads. Harvard University Press, London: England

Rolland RM, Parks SE, Hunt KE et al (2012) Evidence that ship noise increases stress in right whales. Proc Biol Sci 279(1737):2363–2368. https://doi.org/10.1098/rspb.2011.2429

Rosell F (2014) Secrets of the Snout. The University Press of Chicago

Rosell F, Cross HB, Johnsen CB et al (2019) Scent-sniffing dogs can discriminate between native Eurasian and invasive North American beavers. Sci Rep 9:1–9. https://doi.org/10.1038/s41598-019-52385-1

Rosell F, Kniha D, Haviar M (2020) Dogs can scent-match individual Eurasian beavers from their anal gland secretion. Wildlife Biol 2:1–7. https://doi.org/10.2981/wlb.00658

Rutter NJ, Howell TJ, Stukas AA et al (2021a) Can volunteers train their pet dogs to detect a novel odor in a controlled environment in under 12 weeks? J Vet Behav 43:54–65. https://doi.org/10.1016/j.jveb.2020.09.004

Rutter NJ, Howell TJ, Stukas AA et al (2021a) Diving in nose first: the influence of unfamiliar search scale and environmental context on the search performance of volunteer conservation detection dog-handler teams. Animals 11(4):1177. https://doi.org/10.3390/ani11041177

Rutter NJ, Mynott JH, Howell TJ et al (2021c) Buzzing with possibilities: training and olfactory generalization in conservation detection dogs for an endangered stonefly species. Aquat Conserv 31(4):984–989. https://doi.org/10.1002/aqc.3531

Savidge JA, Stanford JW, Reed RN et al (2011) Canine detection of free-ranging brown treesnakes on Guam, New Zealand. J Ecol 35(2):174–181. http://www.jstor.org/stable/24060665

Schellinck H, Brown RE, Slotnick B (1991) Training rats to discriminate between the odors of individual conspecifics. Anim Learn Behav 19:223–233. https://doi.org/10.3758/BF03197880

Schellinck H, Slotnick B, Brown RE (1997) Odors of individuality originating from the major histocompatibility complex are masked by diet cues in the urine of rats. Anim Learn Behav 25:193–199. https://doi.org/10.3758/BF03199058

Schoon GA, Haak R (2002) K9 suspect discrimination: training and practicing scent identification line-ups. Detselig Enterprises, Calgary, Alberta: Canada

Shuttleworth CM, Cripps R, Everest DJ (2020) Using dogs in wildlife surveillance. Vet Rec 187(11):452. https://doi.org/10.1136/vr.m4539

Sims DW, Humphries NE, Hu N et al (2019) Optimal searching behaviour generated intrinsically by the central pattern generator for locomotion. eLife 8:e50316. https://doi.org/10.7554/eLife.50316

Slotnick B, Restrepo D (2005) Olfactometry with Mice. Curr Protoc Neurosci 33.1:8.20.1–8.20.24. https://doi.org/10.1002/0471142301.ns0820s33

Slotnick B, Schellinck H (2002) Methods in olfactory research with rodents. In: Simon SA, Nicolelis M (eds) Frontiers and methods in chemosenses. CRC Press, Boca Raton, FL, pp 21–61

Smallwood KS, Bell DA, Standish S (2020) Dogs detect larger wind energy effects on bats and birds. J Wildl Manage 84(5):852–864. https://doi.org/10.1002/jwmg.21863

Smith DA, Ralls K, Cypher BL et al (2005) Assessment of scat-detection dog surveys to determine kit fox distribution. Wildl Soc Bull 33:897–904. https://doi.org/10.2193/0091-7648(2005)33%5B897:AOSDST%5D2.0.CO;2

Smith DA, Ralls K, Cypher BL et al (2006) Relative abundance of endangered San Joaquin kit foxes (Vulpes macrotis mutica) based on scat-detection dog surveys. Southwest Nat 51:210–219. https://doi.org/10.1894/0038-4909(2006)51[210:RAOESJ]2.0.CO;2

Smith DA, Ralls K, Davenport B et al (2001) Canine assistants for conservationists. Science 291:435

Smith DA, Ralls K, Hurt A et al (2003) Detection and accuracy rates of dogs trained to find scats of San Joaquin kit foxes (Vulpes macrotis mutica). Anim Conserv 6:339–346. https://doi.org/10.1017/S136794300300341X

Stanhope K (2015) Ecological monitoring using wildlife detection dogs: Bat carcass searches at the Wanlip wind turbine. Pract Monit Meas Mitigat Succ 88:29–32

Statham MJ, Woollett DA, Fresquez S et al (2019) Non-invasive identification of herpetofauna: pairing conservation dogs and genetic analysis. J Wildl Manage 84(1):66–74. https://doi.org/10.1002/jwmg.21772

Suma P, La Pergola A, Longo S et al (2014) The use of sniffing dogs for the detection of Rhynchophorus ferrugineus. Phytoparasitica 42(2):269–274. https://doi.org/10.1007/s12600-013-0330-0

Terrace HS (1963a) Discrimination learning with and without errors. J Exp Anal Behav 6:1–27

Terrace HS (1963b) Errorless transfer of a discrimination across two continua. J Exp Anal Behav 6:223–232

Terrace HS (1964) Wavelength generalization after discrimination learning with and without errors. Science 144:78–80

Terrace HS (1966) Stimulus control. In: Honig WK (ed) Operant behavior: areas of research and application. Appleton-Century-Croft, New York, pp 271–344

Thomas ML, Baker L, Beattie JR et al (2020) Determining the efficacy of camera traps, live capture traps, and detection dogs for locating cryptic small mammal species. Ecol Evol 10(2):1054–1068. https://doi.org/10.1002/ece3.5972

Thompson CM, Royle JA, Garner JD (2012) A framework for inference about carnivore density from unstructured spatial sampling of scat using detector dogs. J Wildl Manage 76(4):863–871. https://doi.org/10.1002/jwmg.317

Thompson SA, Thompson GG, Withers PC et al (2020) Conservation detection dog is better than human searcher in finding bilby (Macrotis lagotis) scats. Aust Zool 41(1):86–93. https://doi.org/10.7882/AZ.2020.012

Vesely DG (2008) Training of conservation detection dogs to locate Kincaid's Lupine (Lupinus sulphureus ssp. Kincaidii). U.S. Fish & Wildlife Service, Oregon State Office. http://www.oregonwildlife.org/wp-content/uploads/2012/06/PlantsNDogs_2008d.pdf

Vice DS, Engeman RM (2000) Brown tree snake discoveries during detector dog inspections following supertyphoon paka. Micronesica 33(1/2):105–110

Villablanca FX, Kofron CO, Oliver L et al (2021) Survey for Morro Bay kangaroo rat: a rare mammal of uncertain status. J Fish Wildl Manag 12:263–270

Vynne C, Skalski JR, Machado RB et al (2011a) Effectiveness of scat-detection dogs in determining species presence in a tropical savanna landscape. Conserv Biol 25(1):154–162

Vynne C, Keim JL, Machado RB et al (2011b) Resource selection and its implications for wide-ranging mammals of the Brazilian Cerrado. PLoS ONE 6(12):e28939. https://doi.org/10.1371/journal.pone.0028939

Vynne C, Baker MR, Breuer ZK et al (2012) Factors influencing degradation of DNA and hormones in maned wolf scat. Anim Conserv 15(2):184–194. https://doi.org/10.1111/j.1469-1795.2011.00503.x

Wasser SK, Davenport B, Ramage ER et al (2004) Scat detection dogs in wildlife research and management: application to grizzly and black bears in the Yellowhead Ecosystem, Alberta, Canada. Can J Zool 82:475–492. https://doi.org/10.1139/z04-020

Wasser SK, Hayward LS, Hartman J et al (2012) Using detection dogs to conduct simultaneous surveys of northern spotted and barred owls. PLoS ONE 7(8):e42892. https://doi.org/10.1371/journal.pone.0042892

Wasser SK, Lundin JI, Ayres K et al (2017) Population growth is limited by nutritional impacts on pregnancy success in endangered Southern resident killer whales (Orcinus orca). PLoS ONE 12(6):1–22. https://doi.org/10.1371/journal.pone.0179824

Wasser SK, Smith H, Madden L et al (2009) Scent-matching dogs determine number of unique individuals from scat. J Wildl Manag 73(7):1233–1240. https://doi.org/10.2193/2008-530

Waters J, O'Connor S, Park KJ et al (2011) Testing a detection dog to locate bumblebee colonies and estimate nest density. Apidologie 42:200–205. https://doi.org/10.1051/apido/2010056

Wilbert T, Woollett DA, Whitelaw A et al (2015) Non-invasive baseline genetic monitoring of the endangered San Joaquin kit fox on a photovoltaic solar facility. Endanger Species Res 27(1):31–41. https://doi.org/10.3354/esr00649

Wilbert TR, Woollett DA, Westphal MF et al (2019) Distribution, fine-scale subdivision, and population size of San Joaquin kit foxes in the Ciervo-Panoche Natural Area, California. Conserv Genet 20:405–417. https://doi.org/10.1007/s10592-018-1122-3

Wultsch C, Waits LP, Kelly MJ (2014) Non-invasive individual and species identification of Jaguars (Panthera onca), Pumas (Puma concolor), and Ocelots (Leopardus pardalis) in Belize, Central America using cross-species microsatellites and fecal DNA. Mol Ecol Resour 14(6):1171–1182. https://doi.org/10.1111/1755-0998.12266

Wysocka NK (2021) Using sniffer dogs for non-invasive detection of Heterobasidion root rot from scent stimuli derived from Norway spruce trees. [Master's Thesis] Swedish University of Agricultural Sciences, Southern Swedish Forest Research Centre

Zwickel FC (1980) Use of dogs in wildlife biology. In: Schemnitz SD (ed) Wildlife Management Techniques. The Wildlife Society, USA

MIX
Papier aus verantwortungsvollen Quellen
Paper from responsible sources
FSC® C105338

If you have any concerns about our products,
you can contact us on
ProductSafety@springernature.com

In case Publisher is established outside the EU,
the EU authorized representative is:
Springer Nature Customer Service Center GmbH
Europaplatz 3, 69115 Heidelberg, Germany

Printed by Libri Plureos GmbH
in Hamburg, Germany